STABILITY AND SAFETY OF SHIPS

Risk of Capsizing

VADIM BELENKY

NIKITA B. SEVASTIANOV

EDITORS

R. Bhattacharyya

M. E. McCormic

Second Edition

2007
The Society of Naval Architects and Marine Engineers (SNAME)
www.sname.org

To Wes Wheeler
with great pleasure and wishes
of success in Naval architecture
from the author
Vadim Belenky 16 - Nov - 2007

STABILITY AND SAFETY OF SHIPS: RISK OF CAPSIZING
Vadim Belenky and *Nikita B. Sevastianov*

Editors: *R. Bhattacharyya* and *M. E. McCormic*

Cover design: *Susan Evans Grove*

Published by
The Society of Naval Architects and Marine Engineers
601 Pavonia Ave.
Jersey City, NJ 07306
www.sname.org

First Edition 2003 by Elsevier
Second Edition 2007

Printed in the United States of America
by Automated Graphics Systems, Inc. (AGS)

Bibliographical Note

This SNAME edition, first published in 2007 is republication of the work originally published by Elsevier, Amsterdam; Boston in 2003. A new preface, Appendix 2, and a list of additional references have been prepared especially for this edition.

Library of Congress Cataloging-in-Publication Data
A catalog record from the Library of Congress has been applied for.

ISBN 0-939773-61-9

This book is dedicated to the seafarers that lost their lives at sea.

Preface to the Second Edition

The first edition of this book was published by Elsevier as Volume 10 of Elsevier Ocean Engineering Book series edited by R. Bhattacharyya and M. E. McCormick.

Originally, this book was the second part of the two-volume monograph united under the common title ***Stability and Safety of Ships.*** The first part was published as Volume 9 of Elsevier Ocean Engineering Book series with the subtitle "Regulation and Operation" authored by Kobylinski and Kastner. It described the state of the art and historic perspective of intact stability regulations as well as covered the operational aspect of ship stability. Volume 10, subtitled "Risk of Capsizing," contained descriptions of contemporary approaches and solutions for evaluation of dynamic stability as well as a detailed review of the research results in the field and was meant to serve as an extended reference source for the development of future intact stability regulations. Both parts were written with the same philosophy but could be read separately.

The appearance of new types of naval and commercial vessels with unconventional dynamics in waves made conventional methods of evaluation of dynamic stability unreliable for the most part, as these methods are based on previous experience and statistics. It is well known that the best approach is to use the physically sound solution for ship motion in waves employing Nonlinear Dynamics and theory of stochastic processes. This allows developing new views on different types of stability failures including capsizing in dead ship conditions, surf-riding and broaching, parametric resonance and pure loss of stability on the wave crest. The above approach has defined the increased interest of maritime industry to the problems of ship dynamics. Understanding the importance of these problems motivated IMO to resume discussion on new approaches to intact stability regulations in 2002.

Among the naval architects whose research results and organizational efforts determined these new views in recent years, I would like to mention: P. R. Alman, H. P. Cojeen, J. O. de Kat, A. Francescutto, Y. Ikeda, L. Kobylinski, M. A. S. Neves, J. R. Paulling, L. Peres Rojas, P. Purtell, A. M. Reed, R. Sheinberg, K. Spyrou, A. W. Troesch, N. Umeda and D. Vassalos.

This list, of course, is far from being complete, so I would like to ask those colleagues who were not mentioned in this list to accept my sincere apology.

To assist this development the Society of Naval Architects and Marine Engineers (SNAME) decided to publish a second edition of Volume 10 since the first edition is out of print. Volume 9 remains available. I am very grateful to R. Bhattacharyya, W. France, S. Evans Grove and R. Tagg for their help with organization of the second edition.

My special thanks are due for William Belknap, Michael Hughes and Arthur Reed for their detailed review of Chapter 3, for Marcello Neves for his thorough review of Chapter 4 and for Yury Nechaev for additional corrections to regression coefficients in Appendix I.

The second edition is almost an exact reproduction of the first edition with the exception of corrected typographical errors, updated text for some chapters to account for the most recent development in parametric roll and numerical simulation of irregular roll motions. Corresponding updates were made in the list of references.

I am grateful to all my colleagues, discussions with whom were very helpful in updating the book, in particular: G. Bulian, A. Degtyarev, P. Handler, B. Hutchison, B. Johnson, W. M. Lin, L. McCue, K. Metselaar, W. Peters, and K. Weems.

The author considers it as a pleasant duty to thank management and employees of the American Bureau of Shipping and first of all: G. Ashe, R. I. Basu, A. J. Breuer, C. J. Dorchak, T. Gruber, T. Ingram, B. Menon, D. Novak and H. Yu – all of whom shared the author's interest to the problems of ship dynamics and made possible for the author to continue working in this direction, including publication of the second edition of this book.

Language editing of the second edition was performed by Robert M. Conachey, whose efforts are greatly appreciated.

V. Belenky

April 2007

Series Preface

In this day and age, humankind has come to the realization that the Earth's resources are limited. In the 19th and 20th Centuries, these resources have been exploited to such extent that their availability to future generations is now in question. In an attempt to reverse this march towards self-destruction, we have turned our attention to the oceans, realizing that these bodies of water are both sources for potable water, food and minerals and are relied for upon for World commerce. In order to help engineers more knowledgeably and constructively exploit the oceans the **Elsevier Ocean Engineering Book Series** has been created

The **Elsevier Ocean Engineering Book Series** gives experts in various areas of ocean technology the opportunity to relate to others their knowledge and expertise. In a continuous process, we are assembling world-class technologists who have both the desire and the ability to write books. These individuals select the subjects for their books based on their educational backgrounds and professional experience

The series differs from other ocean engineering book series in that the books are directed more towards technology than science, with a few exceptions. Those exceptions we judge to have immediate applications to many of ocean technology fields. Our goal is to cover the broad areas of naval architecture, coastal engineering, ocean engineering acoustics, marine systems engineering, applied oceanography, ocean energy conversion, design of offshore structures, reliability of ocean structures and systems and many others. The books are written so that readers entering the topic fields can acquire a working level of expertise from their readings.

We hope that the books in the series are well received by the ocean engineering community.

Rameswar Bhattaharyya

Michael E. McCormic

Series Editors

Foreword

Naval Architects have recognized since the earliest times the importance of stability knowing well that ships have to survive the perils of the sea. The amount of knowledge on stability accumulated over the years is now enormous. There is, however, a lack of publications containing the review of the knowledge in this particular field.

The intention of this book is to fill this gap and to present, as far possible, the state-of-the-art focused on the regulatory, operational and future development aspects of intact stability. This book is addressed to readers who are interested in promoting safety against capsizing, who are involved in research on, and practical application of stability regulations on an international or national level, to ship operators and designers and members of the maritime administrations.

The book is divided in two volumes, 1 and 2. (These correspond to Volumes 9 and 10 of the Elsevier Ocean Engineering Series). The first volume (authored by L. Kobylinski and S. Kastner with subtitle "Regulation and Operation") describes the state of the art in the field of intact ship stability. It is focused on how intact safety has been promoted; it considers stability regulations: their current state is given an historical perspective. The methods used to develop these regulations are carefully examined. Besides regulations, the first volume addresses the operational aspect of stability: ocean environment and ship behavior are described as they are seen from the bridge. Special attention is paid to on-board stability measurements and their accuracy.

The second volume (authored by V.L. Belenky and N.B. Sevastianov with the subtitle "Risk of Capsizing") is a gathering of today's knowledge for tomorrow's development. It is focused on how risk and reliability can be applied for evaluation and regulation of the intact stability. First, the framework of future risk-based stability regulations is considered. Then the book examines physical phenomena associated with stability loss including broaching, greenwater influence and breaking waves action in order to evaluate the risk of capsizing in different situations. Much attention is paid to the mechanics of capsizing: as it seems to be necessary for interpretation and validation of numerical simulation results. The latter is meant to facilitate confident applicability of these methods for new ship designs.

Volume 9: L. Kobylinski and S. Kastner

Stability and Safety of Ships. Volume 1: Regulation and Operation

ISBN – 0-08-043001-5

Volume 10: V. L. Belenky and N. B. Sevastianov

Stability and Safety of Ships. Volume 2: Risk of Capsizing

ISBN – 0-08-044354-0

Preface

Problems of ship stability concern a wide circle of professionals related to design, production and operation of ships. This book, addressing contemporary and future problems of stability safety of ships, is written for these professionals including graduate students seeking a career in the maritime field.

Providing a ship with sufficient stability is still a problem; and there are two reasons why it is so. First of all, ship design is constantly evolving responding to ever-changing challenges of the markets. This makes the previous experience partially obsolete and only conditionally applicable. New ships have size and hull geometries different from the past. Modes of operation have also changed significantly with increasing shipping traffic and exploration of ocean mineral resources.

Secondly, the loss of stability or capsizing itself is a complicated physical phenomenon, which does not have a complete mathematical description for all the cases that might be encountered in practice. Increasing computing power facilitates numerical simulation as the principal tool for engineering analysis and decision-making. The amount and quality of the data delivered by the simulation approaches those offered by the model experiments. Interpretation of these data becomes a priority, in order to utilize these opportunities in an efficient and confident manner.

In addition to the above, the industry demand for risk-based rules and regulations has risen in recent years. Wide circles of professionals involved in all aspects of maritime business realise that there is inevitable risk associated with seafaring and they would like to know how large this risk is and how to control it.

These are the questions the authors try to address here by assembling knowledge that would be helpful for future development of risk-based stability regulations.

There are two parts in this volume. The first part contains application of reliability theory adapted for stability regulations and fundamentals for new risk-based stability criteria. These criteria, however, require a more sophisticated description of ship dynamics in irregular seas and gusty wind.

Such a description is given in the second part of this volume. It starts from the very basic hydrodynamic review in order to examine the background of the mathematical model of nonlinear rolling and capsizing. Then this model is applied for different situations, known to be hazardous for stability: beam seas, following seas, action of the sudden gust of wind, greenwater shipping, breaking wave action, etc. First, all these phenomena are considered in a deterministic sense, assuming that the wave is regular, wind is constant and so on. Most of deterministic phenomena are examined based on Nonlinear Dynamics (the latest development in applied mathematics allowing the treatment of nonlinear systems without limitation). The reader, however, is not required to be familiar with it: all information, necessary for understanding is given there in visual form with examples.

Then stochastic factors are introduced into the consideration and the above problems become probabilistic, solutions result ultimately in methods for estimation of capsizing risk. Despite the fact these methods are based on theory of stochastic processes and probabilities of rare events, the reader is required to have only a basic knowledge of probability, all other information, which is necessary to understand complicated concepts of the risk-based approach, is given. In other words, the second part is written to bridge the gap between the level of undergraduate student and contemporary scientific level.

The idea to write the book, which could be used as a comprehensive references source and systematic gathering of knowledge about stability, belonged to the late Professor Nikita B. Sevastianov. Another author had the honour to be his student and close co-worker at the Department of Naval Architecture and National Laboratory of Seakeeping at Kaliningrad Institute of Technology in Russia. Professor Sevastianov was a wonderful teacher, engineer and scientist. His ideas of risk and reliability based safety regulations ran well ahead of the time, being first formulated as early as 1963. The author was heavily influenced by Sevastianov's great personality and is indebted to him for professional and scientific growth.

This book was meant to present the subject from different perspectives, and there were four authors originally: N. Sevastianov, L. Kobylinski, S. Kastner and V. Belenky. The decision was then taken to present the work in two volumes. The first volume of the book is published under the names of Kobylinski and Kastner with the subtitle "Regulation and Operation". It generalises past experience and describes present state of the art of stability safety from regulatory and operational points of view.

This volume is the second one. It was written so as to be independent of the first volume, however, it is still considered to be a continuation of the first volume: their combination is meant to give a complete coverage of the subject on its past experience, present status and future development. Since this volume is focused on the future, it will be especially interesting for graduate students in Naval Architecture, consultants and scientists working for R&D centers and departments of maritime administrations, classification societies and other bodies responsible for the development of stability regulations and operation guides.

It can be also useful for professionals in the neighbouring fields interested in engineering applications of reliability, nonlinear dynamics, stochastic processes and random dynamical systems. The authors of both volumes are confident that the span of material covered in this new book on stability and safety of ships can be widely used for reference, study and training in the maritime community, from ship designers to ship operators, from administrators to researchers.

N. Sevastianov was the author of Part 1 on the probabilistic approach to stability criteria that was the subject of his particular interest, but which unfortunately was not completed because of his premature death. He finalised only Chapter 1 personally. Chapter 2 was completed by V. Belenky based on materials selected by N. Sevastianov. Part 2 of this volume was written by V. Belenky.

I am very grateful to Professor Lech Kobylinski, who is an author of Part 1 of Volume 1 and with whom I had the pleasure to work for a number of years, performing research in the field of probabilistic approach to stability regulations, while staying at the Technical University of Gdansk in Poland. This cooperation was very important for creating the general book structure. I am also grateful to Professor Sigismund Kastner (University of

Applied Sciences Bremen), who is the author of Part 2 of Volume 1, for his help and cooperation, especially in the organization of a series of meetings of the authors in Bremen, which facilitated this book's progress.

All the authors are grateful to series editors Professor R. Bhattacharyya and Professor M. E. McCormick for a wonderful publication opportunity under the Elsevier Ocean Engineering Book Series. The aid of Rameswar Bhattacharyya is especially appreciated: without his continuous support and encouragement, this book probably would never have been completed.

It took almost eleven years to complete this book and many people from different countries helped me in this task. First of all, I would like to acknowledge my colleagues, who directly contributed to the contents of this volume, participating in writing certain subchapters: A. V. Boukhanovsky, A. B. Degrtyarev, S. Grochowalski, S. V. Mordachev, K. Spyrou and N. Umeda.

I am very grateful to J. Falzarano, A. Francescutto, Yu. I. Nechaev, J. W. Kim, Y. S. Shin, A. W. Troesch and K. M. Weems for discussions and review of materials in some of the chapters and subchapters.

My special thanks also go to all the colleagues whose help and support was quite valuable during these years: P. R. Alman, J. C. Card, W. A. Cleary, K. McCreight, H. P. Cojeen, T. Gruber, W. B. Morgan, W. M. Lin, W. L. Thomas, J. S. Spencer (USA); A. Matsuda, Sh. Suzuki, Yu. Yamakoshi (Japan); H. Hormann, J. O. deKat, M. Renilson, and D. Vassalos (EU); J. Bielanski and M. K. Gerigk (Poland); V. V. Garkavy, D. V. Jurko, K. V. Khabaznia, V. A. Lapin, V. A. Mamontov, P. Yu. Markov and K. V. Sokolov (Russia).

I would also like to thank the institutions and organizations I have been working for during these years, their management and employees for help and support, namely American Bureau of Shipping (Houston, USA), National Research Institute for Fisheries Engineering (Hasaki, Japan), Technical University of Gdansk (Poland) and my *alma mater* Kaliningrad Institute of Technology (now known as of Kaliningrad University of Technology, Russia)

The author of this volume appreciates language aid and preliminary proofreading of certain chapters done by J. Philip Brindley. The help of Professor Leonid M. Belenkiy is greatly acknowledged for providing myself with much valuable advice on general composition of the material and everyday support on various issues on writing technique. I am especially grateful to Robert M. Conachey, who performed language editing of the whole manuscript and gave much advice and suggestions on clarity of the text.

I wish to express my most sincere thanks to my wife Alina Belenky for her understanding and patience during all these long years of my work with this book.

V. Belenky

April, 2003

The views and opinions expressed in this book are solely and strictly those of the authors and do not necessarily reflect those of American Bureau of Shipping, its affiliates, subsidiaries or any of their respective officers, employees or agents or Kaliningrad University of Technology

Table of Contents

Part 1

Probabilistic Approach to Stability and Risk Assessment

Chapter 1

Philosophy of Probabilistic Evaluation of Stability and Safety

1.1 General Concepts of Probabilistic Evaluation of Stability, Safety and Risk at Sea

It is clear from the review [Blagoveshchensky, 1932, 1951; Lugovsky, 1971, Rahola, 1935, 1939] of the existing national and international practices of stability standardization that the development of stability standards was initiated by experts in the theory of naval architecture. Therefore, this problem is usually considered by naval architects and seamen as an aspect of the theory of ships. However, it is not quite so.

The final goal of setting ships' stability standards is to ensure their safe operation without fatal capsizing casualties during their service lives. Similar tasks are considered practically concerning any other products of technology, see [Sevastianov, 1982]. Indeed, if we substitute in the former sentence the word "stability" for the word "strength" and the word "capsizing" for the word "failure" then the aim of standardization will be changed in a specific technical aspect. However, the very essence of setting standards will remain the same: to ensure operation of some object (product, system, etc.) without failure during a given time. Only the nature, the causes and the form of the failure are specific.

Failures in a complicated system may occur both in the system as a whole and in some of its elements. This should be especially borne in mind when we are developing the standards of sea-keeping qualities. Such failures as capsizing, foundering, loss of longitudinal strength practically mean the total loss of the system called a vessel. Damage to the propeller, steering engine or hatch cover does not mean the immediate total loss of a vessel, as a rule, although sometimes they may happen to be an important link in the chain of events which result in the total loss of the vessel. In our further consideration of stability we shall deal only with such failures as capsizing or catastrophic heeling.

First of all, it is necessary to consider the general plan of any standardization. It can be represented as a set of the following four sub-problems:

1. The definition of the aims of standardization,
2. The choice of criteria,
3. The setting of the norm (standard) for each criterion,
4. The evaluation of the likelihood of achieving the goal and the technical and economical consequences of implementing standards.

Defining the aims of standardization for a specific regulation (stability, strength, noise suppression of radio equipment, etc.) is connected first of all with social requirements.

It may be narrowed to mentioning only such dangerous or undesirable events, the risk of which should be lowered to zero or to an acceptable level. We shall call such dangerous events "casualties" or "capsizing".

The goal of any standard is preventing certain types of casualties with certain objects during a certain time interval. Such an interval may be determined as the duration of a missile's flight to the target or the navigation period up and down the rivers which become frozen in winter, or lifetime of a vessel from its launch until it is sold for scrap metal.

The choice of criteria is a specific task for each type of casualty. The word "criterion" in Greek means "an instrument for judgment". We shall examine and use the criteria for estimating how great the risk of a casualty is. It has become a custom in modern practice to judge the possibility of casualties not by a single criterion but by a set of some criteria. Apparently, the characteristics chosen as criteria should be functions of properties of the object itself. Besides the arguments determining the criterion, its value should take into consideration the external condition parameters under which the object operates. As far as this concerns stability such conditions should include the forces affecting the vessel, their orientation, dynamic or static application as well as the ability of the crew to maintain the necessary safety level. Simple and effective stability control methods are very important in this respect. Therefore, it is clear that the sub-task of criteria choice necessitates participation of such experts who know in detail the properties of the corresponding objects and their operational conditions.

The setting of the norms is the indication of some conventional boundary between the permissible and impermissible values of the criteria, that is between the points "good" and "bad" for providing stability.

It is obvious that the risk of capsizing will change continuously with continuous alteration of the usual stability criteria. Unfortunately, the actual interdependence between the risk and practical stability criteria cannot be expressed in explicit form by a simple formula. It exists as a rule in a latent, implicit form. This problem will be discussed in detail below.

We will consider a simple example explaining the nature of usual stability norms.

Let us imagine an ordinary deckless vessel or a boat affected by a static heeling moment, M_h .

It is clear that the full loss of stability may occur after flooding the internal volume of the boat whilst heeling. It is possible to suggest an angle of heel ϕ, as a criterion of stability. In the framework of initial stability theory this angle may be expressed by the elementary metacentric formula:

$$\phi = \frac{M_h}{W \cdot GM} ,$$

Where W is the vessel's displacement and GM is the initial metacentric height.

Would it be reasonable to adopt as a norm the angle of inclination, ϕ_f, corresponding to the immersion of the lowest point of the vessel's freeboard? A "pure" theoretician would probably agree with such a suggestion. Really, at the angle of heel $\phi < \phi_f$ the boat has some residual freeboard and water cannot enter into the vessel. Nevertheless it is not necessary to be a seaman to have some doubts about the practicality of such a norm. We shall not criticize this norm for the inaccuracy of the metacentric formula at large angles of heel. Indeed, it is possible to find the angle ϕ more precisely on the basis of a stability diagram if the heeling moment, M_h is known. But who can guarantee proper accuracy of the assigned moment M_h or of the initial value of the actual freeboard? Are we sure of the fact that all the mistakes will always increase safety? And what is the practical validity of the implicit assumption about heeling under the condition of absolutely still water? That's why even in the simplest example it is impossible to accept the value ϕ_f as a practical norm though it is theoretically an indisputable boundary between zero risk and the situation when the boat is doomed to perish. It would be reasonable to introduce some reserve into the norms foreseeing the unpredictable random external conditions (sea state, ship motion, etc.) and inaccuracy of the given initial data. In other words, we have to insure ourselves against our own ignorance of all the actual circumstances of the situation being considered. That is another matter that we can introduce such a reserve in an explicit or in an implicit form, and it is clear that the magnitude of this reserve depends on the factors and circumstances which lie beyond the framework of ordinary theory with its deterministic approach.

The last sub-task of setting the safety standard concerns the evaluation of the guarantee of safety, which should be ensured by the standards being suggested. It also concerns foreseeing technical and economic consequences of introducing these standards.

It should be noted that practical implementation of various seakeeping standards has not yet raised the problem of guarantee in its explicit form.

Probably the experts faced this problem for the first time while developing the IMO Stability Recommendations. This problem is rather new and practically important. Therefore, we shall consider it and its history in detail.

The draft of the IMO Stability Recommendations was developed in 1967. The interested countries tried to evaluate the acceptability of this draft on the basis of the numerous stability calculations made for their ships. These were in accordance with the criteria and norms suggested by IMO [1967a], resolutions A-167 and A-168. Maritime Authorities of countries, which had already used their own national stability regulations, compared the results of such calculations with the stability estimations made in compliance with existing regulations. They found out that the stability of some vessels, which were considered safe, was deemed insufficient according to the IMO draft. At the same time, in the Intact Stability Casualty Records, there were found several vessels which perished though their stability was considered sufficient according to the IMO draft. The comparison of the results obtained in different countries did not give any definite answer even to a rather trivial question: which of the existing stability standards are more strict and which ones are less so? The answers obtained by different countries did not coincide in some cases.

No doubt the clear understanding of the contradiction in the balance of merits and demerits caused by introducing new stability standards was a significant achievement of IMO. The economic cost and other difficulties were evident. First, they were connected with the ballasting of many vessels (decreasing their dead-weight); recalculations of stability according to new requirements; a greater number of inclining tests; the issue of the new information on stability properties for many vessels. All these consequences require rather significant expenses. For many developing countries these expenses would increase still more due to the necessity to invite foreign experts to carry out the corresponding calculations and tests. However, there has never existed any method for evaluating how far the risk of capsizing can be reduced by introducing new stability standards.

The deadlock, which arose in further IMO activities for developing Stability Recommendations, was overcome due to the coincidence of three circumstances. First, some dramatic casualties took place late in the 1960's with Russian, English, Japanese, Canadian and other countries fishing vessels, particularly under the conditions of severe icing. The 19th of January 1965, was a dark day in the history of the Russian fishing fleet. Four medium-sized fishing trawlers were lost during one night in the Bering Sea, where more than one hundred vessels were catching herring. The catastrophe was caused by heavy icing.

These and similar accidents were a powerful incentive for the Maritime Administrations to concentrate their efforts on stability problems and on the development of reliable stability standards.

The second important reason for further IMO activities was the fact that by 1968 IMO had completed an unprecedented five-year collection of Intact Stability Casualty Records. The majority of IMO members were engaged in this work. No separate country would be able to collect such a representative set of statistical data concerning the stability characteristics of lost vessels and the circumstances of their loss. The stability casualty is generally fatal not only for the ship itself, but also for the crew and the passengers. As a rule, there are no surviving witnesses. In the case of the four Russian trawlers, only one man was taken alive from the bottom of the capsized trawler "Boxitogorsk" which was floating with its keel up. The other 96 crew members of these vessels perished with their ships.

These Casualty Records gave the basic data for collective investigations by German and Polish experts and for their suggestions on the stability criteria and norms. These suggestions were then submitted in 1967 as the first draft of the IMO Stability Recommendations for cargo, passenger and fishing vessels, [IMO, 1967], IMO resolutions A-167 and A-168.

In 1968, the Russian delegation in the IMO Working Group on Safety of Fishing Vessels suggested its method for comparison of rigidity and effectiveness of various stability criteria and norms [Sevastianov, 1968a; IMO, 1967a]. The method uses the concept of the so-called "critical" height of the vessel's centre of gravity above a baseline. This concept is very convenient for the comparison with the results achieved in accordance with different sets of stability criteria and norms. The same unified measure of stability was applied to compare the actual stability of capsized vessels with that stability level which would be a sufficient minimum for the same vessels in accordance with various stability standards. The detailed description and foundations of these vessels were published in Russian journals and books [Sevastianov, 1968a, 1970]. As far as we know, some concepts were introduced there for the first time. One of these concepts was an average reserve provided by the given stability standard

$$\bar{x} = \frac{\sum_{i=1}^{n} x_i}{n} \tag{1.1}$$

Where: i=1,2,3,... n is an index of a certain vessel in the list of the lost vessels; n is the total number of the vessels in the list;

$$x_i = \frac{KG_i - KG_{cri}}{KG_{cri}} \tag{1.2}$$

The sign of the f_i -value is to be determined by the numerator of this fraction.

It is obvious that, if $x_i > 0$ the lost vessel's stability would be recognized insufficient in accordance with the given standard ($KG_i > KG_{cri}$) and such a conclusion would be justified by the fact of the loss. But the inequality $x_i < 0$ means an error of the given standards because the stability of this capsized vessel would be evaluated as quite sufficient.

It is evident that those standards which give less scatter (variance) of their values x_i are more relevant to the real circumstances of vessel loss. That's why it was possible to suggest a special measure of inadequacy for any stability standards. This measure is the standard deviation

$$\sigma_x = \sqrt{\frac{\sum_{i=1}^{n} (x_i - \bar{x})^2}{n-1}} \tag{1.3}$$

The introduction of the concepts of a mean stability reserve $\bar{x}$ and σ_x lead to one more concept, namely a concept of the ideal stability standard. Such an imaginary standard would be able to predict for all registered stability casualties the true "critical" KG -values, that is just those values which actually took place at the moment of loss of the perished vessels. These predicted KG_{cri} -values would exactly correspond to the circumstances of each casualty. In other words, such ideal standards applied to the lost vessels would give the values:

$$x_i = \frac{KG_i - KG_{cri}}{KG_{cri}} \equiv 0 \tag{1.4}$$

$$\bar{x} = \frac{\sum_{1}^{n} x_i}{n} \equiv 0 \tag{1.5}$$

$$\sigma_x = \sqrt{\frac{\sum_1^n (x_i - \bar{x})^2}{n-1}} \equiv 0 \tag{1.6}$$

Of course, such ideal standards should take into account the smallest variations of any arguments and parameters influencing stability in any imaginable situation. But we must admit that mankind will hardly be able to compile such precise equations of ship motion, such detailed descriptions of external forces and in addition to have at its disposal a sufficient number of absolutely precise and detailed stability casualty records. This is why it is impossible to achieve such an ideal standard in practice. However, it is equally true for any other ideal concepts; for example, for the ideal fluid, ideal propeller or ideal thermodynamic cycle in combustion engines. Nevertheless, this consideration does not make such ideal concepts useless because they indicate those limits, which cannot be exceeded by efforts of human wisdom and inventiveness. In any case, the ideal models permit comparison of the suitability of different approximate standards, mathematical models of a fluid, construction of machines and propulsive devices.

It was found that, in particular, that the set of x_i - values taken from a sufficient number of capsized vessels may give the evaluation of some conventional guarantee of safety, which is provided by the given set of stability criteria and norms.

Such a conventional guarantee is the average probability of safe navigation of a vessel belonging to a fleet. This fleet consists of vessels of the same type, which are included in the list of casualty records. It is also supposed that the vessels of this fleet keep - permanently - the position of their centers of gravity at the height which is critical for each of them in accordance with the given stability standard and its criteria and norms.

Such a conventional guarantee can be calculated if one knows the law of distribution of the x_i values besides the $\bar{x}$ and σ_x values. In the terms of the theory of probability and mathematical statistics it means that we consider the total number of casualties as a set of test results in which the random continuous variable f was realized in the values of x_i . A special checking procedure was carried out in accordance with so called χ^2-criterion of compliance. It confirmed that normal (Gaussian) distribution law might be used with an acceptable degree of accuracy to express the distribution of x. The density of distribution in accordance with the normal law may be written in the form

$$f(x) = \frac{1}{\sigma_x \sqrt{2\pi}} e^{-\frac{(x-\bar{x})^2}{2\sigma_x^2}} \tag{1.7}$$

The curves in Fig. 1.1 show the normal distribution obtained on the basis of IMO Intact Stability Casualty Records [Sevastianov, 1970] for fishing vessels. Curve N 1 corresponds to IMO Stability Recommendations for Fishing Vessels, [IMO, 1968]. Curve N 3 is drawn for a simplified stability criterion (see formula 1.19 below), which was discussed at an early stage of IMO activity. Curve N 2 is calculated according to the Rules for Classification and Construction of Seagoing Steel Vessels issued by the Russian Register in 1967.

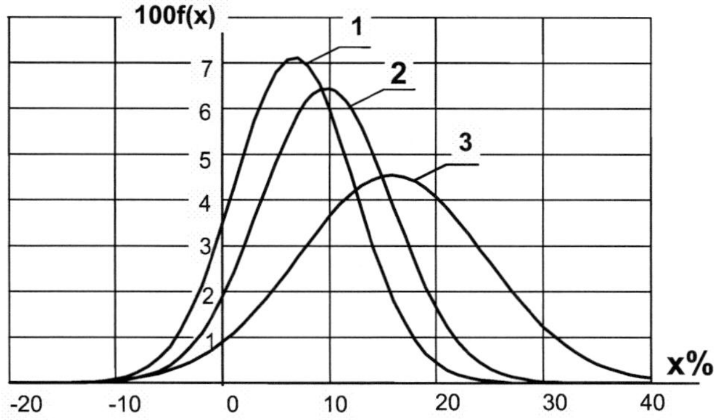

Fig 1.1 Normal distribution of accidental value X_i:

1.- IMO Stability recommendation, 1968. 2.- Draft Stability Standards of the Russian Register, 1967. 3.- Simplified criterion, formula (1.19).

It is useful to recall that the law of distribution of random variables may be written in an integral form:

$$P(x) = \int_{-\infty}^{x} f(x)dx \tag{1.8}$$

If the probability density $f(x)$ corresponds to the normal law (1.7) then:

$$P(x) = \frac{1}{\sigma_x \sqrt{2\pi}} \int_{-\infty}^{\varphi} e^{-\frac{(t-\bar{x})^2}{2\sigma_x^2}} dt \tag{1.9}$$

The function $P(x)$ may be interpreted as the probability of such an accidental event, which is to be determined by the double inequality:

$$-\infty < x_i < x \tag{1.10}$$

Here x_i is any x -value chosen at random and:

$$P(x) = P(-\infty < x_i < x) \tag{1.11}$$

The integral (1.9) may be calculated by numerical methods or by application of so called Laplace function or by the integral of probability tabulated in the mathematical manuals.

$$\Phi(y) = \frac{2}{\sqrt{2\pi}} \int_0^y e^{-\frac{z^2}{2}} dz \ ; \tag{1.12}$$

Where

$$y = \frac{|x - \bar{x}|}{\sigma_x} \tag{1.13}$$

Consequently,

$$\begin{aligned} y &= \pm\infty \quad at \quad x = \pm\infty \\ y &= -\frac{\bar{x}}{\sigma_x} \quad at \quad x = 0 \\ y &= 0 \quad at \quad x = \bar{x} \end{aligned} \tag{1.14}$$

Thus the value of $P(x=0)$ expresses the probability of an event that the value of x will be within the interval $-\infty \le x_i \le 0$, that is x_i chosen at random will be negative.

The value of $P(x=0)$ is given by the formula:

$$P(x=0) = \frac{1}{2}\left[-\Phi(x=-\infty) + \Phi\left(x = -\frac{\bar{x}}{\sigma_x}\right)\right] = \frac{1}{2}\left[1 - \Phi\left(\frac{\bar{x}}{\sigma_x}\right)\right] \tag{1.15}$$

We have already come to the conclusion that a value $x_i < 0$ means an error for the given stability standard. Such a standard would consider the corresponding capsized vessel quite stable since its actual height of the centre of gravity was below the critical value at the time of loss. Therefore, the probability of $P(x=0)$ is a measure of unreliability for such a standard. Then the probability of correct sign of x_i

$$\Gamma = 1 - P(x=0) \tag{1.16}$$

It may be called a conventional guarantee of safety provided by the given stability standard. By using the word "conventional" we must bear in mind some special meaning. We assume that the master of a ship thoroughly controls the height of the actual centre of gravity of his vessel to prevent its shifting above the critical level KG_{cr} determined by the given stability standard. It means that at any moment the following inequality takes place:

$$KG \le KG_{cri} \tag{1.17}$$

If we assume that the real guarantee would be greater than the conventional one, then the conventional guarantee may be considered as an estimation of the minimal value of a real guarantee which is provided by the given stability standard. Fig. 1.2 shows the curves $P(x)$ for the stability requirements of the Russian Register (curve 1), for the IMO Recommendations, 1968 (curve 2, fishing vessels) and for two rather simple but primitive standards N1 and N2 which were considered at the very beginning of the discussion on Stability Recommendations. These two standards may be reduced to the setting of the critical initial metacentric height values:

$$GM_{cr1} = 0.20 + 0.035\frac{B}{f}, \quad \text{m} \tag{1.18}$$

$$GM_{cr2} = 0.50 + 0.035\frac{B}{f}, \quad \text{m} \tag{1.19}$$

Here: GM_{cr1} is the critical (the least permissible) initial metacentric height in accordance with the first variant of the standard (1.18); GM_{cr2} is the same for the second variant (1.19); B is breadth of a vessel amidships; f is the least actual freeboard value in the given loading conditions. Other particulars of size and form of a vessel are not taken into account by these variants of suggested standards. Table 1.1 contains calculated numerical characteristics $\bar{x}$,σ_x , $P(x=0)$ and Γ .

Fig. 1.2 clearly shows that a high value of guarantee, Γ, may be achieved by two alternative methods.

The first method suggests improving the stability standard by the minimization of its measure of inadequacy σ_x, that is by reducing σ_φ as close to zero as is possible (see formula 1.15). For a constant value $\bar{x}$, σ_x decreases, the slope of the curves $P(x)$ increase and the conventional safety guarantee Γ is tending to the ideal value Γ=1 (Fig. 1.2).

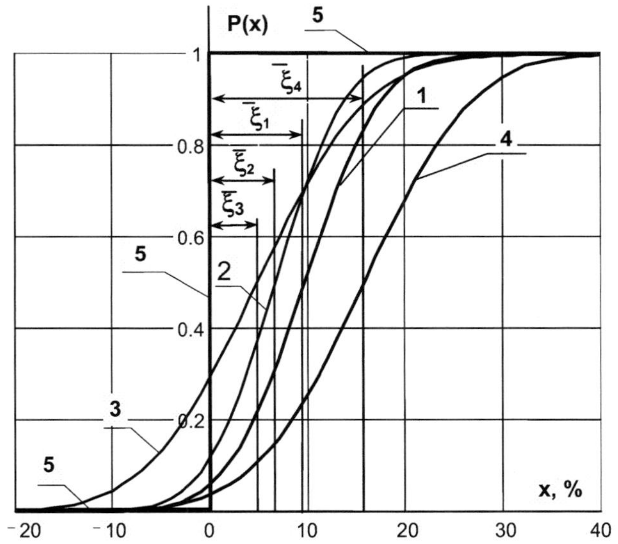

Fig. 1.2 Integral distribution law of the accidental value x

1.- Draft Stability Standards of the Russian Register 2 .- IMO Stability recommendation, 1968., 1967. 3.- Simplified criterion, formula (1.18). 3.- Simplified criterion, formula (1.19). 5.-"Ideal" Standards

The second advantage of this method is that the average reserve $\bar{x}$ may be decreased with decreasing σ_x-value at a constant or even increased value Γ.

These advantages make the first method of improving stability standards the most rational.

The second method is much simpler, but not so rational. It may be implemented by assigning more conservative norms for each criterion. Let us make this point clear by the example of standards N1 and N2. They differ from each other only in the critical values of GM_{cr} which for standard N2 is 0.3 m larger than N1. It means that the critical height of the centre of gravity KG_{cr2} of a certain vessel would be smaller than KG_{cr1} by the same value of 0.3 m. Consequently the average stability reserve $\bar{x}_2$ should be larger than $\bar{x}_1$. Therefore, the curve $P_2(x)$ will be shifted along the x -axis towards the greater x -values. But the σ_x -value remains almost constant. That's why the slope of the curve $P_2(x)$ is practically equal to the slope of curve $P_1(x)$. It is clear that standard N2 provides a greater guarantee as a whole than N1. However, it is achieved by more severe requirements for all vessels, not only for the vessels with an inherently worse stability. To meet these requirements it would be necessary to lower the centre of gravity even for those types of vessels, which have successfully operated for many years. Therefore, the seductive simplicity of the standards N1 and N2 might be extremely ineffective though the standard N2 has the highest guarantee among other considered stability standards. At the same time, IMO Recommendations [IMO, 1968], have the least measure of inadequacy σ_x and ensure rather a high conventional guarantee Γ. To achieve the same guarantee, which is provided by the more complicated Rules of the Russian Register, it would be enough to increase the average reserve $\bar{x}$ from 6.7% up to 8.9%.

These considerations and estimations of various suggested stability standards were to become the first international stability standard, which is known all over the world as IMO Stability Recommendations, [IMO, 1968]. Although these Recommendations have not yet become a part of the ratified International Conventions, they are widely applied in practice and have even became a component of national stability requirements of many countries*.

Table 1.1 Reliability Characteristics of Four Different Stability Standards for Fishing Vessels

Stability standards	Measure of mean reserve $\bar{x}$ %	Measure of inadequacy σ_x %	Measure of unreliability $P(x=0)$ %	Conventional guarantee Γ %
1. Russian Register of Shipping (1967 draft)	9.7	6.2	5.6	94.3
2. IMO Stability Recommendations for Fishing vessels, 1968	6.7	5.6	11.5	88.5
3. Simplified Standard N1	4.8	8.9	29.5	70.5
4. Simplified Standard N2	15.8	8.8	3.6	96.4

1.2 Vectors of Assumed Situations and Loading Conditions. Risk Function

Considering the general philosophy of stability standardization, we came across a number of concepts, which are closely connected with random events and values. They include such data as weather conditions at certain moments or time intervals given in advance; the rolling characteristics in irregular seas; fluctuation of the height of the vessel's centre of gravity and of the draft or displacement, etc. We can state that these events may either occur or not occur with some probability. But we cannot predict them in a deterministic sense. While analyzing statistical data, we applied the apparatus of mathematical

* As of March, 1993

statistics based on the theory of probability. Comparing different stability standards, their criteria and norms, we had to introduce the unified stability measure, which is the relative height of the vessel's centre of gravity. It fluctuates by being influenced by random circumstances. So it is a random process.

Then, a natural question arises: is it possible to consider the whole chain of sequential events during the lifetime of a vessel as a flow of random impulses or processes which can capsize or not capsize the vessel at some moment of time with some probability? In this sense, capsizing itself might be considered as a random event which can be realized or not realized under certain conditions.

The problem of setting stability standards thus formulated has been investigated for more than thirty years. But it does not mean that any use of the theory of probability or statistics is equivalent to the probabilistic approach in the true sense of the word. For a long time many attempts have been made to take into account the irregular rolling amplitudes and the gusts of wind by setting such amplitude values or such a value of wind velocity in a gust which have a rather small probability of being exceeded.

Nevertheless, it would not be quite correct to state that the figures and formulae given by such standards provide for such rolling amplitude and such wind gust force which cannot ever be exceeded. The form of criteria and norms does not differ from the deterministic ones. And the guarantee of safety still remains conventional.

In further analysis we shall use the term "probabilistic approach" only for such an approach for setting stability standards where the probability of capsizing itself is used as a universal criterion or may be expressed simply and solely through other functions. In many cases such a replacement is more convenient. For instance, we can consider a random event of non-capsizing instead of an event of capsizing. But these two events are opposite in their meanings. Therefore their probabilities are related by the equality:

$$P(X) + P(\overline{X}) = 1 \qquad (1.20)$$

We use here the notations X for capsizing and $\overline{X}$ for the opposite event, that is, for non-capsizing. Therefore:

$$P(X) = 1 - P(\overline{X}) \qquad (1.21)$$

Later, we shall analyze some other values and functions, which may be expressed by $P(\overline{X})$ and vice versa.

The main advantages of the probabilistic approach will be considered in comparison with the deterministic one. Let us introduce some elementary concepts whilst not aiming to apply strict definitions in the beginning. It is intuitively clear that one can talk about more or less dangerous situations at any moment of the vessel's life. Stormy conditions are usually connected with a more significant risk than conditions of still weather if the loading conditions in both situations are the same. Nevertheless, some loading conditions can be very unfavorable for stability. Then, the calm sea state in a harbor may be more dangerous than stormy conditions at sea for the same vessel, if the stowage of its cargoes and stores are especially unfavorable in the port. Hence, it is necessary to define two concepts: the situations and the loading conditions which should be taken into account while calculating stability (assumed situations and assumed loading conditions).

The concept of an assumed situation was partially discussed in subchapter 1.1. It includes weather characteristics: wind, sea state, heading of the vessel relative to the wind velocity and to the general direction of the waves propagation, speed of the vessel, additional forces caused by the specific intended service of the vessel (pull of the towing rope, forces caused by fishing gear, reaction of ballasting devices, etc.). Usually, it is possible to determine the situation by a set of parameters, such as:

- Mean wind velocity u_{Am};
- Standard deviation of pulsation components of the wind velocity σ_u ;
- Height of significant waves h_s;
- Mean period of visible waves T_m ;
- Speed of the vessel v;
- Heading to waves ψ;
- Heading to wind ψ_A;
- Additional external forces (or resultant vector of such forces $\vec{F}_e$).

One may consider any set of many parameters as a multidimensional vector. Therefore we shall denote a vector of a certain situation as:

$$\vec{S} = \vec{S}\left(u_{Am}, \sigma_u, h_s, T_m, v, \psi, \psi_A, \vec{F}_e\right). \tag{1.22}$$

The loading conditions may also be represented as vector $\vec{L}$. For the purpose of stability estimation it is usually sufficient to take into account the following components of this vector: vessel's displacement W, the height of the centre of gravity KG, and radius of gyration around the central longitudinal axis R_x. Then one can write

$$\vec{L} = \vec{L}(W, KG, R_x) \tag{1.23}$$

For cases when the dynamic stability is to be considered, it is more convenient to use another system of arguments, which is fully equivalent to (1.23):

$$\vec{L} = \vec{L}(W, M_{xoy}, \omega_\phi) \tag{1.24}$$

Here, M_{xoy} is a static moment of displacement relating to the base plane of a vessel: $M_{xoy} = W \cdot KG$; ω_ϕ is a natural rolling frequency.

If we are ready to agree that to evaluate stability it is enough to know two vectors $\vec{S}$ and $\vec{L}$ and in addition to have the vessel's lines drawing and its general arrangement drawings, then we can introduce another important concept called "risk function". We can measure the degree of risk at any situation and any loading conditions with the help of such a function:

$$\lambda = \lambda(\vec{S}, \vec{L}) \tag{1.25}$$

Let us bear in mind that all arguments in these formulas undergo significant alterations: weather conditions, heading, speed, loading conditions, etc. In other words, they all are functions of time:

$$\vec{S} = \vec{S}(t)\,; \quad \vec{L} = \vec{L}(t)$$

Consequently the function λ itself is:

$$\lambda = \lambda(t) \tag{1.26}$$

It is rather a difficult task to write this dependence in an explicit form. So we shall discuss the possible methods of calculation of the risk function in other parts of our analysis. Now, let us note that, if the vectors $\vec{S}(t)$ and $\vec{L}(t)$ are stationary at some time interval, then the risk function $\lambda(t)$ at the same interval is a constant number. A stationary situation and stationary loading mean that the components of these vectors are also expressed by constant numbers including such parameters as statistical characteristics of irregular sea, gusty wind, probabilities of various values of displacement, heights of the centre of gravity, etc.

Let us suppose that we know function $\lambda(t)$ for a vessel of certain type - A - and for the other type - B (Fig. 1.3). In addition, we suppose that the maximal and the minimal values of these two functions are equal during some large time interval T, that is:

$$\begin{cases} \lambda_{A\min} = \lambda_{B\min} \\ \lambda_{A\max} = \lambda_{B\max} \end{cases}$$

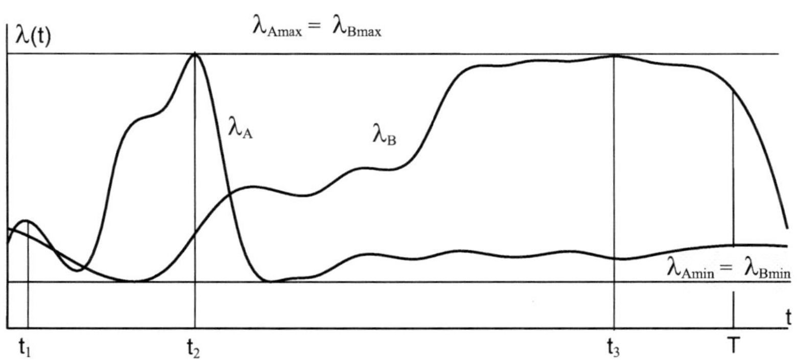

Fig. 1.3 Estimation of safety level on the basis of risk functions for vessels of type A and type B

But the run of these functions on the whole is different. One may wonder: which type of vessels, A or B, is safer?

At least three answers may be given to this question depending on what moment or what interval of time this question refers to.

Indeed, it is possible to compare the safety of these vessels:

- At the same time moment t_1 ;
- At different time moments t_1 and t_2 which correspond to conditions $\lambda_A = \lambda_{A\max}$ and $\lambda_B = \lambda_{B\max}$;
- During the given interval of time T as a whole.

Let us note that from the viewpoint of existing deterministic stability standards, there is no difference between the second and the third answers since both of the variants would be considered equally dangerous (and the most dangerous!) during the interval *T*.

But type A is much more reliable than type B from the viewpoint of people who are working or traveling aboard the vessel and from the viewpoint of the ship owner or insurance company. One can assert this because the critical state $(\lambda \approx \lambda_{max})$ lasts only during a small part of the time interval *T* for vessel A while for vessel B it lasts much longer. This fact increases the chance of vessel A avoiding capsizing.

The probabilistic approach to the evaluation of stability is in full agreement with the last point of view. One may say that the deterministic approach gives a definite answer to the question whether the vessel will capsize or not when being affected by the external forces caused by determined vector $\vec{S}$ under determined loading conditions $\vec{L}$. This answer may be an unconditional "Yes" or "No" without any stipulations.

The probabilistic answer is also "Yes" or "No" but, with two principal stipulations: during the given time interval and with a certain probability.

For a better understanding of the difference between the deterministic and the probabilistic approach, it would be useful to think over an example taken from quite a different field. Let us imagine some advanced fortifications, which are under fire from the enemy at random moments of time. And we can see a soldier standing at the front line near a high-ranking general who came from the headquarters to inspect the regiment. The risk of being killed during the shooting is the same for the soldier as for the general. But an important question arises: who of them has a greater chance to survive until the end of the whole war? The answer is clear because such dangerous shooting is a constant or in any case a frequent situation for the soldier but, it is only a short episode in the general's service.

So, the heart of the matter in the development of a probabilistic approach to setting stability standards is the discovery of links between the probability of capsizing and a certain time interval given in advance of the capsizing event.

1.3 The Probability of Survival and Its Interpretation in the Task of Stability Estimation

The simple considerations and concepts given above may be arranged in the strict mathematical formulas. In the papers [Sevastianov, 1963, 1970, 1978, 1979] the risk function $\lambda(t)$ has another definition, which is necessary for our stability investigation. It is the probability of capsizing during the time unit adjacent to the moment *t*. This conditional probability should be determined with the assumption that the capsizing has not yet occurred until the current moment *t*. In every infinitely small interval dt only an infinitely small probability of capsizing exists and it is proportional to dt :

$$dP_{t,dt}(X) = \lambda(t)dt \qquad (1.27)$$

The probability of non-capsizing during the same interval is equal to:

$$dP_{t,dt}(\overline{X}) = 1 - \lambda(t)dt$$

In 1963 it was proved that the probability of non-capsizing during any finite interval might be expressed by the formula [Sevastianov, 1963]:

$$P_T(\overline{\mathrm{X}}) = \exp\left[-\int_0^T \lambda(t)dt\right] \tag{1.28}$$

Here the symbol exp[x] means e^x.

The same expression is used in the contemporary general theory of the reliability to calculate the probability of the operation of an element without a failure or a system during the given time interval. In English scientific literature this function has its own impressive personal name, "probability of survival". We shall use this term the following discussions.

The functions $\lambda(t)$ and $P_T(\overline{\mathrm{X}})$ have a number of properties which will be analyzed.

1. It is evident from (1.27) that

$$\lambda(t) = \frac{dP_{t,dt}(X)}{dt} \tag{1.29}$$

2. Being a probability, the risk function cannot be negative:

$$\lambda(t) \geq 0 \tag{1.30}$$

Therefore the integral:

$$F = \int_0^T \lambda(t)dt \geq 0 \tag{1.31}$$

And

$$F = 0 \tag{1.32}$$

Only under the condition $\lambda(t) \equiv 0$

3. $$F \to \infty \tag{1.33}$$

Under the condition $t \to \infty$ if $\lambda(t)$ does not become zero from some moment t till $t = \infty$.

4. Consequently in this case:

$$P_T(\overline{\mathrm{X}}) = e^F \to 0 \quad \text{if} \quad F \to \infty \tag{1.34}$$

It means that the probability of survival becomes less and less in the course of time t: as T increases, $P_T(\overline{\mathrm{X}})$ decreases.

5. According to (1.28), the absolute value of the power of the exponential is expressed by the area between the curve $\lambda(t)$ and the axis of the abscissas within the interval $0 \leq t \leq T$. The same area may be calculated by the formula:

$$F = \int_0^T \lambda(t)dt = \lambda_a T \tag{1.35}$$

Where λ_a is an average value of the risk function within the same interval T. Therefore, this value may be represented graphically by the height of the rectangle with the area equal to F :

$$\lambda_a = \frac{F}{T} \tag{1.36}$$

This value at the given area F does not depend on the form of the risk function curve within the interval T.

6. The probability of survival during the time unit (T=1) is equal to:

$$P_{T=1}(\overline{X}) = e^{-\lambda a} = 1 - \lambda_a + \frac{\lambda_a^2}{2!} - \frac{\lambda_a^3}{3!} + \ldots \tag{1.37}$$

There are statistical data and the results of special calculations [Sevastianov, 1970, 1978] which confirm that in different countries:

$$10^{-4} \le \lambda_a \le 3 \cdot 10^{-3}, \quad 1/\text{year} \tag{1.38}$$

The above range is for small and medium-sized fishing vessels. For larger modern cargo and passenger vessels the value of λ_a is less. Therefore if we choose the time unit equal to 1 year, we may state that λ_a is a very small value and that is why

$$P_{T=1}(\overline{X}) = 1 - \lambda_a \tag{1.39}$$

Formula (1.39) is accurate up to the error of the second order. Then the probability of capsizing per year is equal to

$$P_{T=1}(X) = 1 - P_{T=1}(\overline{X}) = \lambda_a \tag{1.40}$$

7. But it means that average number of capsized vessels per year will be:

$$N_c = \lambda_a N \tag{1.41}$$

Where N is a total number of vessels of the same type, which were in operation during the given year. But in the majority of marine countries there are given state statistics for different kinds of casualties including stability casualties. The main index for each kind of casualty is usually an annual frequency of accidents:

$$n_c = \frac{N_c}{N}, \quad 1/\text{year}$$

Consequently the average value of the risk function is equal to an average annual index which is collected and systematized by the maritime administrations and by the insurance companies.

8. The time t_c from the launching of a vessel until she capsizes is a typical continuous random variable. The probability of such values is governed by the distribution law which is expressed by the density of probability $f(t_c)$.

In some cases it is more convenient to use the integral form of distribution law which is linked with the density of probability $f(t_c)$ by the formula:

$$P(t_c) = \int_{-\infty}^{t_c} f(t_c)dt_c \tag{1.42}$$

Here: t - is a current variable time within the limits of the integral; $P(t_c)$ - is a probability of the inequality $t \le t_c$. It means that the random variable will lie within the indicated limits.

The integrand, function $f(t_c)$, may be determined by the first derivative of:

$$f(t_c) = \frac{dP(t_c)}{dt_c} \tag{1.43}$$

One of the important properties of the curve $f(x)$ for any variable lies in the fact that the first moment of the area under the curve is taken relatively. The origin of co-ordinates is equal to the average value of a variable. In our case it is the average time from the launching of the vessel until its capsizing, that is, the average lifetime of the vessel:

$$t_{ca} = \int_0^{\infty} t_c f(t_c)dt_c \tag{1.44}$$

9. Let us use these properties to determine the function, $f(t_c)$ and the average lifetime, t_{ca}. To do so, it is necessary to understand what is the meaning of the event which we bear in mind stating that the first capsizing will happen at some moment t_c. This event is a composition of two other events. The first of them - $\overline{X}_{ta}$ - is expressed by the fact that there was no capsizing event during the time interval t_c. The second event is the fact of the vessel's capsizing during the infinitely small interval adjacent to the moment t_c.

The probability of the first event being equal to the known value is (see (1.28)):

$$P_f = P_{tc}(\overline{X}) = \exp\left[-\int_0^{t_c} \lambda(t)dt\right]$$

The probability of the second event is an infinitely small value dP_s as this event would happen during an infinitely small interval dt. This probability is expressed by formula (1.27):

$$dP_s = dP_{tc,dt}(X) = \lambda(t_c)dt_c$$

According to the well-known probabilities product rule, the probability of composition of these two events is:

$$dP(\overline{X}_{tc}, X_{tc,dt}) = P_f(\overline{X}_{tc}) \times dP_s(X_{tc,dt}|\overline{X}_{t_c}) = \lambda_{tc} \exp\left[-\int_0^{t_c} \lambda(t)dt\right] \times dt$$

But the same elementary probability may be obtained by the formula (1.43):

Taking into account that it may coincide with any current moment of time, we'll omit the subscript "c" and write:

$$f(t) = \lambda(t)\exp\left[-\int_0^t \lambda(t)dt\right] \qquad (1.45)$$

Hence, the average duration of the lifetime of the vessel:

$$t_a = \int_0^\infty t \cdot \lambda(t)\exp\left[-\int_0^t \lambda(t)dt\right]dt \qquad (1.46)$$

This expression may be significantly simplified if it is assumed that the average value of λ does not alternate under some stationary conditions of operation. Then $\lambda(t) = \lambda_a$ and therefore:

$$t_a = \lambda_a \int_0^\infty te^{-\lambda_a t}dt$$

But the definite integral in this formula is equal to $1/\lambda_a^2$. Readers can verify this statement by elementary integration. So finally in this case:

$$t_a = \frac{1}{\lambda_a} \qquad (1.47)$$

In accordance with these results the distribution laws for the lifetime may be written in the following manner:

$$f(t) = \lambda_a \exp(-\lambda_a t) \qquad (1.48)$$

And:

$$P(t) = \int_0^t f(t)dt = 1 - \exp(-\lambda_a t) \qquad (1.49)$$

Besides:

$$P_T(\overline{X}) = \exp(-\lambda_a t) \qquad (1.50)$$

10. Let us imagine the risk function $\lambda(t)$ as it is shown in Fig. 1.4. Divide the maximal ordinate into n equal intervals and draw lines parallel to the axis t. Then draw perpendicular lines upon the axis t from the points M_1, M_2, M_3,..., M_k. Thus, we have divided the interval T into k sub-intervals. Each of them has its own duration, τ_i, which is to be determined by the corresponding instants of time, $t_0, t_1, ..., t_{k-1}, t_k$. Table 1.2 gives various probabilistic characteristics of the sub-intervals τ_i, the approximate λ_{ai} - values and probabilities $P_{\tau i}(\overline{X})$ being among them.

Such a procedure makes it possible to calculate the probability $P_T(\overline{X})$ over the whole finite large interval T:

$$P_T(\overline{X}) = e^{-(\lambda_{a_1}\tau_1+\lambda_{a2}\tau_2+\cdots+\lambda_{ak}\tau_k)}$$

Or:

$$P_T(\overline{X}) = \exp\left[-\sum_{i=1}^{k}\lambda_{a_i}\tau_i\right] \quad (1.51)$$

Evidently, the magnitude of sub-intervals may be assigned by some other or even arbitrary manner. In any case, the less the duration's τ_i are, the less the error of such an approximation will be.

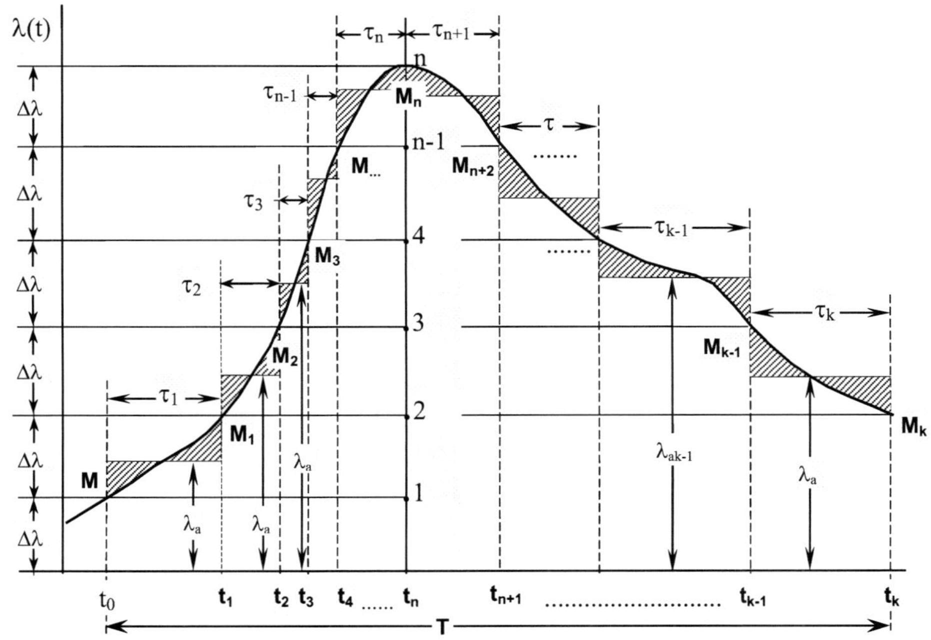

Fig. 1.4 Illustration of an algorithm for determining "probability-of-survival" in accordance with the risk function curve

The average value λ_{ai} for each sub-interval may be found by various procedures. For instance, it may be obtained by the equalization of the areas, which are hatched in fig. 1.4 or by any appropriate method.

It is important for further analysis to emphasize that the probability in the formula (1.51) does not depend on the permutations of the components in the sum $\sum_{i=1}^{k}\lambda_{a_i}\tau_i$.

Table 1.2 Characteristics of sub-intervals

Sub-intervals	τ_1	τ_2	τ_3	...	τ_k
Duration of sub-intervals	$\tau_1 = t_1 - t_0$	$\tau_2 = t_2 - t_1$	$\tau_3 = t_3 - t_2$	...	$\tau_k = t_k - t_{k-1}$
Ordinates at the ends	$\lambda(t_0)$ $\lambda(t_1)$	$\lambda(t_1)$ $\lambda(t_2)$	$\lambda(t_2)$ $\lambda(t_3)$	...	$\lambda(t_{k-1})$ $\lambda(t_k)$
Mean values	λ_{a1}	λ_{a2}	λ_{a3}	...	λ_{a4}
Probability of survival during the sub-interval	$P_{\tau 1}(\overline{X}) = = e^{-\lambda_{a1} t_1}$	$P_{\tau 2}(\overline{X}) = = e^{-\lambda_{a2} t_2}$	$P_{\tau 3}(\overline{X}) = = e^{-\lambda_{a3} t_3}$	...	$P_{\tau k}(\overline{X}) = = e^{-\lambda_{ak} t_k}$

1.4 The Problems of Criteria and Norms in the Probabilistic Approach to Stability Standards

Let the symbol T signify the time interval given in advance. The quantity $P_T(\overline{X})$ could evidently be suggested as a universal stability criterion. Indeed, the vessels which may be characterized by the numerical value of $P_T(\overline{X})$ significantly below $P_T(\overline{X}) = 1$ may hardly be recognized sufficiently stable and vice versa, the value $P_T(\overline{X})$ being very close to 1 under the severe weather and loading conditions, might be considered completely safe. Moreover, the usage of $P_T(\overline{X})$ -value as a stability criterion fully removes the separate problem of guarantee since this value is at the same time a guarantee itself.

It should also be noted that it is a real, but not a conventional guarantee since it takes into account, in principle, both the real parameters of numerous assumed situations in explicit form and the real fluctuations of loading conditions.

Besides $P_T(\overline{X})$ some other alternative criteria may be suggested, namely:

- Risk function averaged over all possible situations and loading conditions or, what is the same, over the whole lifetime of a vessel;
- The average duration of lifetime of a vessel t_a from the launching until the capsizing.

These alternative kinds of criterion are closely connected with each another:

$$P_T(\overline{X}) = e^{-\lambda_a t} = e^{-T/t_a} \tag{1.52}$$

$$\lambda_a = -\frac{\ln P_T(\overline{X})}{T} = \frac{1}{t_a} \tag{1.53}$$

$$t_a = \frac{1}{\lambda_a} = -\frac{T}{\ln P_T(\overline{X})} \tag{1.54}$$

Thus, these possible criteria may be recognized equivalent. But they are not equally convenient because of the different possibilities to assign corresponding norms for each of these criteria.

Let us start our analysis from the criterion t_a. Taking into account the double inequality (1.38), we see that for the present time such an average value t_a for modern small and medium-sized fishing vessels is to be within the bounds:

$$330 < t_a < 10{,}000 \text{ years}$$

For large cargo and passenger vessels, this value may be significantly larger. It is absolutely clear that no vessel has a chance to remain in operation during such a "theoretically possible" lifetime. And no one naval architect could justify the choice of such a stability criterion though the same criterion of reliability is successfully applied to other objects with a short lifetime, for instance, to electric light bulbs. Thus, we must reject this criterion because of the impracticality of its norm for estimating the vessels' stability.

The next alternative criterion $P_T(\overline{X})$ leads to difficulties of another kind. It might be acceptable for evaluation of safety in the cases of less dangerous accidents caused, for instance, by collisions or by some failure of the main engine. In such circumstances the accident usually lasts for hours. In most cases it is sufficient enough to save the crew and, sometimes, the vessel itself. But stability casualties last often for only some dozen seconds. Therefore, the safety of the crew can only be ensured against the threat of capsizing by a vessel having sufficient stability. That's why it is possible to accept the norm of probability of vessel's survival in the situation of collision within the bounds:

$$0.6 < P < 0.8$$

But if we are ready to adopt the existing level of safety against the threat of capsizing and to keep the statistical values λ_a (see 1.38) then we have to ensure:

$$0.942 < P_T(\overline{X}) < 0.998$$

However, the lower bound is hardly sufficient nowadays. Indeed, such a norm would mean that approximately six vessels out of one hundred will, on the average, capsize during twenty years of operation. Even the norm 0.99 is not indisputable since the probability of survival of each one hundred vessels is only:

$$P_{100T}(\overline{X}) = 0.99^{100} \approx 0.37$$

Consequently, a sufficiently conservative norm for criterion $P_T(\overline{X})$ should be between 0.99 and 1. But the absolutely stable vessel with $P_T(\overline{X})=1$ is nothing but an unattainable theoretical image. Thus, we have a very narrow range to choose a sufficient and achievable norm.

That narrowness gives rise to doubt: is the criterion $P_T(\overline{X})$ sensitive enough for the usual measures of stability improvement?

Indeed, let us consider the practical question: if the value of $P_T(\overline{X}) = 0.99$ is insufficient, but the value $P_T(\overline{X}) = 1$ is unattainable, then how much ballast should be laid onto the vessel's bottom to increase the stability criterion from 0.990 up to 0.991 and to ensure the observance of such a stability standard?

It is necessary to carry out rather lengthy calculations to give the exhaustive answer to this simple question. But it is clear without any calculations that by lowering the centre of gravity as much as possible we can increase the existing probability of survival only by a very small quantity. Fig. 1.5 clarifies this statement. The abscissa of this graph is the relative lowering of the risk function. The initial magnitude of $P_T(\overline{X})$ is read along the second axis. The curve on this graph shows how significant the relative decrement of the risk function should be to increase the initial value of probability $P_T(\overline{X})$ by 0.001.

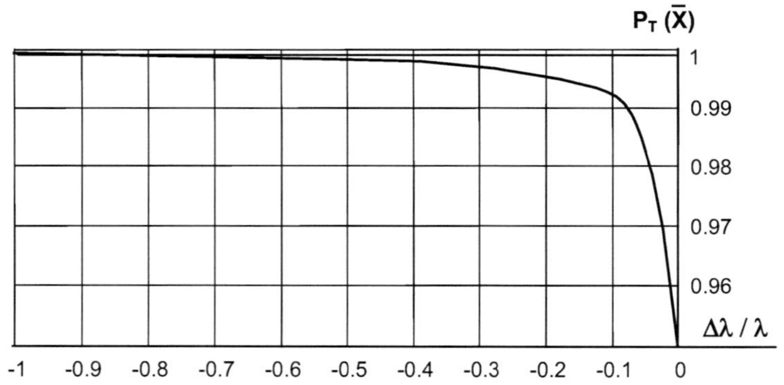

Fig. 1.5 Sensibility of probability of survival relating to the risk function

The reader can see that the decrement of λ_a is rather small at $P_T(\overline{X}) < 0.9$. But if the initial value $P_T(\overline{X})$ is more than 0.95, the required decrease in the λ_a - value becomes drastically large.

The same conclusion may be obtained analytically with the help of the first derivative:

$$\frac{dP_T(\overline{X})}{d\lambda_a} = \frac{P_T(\overline{X}) \times \ln P_T(\overline{X})}{\lambda_a}$$

This relationship might be considered as a sensitivity measure of the probability $P_T(\overline{X})$ to the alterations of the risk function λ_a.

Let us write it in a new form:

$$\frac{dP_T(\overline{X})}{P_T(\overline{X})} = \ln P_T(\overline{X}) \times \frac{d\lambda_a}{\lambda_a} \qquad (1.55)$$

Hence, it is clear that the increment $dP_T(\overline{X}) \to 0$ under the condition $P_T(\overline{X}) \to 1$ with any alterations of risk function λ_a.

Such a small sensitivity of the above-analyzed criterion in relation to the large sensitivity of the risk function (and of the usual stability characteristics) is the main reason to refute this criterion.

The third type of criterion suggested above is the value that indicates a risk function is averaged over the whole of a vessel's lifetime.

We have discussed its meaning as an index of annual casualty statistics. This casualty index, n_{ca}, averaged over a reasonable number of years may be suggested as a norm of the criterion λ_a and the main stability requirement may be written as follows:

$$\lambda_a \leq \frac{n_{c_a}}{k} \tag{1.56}$$

Here: k is a factor of reserve which meets the social need to keep ($k = 1$) or to increase ($k > 1$) the existing level of safety.

Such a norm seems natural and understandable enough. The acceptance of such a criterion and such a norm would make it possible to attain the general goal of probabilistic stability standards and to ensure equal safety for different vessels working under alternating conditions of weather, loading and intended service.

But the main obstacle to this criterion is the difficulty in finding the quantity λ_a, taking into account the existing (or new if necessary) solutions of the stability tasks in various assumed situations. Now, we will begin our discussion of these problems.

The general idea of such an analysis is as follows. It is necessary to obtain the risk function $\lambda(t)$ for probabilistic assessment of ship's safety against capsizing. It depends, in its turn, on time, since navigation conditions (or a vector of assumed situations $\vec{S}$) and loading conditions (or a vector of loading $\vec{L}$) are subject to time. The set of possible situations is, strictly speaking, an infinite set as the number of its elements (situations) is also infinite. But for practical applications of existing computing systems, it is necessary to substitute the infinite set of continuously changing elements by the finite number of discrete stationary situations and stationary loading conditions. When this is done the task of calculating the risk function may be solved step by step, for each certain situation in combination with a certain loading conditions in the beginning and then for the whole set of such combinations.

In the next subchapter we'll try to get such a value of risk function which is an average value within some sufficiently large time interval T. The probabilities of all possible discrete situations and loading conditions will be taken into account by such an average value.

1.5 Algorithm of Averaging of Risk Function

We have found in previous analysis that any form of probabilistic criteria may be determined by the risk function $\lambda(t)$. This function depends on the vector of assumed situations $\vec{S}(t)$ and the vector of loading conditions $\vec{L}(t)$, which in their turn are dependent on time t (see formula 1.24). The main components of these vectors were briefly considered in subchapter 1.2. We shall discuss the properties of these vectors in detail below. Here, we suggest an algorithm of averaging risk function taking into account all possible situations and all possible loading conditions.

From the mathematical point of view (see subchapter 1.3), the risk function is a time density of capsizing probability or, equivalently, is a probability of capsizing during the time unit adjacent to a time moment t. Function $\lambda(t)$ is to be calculated on the assumption that for a certain situation $\vec{S}(t)$ and a certain loading conditions $\vec{L}(t)$ had occurred by the moment t. This probability is a conditional one because we suppose that the capsizing had not taken place before the moment t.

Then the infinitely small probability of capsizing during the infinitely small time interval dt may be expressed in accordance with (1.27) by the formula:

$$dP_{t,dt}(\mathrm{X}) = \lambda\left(X, \vec{S}(t), \vec{L}(t)\right)dt \tag{1.57}$$

Here, symbols in the brackets may be interpreted as the enumeration of the following accidental events:

- Event A_1 is the occurrence of the situation vector $\vec{S}(t)$ by the moment t;
- Event A_2 is the occurrence of the vector of loading conditions $\vec{L}(t)$ by the same moment t:
- Event $A_3 \equiv X$ is the event of the vessel capsizing during the time interval dt adjacent to the moment t (under the condition that no capsizing had taken place before moment t).

To shorten further intermediate transformations, we can rewrite formula (1.57) as follows:

$$dP_{t,dt}(A_1, A_2, A_3) = \lambda(A_1, A_2, A_3)dt \ . \tag{1.58}$$

For our further analysis, let us recall two important rules of probability theory. The first is the product rule, which concerns the probability of coincidence of several accidental events.

$$\begin{aligned} &dP_{t,dt}(A_1, A_2, A_3, \ldots, A_n) = \\ &= P(A_1) \times P(A_2|A_1) \times P(A_3|A_1, A_2) \times \cdots \times P(A_n|A_1, A_2, \ldots, A_{n-1}) \end{aligned} \tag{1.59}$$

Symbol $P(A_1, A_2, ..., A_n)$ denotes the probability of simultaneous occurrences of all these events $A_1, A_2, ..., A_n$, that is their coincidence;

$P(A_1)$ is an unconditional (absolute) probability of event A_1;

$P(A_2|A_1)$ is a conditional probability of event A_2 which should be found assuming that event A_1 has taken place;

$P(A_3| A_2, A_1)$ is a conditional probability of event A_3 which should be found assuming that the coincidence of events A_1 and A_2 have taken place;

$P(A_n| A_1, A_2, ..., A_{n-1})$ is a conditional probability of event A_n which should be found assuming that the coincidence of all events $A_1, A_2, ..., A_{n-1}$ have taken place.

The order the events are numbered in formula (1.59) is of no significance.

The second rule (addition theorem) is to be applied if we face several incompatible (alternative) outcomes of a stochastic test. Let us imagine that k of different incompatible results R_1 or R_2 or ... R_i or ... R_k may appear in some test. We may consider each of these results as an accidental event which has its own probability $P(R_1)$, $P(R_2)$ ... $P(R_k)$.

The question arises: what is the probability that one of a certain smaller set of the possible incompatible results will appear as an outcome of the test, for example:

The set of events "R_g or R_j or R_h" we'll call a sum of events $R_g + R_j + R_h$.

Then the rule of probability addition should be applied. It gives:

$$R_g + R_j + R_h = P(R_g) + P(R_j) + P(R_h)$$

Or in a brief form:

$$P\left(\sum R_i\right) = \sum P(R_i) \tag{1.60}$$

Where $i= g, j, h$.

Hence, for the sum of all the possible outcomes ("full group of incompatible events" from $i = 1$ to $i = k$):

$$\sum_{i=1}^{k} P(R_i) = 1 \tag{1.61}$$

The important consequence of these two rules is expressed by the composite probability formula. Let us calculate the probability of some event A which can occur only together with one of the set of incompatible events H_1, H_2 ,..., H_g. The composite probability formula may be written as follows:

$$P(A) = \sum_{i=1}^{g} P(H_i) \times P(A|H_i) \tag{1.62}$$

Let us use formulae (1.59), (1.60) and (1.62) to transform the equality (1.57) into a more convenient form. First of all, let us note that the set of possible situations and the analogous set of possible loading conditions are, generally speaking, infinite and uncountable sets. It involves some difficulties in analyzing and computing. To avoid them, it is expedient to substitute these sets by the finite and countable sets of discrete situations S_i ($i = 1, 2, 3,.., n$) and discrete loading conditions L_j($j = 1, 2, 3,..., m$). In other words we shall represent the whole set of possible assumed situations as an enumeration of n different quasi-stationary situations. The same approach may be used for a set of loading conditions. Besides, it should be noted that each discrete situation S_i has its own probability of existence at any time moment chosen at random. This probability may be averaged over the large interval of time T:

$$P(S_i) = \frac{T_i}{T} \tag{1.63}$$

Here, T is a sufficiently large time interval approximately equal to, for example, with the vessels life time; T_i is an average total duration of situation S_i taken over the time interval T.

The analogous statement may be expressed in relation to any discrete loading conditions L_j with its probability:

$$P(L_j) = \frac{T_j}{T} \tag{1.64}$$

Let us choose some combinations of discrete situation S_i and discrete loading conditions L_j. Now we have the right to formulate the task: to express the probability of capsizing in situation S_i under loading conditions L_j during the infinitely small time interval dt. Apparently, it should be in accordance with (1.57) and (1.59):

$$dP_{dt,i,j} = \lambda(\mathrm{X}, S_i, L_j)dt = P(S_i) \times P(L_j|S_i) \times \lambda(\mathrm{X}|S_i, L_j)dt$$

But we can write a similar expression for any combination of a certain assumed situations of any index i (i = 1, 2, ..., n) and a certain loading conditions with any index j (j = 1, 2,..., n). Then the composite probability of capsizing during any time element dt within the interval T may be represented by the double sum:

$$dP_{dt} = \sum_{i=1}^{n}\sum_{j=1}^{m} dP_{dt,i,j} = \sum_{i=1}^{n} P(S_i) \sum_{j=1}^{m} P(L_j|S_i) \times \lambda(\mathrm{X}|S_i, L_j)dt$$

Dividing the left and the right parts of this equality by dt and remembering that $dP_{dt} / dt = \lambda_a$ we get:

$$\lambda_a = \sum_{i=1}^{n} P(S_i) \sum_{j=1}^{m} P(L_j|S_i) \times \lambda(\mathrm{X}|S_i, L_j) \tag{1.65}$$

Let us try to interpret this equality. The expression standing under the symbol $\sum_{j=1}^{m}$ is a conditional risk function calculated for capsizing in the situation S_j under loading conditions L_j and multiplied by the conditional probability of the event $L_j|S_j$. After the operation of addition, such a sum represents the risk function value averaged over the whole set of possible loading conditions in the given situation S_j:

$$\lambda_a(X|S_i) = \sum_{j=1}^{m} P(L_j|S_i) \times \lambda(\mathrm{X}|S_i, L_j) \tag{1.66}$$

Then, the same procedure should be used for all the remaining situations. For this purpose, each value $\lambda_a(X|S_i)$ should be multiplied by the probability of the corresponding situation. Summing up such products (from i =1 to $i = n$) we get the risk function averaged over the whole set of loading conditions and over the whole set of the assumed situations. Formula (1.65) expresses the main result of this subchapter of our analysis.

Determining the λ_a - value in accordance with this formula provides the possibility of calculations of any probabilistic criteria considered in subchapter 1.4 and by Sevastianov [1982, 1982a].

To use this method in practice it is necessary to list all the possible discrete loading conditions. The word - all - means here " all in accordance with replacement of the real infinite uncountable sets of situations and loading conditions by some conventional set of discrete situations and loadings ". It is neither an easy nor a simple task. But tasks of this nature are very common in practice. The habitual characteristics of wind force (wind velocity or wind pressure and so on) are continuous variables, but we often replace them by a scale of discrete numbers and say, for example, "wind force 6 of the Beaufort scale". One can also say about waves: "sea state 8" or " sea of force 5". For many practical needs such discrete characteristics are quite sufficient.

Only in some rare and special cases do we use continuous scales and say: " a day before the casualty the master had been informed that significant waves of 5.47 meters at a mean wind velocity of 22.2 m/s were forecast ".

Therefore, the problem of discrete sets of situations or loading conditions is a rather practical than a principal one. One may even expect that in some time standard, sets of these vectors might be suggested in a form of discrete scales.

But the most complicated problem for using formula (1.65) is the problem of stability of a vessel under the fixed loading conditions in various assumed situations. Purely probabilistic aspects of setting stability standards are fully embraced by the formula (1.65), which combines the contemporary theory of reliability with the classical hydro-mechanical approach to the stability problems.

Thus, it is necessary to obtain all the solutions for calculating the risk function.

It is necessary to emphasize that the appearance of new hydromechanical solutions will require just a simple replacement of figures in formula (1.65) by other figures, which are more relevant to the mechanics of capsizing.

Chapter 2

Probabilistic Evaluation of Environmental and Loading Conditions

2.1 Lightweight Loading Conditions

Two vectors were mentioned in subchapter 1.2: the vector of loading conditions and the vector of assumed situations. We start with the vector of loading conditions including displacement *W*, the height of the centre of gravity *KG,* and radius of gyration around the central longitudinal axis R_x; we have introduced it as:

$$\vec{L} = \vec{L}(W, KG, R_x)$$

All the components of the vector are changing during the voyage of the vessel as well as her whole lifetime. For example, changes in the displacement of a trawler during one voyage can be taken: see fig 2.1, [Rakov and Sevastianov, 1981].

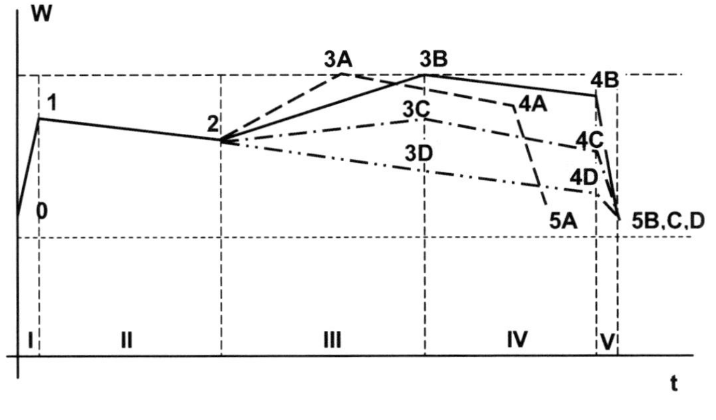

Fig. 2.1 Change of displacement during the voyage of a large fishing vessel

Legend

A – – – – – Big catches
B ———— Planed catches
C –·–·–·· Small catches
D –·–·–·· Failure

Stages of the voyage

I loading in the port (0 - 1)
II heading to fishing ground(1 - 2)
III fishing (2 - 3)
IV heading back (3 - 4)
V unloading in the port (4 - 5)

Möckel [1960] presented stability as a function of time: $GM(t)$. Rakov considered two components: $GM(t)$ and $W(t)$ as functions of time. Sevastianov [1970] proposed probabilistic interpretation of these figures. Moisseyeva [1971] gathered and processed statistical data on changes of loading conditions during the voyage of a large Russian trawler, "Mayakovsky".

Vector presentation of loading condition was proposed by Pavlenko [1949] and developed by Moisseyeva [1976].

The components of the loading condition vector are not independent. There is a correlation between the displacement, KG and mass distribution. The degree of this correlation is different for different vessels and depends on peculiarities of service. Vector presentation of loading makes it possible to take into account such a relationship.

So we shall consider loading as a vector, $\overline{OM}$, in co-ordinate space O, W, KG, R_x. End point M of the vector $\overline{OM}$ is moving in space during the ship's lifetime; see fig. 2.2 [Rakov and Sevastianov, 1981].

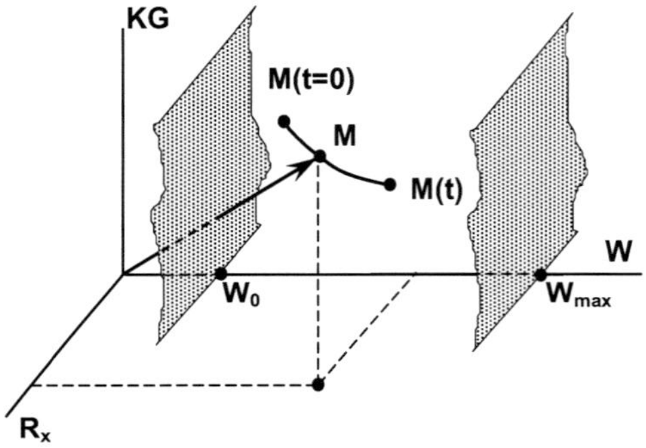

Fig. 2.2 Vector presentation of loading condition. W_0 - light displacement; W_{Max} - maximum displacement

Application of the scheme of probabilistic stability assessment described below in subchapter 2.2 requires knowledge of probabilistic characteristics of loading conditions and their dependence on time.

First, let's consider the structure of components of the vector. Displacement as a measure of the weight of the vessel is conventionally described as a sum of three terms:

$$W = W_0 + D_w + P \qquad (2.1)$$

Where: W_0- lightweight displacement; D_w - deadweight; P - special load (icing, loading caused by green water shipping, etc.) These weights can also act as external forces on the ship. Values for KG and R_x also can be presented in such a form in the same manner.

Let's start from lightweight loading conditions. Experience of long-term operation of vessels showed that after 8 -10 years, stability was considerably changed in the same state of loading [Sevastianov, 1970]. First of all, the lightweight changing was caused by ageing.

It is known that corrosion of a ship's shell plating below the waterline can reach up to 0.1-0.2 mm per year. For example, a medium size trawler might lose about 5 metric tons of weight over 10 years: the underwater area of such ship is about 300 m^2. This would increase KG up to 3-4 cm [Sevastianov, 1970]. More details and practical formulae are available from [Guralnik and Kulagin, 1995].

Another reason is moisture in wooden deck planking and other wooden details. The moisture content of air-dried wood, which is normally used in shipbuilding, is about 18 - 20%. After the first several years of service, this quantity increases up to 30 - 40% and then becomes stable. Such an increase means an additional 1-2 tons of lightweight displacement and about 1 cm of KG for a vessel such as a medium trawler [Sevastianov, 1970].

An even more significant increase can be caused by a rise of moisture content of internal insulation. For example, such an increase in light displacement reached up to 10 metric tons for the Russian medium size trawler "Bologoe". The change in *KG* depends on the location of such spaces in the ship [Sevastianov, 1970].

Repairing, service between voyages and modernization are also important factors. Installation of additional equipment and strengthening of damaged structures during repairs also alters lightweight.

So, lightweight loading conditions are subject to change due to all the above mentioned factors. This change can lead to a non-compliance of stability and free board standards that can be illustrated in a form of an ellipse of probability of loading condition vector $\vec{L}$ shown in fig. 2.3.

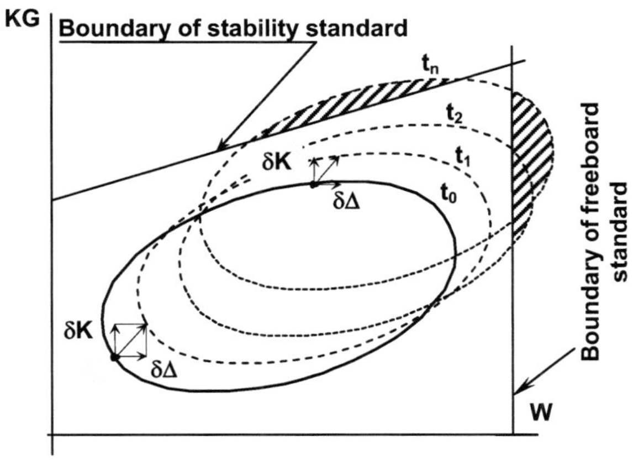

Fig 2.3 Changing of ellipse of probability of loading condition vector due to ageing [Guralnik and Kulagin 1995]

2.2 Time Varying Components of Loading Conditions

Consider records of displacement, *KG* and R_x, made on a daily basis for several voyages of a ship. These records can be considered as stochastic processes with the exception for a few cases when the vessel services the same route and has exactly the same amount of cargo all the time (even though a certain randomness is introduced by weather). Stochastic character of loading conditions is especially strong for fishing vessels. It depends on the quantity of fish in a certain region, organization of fishing, frequency of catching cycle, etc.

Generally, the trajectory of vector end *M* (see fig. 2.2) has to be considered as a particular realization of the stochastic process of loading conditions. The probability of a certain loading conditions appearing in a certain moment can be completely defined by the 4-D probability distribution of stochastic vector $\overline{OM}$. If we know time history of displacement, KG and R_X such a distribution can be easily determined as the following limit:

$$f(\overline{OM}) = \lim_{\substack{\Delta t \to 0 \\ \Delta W \to 0 \\ \Delta KG \to 0 \\ \Delta R_X \to 0}} \frac{p_{lijk}}{\Delta t \cdot \Delta W \cdot \Delta KG \cdot \Delta R_X} \tag{2.2}$$

Where: p_{lijk} -frequency of getting the vector end point in a 4D elementary parallelepiped with size Δt_l, ΔW_i , ΔKG_j, ΔR_{Xk} with co-ordinates t_l, W_i, KG_l and R_{Xk}

The practical algorithm of calculating the distribution, having data on loading conditions varying with time, is described in detail in [Moisseyeva, 1976]. It is shown that it is not necessary to use time as an independent co-ordinate here - a dangerous situation can appear at any moment of time. The algorithm, in general, is based on the same principles as an ordinary histogram calculation: all the axes should be subdivided into elementary ranges. The frequency of hitting the elementary parallelepiped by the end point of the 3D stochastic vector of the loading condition yields the required value of the estimate probability density.

The practical realization of this approach requires detailed knowledge of the time history of loading conditions during a significant number of voyages. Meylunas and Braslavskaya [1976] developed such a statistics-based method for a 2D loading condition vector for cargo vessels. Consider a cargo vessel, which heads from one port to another, having a random quantity of cargo of a random kind and some quantity of fuel and food stores. We characterize the quantity of cargo by special value η, which is the weight cargo coefficient:

$$\eta = P_c / P_{Fc}$$

Were P_c is the current amount of cargo, P_{Fc} is full cargo capacity. The value η should be considered as a random number because the type of cargo is arbitrary.

We assume, when the ship has less than 10% of fuel and food stores ("storm reserve"), she should refill her stores up to full capacity. It is convenient to use a special value ξ for characterizing the quantity of stores. It randomly changes from 0 (only 10% of stores - storm reserve) up to 1 (full stores). So, the current quantity of stores and its moment can be expressed as:

$$P_{St} = P_{St}^0 \cdot \xi \tag{2.3}$$

$$M_{ZSt} = M_{ZSt}^0 \cdot \xi \tag{2.4}$$

Where P_{St}^0 and M_{ZSt}^0 are weight and static moment at 90% ship stores.

Full displacement and moment at any moment of the voyage can be expressed as:

$$W(t) = W_0 + P_{Fc} \cdot \eta + P_{St}^0 \cdot \xi \tag{2.5}$$

$$M_Z(t) = W_0 \cdot KG_0 + P_{Fc} \cdot z_c \cdot \eta + M_{ZSt}^0 \cdot \xi \tag{2.6}$$

Where: W_0 and KG_0 correspond to unloaded ship with 10% of stores on board; z_c is elevation of centre of gravity of the cargo.

Good seamanship practice requires filling of a hold by a cargo, so z_c will be defined only by the volume of cargo. Assuming the sides of a ship are vertical, we can write:

$$z_c = h_{db} + (z_{Fc} - h_{db}) \cdot \frac{\kappa_c}{\kappa_{Fc}} \cdot \eta \tag{2.7}$$

Where: h_{db} is double bottom height or the height of the lowest filling level of cargo in the hold; z_{Fc} is elevation of centre of gravity at full cargo capacity. κ_{Fc} is full volume coefficient (the volume of full cargo capacity per one metric ton of displacement). κ_c is current volume coefficient (the volume of one metric ton of currently loaded cargo): it is a random variable, because the kind of cargo is arbitrary.

Random variable ξ is assumed to be independent of the other two values. It means that the relation between routing and type of cargo is not taken into account. This assumption leads to the introduction of the constant distribution of ξ, because the ship can be at any stage of her route with the same probability:

$$f(\xi) = \begin{cases} 0; & if \quad \xi \notin [0;1] \\ 1; & if \quad \xi \in [0;1] \end{cases} \tag{2.8}$$

Two other random variables η and κ_c are related as follows:

$$\frac{\kappa_c}{\kappa_{Fc}} \cdot \eta \le 1 \tag{2.9}$$

Probability distributions of η and κ are considerably different depending on type of cargo: general or mass [Meylunas, 1971]. We consider the transporting of these two different kinds of cargo as two forms of the shipping operations.

For transporting a mass cargo, the loading coefficient, η, is about unity, so weight capacity is fully used. When transporting a general cargo, the volume capacity coefficient, κ_c, is about unity, so volume capacity is fully used.

Let's start from the first case: using full weight capacity while transporting mass cargo. We have η=1, values ξ and κ_c keep their random character. The purpose is to obtain the joint distribution of displacement and *KG*, which can be expressed as:

$$f(W, KG) = f(W) \cdot f(KG \mid W) \tag{2.10}$$

The distribution of displacement is uniform, taking into account (2.5) and (2.8):

$$f(W) = \begin{cases} 0 \quad if \quad W \notin [W; W + P_{St}^0] \\ 1/P_{St}^0 \quad if \quad W \in [W; W + P_{St}^0] \end{cases} \tag{2.11}$$

To obtain the joint distribution density $f(KG|W)$, we exclude random variable ξ from (2.6) and consider displacement as a parameter:

$$KG = \frac{1}{W}\left[W_0 KG + P_{Fc} h_{db} + P_{Fc}(z_{Fc} - h_{db})\frac{\kappa_c}{\kappa_{Fc}} + \frac{M_{ZSt}}{P^0_{St}}\left(W - W_0 - P_{Fc}\right)\right] \qquad (2.12)$$

Equation (2.12) is a linear function of the random variable κ if W is given. Taking $f(\kappa_c)$ from fig. 2.4, we obtain conditional distribution $f(KG|W)$.

Equation (15.12) allows derivation of conditional distribution density of KG if the displacement is given, which completes consideration of the mass cargo case:

$$f(KG\,|\,W) = f(\kappa_c)\cdot|\,KG(\kappa_c)\,| \qquad (2.13)$$

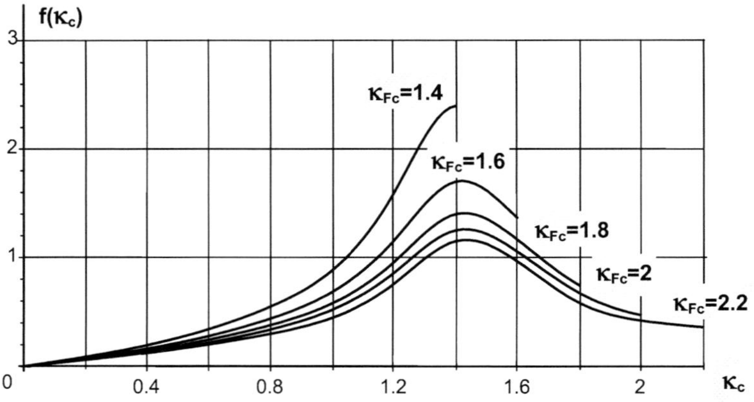

Figure 2.4 Probability density of current volume coefficient for transporting of mass cargo [Meylunas, 1971]

The example of joint distribution is given in fig. 2.5 for a cargo vessel with characteristics, which are given in the table below:

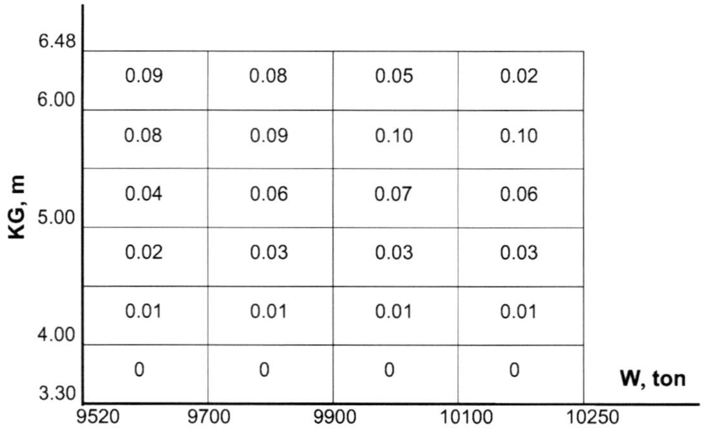

Fig. 2.5 Example of joint distribution density of KG and displacement for ship transporting mass cargo [Meylunas and Braslavskaya, 1976]

Length: 120.5 m	Breadth: 2.8 m	Depth: 9.5m
h_{db} = 1.1 m	W_0 = 3625 MT	KG_0 = 7.3 m
P_{Fc} = 5895 MT	z_{Fc} =5.95 m	κ_{Fc} = 1.5 m³/MT
P^0_{St} =730 MT	M^0_{ZSt} = 928 MT m	

Let's consider the second case, when full volume capacity is used (general cargo). The weight capacity of a ship is not fully used:

$$\eta < 1$$

We have two random variables η and ξ. Distribution of the stores coefficient is evidently the same, as in the previous case. Uniform distribution density is proposed for cargo coefficient η [Meylunas and Braslavskaya, 1976].

$$f(\eta) = \begin{cases} 0; \; if \quad \eta \notin [\eta_{\min}; 1] \\ \dfrac{1}{1-\eta_{\min}}; \; if \quad \eta \in [\eta_{\min}; 1] \end{cases} \tag{2.14}$$

Where η_{min} is the minimum value of the cargo weight coefficient, which is evaluated from statistics. It depends on the full volume capacity coefficient, κ_{Fc}.

It is convenient to express the value of η_{min} through $M\eta$ – mean value of η, which can be determined from the graph in fig. 2.6.

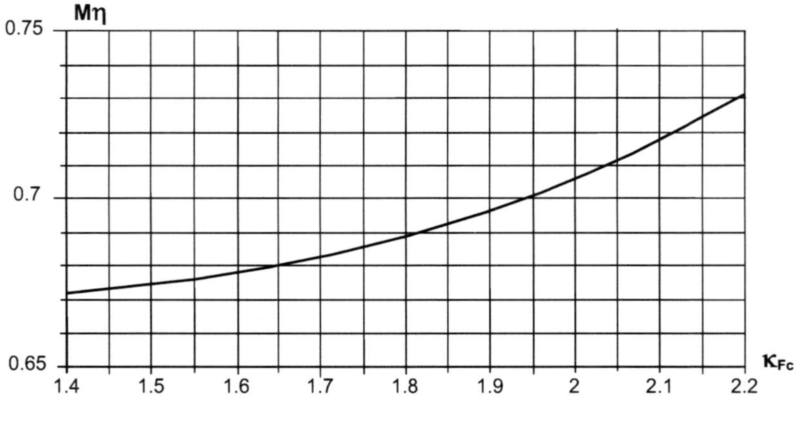

Fig. 2.6 Value of weight coefficient vs. full volume capacity coefficient

Displacement is considered as a deterministic function of two random arguments η and ξ as in equation (2.5). Random variables η and ξ are independent, their distributions are known, so distribution of the displacement can be found. Meylunas and Braslavskaya [1976] presented it in trapezoidal form, see fig. 2.7.

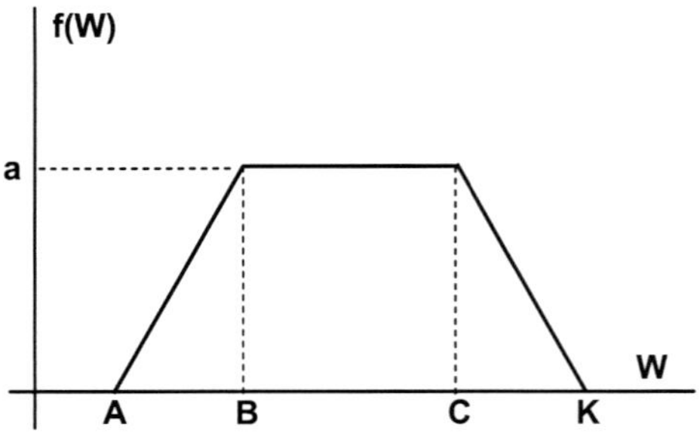

Fig. 2.7 Probability density function of displacement for ship using full volume capacity

Coordinates of the points *A*, *B*, *C*, and *K* as well as the height of trapezoid figure *a* can be calculated as:

For the case: $P_{st} + \eta_{\min} P_c > P_c$

$$\begin{aligned} W_A &= W_0 + P_c \cdot \eta_{\min} \\ W_B &= W_0 + P_c \\ W_C &= W_0 + P_{st} + P_c \cdot \eta_{\min} \\ W_K &= W_0 + P_{st} + P_c \\ a &= 1/P_{st} \end{aligned} \tag{2.15}$$

For the case: $P_{st} + \eta_{\min} P_c < P_c$

$$\begin{aligned} W_A &= W_0 + P_c \cdot \eta_{\min} \\ W_B &= W_0 + P_{st} + P_c \cdot \eta_{\min} \\ W_C &= W_0 + P_c \\ W_K &= W_0 + P_{st} + P_c \\ a &= 1/[P_c \cdot (1 - \eta_{\min})] \end{aligned} \tag{2.16}$$

For deriving a joint distribution, it is sufficient to obtain conditional distribution of *KG*, when the displacement is given, see equation (2.10). For this purpose we exclude ξ from equation (2.6):

$$KG(\eta, W - given) = \frac{1}{W}\left[W_0 \cdot KG + P_c \cdot z_c \cdot \eta + \frac{m_{st}}{P_{st}} \cdot (W - W_0 - P_c \cdot \eta)\right] \tag{2.17}$$

The equation (2.17) is a linear function of η, consequently the conditional distribution density can be expressed as:

$$f(KG \mid W) = f(\eta \mid W) \cdot \frac{d[\eta(KG, W - given)]}{dKG} \tag{2.18}$$

Where: $f(\eta|W)$ is the distribution density of random variable η, when the displacement is given. It is of uniform distribution. A range of changes in random variable η, when the displacement is given can be obtained from equation (2.18) where other random variables change from 0 to 1.

$$\underbrace{\frac{W - W_0 - P_{St}}{P_c}}_{\min} \le \eta(W - given) \le \underbrace{\frac{W - W_0}{P_c}}_{\max} \tag{2.19}$$

So, the distribution is:

$$f(\eta \mid W)=\begin{cases} 0; \quad if \quad \eta(W-given)\notin\left[\dfrac{W-W_0-P_{st}}{P_c};\dfrac{W-W_0}{P_c}\right] \\ \dfrac{P_c}{P_{St}}; \quad if \quad \eta(W-given)\in\left[\dfrac{W-W_0-P_{st}}{P_c};\dfrac{W-W_0}{P_c}\right] \end{cases} \tag{2.20}$$

The derivative of $\eta(KG,\ W\text{ - }given)$, which is an inverse function to equation (2.17) can be written as follows:

$$\frac{d[\eta(KG, W-given)]}{dKG}=\frac{W}{P_c\cdot z_c - P_c\cdot\left(m_{st}/P_{st}\right)} \tag{2.21}$$

A range of change in random variable KG, when the displacement is given can be obtained by substitution of equation (2.17) into equation (2.19):

$$\underbrace{W_0KG_0+m_{st}+z_c(W-W_0)}_{\min}\le KG\cdot W\le \underbrace{W_0KG_0+m_{st}+z_c(W-W_0-P_{st})}_{\max} \tag{2.22}$$

Finally:

$$f(KG\mid\Delta)=\begin{cases} 0; \quad KG\notin\left[\dfrac{W_0KG_0+m_{st}+z_c(W-W_0-P_{st})}{W};\dfrac{W_0KG_0+m_{st}+z_c(W-W_0)}{\Delta}\right] \\ \dfrac{W}{m_{st}-P_{st}z_{st}}; \quad KG\in\left[\dfrac{W_0KG_0+m_{st}+z_c(W-W_0-P_{st})}{W};\dfrac{\Delta_0KG_0+m_{st}+z_c(W-W_0)}{W}\right] \end{cases}$$

The example of the joint distribution for the above mentioned cargo vessel is given in fig. 2.8.

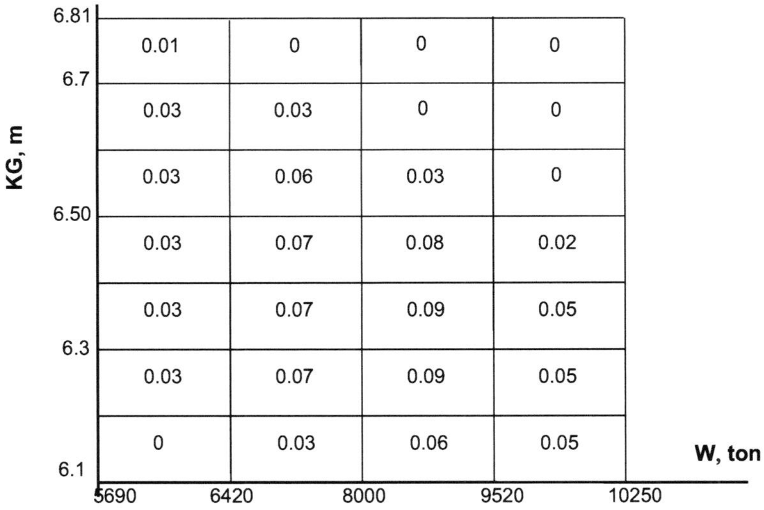

Fig. 2.8 Example of joint distribution density of KG and displacement for ship transporting general cargo [Meylunas and Braslavskaya, 1976]

Such a general solution can only be applied for cargo vessels. Operation of a fishing vessel involves more random values. First of all, a fishing vessel has two basic regimes of operation: heading (to/from fishing ground) and fishing itself.

The fishing regime is more complicated. It can be a sequence of lifting of stochastic weight catches, where time between two neighborhood catches is also random. Moisseyeva [1976] calculated the distribution of loading condition vector $\vec{L} = \vec{L}(\Delta, GM)$ for the large Russian trawler "Mayakovsky". Apollinariev and Sevastianov [1992] obtained the distribution of loading conditions through simulation of a complete fishing operation. The catch was presented as a stochastic flow of singular events associated with random weight of the caught fish; intervals between catches were also random numbers. Detail simulation of fish processing aboard the ship allowed precise prediction of distribution of displacement and *KG*. Characteristics of input catches flow were based on existing fishing statistics.

2.3 Meteorological Components of Assumed Situation[1]

We introduced a concept of a vector of an assumed situation in subchapter 1.2 and indicated its possible components. These components are dependent on time, so the vector $\vec{S}$ has to be considered as a multidimensional stochastic process [Sevastianov, 1982b; Mordachev, et al 1994; Mordachev, 1995].

It is convenient to separate the components of the vector of assumed situation $\vec{S}$ in two auxiliary sub-vectors:

$$\vec{S} = \vec{S}(\vec{S}_1, \vec{S}_2) \qquad (2.23)$$

Where $\vec{S}_1$ is a sub-vector describing waves and wind and $\vec{S}_2$ is a sub-vector describing ship heading and additional forces. The first sub-vector characterizes environmental weather conditions, the second one contains operational information, which is also dependent upon human decision making:

$$\begin{cases} \vec{S}_1 = \vec{S}_1(u_{Am}, \sigma_u, \alpha_u, h_s, T_w, \alpha_{wu}, ...) \\ \vec{S}_2 = \vec{S}_2(v, \psi_A, \delta, P, \varepsilon_{px}, \varepsilon_{py}, ...) \end{cases} \qquad (2.24)$$

With:

u_{Am} - mean wind velocity
σ_u- standard deviation of fluctuating part of wind velocity
α_u- angle between wind direction and geographical North
h_s- height of significant waves
T_m- mean period of waves
α_{uw}- angle between general directions of wind and waves
v - ship heading speed
ψ_A- angle between ship heading and general wind direction

[1] Written in co-authorship with Prof. Alexander B. Degtyarev and Dr. Alexander V. Boukhanovsky of Institute for High Performance Computing and Information Systems (Russia)

δ- rudder angle
P - module of principal vector of additional external forces, related with ship service
ε_{px}, ε_{py}- angles between P and ship co-ordinate axis *Kxyz*

There are dots in formulae (2.24); they mean that they are not closed for addition of possible new components, if they are significant for stability evaluation. For example, it may be necessary, to include geometric characteristics of the ship's hull, angular velocity of ship turns or radius of curvature of her trajectory in sub-vector $\vec{S}_2$ as components. It could be useful to include width of wave spectrum in sub-vector $\vec{S}_1$ and principal external heeling moment in sub-vector $\vec{S}_2$. However, every new component will make analysis more complicated. We are interested in reducing components, the influence of which on the stability evaluation are not so important.

A close look at the components reveals that there is one main argument or component in each of the sub-vectors, a small value of which reduces the importance of all sub-vectors. For example, when $u_{Am} \to 0$, then it is impossible to meet a significant fluctuation of wind velocity so $\sigma_u \to 0$. Therefore, a prolonged period of severe irregular waves is also impossible, so all the components of sub-vector $\vec{S}_2$ loose their significance.

The importance of reducing the number of components can be illustrated by the following example. If it will be assumed that all components of vector $\vec{S}$ are independent and it is enough to have k values on each component, then the total number of assumed situations would be as follows:

$$N_s = k_s^{ns}$$

So, if total number of components in equation (15.24) is n_s=12 and k_s=5, we get:

$$N_s = 5^{12} \approx 2.44 \cdot 10^8$$

The vector of loading condition has 3 components and each component is presented, say, by 5 values. The total number of the loading condition cases, which should be taken into account in each assumed situation, will be determined as follows:

$$N_l = 5^3 = 125$$

The total number of risk function computations in all assumed situations with N_l discrete loading conditions is:

$$N_\lambda = N_s \cdot N_l \approx 3.05 \cdot 10^{10}$$

The number of necessary cases makes calculations quite bulky and leaves small hope for use of this method as an everyday tool.

However, this number can be significantly reduced by taking into account dependence of arguments. Such dependence of arguments of the sub-vector $\vec{S}_1$ (waves and wind) can be found by analyzing meteorological statistical data. Components of the sub-vector $\vec{S}_2$ (ship heading and additional forces) are dependent on human decision makings and circumstances of the vessel's operation. Introducing some models of operator strategies can reveal their interdependency. This subject is considered in the next subchapter.

Further, we consider meteorological components of the assumed situation vector in order to find their interdependence in a form, which can be of use to simplify the problem. Meteorological data today is the result of long term observations averaged and represented by tables of a special type for a certain season and geographical region. These kind of data can be found in "Global Wave Statistics", published by British Maritime Technology Limited (BMT) [Hogben, et al 1967, 1986] (see also chapter 9 of [Kobylinski and Kastner, 2003]). Another source is "Wind and Waves in Seas and Oceans" [Davidan, et al 1974] issued by the Russian Register. These sources divide all world oceans and seas into geographical regions. The data also are divided by seasons. Examples of such tables are given in table 2.1.

Table 2.1 Wind velocity probability by direction, %

Season	Wind velocity m/s	Direction							
		N	NE	E	SE	S	SW	W	NW
1 region of Baltic sea									
Winter	<6	3	2	2	2	3	3	4	3
	6-12	4	4	5	4	6	10	10	6
	12-16	2	2	2	2	2	5	4	2
	>16	1	<0.5	<0.5	<0.5	1	2	2	1
Spring	<6	6	8	7	5	5	8	6	6
	6-12	4	8	7	3	4	7	6	3
	12-16	1	1	1	<0.5	<0.5	1	1	1
	>16	<0.5	<0.5	<0.5	<0.5	<0.5	<0.5	<0.5	<0.5
Summer	<6	6	6	5	4	6	10	9	3
	6-12	4	5	5	4	6	9	10	3
	12-16	<0.5	<0.5	<0.5	<0.5	<0.5	2	1	1
	>16	<0.5	<0.5	<0.5	<0.5	<0.5	<0.5	<0.5	<0.5
Autumn	<6	5	2	3	3	3	4	4	3
	6-12	4	3	5	7	7	11	11	6
	12-16	1	1	1	2	2	3	3	2
	>16	<0.5	<0.5	<0.5	<0.5	<0.5	1	1	1

These tables contain data on statistical frequency of meeting certain characteristics of wave height, mean wave period, mean wind velocity and wind direction. Data of correlation of these random variables can be found for several regions.

A very clear relationship exists between mean value of wind velocity, u, and standard deviation of the fluctuating part of wind velocity, σ_u. In accordance with measurements variance of instant relative wind velocity $\bar{u} = u(t)/u_{Am}$ is equal [Lugovsky, 1976]:

$V_{\bar{u}} = 0.0167$ Caspian sea

$V_{\bar{u}} = 0.0206$ Antarctic

For the first expansion estimate, this data can be generalized: the first case for all closed seas, and the second one for open seas and oceans.

Standard deviation σ_u can be expressed as follows in this case:

$$\sigma_u = u_{Am} \cdot \sqrt{V_{\bar{u}}} \quad (2.25)$$

The next is the relationship between mean value of wind velocity and statistical characteristics of wave height. NATO data sheet for the Open North Atlantic might be used as a sample of such data [Buckley, 1992].

Table 2.2 Significant wave height and sustained wind speed for North Atlantic [Buckley, 1992]

Sea State	Significant Wave Height (m)		Sustained Wind Speed (kn)		Percent-age Pro-bability of Sea State	Modal Wave Period	
Number	Range	Mean	Range	Mean		Range	Most Portable
0-1	0-0.1	0.05	0-6	3	0.7	-	-
2	0.1-0.5	0.3	7-10	8.5	6.8	3.3-12.8	7.5
3	0.5-1.25	0.88	11-16	13.5	23.7	5.0-14.8	7.5
4	1.25-2.5	1.88	17-21	19	27.8	6.1-15.2	8.8
5	2.5-4	3.25	22-27	24.5	20.64	8.3-15.5	9.7
6	4-6	5	28-47	37.5	13.15	9.8-2.2	12.4
7	6-9	7.5	48-55	51.5	6.05	11.8-18.5	15.0
8	9-14	11.5	56-63	59.5	1.11	14.2-18.6	2.4
>8	>14	>14	>63	>63	0.05	18.0-23.7	20.0

There are data available on the relationship between wave height and mean wave period. An example of this relationship is given in fig. 2.9. This chart was used as an industry standard in Russia for calculating ship motions [Boroday and Netsvetaev, 1969, 1982; Boroday, et al 1989]. The upper curve corresponds to maximum observed mean periods and can be considered as values of mean periods of dying waves. The lower curve corresponds to minimum observed mean periods of growing waves. The middle curve can be considered as values of mean periods of fully developed waves. It can be approximated by the following formula [Davidan, et al 1974]:

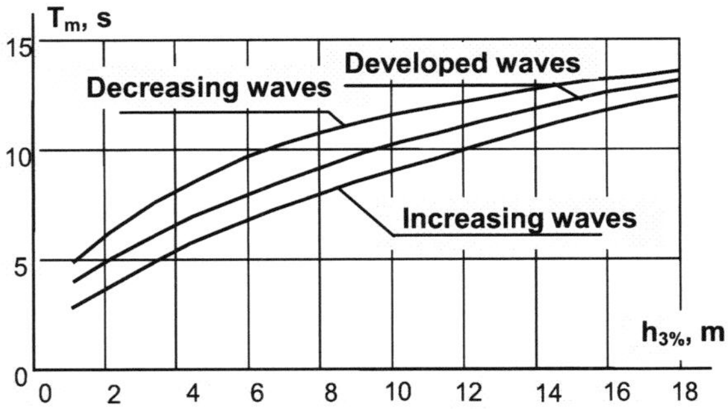

Fig. 2.9 Mean period of irregular waves vs. $h_{3\%}$ wave height with 3% probability of exceeding

$$T_m = 3.3 \cdot \sqrt{h_{3\%}} \tag{2.26}$$

This formula contains wave height with 3% of probability of exceeding ($h_{3\%}$), which is used along significant wave height (mainly in Russian literature):

$$h_s = 1.33 \cdot h_{3\%}\,; \qquad h_{3\%} = 0.75 \cdot h_s$$

The last parameter, α_{uw} is the angle between wind and general direction of waves. This angle may be different from zero. A possible reason for this phenomenon is the influence in neighboring regions, the waves generated there may reach quite far away in a form of a swell. If swells are likely in the region, they are usually included into wave statistics.

Finally, if the ship route is known, statistical relationships between components of the "weather" sub-vector might reduce the problem to one random variable, defined by the region. (Such an assumption is definitely acceptable for the first expansion of the risk analysis). The voyage can be considered as a flow of events associated with a random number describing the weather in the region – usually, significant wave height.

The central problem is how to relate wave statistics with the ship motions: averaged characteristics of waves (including spectrum) do change in time. Therefore, we have to work in a "synoptic" time scale and we cannot assume that significant wave height stays the same. Completely different models are needed for such analysis. These models can be developed by averaging weather states for longer times. It is assumed that a state of weather can be fully defined by a 3D spectrum (taking into account frequency and direction at the same time).

There are two types of waves at sea: wind waves and swell (one or several systems). In fact, they are distinct physical phenomena: mechanics of their generation is different. Each of these systems has its own statistical measures: wave height h, direction β, and period T. For example, significant wave height is a statistical measure of wave height, mean period of wave is a statistical measure of wave period, etc.

Since we deal with a synoptic time scale, the spectrum is no longer constant: in fact, it is a stochastic function. However, it is difficult to use spectral density for statistical analysis [Jonson and Wichern, 1992]. A reasonable alternative is to consider it as a deterministic function of random numbers. We assume wave height h, direction β, period T, and characteristics of wave generation (free run X, duration T_w, water depth H_w, etc.) as a set of random numbers. It depends on geographical co-ordinates (x, y) and time t:

$$S_{RN}(x,y,t) = \{h, T, \chi, X, T_W, H_W, \dots\}$$

Each value of $S_{RN}(x,y,t)$ is mapped to spectral density:

$$S(\omega,\beta) = S(\omega,\beta,S_{RN})$$

To build such a model we have to resolve the following problems:

- Find a set of parameters (parameterization) to map the presentation of the spectrum as a stochastic function onto presentation of spectrum as a deterministic function of random arguments.
- Classification of spectra in the characteristics of S_{RN}.

- Find areas of stability in space and time states of the spectrum.

Parameterization means that we find some typical representatives of every class of spectra. Then the spectra would differ inside the class only by numerical values of the parameters. Such a representative spectrum could be considered as a certain characteristic of the wave climate: (see also subchapter 9.12 of the [Kobylinski and Kastner, 2003]). The term "spectral wave climate" was introduced in the conclusion of the International conference "Provision and Engineering/Operational Application of Ocean Wave Spectra" in 1998.

The first attempt of classification of wave spectra was made by Buckley [1988]. He analyzed more than 2 million of spectra that were generated over 12 years at 13 buoys located in coastal waters of the USA. All the wave situations that the study included were broken down into twelve types, according to their significant wave height. Then all spectra in each class were averaged by each frequency. However, wind waves of the same height could result from various weather conditions (rising sea, decaying wind waves, swell, etc.). They are characterized by different spectra.

A classification of spectra $S_i(\omega, h_s)$ based on a single parameter, h_s or variance V_ξ, would lead to difficulties of interpretation. A more fruitful approach, which also leads to a better understanding of the results, is based on classification with respect of "persistent conditions". For example, Lopatoukhin, et al. [1990] for the tropical Pacific, proposed four types of wave generating conditions leading to the development of common features in their spectra (see table 2.3). Variability of wind wave spectra within each of the four groups can be expressed using quantile (or probability of exceeding a given level) diagrams as shown in fig. 2.10.

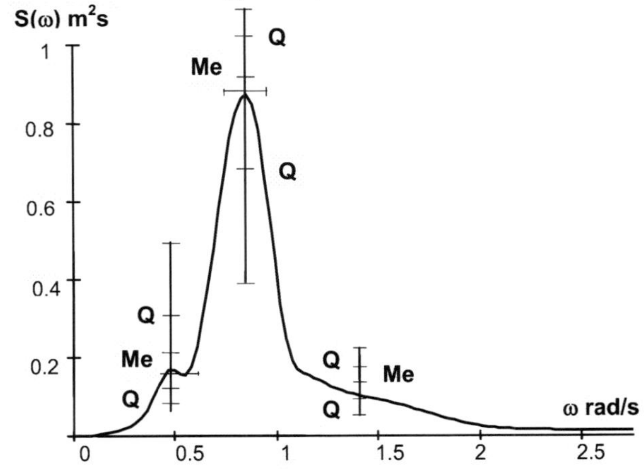

Fig. 2.10 A typical wave spectrum generated under the action of strong trade winds in the tropical part of the North Pacific Ocean

Another classification also not based on formal partitioning of wave height and other statistical characteristics was mentioned in subchapter 9.12 of [Kobylinski and Kastner, 2003]; it is descried completely in [Lopatoukhin, et al, 1999; Boukhanovsky, et al., 2000]. Functional representation of such classes of spectral densities $S(\omega)$ can be made using the following well known approximation:

$$S(\omega) = A\omega^{-k} \exp\left(-B \cdot \omega^{-n}\right) \tag{2.27}$$

Where A, B, k, n are parameters reflecting wave generating conditions.

Table 2.3 Typical frequency spectra for the tropical part of North Pacific Ocean

Type	%	Wind, m/s	Variance, cm^2	Peak No.1, Rad/s	Peak No.2, Rad/s	Spectral shape $S(\omega)$, $(m^2\ s)$
ITCZ	40	<6	650-1300	0.4-0.7	0.8–1.1	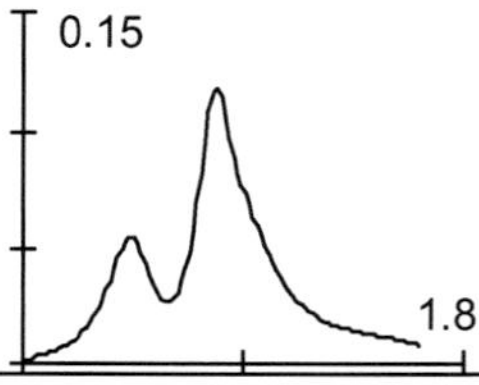
MTW	25	<8	1500-2800	0.4-0.7	0.8–1.1	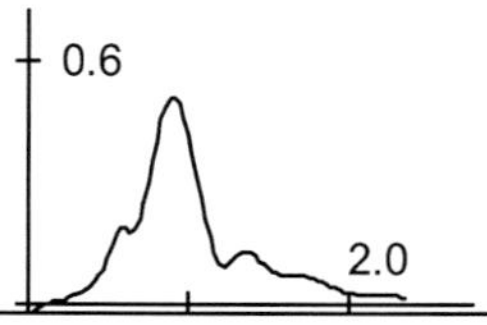
STW	25	8-15	2500-4500	0.4-0.7	0.7–1.0	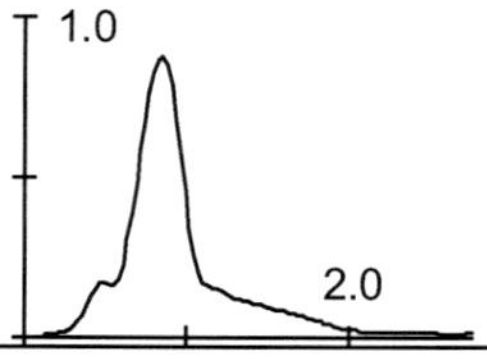
TC	10	>15	>4500	0.4–0.7	—	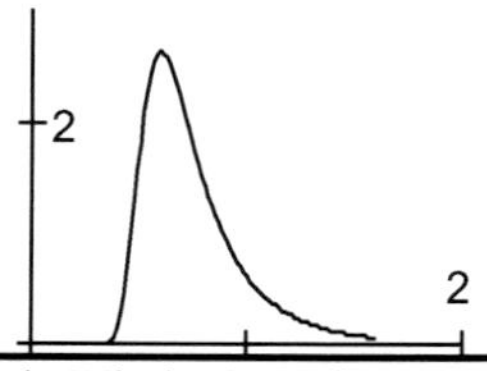

Note: ITCZ is Inter-Tropical Convergence Zone, MTW is moderate Trade Winds, STW is strong Trade Winds, TC is tropical cyclone.

Mean wave height, $\bar{h}$, is the only parameter needed for description of wind sea waves. Mean wave period can be estimated by various relations (see above or [Davidan, et al, 1974]). Single-peaked spectral approximation (2.27) for swell depends on two parameters, namely $\bar{h}$ and T_m. The ratio $\beta = gT_m^2/\bar{h}$ reflects non-dimensional steepness. A complex wave can be expressed, in the first approximation, as the sum of spectra:

$$S_{Complex}(\omega) = S_{Wind}(\omega) + S_{Swell}(\omega) \qquad (2.28)$$

The proposed approximation uses spectral moments and some other related variables. It makes possible to represent any spectral density function $S(\omega)$ as $S(\omega, S_{RN})$. All operations with such functions $S(\omega)$ inside their class are ones with deterministic functions of random arguments S_{RN}. For example, it is possible to define the mean spectrum:

$$\overline{S}(\omega) = S\left(\omega, \overline{S_{RN}}\right) \tag{2.29}$$

The quantile (the value with given probability of exceeding certain level p) spectrum:

$$S_p(\omega) = S\left(\omega, \left[S_{RN}\right]_p\right) \tag{2.30}$$

Here $\overline{S_{RN}}$ is a mean value of the set of random numbers and $\left[S_{RN}\right]_p$ is its p-quantile (the value with given probability of exceeding certain level).

The above examples dealt with frequency wave spectra only. However, direction of propagation of different wave systems is also important. A 3-D pattern of wave spectrum is particularly complicated for moderate wave heights. In the case of a strong storm or "dead" swell (swell caused by a storm far away) the spectrum is characterized by a relatively narrow directional distribution. Fig. 2.11 shows frequency-directed climatic spectrum of complex sea in the north-eastern part of the Black Sea [Boukhanovsky, et al, 2001]. This case requires application of multidimensional statistical analysis of condition of wave generation and helps to find the best classification of spectra, especially if different wave systems exist in the same frequency domain

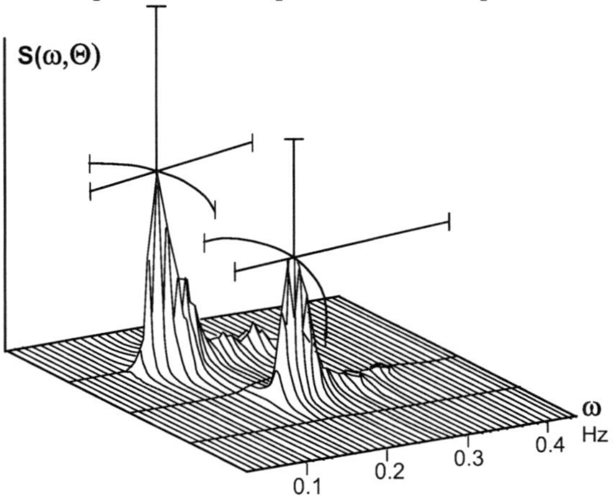

Fig. 2.11 Frequency-directed climatic spectrum of complex sea. North-eastern part of the Black Sea

Alternation of a situation such as "storm" and "weather window" (slack sea) is usual for mid-latitudes regions [Davidan, 1995]. Wave statistics consider "storm" as a phenomenon when process $h(t)$ is greater than a certain threshold. "Weather window" is the time when waves are less than the same threshold. If the threshold is Z (e.g. mean value $h[t]$), then upcrossing could be parameterized as an impulse with maximum height h^+, duration $\Im$, interval Θ between adjacent storms and minimum height h^- (fig.2.12).

Such parameterization allows interpretation of a series of storms and weather windows as a related sequence of stochastic surges or impulses, i.e. impulse random process [Tikhonov, 1982]. A sample can be generated as follows:

$$\xi(t) = \sum_{k=1}^{n} w_k\left(Z, t - \sum_{j=1}^{k-1}\left(\Im_j + \Theta_j\right)\right) \tag{2.31}$$

With:

$$w(Z,t)=\begin{cases} Z+\left(h^{+}-Z\right)\cdot u(t/\Im) & 0\le t\le \Im, \\ Z-\left(h^{-}-Z\right)\cdot u((t-\Im)/\Theta) & \Im\le t\le \Im+\Theta \\ 0 & (t<0)\cup(t>\Im+\Theta) \end{cases}$$

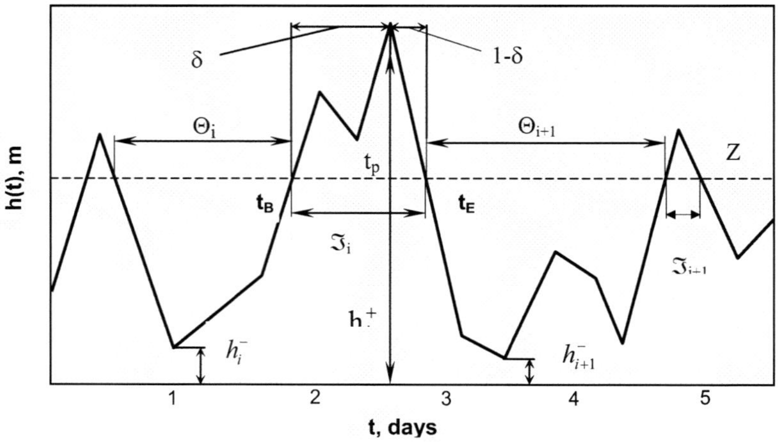

Fig 2.12 Parameterization of "storms" and "weather windows" sequence

Function $u(t)$ describes the shape of the non-dimensional impulse. A triangular shape of this function might be expressed as:

$$u(t)=\begin{cases} t/\delta & 0\le t\le \delta, \\ 1/(1-\delta)-t/(1-\delta) & \delta\le t\le 1 \\ 0 & (t<0)\cup(t>1) \end{cases}$$

This shape serves as a good first approximation. Parameter δ sets asymmetry of function $u(t)$, see fig. 2.12. If $\delta = 0.5$, the function is symmetric.

Analysis of interdependency of these parameters [Lopatoukhin, et al, 2000] shows that pairs of values (h^+, h^-), (h^+, Θ), $(h^-, \Im)$, $(\Im, \Theta)$ can be assumed independent for the first expansion. Therefore, the four-dimensional distribution for these parameters can be simplified:

$$f(h^+,h^-,\Im,\Theta)=f(h^+,\Im)f(h^-,\Theta)$$

Each of these 2D distributions can be presented as:

$$f(h^+,\Im)=f(h^+\mid\Im)f(\Im) \quad ; \quad f(h^-,\Theta)=f(h^-\mid\Theta)f(\Theta)$$

Distributions $f(\Im)$ and $f(\Theta)$ are exponential distributions, since $\Im$ and Θ are duration of upcrossings [Tikhonov, 1982; Leadbetter 1986]. The corresponding cumulative distribution can be taken in a generalized form with parameters α and k:

$$F(x) = 1 - \exp\left[-\left(\alpha \frac{x}{\bar{x}}\right)^k\right] \tag{2.32}$$

Distributions of h^+ and h^- actually are extreme distributions [Leadbetter, 1986], since these values are extremes for storms and weather windows respectively. For the given values of duration $\Im$ and Θ and threshold Z, they can be expressed as:

$$\begin{aligned} h^+ &= \max_{0 \le t \le \Im} \left(h(t) : h(t) > Z\right) \\ h^- &= \max_{0 \le t \le \Theta} \left(h(t) : h(t) \le Z\right) \end{aligned} \tag{2.33}$$

Since wave height has a logarithmic-Gaussian distribution in a synoptic time scale (we mean time duration from several days to several weeks) and section by level Z does not affect distribution [Leadbetter, 1986]:

$$F\left(h^+ \mid \Im\right) = \begin{cases} \exp\left[-\exp\left[-\dfrac{h^+ - A(\Im)}{B(\Im)}\right]\right] & \text{for } \; h^+ \ge Z \\ 0 & \text{for } \; h^+ < Z \end{cases} \tag{2.34}$$

Where A and B are the following functions of statistical moments:

$$B(\Im) = \frac{\sqrt{6}}{\pi}\sigma(\Im)$$
$$A(\Im) = m(\Im) - 0.5772B(\Im)$$

A three-parameter Weibull distribution can serve as a good approximation of the above [Angelides, et al, 1981; Boukhanovsky, et al, 1998]:

$$F(y \mid x) = 1 - \exp\left[-\left(\frac{y - Z}{A(x) - Z}\right)^{B(x)}\right] \tag{2.35}$$

Where the third parameter Z determines the threshold and the first two parameters A and B are estimated using sample data.

The Monte-Carlo approach and use of formulae (2.32)-(2.35) make it possible to reproduce the whole variety of values of function S_{RN}:

$$\begin{aligned} \Im_k &= F_{\Im}^{-1}\left(\gamma_1^{(k)}\right) \\ \Theta_k &= F_{\Theta}^{-1}\left(\gamma_2^{(k)}\right) \\ h_k^+ &= F_{h^+ \mid \Im}^{-1}\left(\gamma_3^{(k)} \mid \Im_k\right) \\ h_k^- &= F_{h^- \mid \Theta}^{-1}\left(\gamma_4^{(k)} \mid \Theta_k\right) \end{aligned} \tag{2.36}$$

Upper index "-1" means here the inversion of the cumulative distribution or in other words, computer generation of random numbers distributed according to a given function. Set $\left\{\gamma_i^{(k)}\right\}$ denotes a system of four computer-generated pseudo random numbers.

The data analysis shows that storm shapes are quite diverse and there are many ways to classify them. The classification results significantly depend on the selection of *Z*. The shapes are more diverse for smaller values of *Z*, while for larger values of *Z* they become more uniform.

One of possible classification shown in subchapter 9.12 of [Kobylinski and Kastner, 2003] suggests 5 classes of storms [Boukhanovsky, et al, 2000]. An increase in the level *Z* leads to zero probability of the storms of 5th class. Another classification was made based on the same data [Lopatoukhin, et al, 2000]. Discriminate analysis was used for this new classification. It resulted with a more detailed classification of the same storms. Eight types of storms for wave heights exceeding the mean seasonal wave height $h=Z$ and four types for wave heights exceeding $h=2Z$ and $h=3Z$ were proposed, see table 2.4.

Here, all storm types are corresponded with dominating meteorological conditions. It is worth mentioning that in spite of differences between the classification methods, the whole set of storm shapes for wave heights exceeding $h=2Z$ fell almost similarly into the same four groups.

Weather windows can also be classified similarly, see table 2.5.

Table 2.4 A classification of storm shapes based on discriminate analysis

Type	Shape	Description	Threshold					
			1h	2h	3h	*1h*	*2h*	*3h*
			Number of storms			%		
I	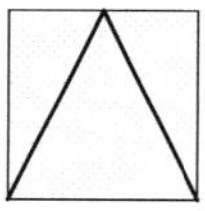	Steady increase and decrease of wind	39	21	14	20.3	23.1	41.2
II	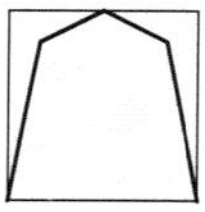	Stable wind at phase of maximal storm development	40	39	4	20.8	42.9	11.8
III	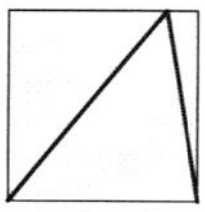	Duration of increase is considerably longer than one of decrease. This type is specific for "slow" storms	33	16	7	17.1	17.6	20.6
IV	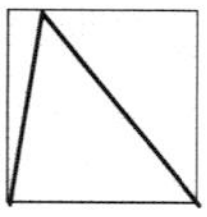	Expressed asymmetry of the shape with domination of the decrease phase. This type is specific for "quick" storms.	37	15	9	19.3	2.4	26.4
V		The discriminate analysis gives a separate type for this storm shape. It bears some similarity to type IV. This shape is typical for fast and deep cyclones.	12	*	*	6.3	*	*

Table 2.4 A classification of storm shapes based on discriminate analysis (continued)

VI		Intermittent increase and decrease of waves caused by instabilities of the atmospheric flow. They are typical for a shallow or a slow moving cyclone.	8	*	*	4.2	*	*
VII		Passage of a deep cyclone with distinct separation of fronts. Depending on the cyclone track wind wave field either develops having swell as its background or generates swell as a residual signal.	19	*	*	9.9	*	*
VIII		A "chain" of storms, which cannot be separated due to small threshold value of Z.	4	*	*	2.1	*	*

Table 2.5 A classification of weather windows

Type	**Shape**	**Description**	**Threshold**					
			1h	**2h**	**3h**	***1h***	***2h***	***3h***
			Number of weather windows			*%*		
I	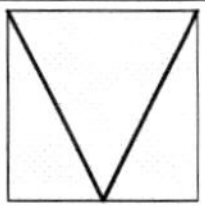	Smooth decrease and then increase of storm activity.	31	22	16	14.9	22.2	47.1
II		Wind waves in the "window" are much weaker than the selected threshold value *h*.	67	17	14	32.2	17.2	41.2
III	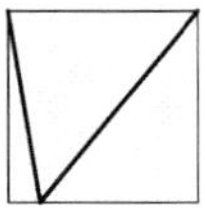	Gradual increase of storm activity or result of passage of a chain of storms with different tracks.	39	14	*	18.8	14.1	*
IV	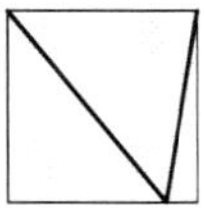	Strong residual wave field that is decaying after storm passage.	49	16	*	23.6	2.2	*
V		Wave heights close to the threshold value *h*.	22	30	4	10.5	30.3	11.7

The calculation of matrix of transitions between storms and weather windows shows that correlation is weaker between certain classes [Lopatoukhin, et al, 2000]. Therefore, it is possible to formulate the most probable scenarios of weather in the give region.

The annual cycle of storms manifests itself in variations of the monthly mean wave height $\bar{h}(t)$ between seasons. Also, synoptic variability is higher in winter than in summer. Such cyclic variations can be expressed as:

$$\bar{h}(t) = m(t) + \sigma(t)(\eta(t) + \xi(t)) \tag{2.37}$$

Where: $m(t)$ is the annually averaged value of wave height, $\sigma(t)$ is the standard deviation of monthly mean wave heights from $m(t)$. Process $\eta(t)$ can be modeled by stochastic models for extra-annual rhythms [Dragan, et al, 1987; Lopatoukhin, et al, 2000]. Finally, $\xi(t)$ is the impulse stochastic process described by (2.31) with parameters (2.36). The approach described above allows simulation of any weather scenario.

2.4 Operational Components of an Assumed Situation [1]

When operating a ship, the master chooses a heading speed and a course angle in absolute (geographical) coordinates. Making such a decision, the master is aiming to reach his destination port in time and to avoid danger. In reality, such a decision takes a balance between safety and efficiency, between financial success and risk to life. A level where such a balance is established is dependent upon a huge number of circumstances, including economic criteria and human judgment. The human factor is most significant in this area of stability safety, and evidently, the probabilistic approach to stability could not be developed without taking the human factor into account [Sevastianov, 1982b; Mordachev, et al, 1994; Mordachev, 1995].

We will not try here to build a detailed model of human behavior while operating a ship. Nevertheless, we can set some very simple assumptions, which make it possible to cover a majority of the master's behavioral patterns that could be imagined. Such variants of behavior or strategies could be defined as follows:

1. A "Careful" strategy assumes that the course and speed should be chosen to minimize risk of capsizing. Only one limitation should be taken into account: the ship should reach her destination after finite time.
2. A "Mercantile" strategy assumes that course is direct between starting point and the destination, speed is maximal and danger of capsizing is not taken into account.

Introducing these two assumed strategies reflects two main motivations of human behavior at sea: willingness to avoid danger and readiness to attain maximum profit. However, the master is able to make a mistake as any other human. To cover the possibility of a mistake, we should introduce the third strategy:

3. A "Fool" strategy assumes that course and speed should be chosen to maximize risk of capsizing. Only one limitation should be taken into account: the ship should reach her destination after finite time.

---

[1] Written in co-authorship with Prof. Sergey V. Mordachev of Kaliningrad Institute of Technology (Russia)

Using the above strategies, mathematical models of heading course choice and speed can be developed. Such models could be based on a detailed analysis of an actual ship operation taking into account recent psychological research or it may be quite simple - a matter of a special study. Again, we are not trying here to find a final solution, we only wish to indicate the point of application of such results in the general scheme of the probabilistic approach to evaluating stability safety. To do this, we shall consider a very simple model, which is based mainly on common sense. We assume that:

- A certain boundary level of capsizing risk exists;
- The level of risk depends on the sea state, heading velocity and course angle;
- The master is aware of the risk level;
- The "careful" master will try to keep his current level less than the boundary. To achieve this goal the "careful" master changes the course in order to avoid dangerous regions;
- The "mercantile" master will not take any risk resulting in a delay in arriving at the destination in time;
- The "fool" master will try to keep his current level above the boundary. To achieve this goal, the "fool " master changes the course in order to go through "dangerous" regions

These strategies can be simulated using available ship routing software and wave databases. Changes in the weather can be modeled as it was described above, in subchapter 2.3. If we imagine that we do have methods to evaluate risk function, then we would receive three values:

- Averaged risk function for a given voyage, assuming a "careful" strategy λ_C,
- Averaged risk function for a given voyage, assuming a "mercantile" strategy λ_M,
- Averaged risk function for a given voyage, assuming a "fool" strategy λ_F.

These values contain important information of the influence of a human factor on operation risk of particular ship:

$$E_M = \lambda_M - \lambda_C \tag{2.39}$$

This estimate describes the ship's ability to resist a human's premeditated mistakes caused by neglect of the danger of capsizing because of intent to gain profit:

$$E_F = \lambda_F - \lambda_C \tag{2.40}$$

This estimate describes a ship's ability to resist human non-premeditated mistakes. This value can be considered as some estimation of the degree of a ship being "foolproof".

To complete our consideration of the application of the probabilistic approach to a ship's stability estimation, we should address different scenarios of capsizing and the probability of their occurrence. We have defined an assumed situation by using all possible data, which describe environmental and human circumstances, because they both are external in relation to a ship. However, different physical phenomena could be responsible for capsizing, in the same assumed situation, see, for example, [Alman, et al 1999]. For example, a ship could capsize in following or quartering seas because of decreasing stability on the wave crest, or because of parametric resonance, or because of broaching. Different physical phenomena require different algorithms for calculation of capsizing probability.

Part 2

Dynamics of Capsizing

Chapter 3

Equations for Nonlinear Motions[1]

Generally speaking, loss of ship stability (capsizing or catastrophic heel) is a phenomenon of ship motions in waves and therefore has to be considered as a result of fluid motion around the ship hull. Although a general formulation of the problem may not always be practical, it is important to start from one, in order to have a clear understanding of the assumptions made to develop more practical models.

This chapter shows how the models for capsizing behavior are built and what assumptions and simplifications are involved. Generally, marine hydrodynamics focuses on a solid body moving on the free surface of water under the action of waves. In this chapter, we consider a ship in the middle of an ocean: so the water is deep and the shores are far away. It means that any influence of the ship motions is negligible at the shores. In other words, there are no boundaries of the fluid domain other than the free surface. Recognizing that the influence of forward speed is very important for ship hydrodynamics, we consider the ship with zero forward speed in this chapter to avoid complex derivations. The effect of forward speed on ship motions is discussed in detail in chapters 6, 8 and 9.

We start with a brief review of hydrodynamics of the ship in waves: subchapters 3.1 through 3.4 present the essence of the matter mostly following Newman [1977]. Therefore, results only are shown, since the derivations are available from the above source. Subchapters 3.5 and 3.6 consider a model of roll motions that will be used for capsizing study in the following chapters.

3.1 General Equations of Fluid Motions

3.1.1 Forces and Stresses in Fluid

Force is a measure of interaction between bodies. To consider forces in a fluid, we mentally extract a volume from the fluid domain: To keep the status quo we have to compensate for the interaction with the rest of the fluid by applying forces all over its boundary, see fig. 3.1.

---

[1] The author is grateful to K. M. Weems and J.W. Kim for review and discussion of materials of this chapter. A.M. Reed, W.F. Belknap and M.J. Hughes performed a thorough review of this chapter which was very helpful for the preparation of the second edition and greatly appreciated.

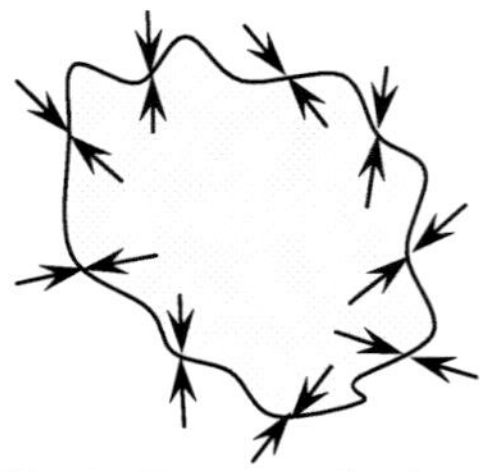

Fig. 3.1 Forces acting in fluid

This allows introducing an important concept of surface forces as internal reactions in fluid. Also, there are mass forces acting in the fluid, and gravity is one the most important of them for the waves and ship motions.

These forces differ from one point in the fluid to another. They also are changing with time. Let us pick an arbitrary point in the vicinity of the boundary of the volume we just extracted and place a small tetrahedron around it, so its slope would be a small part of the boundary surface, see fig. 3.2.

Consider the equation of forces acting on this small tetrahedron:

$$\Delta m \frac{d\vec{v}}{dt} = \vec{F}_M + \vec{P}_B - \vec{P}_x - \vec{P}_y - \vec{P}_z \qquad (3.1)$$

Here Δm is a mass of fluid in the tetrahedron and $\vec{v}$ its instant velocity. $\vec{F}_M$ is the mass force (not shown in fig. 3.2), $\vec{P}_B$ is the surface force acting through the boundary surface element ΔS, $\vec{P}_x$, $\vec{P}_y$ and $\vec{P}_z$ are the surface forces acting through the surfaces ΔS_x, ΔS_y and ΔS_z belonging to the coordinate planes.

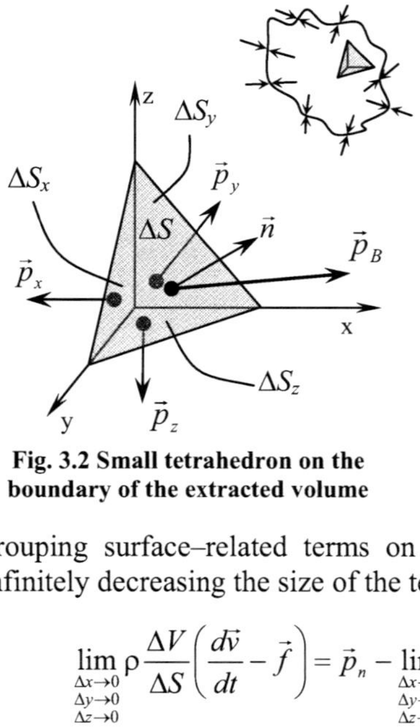

Fig. 3.2 Small tetrahedron on the boundary of the extracted volume

Since the tetrahedron is small, the forces acting on it are small too. Therefore, we can consider them being proportional to corresponding volume and area values: The volume element for mass force and surface elements for the surface forces:

$$\begin{aligned}\rho \Delta V \frac{d\vec{v}}{dt} &= \rho \Delta V \vec{f} + \vec{p}_B \Delta S \\ &- \vec{p}_x \Delta S_x - \vec{p}_y \Delta S_y - \vec{p}_z \Delta S_z\end{aligned} \qquad (3.2)$$

Here, $\vec{f}$, $\vec{p}_x$, $\vec{p}_y$ and $\vec{p}_z$ are the forces' action on the unit volume and are through the unit area. Correspondingly, ρ is density of the fluid, ΔV is the small volume of the tetrahedron. Dividing both parts by the area of boundary surface element ΔS, grouping surface–related terms on the left hand side and considering a limit while infinitely decreasing the size of the tetrahedron.

$$\lim_{\substack{\Delta x \to 0 \\ \Delta y \to 0 \\ \Delta z \to 0}} \rho \frac{\Delta V}{\Delta S}\left(\frac{d\vec{v}}{dt} - \vec{f}\right) = \vec{p}_n - \lim_{\substack{\Delta x \to 0 \\ \Delta y \to 0 \\ \Delta z \to 0}}\left(\vec{p}_x \frac{\Delta S_x}{\Delta S} - \vec{p}_y \frac{\Delta S_y}{\Delta S} - \vec{p}_z \frac{\Delta S_z}{\Delta S}\right) \qquad (3.3)$$

Here Δx, Δy, and Δz are edges of the tetrahedron, having in mind that the volume is proportional to the third degree of the linear size and the area to the second, we see that:

$$\lim_{\substack{\Delta x \to 0 \\ \Delta y \to 0 \\ \Delta z \to 0}} \frac{\Delta V}{\Delta S} = 0 \qquad (3.4)$$

Also:

$$\frac{\Delta S_x}{\Delta S} = \vec{n} \cdot \vec{x}; \quad \frac{\Delta S_y}{\Delta S} = \vec{n} \cdot \vec{y}; \quad \frac{\Delta S_z}{\Delta S} = \vec{n} \cdot \vec{z} \tag{3.5}$$

where $\vec{n}$ is a normal vector to the boundary surface element of unit length as it is shown in fig. 3.2, while $\vec{x}, \vec{y}$ and $\vec{z}$ are vectors of unit length directed along coordinate axes (orts). So, the scalar product of a normal vector and an ort represent the cosine of the angle between two vectors. The equation (3.2) now can be re-written as:

$$\vec{p}_B = \vec{p}_x \cdot (\vec{n} \cdot \vec{x}) + \vec{p}_y \cdot (\vec{n} \cdot \vec{y}) + \vec{p}_z \cdot (\vec{n} \cdot \vec{x}) \tag{3.6}$$

We remember here that $\vec{p}_x$, $\vec{p}_y$ and $\vec{p}_z$ are the forces acting on the unit area. Each of these vectors can be presented with their own components as:

$$\begin{aligned} \vec{p}_x &= p_{xx}\vec{i} + p_{xy}\vec{j} + p_{xz}\vec{k} \\ \vec{p}_y &= p_{yx}\vec{i} + p_{yy}\vec{j} + p_{yz}\vec{k} \\ \vec{p}_z &= p_{zx}\vec{i} + p_{zy}\vec{j} + p_{zz}\vec{k} \end{aligned} \tag{3.7}$$

These components are known as stresses: the stress is defined as the force acting on the specifically oriented surface of the unit area. Therefore, the stress depends on the force and orientation of the surface. A value defined by the vector and the oriented surface is known in mathematics as a tensor, see for example [Bronshtein and Semendyayev, 1997]. A tensor can also be written in matrix form (τ is a standard nomenclature for stress):

$$\boldsymbol{\tau} = \begin{pmatrix} p_{xx} & p_{xy} & p_{xz} \\ p_{yx} & p_{yy} & p_{yz} \\ p_{zx} & p_{zy} & p_{zz} \end{pmatrix} = \begin{pmatrix} \tau_{xx} & \tau_{xy} & \tau_{xz} \\ \tau_{yx} & \tau_{yy} & \tau_{yz} \\ \tau_{zx} & \tau_{zy} & \tau_{zz} \end{pmatrix} = \begin{pmatrix} \tau_{11} & \tau_{12} & \tau_{13} \\ \tau_{21} & \tau_{22} & \tau_{23} \\ \tau_{31} & \tau_{32} & \tau_{33} \end{pmatrix} \tag{3.8}$$

We also used numerical nomenclature for the coordinate axis: x-coordinate as number 1, y as 2 and z as 3. The matrix of a stress tensor is symmetric relative to the main diagonal [Newman, 1977]:

$$\tau_{ij} = \tau_{ji} \tag{3.9}$$

Using a similar nomenclature for the normal vector $\vec{n} = (n_1, n_2, n_3)$, the stress on the boundary can be written as:

$$p_{Bi} = \sum_{j=1}^{3} \tau_{ij} n_j = \tau_{ij} n_j \tag{3.10}$$

The sign for summation can be skipped: the presence of index j ONLY on the right hand side of the formula is the indication that the summation is done by this index. The entire surface force then can be expressed with the integration of (3.10) over the surface S covering the volume V:

$$\begin{aligned} \vec{F}_B &= \iint_S \vec{p}_B dS = \iint_S \left(p_{B1}\vec{i} + p_{B2}\vec{j} + p_{B3}\vec{k} \right) dS = \\ &= \iint_S \left(\vec{i} \sum_{j=1}^{3} \tau_{1j} n_j + \vec{j} \sum_{j=1}^{3} \tau_{2j} n_j + \vec{k} \sum_{j=1}^{3} \tau_{3j} n_j \right) dS \end{aligned} \tag{3.12}$$

In the tensor notation:

$$\vec{F}_B = \iint_S \left(\vec{i}\cdot\tau_{1j}n_j + \vec{j}\cdot\tau_{2j}n_j + \vec{k}\cdot\tau_{3j}n_j\right)dS$$

The same force, but expressed by components is presented in short record using tensor notation:

$$F_{Bi} = \iint_S \left(\tau_{ij}n_j\right)dS \tag{3.13}$$

Mass forces are express with the integration of the fluid density over the volume V:

$$F_{Mi} = \iiint_V f_i dV \tag{3.14}$$

3.1.2 Relationship of Volume and Surface Integrals. Transport Theorem

Before proceeding further, let us review the theorems on the relationship of surface and volume integrals [Bronshtein and Semendyayev, 1997]. If we have a space with a differentiable vector defined in each point (continuous differentiable vector field), say $\vec{Q}(x,y,z)$, then:

$$\iiint_V \left[\frac{\partial\vec{Q}}{\partial x}\cdot\vec{i} + \frac{\partial\vec{Q}}{\partial y}\cdot\vec{j} + \frac{\partial\vec{Q}}{\partial z}\cdot\vec{k}\right]dV = \iint_S \vec{Q}\cdot\vec{n}dS \tag{3.15}$$

Where S is surface "covering" volume V and $\vec{n}$ is an outer normal vector to S with the unit length. This is the divergence theorem. The divergence itself is the vector operator defined as:

$$\mathrm{div}\left(\vec{Q}\right) = \frac{\partial\vec{Q}}{\partial x}\cdot\vec{i} + \frac{\partial\vec{Q}}{\partial y}\cdot\vec{j} + \frac{\partial\vec{Q}}{\partial z}\cdot\vec{k} \tag{3.16}$$

If we have a space with a differentiable scalar defined in each point (continuous differentiable scalar field), say $\Phi(x,y,z)$, then:

$$\iiint_V \left[\frac{\partial\Phi}{\partial x}\vec{i} + \frac{\partial\Phi}{\partial y}\vec{j} + \frac{\partial\Phi}{\partial z}\vec{k}\right]dV = \iint_S \Phi\vec{n}dS \tag{3.17}$$

With the S, V and $\vec{n}$ defined as above. This is Gauss's theorem. The integrand of the volume integral is known as the gradient vector operator or just gradient:

$$\overrightarrow{\mathrm{grad}}\left(\Phi\right) = \frac{\partial\Phi}{\partial x}\vec{i} + \frac{\partial\Phi}{\partial y}\vec{j} + \frac{\partial\Phi}{\partial z}\vec{k} \tag{3.18}$$

Similarity between the formulae (3.15) and (3.16) allows the introduction of vector operator ∇ -"nabla", which is defined as:

$$\nabla = \frac{\partial}{\partial x}\vec{i} + \frac{\partial}{\partial y}\vec{j} + \frac{\partial}{\partial z}\vec{k} \tag{3.19}$$

It can be treated like a vector. If the operand is a vector, the operator results in divergence (3.16), if the operand is a scalar the result will be the gradient vector (3.18). It allows presenting both theorems in one formula:

$$\iiint_V (\nabla \cdot A)\, dV = \iint_S (A \cdot \vec{n})\, dS \tag{3.20}$$

It expresses the divergence theorem if A is a vector and Gauss's theorem if A is a scalar value.

The Transport theorem is another tool necessary to proceed. It helps to find the derivatives of volume integrals. Consider a differentiable scalar function defined in a vector field that changes with time: $f(\vec{x},t)$. The transport theorem states, that for fluid volume $V(t)$:

$$\frac{d}{dt}\iiint_{V(t)} f(\vec{x},t)\, dV = \iiint_{V(t)} \left[\frac{\partial f(\vec{x},t)}{\partial t} + \frac{\partial}{\partial x_i}\left(f(\vec{x},t)\cdot u_i\right) \right] dV \tag{3.21}$$

Were u_i is i-th component of fluid velocity vector in the given point. Note, that formula (3.21) is written in tensor notation, so sign of summation is omitted. Proof of the transport theorem can be found in [Newman, 1977].

3.1.3 Conservation of Mass and Momentum

The law of mass conservation for the fluid can be expressed quite straight forward:

$$\frac{dm}{dt} = \frac{d}{dt}\iiint_{V(t)} \rho\, dV = 0 \tag{3.22}$$

It could be interpreted as the fluid volume $V(t)$ which still consists of the same fluid particles despite their changing position with time.

Conservation of momentum requires that the derivative of momentum equals the sum of the forces acting on the system of bodies:

$$\frac{d}{dt}\iiint_V \rho \cdot u_i\, dV = \iint_S \tau_{ij} n_j\, dS + \iiint_V f_i\, dV \tag{3.23}$$

Momentum of the fluid volume consists of individual momentum of each fluid particle. The right side of the equation (3.23) contains surface (surface integral) and mass forces (volume integral) acting on volume $V(t)$, see Subchapter 3.1.1.

Using the divergence theorem (3.15), it is possible to express momentum conservation in volume integrals only:

$$\frac{d}{dt}\iiint_V \rho \cdot u_i\, dV = \iiint_V \left(\frac{\partial \tau_{ij}}{\partial x_i} + f_i \right) dV \tag{3.24}$$

3.1.4 Continuity Equation. Euler's Equations

Application of the Transport theorem (3.21) to mass conservation equation (3.22) and replacing integration with a partial differential equation yields the Continuity equation, which express the mass conservation law in simpler form, provided that the fluid density does not change with time:

$$\text{Tensor notation}: \frac{\partial(\rho u_i)}{\partial x_i} = 0 \qquad \text{Vector notation}: \nabla \cdot \left(\rho \vec{V}\right) = 0 \quad (3.25)$$

Here $\vec{V}$ is velocity of fluid particle and u_i are its components:

$$\vec{V} = u_1\vec{i} + u_2\vec{j} + u_3\vec{k}$$

Assuming constant fluid density, which, indeed, is the case for surface ships equation (3.25) is simplified even more:

$$\nabla\left(\vec{V}\right) = 0 \tag{3.26}$$

An analogous procedure (use of the Transport theorem with further transition to partial differential equation) applied to the equation of conservation of momentum (3.24) yields Euler's equation:

$$\frac{\partial u_i}{\partial t} + u_j \frac{\partial u_i}{\partial x_j} = \frac{1}{\rho}\left(\frac{\partial \tau_{ij}}{\partial x_j} + F_{Mi}\right) \tag{3.27}$$

Here, it is also assumed the flow is incompressible $(\partial\rho/\partial t = \partial\rho/\partial x_i = 0)$, which is a fair assumption for water. Taking into account tensor notation used for equation (3.27), we find that the left hand side is actually a substantial derivative of the fluid velocity by time, which is the acceleration of the fluid particle:

$$\frac{du_i}{dt} = \frac{1}{\rho}\left(\frac{\partial \tau_{ij}}{\partial x_j} + F_{Mi}\right) \tag{3.28}$$

More details on Continuity and Euler's equations are available in [Newman, 1977].

3.1.5 Navier-Stokes Equations

There is a large class of fluids for which stress is linearly dependent on strain rates. Such fluids typically have small viscosity and are called "Newtonian" liquids; water is one of them. For the Newtonian Fluid this relationship is expressed as:

$$\tau_{ij} = \mu \cdot \left(\frac{\partial u_i}{\partial x_j} + \frac{\partial u_j}{\partial x_i}\right) \tag{3.29}$$

Here, μ is the viscous shear coefficient. Equation (3.29) expresses the relation between shear stresses (which are forces - dynamic parameters) with spatial derivative of the velocities (kinematic parameters).

This equation also implies that the fluid is isotropic and its qualities are the same in all directions. It is in line with our previous assumption that the density is a constant and it is definitely adequate for water when considering ship motions.

To derive equation of motions of a Newtonian fluid we need to substitute expression (3.29) in Euler's equation. Since (3.29) includes only shear stresses, it would be necessary to include normal ones too:

$$\tau_{ij} = -p\delta_{ij} + \mu \cdot \left(\frac{\partial u_i}{\partial x_j} + \frac{\partial u_j}{\partial x_i} \right) \tag{3.30}$$

Here p is pressure, which is the value numerically equal to normal stress, but taken with the opposite sign, δ_{ij} is Kronecker delta function:

$$\delta_{ij} = \begin{cases} 1; & i = j \\ 0; & i \neq j \end{cases} \tag{3.31}$$

Substitution of (3.30) into Euler's equation and taking into account continuity equation (3.26) leads to the well-known Navier–Stokes equations (tensor notion, summation by index j):

$$\frac{\partial u_i}{\partial t} + u_j \frac{\partial u_i}{\partial x_j} = -\frac{1}{\rho}\frac{\partial p}{\partial x_i} + \frac{\mu}{\rho}\frac{\partial^2 u_i}{\partial x_j \partial x_j} + F_{Mi} \tag{3.32}$$

The same in the vector notation:

$$\frac{\partial \vec{V}}{\partial t} + \left(\vec{V} \cdot \nabla\right) \cdot \vec{V} = -\frac{1}{\rho}\nabla p + \nu \nabla^2 \vec{V} + \frac{1}{\rho}\vec{F}_M \tag{3.33}$$

Here ν is a coefficient of kinematic viscosity:

$$\nu = \frac{\mu}{\rho} \tag{3.34}$$

3.1.6 Boundary Conditions

Expressions (3.32) or (3.33) represent a system of three partial differential equations. The unknowns are a vector-valued function of velocity (or three velocity components) of the fluid and scalar pressure. This system has an infinite number of solutions, to choose one of them, which describes flow during ship motions in waves, we have to apply certain limitations: set of initial conditions defining the system state at the initial moment and boundary conditions defining values for the unknown function (provided such a solution exists and it is stable). In the case of ship motion the fluid domain is limited by:

- Solid body – ship hull;
- Free surface;
- Far field conditions (consequences of ship motions cannot have serious influence on the flow far away from the ship).

Physical boundary conditions might be dynamic and kinematic: dynamic boundary conditions provide equality of stresses on the boundary, for example: pressure at the free surface should be equal to atmospheric pressure. Kinematic boundary conditions apply limitations on fluid velocity: for example, at the body boundary, the fluid velocity should be equal to the velocity of this particular point of the body.

Once initial and boundary conditions are formulated, the system is theoretically ready for solution. The problem is that there are no analytical solutions for practical cases; such solutions are available for the very simple theoretical cases.

A great deal of effort has been devoted to the development of numerical solutions to the Navier-Stokes equations using a wide variety of methods. Indeed, Navier-Stokes methods make up the core of the discipline of Computational Fluid Dynamics (CFD). In the field of marine hydrodynamics, computational Navier-Stokes methods have been applied to a number of aspects of ship flow problems, including the characterization of viscous roll damping, prediction of ship maneuvering coefficients, and calculation of propeller inflow and performance [Gorski, 2002], [Stern, *et al* 2006]. However, the calculation of the complete problem of a ship moving in waves is outside the practical capabilities of such methods at this time and is likely to remain so for the near future.

3.2 Motions of Ideal Fluid

3.2.1 Model of Ideal Fluid

While a solution for Navier-Stokes equations poses significant numerical challenges, especially in the presence of free surface, we are in need for a simpler model to apply. At this moment we assumed that the fluid is a continuum, it has constant density and no compressibility, its physical qualities are the same in all directions (isotropic quality) and shear stress linearly depends on transverse velocity difference (Newtonian fluid). What kind of additional assumption could we make to simplify the model without giving away something significant for a correct description of ship motions and capsizing in waves?

Let us think about viscosity: it is known from experiments that for water, viscosity plays a significant role in fluid motions only within the thin layer around the body. This layer, known as the boundary layer, is important for the correct prediction of drag force and propulsion performance. However, this is not very significant for the ship motions, because the latter is primarily influenced by surface wind waves, which are not so dependent on viscosity. The only exception is roll: it is known that roll damping does have a viscous contribution.

At the same time the roll damping can be externally characterized and then added to the potential-flow based ship motions calculations. Viscous flow calculations using Reynolds-Averaged Navier-Stokes (RANS) are beginning to be used to build such roll damping models [Korpus and Falzarano, 1996], but model tests remain the most common, still fairly expensive, method of evaluating roll damping. In early stage design or in studies not related to particular ship hulls, an estimate of roll damping coefficients can typically be obtained from experience with similar ships or the published results of systematic hull form studies [Ikeda, *et al*, 1982, Ikeda and Kawahara, 1993, Ikeda, 2002].

Therefore, assumption of a small viscosity influence seems to be acceptable and we further deal with inviscid or ideal fluid:

$$\frac{\partial u_i}{\partial t}+u_j\frac{\partial u_i}{\partial x_j}=-\frac{1}{\rho}\frac{\partial p}{\partial x_i}+F_{Mi} \tag{3.35}$$

It is also important to note that the absence of the viscosity leads to the absence of shear stresses: there are no more interactions between layers.

The next assumption is concerning mass forces. It is quite evident that the only mass force in the fluid that influences ship motions is gravitation. Other mass forces (such as electro-magnetic) might be important for other types of flow, but not for ship motion in waves. The important quality of gravity forces is that they are conservative: work of these forces along the closed line is zero. Taking into account also that the gravity acts downwards and it is in the negative direction of the third coordinate axis $x_3=z$, Euler's equations of motions (in tensor notation) are of the form:

$$\frac{\partial u_i}{\partial t}+u_j\frac{\partial u_i}{\partial x_j}=-\frac{1}{\rho}\left(\frac{\partial p}{\partial x_i}+\rho g h\right);\quad h=\begin{cases}0: & i=1,2\\ z: & i=3\end{cases} \tag{3.36}$$

Kelvin's theorem states that for an inviscid or ideal fluid the amount of rotational motion is constant. Physically it is very clear, since there are no shear stresses, that it is impossible to change characteristics of rotational motion. This theorem allows us to introduce the next assumption that there is no rotational motion in the considered fluid domain.

What consequences does this simplification lead to? Forces caused by rotational motions of fluid are principal for lifting surfaces, like rudders, bilge keels and other fin-like appendages. As we shall see shortly, simplifications achieved for the price of irrotational assumptions are principal, so it is worthwhile to accept this assumption and include the influence of the lifting surface as an external force dependent on local field of fluid velocities.

Absence of rotational motions is expressed as:

$$\frac{\partial u_i}{\partial x_j}-\frac{\partial u_j}{\partial x_i}=0 \qquad \text{with} \quad i\neq j \tag{3.37}$$

Using the above formula makes Euler's equations simpler:

$$\frac{\partial u_i}{\partial t}=-\frac{1}{\rho}\left(\frac{\partial p}{\partial x_i}+\rho g h\right);\quad h=\begin{cases}0: & i=1,2\\ z: & i=3\end{cases} \tag{3.38}$$

The system of partial differential equations (3.38) does not contain any spatial derivatives of the fluid velocities. Further consideration shall show how significant these simplifications are.

3.2.2 Potential. Laplace and Bernoulli Equations. Green's Theorem

Assuming the irrotational character of fluid motions allows presenting the vector field of fluid velocity with a scalar field called potential. The potential is related with the fluid velocity as follows:

$$\vec{V}(x,y,z,t) = \text{grad}(\varphi(x,y,z,t)) = \frac{\partial\varphi}{\partial x}\cdot\vec{i} + \frac{\partial\varphi}{\partial y}\cdot\vec{j} + \frac{\partial\varphi}{\partial z}\cdot\vec{k} \tag{3.39}$$

The same in the tensor notation:

$$u_i = \frac{\partial\varphi}{\partial x_i}$$

This is a significant simplification since instead of three values at each point of the fluid domain and at each moment of time we can address only one value.

Let us consider how the principal equations expressing conservation of mass and momentum will look like for the new unknown function – potential.

The Continuity equation expresses the mass conservation law as:

$$\nabla^2\varphi = \frac{\partial^2\varphi}{\partial x^2} + \frac{\partial^2\varphi}{\partial y^2} + \frac{\partial^2\varphi}{\partial z^2} = 0 \tag{3.40}$$

The same in the tensor notation:

$$\frac{\partial^2\varphi}{\partial x_i \partial x_i} = 0$$

This is a linear partial differential equation of the second order called the Laplace equation. This equation belongs to elliptical types of partial differential equations; it can be solved with initial and boundary conditions.

A mathematical problem defined with an elliptic equation with an appropriate set of initial and boundary conditions belong to a class of Boundary Value Problems. Also, the Laplace equation is linear and it is generally easier to solve than nonlinear Navier-Stokes equations. The solution of the Laplace equation belongs to a certain class of harmonic functions.

For numerical solutions, the fact that the boundary value problem is easier to solve than Navier-Stokes translates into a significantly less amount of required computational resources. A mesh has to be defined at the boundaries only and the solution is much less mesh dependent. A dramatic simplification of the solution procedure actually justifies the inviscid fluid and irrotational motion assumption made above.

Now we need to look at boundary conditions. Let us look again at Euler's equation that expresses law of momentum conservation. Substitution of potential transforms it into the following form (in tensor notation):

$$\frac{\partial\varphi}{\partial t} + \frac{1}{2}\frac{\partial\varphi}{\partial x_i}\frac{\partial\varphi}{\partial x_i} = -\frac{1}{\rho}(p + \rho g h) + C(t); \quad h = \begin{cases} 0: & i = 1,2 \\ z: & i = 3 \end{cases} \tag{3.41}$$

Here $C(t)$ is an arbitrary function, which does not depend on coordinates, but might be dependent on time. This formula, known as Euler's Integral, is frequently called Bernoulli's equation. It provides us with the important relationship between potential and pressure. This is especially important since we need to know the forces acting on a ship in waves to predict her motion.

Bernoulli's equation also will be needed to formulate the boundary condition on the free surface, but we will address this later. Generally, the ideal fluid problem requires less boundary conditions, because there are no shear stresses.

Body boundary conditions also can be expressed in terms of the potential:

$$\frac{\partial \varphi}{\partial n} = \vec{V}_B(x, y, z, t) \cdot \vec{n}(x, y, z, t) \qquad (3.42)$$

Here $\vec{V}_B(x, y, z, t)$ is a velocity of a body surface defined with coordinates x, y and z and the time moment t; $\vec{n}(x, y, z, t)$ is an outer normal vector to the body surface. The body boundary condition in the form (3.42) requires that the normal component of fluid velocity would be equal to the velocity of the corresponding body point at the place of contact. There might be however, a difference in tangential velocities, since shear stresses in the fluid are absent and we do not model the boundary layer, assumed to be very thin here. We will continue our consideration of the boundary condition later when discussing free surface and waves.

Our brief review of potential and its properties would not be complete without mentioning Green's theorem (to be exact, this is just one of its forms):

$$\iint_S \left[\varphi \frac{\partial \psi}{\partial n} - \psi \frac{\partial \varphi}{\partial n} \right] dS = 0 \qquad (3.43)$$

Here S is a closed surface, ψ and φ are solutions for Laplace equations not necessarily matching any boundary conditions. More details can be found in [Newman, 1977].

3.2.3 Hydrodynamic Pressure Forces

Now, consider forces acting on the body caused by hydrodynamic pressures, which being applied along the body surface S_B produce the following force and moment:

$$\vec{F} = \iint_{S_B} p\vec{n}\, dS \qquad (3.44)$$

$$\vec{M} = \iint_{S_B} p(\vec{r} \times \vec{n}) dS \qquad (3.45)$$

Here $\vec{n}$ is the normal vector of unit length at the body surface S_B, $\vec{r}$ is radius vector of this point. Vector $\vec{n}$ points into the body.

To facilitate the use of Gauss's theorem (which is necessary for this derivation) we have to introduce matching surface S_M to "enclose" the body and some fluid volume. Gauss's theorem can only be applied to a finite volume "covered" with the surface, which now consists of matching surface S_M and body surface S_B.

Considering the rate of change of momentum in the fluid volume (which is equal to the sum of all forces), using the Transport theorem and Bernoulli's equation to express pressure through potential, we can express hydrodynamic force and moment in the following form:

$$\vec{F} = -\rho \frac{d}{dt} \iint_{S_B} \varphi \cdot \vec{n}\, dS - \rho \iint_{S_M} \left(\frac{\partial \varphi}{\partial n} \cdot \nabla \varphi - \frac{1}{2} \vec{n} \cdot \nabla \varphi \cdot \nabla \varphi \right) dS \tag{3.46}$$

$$\vec{M} = -\rho \frac{d}{dt} \iint_{S_B} \varphi \cdot (\vec{r} \times \vec{n}) dS - \rho \iint_{S_M} \left[\vec{r} \times \left(\frac{\partial \varphi}{\partial n} \cdot \nabla \varphi - \frac{1}{2} \vec{n} \cdot \nabla \varphi \cdot \nabla \varphi \right) \right] dS \tag{3.47}$$

Details of the derivation can be found in [Newman, 1977].

3.2.4 Forces on Moving Body in Unbounded Fluid. Added Masses

Formulae (3.46) and (3.47) can be simplified if we consider the fluid domain to be unbounded. Naturally, the influence of the matching surface becomes very small and can be excluded:

$$\vec{F} = -\rho \frac{d}{dt} \iint_{S_B} \varphi \cdot \vec{n}\, dS \tag{3.48}$$

$$\vec{M} = -\rho \frac{d}{dt} \iint_{S_B} \varphi \cdot (\vec{r} \times \vec{n}) dS \tag{3.49}$$

The body is involved in arbitrary motion: nevertheless, it is always possible to present this motion as a composition of translational motion together with a certain point called centre of rotation and rotational motion about this point:

$$\vec{r}_M(x,y,z,t) = \vec{r}_0 + \vec{r}' \quad \text{and} \quad \frac{d\vec{r}_0}{dt} = \vec{V}_B \tag{3.50}$$

Here $\vec{V}_B$ is the velocity of translational motion of the body; $\vec{r}'$ is the radius vector of an arbitrary point originating from the mentioned above centre of rotation, say *O'*. Then, the velocity of an arbitrary point on the body surface, can be expressed as:

$$\vec{V}_M(x,y,z,t) = \vec{V}_B + \vec{\Omega} \times \vec{r}' \tag{3.51}$$

Here $\vec{\Omega}$ is the vector of angular velocity of rotation about the centre of rotation *O'*;

Then the body boundary condition, which is a normal component of velocity of an arbitrary point on the body surface, can be expressed as:

$$\frac{\partial \varphi}{\partial n} = \vec{V}_B \cdot \vec{n} + \vec{\Omega} \cdot (\vec{r}' \times \vec{n}) \tag{3.52}$$

We introduce a new notation of components of vectors $\vec{V}_B$ and $\vec{\Omega}$:

$$\vec{V}_B = u_1\vec{i} + u_2\vec{j} + u_3\vec{k}$$
$$\vec{\Omega} = u_4\vec{i} + u_5\vec{j} + u_6\vec{k} \tag{3.53}$$

Let us also present the potential as:

$$\varphi = \sum_{i=1}^{6} \varphi_i u_i \tag{3.54}$$

So, the potential consists of six components. Each component represents fluid velocities caused by the translational motion of the body along the coordinate axes (components 1, 2 or 3) or rotation about these axes (components 4, 5 or 6) with the unit velocity. Each of these components is also a solution of the Laplace equation with the boundary conditions presented as (compare to 3.42):

$$\frac{\partial \varphi_i}{\partial n} = n_i \qquad i = 1,2,3$$
$$\frac{\partial \varphi_i}{\partial n} = \left(\vec{r}'\times\vec{n}\right)_{i-3} \qquad i = 4,5,6 \tag{3.55}$$

Each of these potentials also has to satisfy the far-field condition: it has to be zero at an infinite distance from the body.

Substitution of the new potential presentation (3.54) into formulae for force (3.48) and moment (3.49) allows us to evaluate the derivative taking into account that:

$$\frac{d\vec{n}}{dt} = \vec{\Omega}\times\vec{n}; \quad \frac{d\left(\vec{r}'\times\vec{n}\right)}{dt} = \vec{\Omega}\times\left(\vec{r}'\times\vec{n}\right) \tag{3.56}$$

With the final result becoming:

$$\vec{F} = -\rho\dot{V}_{Mi}\iint_{S_B}\varphi_i\vec{n}\,dS - \rho V_{Mi}\cdot\vec{\Omega}\times\iint_{S_B}\varphi_i\vec{n}\,dS \tag{3.57}$$

$$\vec{M} = \vec{r}_0\times\vec{F} - \rho\vec{V}_M V_{Mi}\times\iint_{S_B}\varphi_i\vec{n}\,dS - \rho\dot{V}_{Mi}\iint_{S_B}\varphi_i\left(\vec{r}'\times\vec{n}\right)dS -$$
$$-\rho V_{Mi}\cdot\vec{\Omega}\times\iint_{S_B}\varphi_i\left(\vec{r}'\times\vec{n}\right)dS \tag{3.58}$$

Since we are free to choose an origin for the coordinate system, we can assume that at the present moment it coincides with the centre of rotation, so $\vec{r}_0 = 0$.

Consider the components of force and moment. To avoid bulky records of the vector product, we introduce an auxiliary three-dimensional array ε_{jkl} defined as:

$$\varepsilon_{jkl} = \begin{cases} 1 & \text{if} \quad \{j,k,l\}\in\{[1,2,3],[2,3,1],[3,1,2]\} \\ 0 & \text{if} \quad j=k \text{ or } j=l \text{ or } k=l \\ -1 & \text{if} \quad \{j,k,l\}\in\{[3,2,1],[1,3,2],[2,1,3]\} \end{cases} \tag{3.59}$$

In other words, it equals 1 if the indexes are in cyclic order, -1 if vice versa and 0 if any of two indexes are equal. It allows writing components of the vector product in a concise manner.

Using the auxiliary array ε_{jkl} as defined above and boundary conditions (3.55), the components of the force and moment vectors can be presented as:

$$F_j = -\rho \dot{V}_{Mi} \iint_{S_B} \varphi_i \frac{\partial \varphi_j}{\partial n} dS - \rho \varepsilon_{jkl} V_{Mi} \Omega_k \iint_{S_B} \varphi_i \frac{\partial \varphi_l}{\partial n} dS \tag{3.60}$$

$$\begin{aligned} M_j = & -\rho \varepsilon_{jkl} V_{Mk} V_{Mi} \iint_{S_B} \varphi_i \frac{\partial \varphi_l}{\partial n} dS - \rho \dot{V}_{Mi} \iint_{S_B} \varphi_i \frac{\partial \varphi_{j+3}}{\partial n} dS - \\ & -\rho \varepsilon_{jkl} V_{Mi} \cdot \Omega_k \iint_{S_B} \varphi_i \frac{\partial \varphi_{l+3}}{\partial n} dS \end{aligned} \tag{3.61}$$

Here, we used tensor notation, so summation symbols are not shown. i defines the index of the potential as in formula (3.54) so it changes from 1 to 6, index j defines the force and moment vector component, it runs from 1 to 3; indexes k and l are also related to vector components, but used to define elements of the vector product, so they are also in the range from 1 to 3.

Both formulae (3.60) and (3.61) contain remarkably similar integrals:

$$m_{ij} = \rho \iint_{S_B} \frac{\partial \varphi_i}{\partial n} \varphi_j dS \tag{3.62}$$

These integrals are actually coefficients at the acceleration-related terms and they form a 6 by 6 matrix. The values are known as added masses and they describe inertial properties of a body moving in a fluid other than ones of a gravitational nature.

Added mass depends on the body's geometry. They express the influence of acceleration in the j-th mode on motion in i-th mode (both indexes are counted from 1 to 3 for translational and from 4 to 6 for rotational motions). As a result, added masses have different units, depending on the indexes:

$$[m_{ij}] = \begin{vmatrix} Kg & Kg & Kg & Kg\cdot m & Kg\cdot m & Kg\cdot m \\ Kg & Kg & Kg & Kg\cdot m & Kg\cdot m & Kg\cdot m \\ Kg & Kg & Kg & Kg\cdot m & Kg\cdot m & Kg\cdot m \\ Kg\cdot m & Kg\cdot m & Kg\cdot m & Kg\cdot m^2 & Kg\cdot m^2 & Kg\cdot m^2 \\ Kg\cdot m & Kg\cdot m & Kg\cdot m & Kg\cdot m^2 & Kg\cdot m^2 & Kg\cdot m^2 \\ Kg\cdot m & Kg\cdot m & Kg\cdot m & Kg\cdot m^2 & Kg\cdot m^2 & Kg\cdot m^2 \end{vmatrix} \tag{3.63}$$

The matrix of added masses is symmetrical. It can easily be seen from Green's theorem (3.43), as both φ_i and φ_j are solutions of the Laplace equation:

$$m_{ij} = \iint_{S_B} \frac{\partial \varphi_i}{\partial n} \varphi_j dS = \iint_{S_B} \frac{\partial \varphi_j}{\partial n} \varphi_i dS = m_{ji} \tag{3.64}$$

As a result, only 21 of the added masses are independent. Another feature is symmetry of the body: if it has one plane of symmetry, there are only 12 non-zero added masses, because all added mass having only one even index are equal to zero:

$$m_{12} = m_{14} = m_{16} = m_{23} = m_{25} = m_{34} = m_{36} = m_{45} = m_{56} = 0 \quad (3.65)$$

This case can be applied to the majority of ships in undamaged conditions. If the body has two planes of symmetry, it has only eight nonzero added masses, because:

$$m_{13} = m_{15} = m_{24} = m_{46} = m_{25} = 0 \quad (3.66)$$

Sometimes the ship can be considered as a symmetrical body relative to the midship section as well. It could help to simplify some problems of ship hydrodynamics, where the details of her shape are not that important such as a preliminary study of controllability. If there were three planes of symmetry, only the principal diagonal of the matrix would be non-zero. Such a simplification could be used for qualitative theoretical research only with the exception of several very specific cases like buoys, etc. Numerical values of the coefficients can be obtained using potential theory. Analytical results are available for some simple shapes, while the general case requires applications of numerical methods, so these symmetry considerations are rather a matter of convenience. As we indicated earlier, we have followed [Newman, 1977], where more information is available.

3.3 Waves

3.3.1 Free Surface Boundary Conditions

We have completed a very brief review of arbitrary body motion in boundless fluid. Now let us introduce a free surface and assume that the fluid is involved in wave motions. Here we will not look at how the ocean waves are generated: this is a separate subject; see [Phillips, 1977] or [Mei, 1983]. A general review of waves on the surface of water is also given in the Chapter 9 of [Kobylinski and Kastner, 2003]. Here we focus on very simple models that would lead to a model of ship motion and capsizing in waves.

First, let us introduce the coordinate system. At this moment we consider the direction of wave propagation would make angle β with positive direction abscissa and z-coordinate is directed up, as is shown in fig. 3.3 (our coordinate system will differ from that used in [Newman, 1977]; as we have to be consistent with our further analyses).

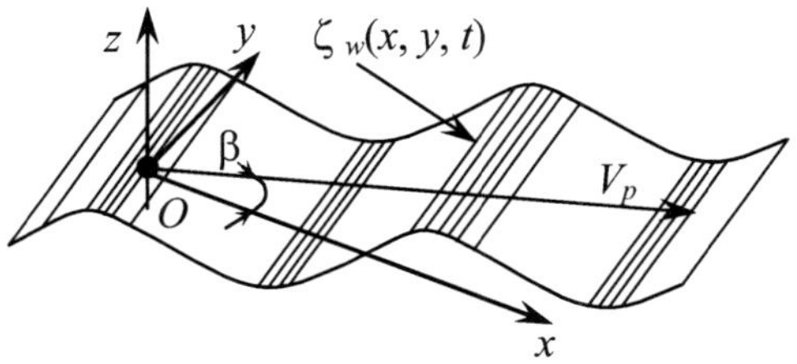

Fig. 3.3 Coordinate system for waves

Now we try to formulate boundary conditions on the free surface. First of all, the particles should not leave the fluid domain. This is the kinematic boundary condition, which can be expressed as:

$$\frac{d}{dt}\left[z - \zeta_w(x,y,t)\right] = 0 \quad (3.67)$$

Taking into account that the coordinates are orthogonal ($\partial z / \partial x = \partial z / \partial y = 0$) and existence of the potential of fluid velocities, the final kinematic free surface boundary condition can be expressed as:

$$\frac{\partial \varphi}{\partial z} - \frac{\partial \zeta_w}{\partial t} - \frac{\partial \zeta_w}{\partial x}\frac{\partial \varphi}{\partial x} - \frac{\partial \zeta_w}{\partial y}\frac{\partial \varphi}{\partial y} = 0 \qquad (3.68)$$

Another boundary condition to be satisfied is that the pressure on the free surface should be equal to the atmospheric one:

$$p - p_a = 0 \qquad (3.69)$$

This is the dynamical boundary condition, using Bernoulli's equation (3.41). It can be written as (choosing the initial time moment so that arbitrary constant *C*(*t*) equals zero):

$$\zeta_w(x, y, t) = -\frac{1}{g}\left(\frac{\partial \varphi}{\partial t} + \frac{1}{2}\nabla\varphi \cdot \nabla\varphi\right) \qquad (3.70)$$

3.3.2 Linearized Free Surface Boundary Conditions. Theory of Small Waves

The expressions (3.68) and (3.70) are nonlinear partial differential equations. In order to derive equations of ship motions in waves, we will have to linearize it. Actually, this is not completely relevant for the ultimate purpose of this study: steep waves, indeed, are the most dangerous for ship stability. However, a completely consistent analytical model is not known at this time and WE HAVE to make this kind of assumption to proceed, especially having in mind applicability of the superposition principle, which will be necessary to consider for irregular waves (Chapter 8). The next best thing, though, is to get a clear understanding of what is assumed and how it might affect the final results.

The linearization of the kinematic boundary conditions leads to the following assumptions:

1. The wave elevation is small, therefore its slopes $\partial \zeta_w / \partial x$ and $\partial \zeta_w / \partial y$ have the same order as that of $\partial \zeta_w / \partial t$
2. The water velocities $\partial \varphi / \partial x$, $\partial \varphi / \partial y$ and $\partial \varphi / \partial z$ are also small values; they are proportional to wave motion and have the same order as the wave slopes.
3. Since the wave elevation is small, the dynamical boundary condition can be considered on $z = 0$, instead of $\zeta_w = 0$

These assumptions effectively mean that the wave is small and allows dropping nonlinear terms both in (3.58) and (3.60). The linearized kinematic free surface boundary conditions are formulated as:

$$\frac{\partial \varphi}{\partial z} = \frac{\partial \zeta_w}{\partial t} \qquad (3.71)$$

The linearized dynamical boundary condition can be expressed as:

$$\zeta_w(x, y, t) = -\frac{1}{g}\frac{\partial \varphi}{\partial t} \quad \text{at} \quad z = 0 \qquad (3.72)$$

Differentiating both sides of (3.72) by t with further substitution into (3.71) allows deriving the combined boundary condition:

$$\frac{\partial^2\varphi}{\partial t^2}+\frac{1}{g}\frac{\partial\varphi}{\partial z}=0 \tag{3.73}$$

3.3.3 Plane Progressive Small Waves

We shall limit ourselves to plane progressive waves here. Later, in Chapter 8 we will consider different models for irregular waves.

The free surface is assumed to have a form of cosine function as:

$$\zeta_w(x,y,t)=\zeta_{Aw}\cdot\cos(kx\cos\beta+ky\sin\beta-\omega t+\varepsilon) \tag{3.74}$$

Where ζ_{Aw} is the wave amplitude (one half of wave height), ω is wave frequency, k is the wave number (spatial frequency) β is angle of wave propagation relative to the axis of the abscissa and ε is initial phase. Here, we can always choose initial moment for ε=0.

Wave frequency has evident relation with the wave period T_w. The wave number is related to the wave length λ.

$$\omega=\frac{2\pi}{T_w};\qquad k=\frac{2\pi}{\lambda} \tag{3.75}$$

The velocity of the wave profile (phase velocity):

$$V_p=\frac{\lambda}{T_w}=\frac{\omega}{k} \tag{3.76}$$

The flow velocity potential is searched in the form, similar to wave profile:

$$\varphi(x,y,z,t)=\mathrm{Re}\left[Z(z)\cdot\exp(-ik(x\cos\beta+y\sin\beta)+i\omega t)\right] \tag{3.77}$$

Here $i=\sqrt{-1}$. To be a solution, the potential (3.77) has to satisfy the Laplace equation and free surface boundary condition. If we substitute the potential (3.77) into the Laplace equation (3.40), the following ordinary differential equation appears:

$$\frac{d^2Z}{dz^2}+k^2Z=0 \tag{3.78}$$

Once we find the solution $Z(z)$ for the equation (3.78), the Laplace equation is satisfied. Since the equation (3.78) is a linear ordinary differential equation of the second order with a constant coefficient, it does not present a problem (however, we will be reviewing ordinary linear differential equations of the second order because of their special importance for the subject in subchapter 3.4). The solution of (3.78) exists in a form:

$$Z(z)=C\cdot\exp(kz)+D\cdot\exp(-kz) \tag{3.79}$$

Arbitrary constants C and D are to be defined by boundary conditions. If we consider deep water, influence of surface wave motion should decay with increasing depth.

$$\lim_{z\to-\infty}\varphi(x,y,z,t)=0\qquad\Rightarrow\qquad D=0 \tag{3.80}$$

Substitution of potential (3.77) with (3.79) and (3.80) into a linearized free surface boundary condition in the form (3.72) delivers the second arbitrary constant:

$$C = -g\zeta_{Aw}\frac{1}{i\omega} \tag{3.81}$$

The potential itself now is presented as:

$$\begin{aligned}\varphi &= \mathrm{Re}\left[\frac{g}{\omega}\zeta_{Aw} i \exp\left(kz - ik(x\cos\beta + y\sin\beta) + i\omega t\right)\right] \\ &= \frac{g}{\omega}\zeta_{Aw} e^{kz}\sin\left(kx\cos\beta + ky\sin\beta - \omega t\right)\end{aligned} \tag{3.82}$$

Substituting (3.82) into the combined boundary condition of the form (3.73) reveals another relationship between wave number and wave frequency:

$$k = \frac{\omega^2}{g} \tag{3.83}$$

Formula (3.73) is called the dispersion relation and, in this form, it is true only for small waves. Partial derivatives of the potential are components of the velocity of fluid particles:

$$\begin{aligned} v_x &= \frac{\partial\varphi}{\partial x} = g\omega\zeta_{Aw} e^{kz}\cos\beta\cos(kx\cos\beta + ky\sin\beta - \omega t) \\ v_y &= \frac{\partial\varphi}{\partial y} = g\omega\zeta_{Aw} e^{kz}\sin\beta\cos(kx\cos\beta + ky\sin\beta - \omega t) \\ v_z &= \frac{\partial\varphi}{\partial z} = g\omega\zeta_{Aw} e^{kz}\sin\left(kx\cos\beta + ky\sin\beta - \omega t\right)\end{aligned} \tag{3.84}$$

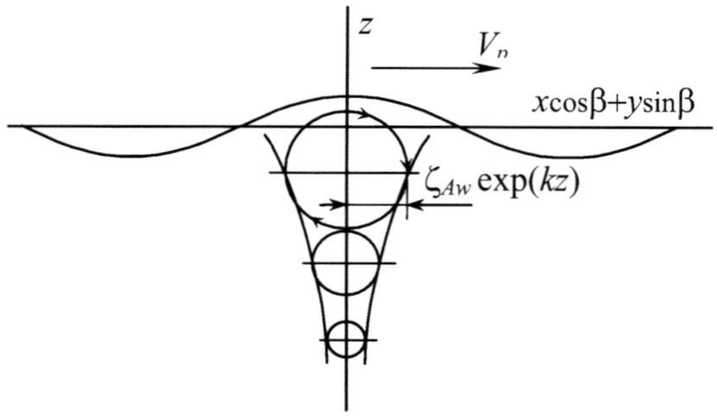

Fig. 3.4 Motions of fluid particles in small progressive plane wave

These formulae suggest an orbital character of fluid and a partial motion in the plane defined $x\cos\beta + y\sin\beta$ that is a well-known outcome of the theory of small waves - see fig. 3.4.

According to this scheme, there is no mass transfer – all fluid particles move with closed trajectories and their averaged velocities are zero. Theory of small waves also allows taking into account finite depth of the fluid domain altering accordingly boundary condition (3.80) see [Newman, 1977].

Completing our consideration of waves here, we would like once again to emphasize that application of small waves theory for large amplitude roll and capsizing of ships is an approximation. Review of the background of nonlinear waves can be found in [Newman,

1977], also [Whitham, 1974] can be recommended. Several nonlinear wave phenomena could be actually captured with the theory for ideal fluid.

3.4 Ship Response in Regular Small Waves

3.4.1 System of Coordinates

First, we have to determine a system of coordinates. This choice is a matter of convenience and tradition. While considering hydrostatics and stability in a static sense only one coordinate system was sufficient. Here, (as in subchapter 10.1 of [Kobylinski and Kastner, 2003]) we need three systems (fig. 3.5):

- A global system ξ η ζ is fixed to the Earth. Position of its origin O is arbitrary. Sometimes, however, it is convenient to substitute it with the semi-movable system that moves with constant forward speed.
- "Main" semi-movable system x y z; in most of the cases, its origin is located in the centre of mass G, which plays a role in the centre of rotation, however sometimes it is more convenient to have in the waterplane (point O'). Motion of this system coincides with translational motion of the ship.
- Ship-fixed system x' y' z'; it is the same as is traditionally used in ship hydrostatics with the only difference that the origin is placed in the waterplane point O' or centre of mass G instead of the point K.

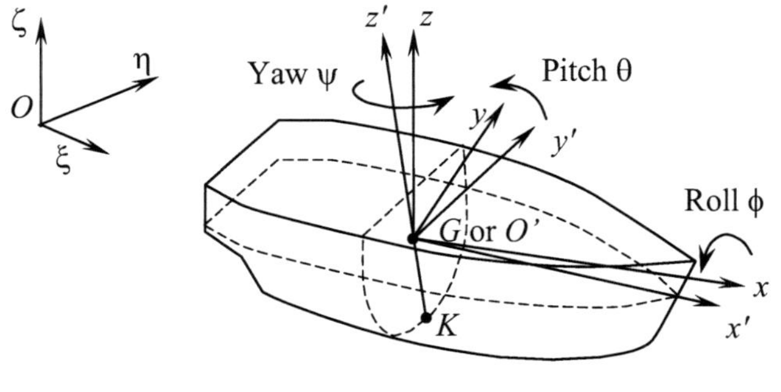

Fig. 3.5 Coordinate systems

The relationship between global and a ship-fixed coordinate system are defined with the following set of formulae [Voitkunsky, 1985]:

$$\xi = \xi_G + x'\cos\psi\cos\theta + y'(\sin\phi\cos\psi\sin\theta - \cos\phi\sin\psi) + \\ + z'(\cos\phi\cos\psi\sin\theta + \sin\phi\sin\psi) \tag{3.85}$$

$$\eta = \eta_G + x'\sin\psi\cos\theta + y'(\sin\phi\sin\psi\sin\theta + \cos\phi\cos\psi) + \\ + z'(\cos\phi\sin\psi\sin\theta - \sin\phi\cos\psi) \tag{3.86}$$

$$\zeta = \zeta_G - x'\sin\theta + y'\sin\phi\cos\theta + z'\cos\phi\cos\theta \tag{3.87}$$

3.4.2 Formulation of the Problem

Consider a ship with zero forward speed under the action of a small wave of sinusoidal form we discussed in subchapter 3.3. We expect the resulting potential of fluid velocities to be contributed from flow caused by wave and body motion:

$$\varphi(\xi,\eta,\zeta,t) = \varphi_B + \varphi_W \tag{3.88}$$

Here φ_B is the potential of fluid velocities induced by the body motions; this potential can be presented in the form (3.54) as the sum of unit potentials, while φ_W is the potential of flow caused by the wave including the interference with the wave diffracted from the body as an obstacle.

We have seen that the wave that has a sinusoidal form also has a similar form analogous to potential: compare equations (3.74) and (3.82). So, it will be quite logical to search the potential of fluid velocities caused by ship motions in a similar form.

$$\varphi(\xi,\eta,\zeta,t) = \mathrm{Re}\left\{\left[\sum_{j=1}^{6}\xi_{Aj}\varphi_j(\xi,\eta,\zeta) + \zeta_{Aw}\varphi_{Aw}(\xi,\eta,\zeta)\right]\cdot\exp(i\omega t)\right\} \tag{3.89}$$

Here φ_j is fluid velocity potential caused by the j-th component of body motion and it does not include any wave influence, while φ_{wA} is the potential of fluid velocities expressing influence of incident and diffracted waves. We have assumed here a sinusoidal form for all the potential; this allows taking the term $\exp(i\omega t)$ out of the inner parentheses. What is left within these parentheses is, indeed, the amplitude. So, actually, we consider further the potential caused by the motion with the amplitude equal to unity. Then, ξ_{Aj} are amplitudes of ship motions for j-th degrees of freedom.

As we have seen from the case of body motion in an unbounded fluid, unit potentials have to satisfy body boundary conditions (3.55), which have to be re-written in terms of amplitude for our case:

$$\begin{aligned} \frac{\partial\varphi_j}{\partial n} &= i\omega\cdot n_j & j &= 1,2,3 \\ \frac{\partial\varphi_j}{\partial n} &= i\omega\cdot(\vec{r}'\times\vec{n})_{j-3} & j &= 4,5,6 \end{aligned} \tag{3.90}$$

The unit amplitude velocity potentials φ_j caused by the body motions are also called radiation potentials referring to the fact that a moving body on the surface of the ideal fluid generates waves.

As we mentioned above, φ_{Aw} is the potential of fluid velocities expressing the influence of incident and diffracted waves:

$$\varphi_{Aw} = \varphi_0 + \varphi_7 \tag{3.91}$$

Where φ_0 is the potential of incident waves (3.82) and φ_0 is the potential of the fluid velocities caused by the diffracted wave. Since the body motion influence is expressed with radiation potentials, we can consider wave diffraction as on an unmovable obstacle. Then, the body boundary condition for φ_{wA} can be expressed as:

$$\frac{\partial \varphi_{Aw}}{\partial n} = 0 \tag{3.92}$$

Let us assign:

$$\xi_{A0} = \xi_{A7} = \xi_{Aw} \tag{3.93}$$

This definition allows rewriting (3.89) in a more compact form:

$$\varphi(\xi,\eta,\zeta,t) = \mathrm{Re}\left[\exp(i\omega t)\cdot \sum_{j=0}^{7} x_{Aj}\varphi_j(\xi,\eta,\zeta)\right] \tag{3.94}$$

All the potentials included in the formula (3.81) must be solutions of the Laplace equation:

$$\nabla\varphi_j = 0\,, \qquad j = 0, 1, ..., 7 \tag{3.95}$$

Besides body boundary conditions, these potentials have to satisfy linearized free surface boundary conditions as defined with equations (3.72)-(3.73) taking into account that we are working with unit amplitude potentials:

$$-\frac{\omega^2}{g}\varphi_j + \frac{1}{g}\frac{\partial \varphi_j}{\partial \zeta} = 0\,, \quad z = 0, \quad j = 0, 1, ..., 7 \tag{3.96}$$

Finally, to ensure uniqueness of solutions, we have to introduce far field conditions that are very far away from the body where only incident and radiated waves exist (radiation condition), see [Newman, 1977, 1978] for more information.

Now we use Bernoulli's equation (3.41) keeping only the linear terms and chose the initial moment to have $C(t)$=0:

$$\frac{\partial \varphi}{\partial t} = -\frac{1}{\rho}(p + \rho g \zeta) \tag{3.97}$$

This equation allows expressing pressure through fluid velocity potential:

$$p = -\rho\left(\frac{\partial \varphi}{\partial t} + g\zeta\right) = -\rho\cdot\left\{\mathrm{Re}\left[i\omega\exp(i\omega t)\cdot\sum_{j=0}^{7}\xi_{Aj}\varphi_j(\xi,\eta,\zeta)\right] + g\zeta\right\} \tag{3.98}$$

Integration of these pressures over the surface of the ship hull produces hydrodynamic forces and moments:

$$\vec{F}(t) = \iint_{S_B} p\cdot\vec{n}\,dS\,; \qquad \vec{M}(t) = \iint_{S_B} p\cdot(\vec{r}\times\vec{n})dS \tag{3.99}$$

Substitution of wave-induced velocity potential allows easy classification of forces:

$$\vec{F} = -\rho g \iint_{S_B} \zeta \cdot \vec{n}\, dS - \qquad \text{– Hydrostatics}$$

$$-\rho \iint_{S_B} \mathrm{Re}\left[i\omega e^{i\omega t} \sum_{j=1}^{6} \xi_{Aj} \varphi_j \right] \vec{n}\, dS - \qquad \text{– Radiation forces} \qquad (3.100)$$

$$-\rho \iint_{S_B} \mathrm{Re}\left[i\omega e^{i\omega t} \zeta_{Aw} (\varphi_0 + \varphi_7) \right] \vec{n}\, dS \qquad \text{– Wave forces}$$

$$\vec{M} = -\rho g \iint_{S_B} \zeta \cdot (\vec{r} \times \vec{n})\, dS - \qquad \text{– Hydrostatics}$$

$$-\rho \iint_{S_B} \mathrm{Re}\left[i\omega e^{i\omega t} \sum_{j=1}^{6} \xi_{Aj} \varphi_j \right] (\vec{r}' \times \vec{n})\, dS - \qquad \text{– Radiation moments} \qquad (3.101)$$

$$-\rho \iint_{S_B} \mathrm{Re}\left[i\omega e^{i\omega t} \zeta_{Aw} (\varphi_0 + \varphi_7) \right] (\vec{r}' \times \vec{n})\, dS \qquad \text{– Wave moments}$$

The classification becomes evident since we know what the nature of contribution of each potential is. Now we can consider them separately.

3.4.3 Hydrostatic Forces

Hydrostatic force and moment are defined by the first component in the formulae (3.100) and (3.101):

$$\vec{F}_{Hs} = -\rho g \iint_{S_B} \zeta \cdot \vec{n}\, dS \qquad (3.102)$$

$$\vec{M}_{Hs} = -\rho g \iint_{S_B} \zeta \cdot (\vec{r} \times \vec{n})\, dS \qquad (3.103)$$

Integration of formulae (3.102) and (3.103) has to be done by the instant wetted surface of the ship hull. It is more convenient to use the ship-fixed coordinate system for this purpose. Following assumptions that motions are small, we consider cosine functions in (3.85)-(3.87) being equal to 1, sine functions being equal to the value of the angle, expressed in radians and we also drop all the terms of the second order. This allows significant simplification of the transition to the ship-fixed coordinate system, so linearized formulae are as follows:

$$\xi = \xi_0 + x' - y'\psi + z'\theta \qquad (3.104)$$

$$\eta = \eta_0 + x'\psi + y' - z'\phi \qquad (3.105)$$

$$\zeta = \zeta_0 - x'\theta + y'\phi + z' \qquad (3.106)$$

Formulae (3.104)-(3.106) can be presented in the vector form too:

$$\vec{r} = \vec{r}' + \vec{r}_0 + \vec{\xi}_R \times \vec{r}' \qquad (3.107)$$

Here: $\vec{r}(\xi, \eta, \zeta)$ is the radius vector of a point on the ship surface in global coordinate system, $\vec{r}(\xi_0, \eta_0, \zeta_0)$ is radius vector of the origin of ship-fixed coordinate system,

$\vec{r}'(x',y',z')$ is radius vector of a point at ship surface defined in the ship-fixed coordinate system and $\vec{\xi}_R(\phi,\theta,\psi)$ is the vector describing rotation of a ship.

Following the logic of small motion assumptions, we consider changes in the wetted surface due to ship motion as a small value; so we shall further count the submerged area as the still water waterline.

It is also more conventional and convenient to consider all the moments relative to the centre of gravity, so instead of (3.103) we are going to work with:

$$\vec{M}_{Hs} = -\rho g \iint_{S_B} \zeta \cdot ((\vec{r} - \vec{r}_0) \times \vec{n}\,) dS \tag{3.108}$$

To convert surface integrals in formulae (3.102) and (3.108) into volumes we need the surface to be closed; at the same time integrands for both of them equal on the instant waterplane and due to the previously made assumption, this waterplane coincides with the calm water surface, so

$$\vec{F}_{Hs} = -\rho g \iint_{S_B+S_0} \zeta \cdot \vec{n}\, dS + \rho g \iint_{S_0} \zeta \cdot \vec{n}\, dS \tag{3.109}$$

$$\vec{M}'_{Hs} = -\rho g \iint_{S_B+S_0} \zeta \cdot ((\vec{r} - \vec{r}_0) \times \vec{n}\,) dS + \rho g \iint_{S_0} \zeta \cdot ((\vec{r} - \vec{r}_0) \times \vec{n}\,) dS \tag{3.110}$$

Where S_0 is instant/calm water waterplane, where $\zeta=0$, so the second integral also equals zero. Now we can re-write these formulae with volume integrals, using Gauss's theorem (3.17):

$$\vec{F}_{Hs} = -\rho g \iiint_{V(t)} \nabla \zeta\, dV = \vec{k} \cdot \rho g \iiint_{V(t)} dV = \vec{k} \cdot \rho g V(t) \tag{3.111}$$

Were $V(t)$ is the instant submerged volume.

$$\begin{aligned} \vec{M}'_{Hs} &= -\rho g \iiint_{V(t)} [\nabla \times (\vec{r} - \vec{r}_0)] \cdot \zeta\, dV = \\ &= -\rho g \iiint_{V(t)} \left[(\xi - \xi_0) \cdot \vec{j} - (\eta - \eta_0) \cdot \vec{i}\,\right] dV \end{aligned} \tag{3.112}$$

Our goal is to relate hydrostatic force and moment with geometrical characteristics of the ship hull, so we switch to ship-fixed coordinate system using formulae (3.104)-(3.106):

$$\vec{M}'_{Hs} = -\rho g \iiint_{V(t)} \left[(x' - y'\chi + z'\theta) \cdot \vec{j} - (x'\chi + y' - z'\phi) \cdot \vec{i}\,\right] dV \tag{3.113}$$

The instant submerged volume can be presented as the volume displacement (in calm water, of course) and additional volume, which submerges or emerges due to ship motion. The difference of these volumes could be obtained by surface integration along the waterplane as the volume between planes $\zeta=0$ and $z'=0$:

$$V(t) = V_0 - \Delta V(t) = V_0 - \iint_{S_0} (\zeta - z')\, dS = V_0 - \iint_{S_0} (\zeta_0 - x'\theta + y'\phi')\, dS \tag{3.114}$$

This presentation is approximate, but it is in line with our small motion assumption. The domain of integration of (3.112) also can be presented in this form:

$$\vec{M}'_{Hs} = -\rho g \iiint\limits_{V_0-\Delta V(t)} \left[(x' - y'\psi + z'\theta)\cdot\vec{j} - (x'\psi + y' - z'\phi)\cdot\vec{i}\right]dV =$$
$$= -\rho g \iiint\limits_{V_0} \left[(x' - y'\psi + z'\theta)\cdot\vec{j} - (x'\psi + y' - z'\phi)\cdot\vec{i}\right]dV + \quad (3.115)$$
$$+ \rho g \iiint\limits_{\Delta V(t)} \left[(x' - y'\psi + z'\theta)\cdot\vec{j} - (x'\psi + y' - z'\phi)\cdot\vec{i}\right]dV$$

Using expression (3.114), the second integral in (3.115) can be presented as:

$$\iiint\limits_{\Delta V(t)} \left[(x' - y'\psi + z'\theta)\cdot\vec{j} - (x'\psi + y' - z'\phi)\cdot\right]dV =$$
$$= \iint\limits_{S_0} \int\limits_{z'}^{\zeta_{0'} - x'\theta + y'\phi + z'} (x' - y'\psi + z'\theta)\cdot\vec{j} - (x'\psi + y' - z'\phi)\cdot d\zeta\, dS \quad (3.116)$$

After the internal integral in (3.116) can be evaluated analytically, formulae for hydrostatic force and moment expressed in ship-fixed coordinates take the following form:

$$\vec{F}_{Hs} = \vec{k}\cdot\rho g V_0 - \vec{k}\cdot\rho g \iint\limits_{S_0} (\zeta_0 - x'\theta + y'\phi)\, dS \quad (3.117)$$

$$\vec{M}'_{Hs} = -\rho g \iiint\limits_{V_0} \left[(x' - y'\psi + z'\theta)\cdot\vec{j} - (x'\psi + y' - z'\phi)\cdot\vec{i}\right]dV +$$
$$+ \rho g \iint\limits_{S_0} \left[\left(x'\cdot\vec{j} - y'\cdot\vec{i}\right)\cdot(\zeta_0 - x'\theta + y'\phi)\right]dS \quad (3.118)$$

Now, we can consider integration of each term of (3.107) and (3.108) and determine its relationship with ship hull geometry:

Area of the waterline and its static moments relative to axes x' and y':

$$S_0 = \iint\limits_{S_0} dS; \qquad S_1 = \iint\limits_{S_0} x'dS; \qquad S_2 = \iint\limits_{S_0} y'dS \quad (3.119)$$

The above formulae are related with the location of centre of flotation as:

$$x_f = \frac{S_1}{S_0}; \quad y_f = \frac{S_2}{S_0} \quad (3.120)$$

Moments of inertia of the waterplane area:

$$S_{11} = \iint\limits_{S_0} x'^2\, dS; \qquad S_{22} = \iint\limits_{S_0} y'^2\, dS; \qquad S_{12} = \iint\limits_{S_0} x'y'dS \quad (3.121)$$

Location of buoyancy centre:

$$x_B = \frac{1}{V_0}\iiint\limits_{V_0} x'dV; \quad y_B = \frac{1}{V_0}\iiint\limits_{V_0} y'dV; \quad x_B = \frac{1}{V_0}\iiint\limits_{V_0} z'dV \quad (3.122)$$

The above formulae allow presenting hydrostatic force and moment in a form:

$$\vec{F}_{Hs} = \vec{k} \cdot \rho g V_0 - \vec{k} \cdot \rho g(\zeta_0 S_0 - \theta S_1 + \phi S_2) \tag{3.123}$$

$$\begin{aligned}\vec{M}'_{Hs} = &-\rho g V_0 \left[(x_B - y_B\psi + z_B\theta) \cdot \vec{j} - (x_B\psi + y_B - z_B\phi) \cdot \vec{i} \right] + \\ &+ \rho g \left[(\zeta_0 S_1 - S_{11}\theta + S_{12}\phi) \cdot \vec{j} - (\zeta_0 S_2 - S_{12}\theta + S_{22}\phi) \cdot \vec{i} \right]\end{aligned} \tag{3.124}$$

As it is known from ship hydrostatics and as it is evident from formulae (3.119), if the waterplane is symmetrical relative to the centerline, which is true for most ships, then:

$$S_2 = 0 \quad \Rightarrow \quad y_f = 0 \tag{3.125}$$

If a ship is symmetrical relative to its center-plane, which, again, is the case for most of the ships (in undamaged conditions) combined moment of inertia, S_{12} becomes zero and the same can be said for the ordinate of buoyancy centre

$$S_{12} = 0 \quad \text{and} \quad y_B = 0 \tag{3.126}$$

We also have a freedom to place the origin of the ship-fixed coordinate system in the centre of floatation: it will illuminate another static moment relative axes of abscissa:

$$S_1 = 0 \quad \Rightarrow \quad x_f = 0 \tag{3.127}$$

The considerations (3.125) – (3.127) offer some simplifications in formulae for hydrostatic force and moment (for certain types of hull geometry only!):

$$\vec{F}_{Hs} = \vec{k} \cdot \rho g(V_0 - \zeta_0 S_0) \tag{3.128}$$

$$\vec{M}'_{Hs} = -\rho g V_0 \left[\left(x_B + \left(\frac{S_{11}}{V_0} + z_B \right)\theta \right) \cdot \vec{j} - \left(x_B\psi - \left(\frac{S_{22}}{V_0} + z_B \right)\phi \right) \cdot \vec{i} \right] \tag{3.129}$$

As it is clearly seen from the formulae (3.128) and (3.129) that within our assumptions for small waves, small motions and the ship hull geometry with one vertical symmetry plane, there is only the vertical hydrostatic force; and the hydrostatic moment exists only about the x and y axes.

3.4.4 Added Mass and Wave Damping

Let us consider now the second term in formulae (3.100) and (3.101) that describes forces related with radiation potentials:

$$\vec{F} = -\rho \iint_{S_B} \mathrm{Re}\left[i\omega e^{i\omega t} \sum_{j=1}^{6} \xi_{Aj}\varphi_j \right] \vec{n}\, dS \tag{3.130}$$

$$\vec{M} = -\rho \iint_{S_B} \mathrm{Re}\left[i\omega e^{i\omega t} \sum_{j=1}^{6} \xi_{Aj}\varphi_j \right] (\vec{r} \times' \vec{n})\, dS \tag{3.131}$$

Re-writing these equations as vector components, and substituting body boundary conditions (3.90) allows presenting both force and moment in unified form (we also swap integration and summation for convenience sake):

$$F_k = -\rho \operatorname{Re}\left[i\omega e^{i\omega t} \sum_{j=1}^{6} \xi_{Aj} \iint_{S_B} \varphi_j \frac{\partial \varphi_k}{\partial n} dS \right] \tag{3.132}$$

Here, we use index k to identify the component of force (k=1,2,3) and moment (k=4,5,6) instead of index i as we did before; to avoid possible confusion with imaginary unit $i = \sqrt{-1}$. We try not to use i as an index while working with complex values.

The above formula is somewhat similar to equations (3.60) and (3.61) that represent force and moment action on the body in an unbounded fluid. Following this pattern we introduce the following coefficient:

$$f_{jk} = -\rho \iint_{S_B} \varphi_j \frac{\partial \varphi_k}{\partial n} dS \tag{3.133}$$

The major difference between the coefficient introduced above and the one defined by (3.62) is that (3.133) is a complex value. It has to be complex, because we have free surface here, which is different from the case of unbounded fluid. Motion of the body in the vicinity of free surface would make waves. The potential for the wave motions must have an imaginary term to describe periodic character of the fluid flow, as we have seen from subchapter 3.3 where the example with a plane progressive wave was considered.

As a result, the coefficient (3.133) is presented in the following form:

$$f_{jk} = \omega^2 a_{jk} + i\omega b_{jk} \tag{3.134}$$

Substitution of the definition (3.134) into equation (3.132) lead to the following for the force and moment:

$$F_k = -\sum_{j=1}^{6} \left(a_{jk} \ddot{\xi}_j + b_{jk} \dot{\xi}_j \right) \tag{3.135}$$

Here $\ddot{\xi}_j$ and $\dot{\xi}_j$ are acceleration and velocities for all six degrees of freedom of ship motions, defined correspondingly as:

$$\begin{aligned} \xi_j &= \operatorname{Re}\left[\xi_{Aj} \exp(i\omega t)\right] \\ \dot{\xi}_j &= \operatorname{Re}\left[\xi_{Aj} i\omega \exp(i\omega t)\right] \\ \ddot{\xi}_j &= -\operatorname{Re}\left[\xi_{Aj} \omega^2 \exp(i\omega t)\right] \end{aligned} \tag{3.136}$$

The coefficient a_{kj} is multiplied with acceleration, and therefore plays a role of measure of inertia. It is defined as added mass for body motion in the vicinity of free surface. Generally it does not equal the added mass for the same body moving in unbounded fluid:

$$a_{jk} \neq m_{kj} \tag{3.137}$$

Coefficient b_{kj} is multiplied with body velocities, so it takes the place of the damping term in the equation of a mechanical oscillator (we study that equation later in subchapter 3.5). It is defined as the wave damping coefficient that describes transfer of energy from ship motions to surface waves.

Both the added mass and wave damping coefficient depend on wave frequency. For the case of zero forward speed, both these figures are represented with symmetric matrixes:

$$a_{jk} = a_{kj}\,; \quad b_{jk} = b_{kj} \tag{3.138}$$

More information on the properties of these values can be found in [Newman, 1977, 1978].

3.4.5 Wave Forces: Formulation of the Problem

Consider the third term of equations (3.100) and (3.101) describing wave action on the ship. As we pointed out above, in subchapter 3.4.2, there are two potentials φ_0 and φ_7 representing fluid velocities caused by incident and diffracted waves:

$$\vec{F}_W = -\rho \iint_{S_B} \mathrm{Re}\left[i\omega e^{i\omega t} \zeta_{Aw} (\varphi_0 + \varphi_7)\right] \vec{n}\, dS \tag{3.139}$$

$$\vec{M}_W = -\rho \iint_{S_B} \mathrm{Re}\left[i\omega e^{i\omega t} \zeta_{Aw} (\varphi_0 + \varphi_7)\right] (\vec{r}' \times \vec{n})\, dS \tag{3.140}$$

Correspondingly, there are two components present in the wave force: one is the result of pressures related with incident wave potential and another one related with diffracted wave potential.

The first component could be considered as wave force (and moment) acting on a ship in waves, but calculated with the assumption that the ship does not affect fluid velocities by her presence. This assumption is known as the Froude-Krylov hypothesis; this wave's force component sometimes is called Froude-Krylov force.

The second component, then, represents the disturbance that a ship makes to the fluid velocities in an incident wave by her presence. This disturbance here is understood as a reflection only, since making of waves by a moving ship is already taken into account with radiation potentials, which result in added mass and wave damping. Therefore, we have to consider a ship as an unmovable obstacle for the purpose of calculation of this force. This wave force component is called the hydrodynamic or diffraction force (and moment).

3.4.6 Froude-Krylov Forces

Consider Froude-Krylov force first:

$$\vec{F}_W^{FK} = -\rho \iint_{S_B} \mathrm{Re}\left[i\omega e^{i\omega t} \zeta_{Aw} \varphi_0\right] \vec{n}\, dS = \mathrm{Re}\left[i\omega e^{i\omega t} \zeta_{Aw} \vec{F}_{Aw}^{FK}\right] \tag{3.141}$$

With $\vec{F}_{Aw}^{FK}$ playing the role of amplitude of the Froude-Krylov force:

$$\vec{F}_{Aw}^{FK} = -\rho \iint_{S_B} \varphi_0\, \vec{n}\, dS \tag{3.142}$$

We use Gauss's theorem to perform transition from the surface to volume integral, adding and subtracting the integral on the waterplane to "close" the surface:

$$\vec{F}_{Aw}^{FK} = -\rho \iint_{S_B+S_0} \varphi_0 \, \vec{n} \, dS + \rho \iint_{S_0} \varphi_0 \, \vec{n} \, dS = -\rho \iiint_{V_0} \nabla \varphi_0 dV + \rho \iint_{S_0} \varphi_0 \, \vec{n} \, dS \tag{3.143}$$

To facilitate further derivation and reveal the structure of equations of ship motions in waves, we assume that the waves are not only small (in a sense of amplitude), but also long in comparison with the ship. This would allow expanding potential φ_0 in a Taylor series in the location of the ship:

$$\varphi_0(\xi, \eta, \zeta) = \sum_{j=0}^{\infty} \sum_{k=0}^{\infty} \sum_{l=0}^{\infty} \frac{1}{j! \, k! \, l!} \cdot \frac{\partial^{j+k+l} \varphi_0(\xi_0, \eta_0, \zeta_0)}{\partial \xi^j \, \partial \eta^k \, \partial \zeta^l} \times \\ \times (\xi - \xi_0)(\eta - \eta_0)(\zeta - \zeta_0) \tag{3.144}$$

Since we consider the ship as an unmovable obstacle, we can place the origin of the global coordinate system into the centre of floatation – therefore:

$$\xi_0 = 0; \quad \eta_0 = 0; \quad \zeta_0 = 0 \tag{3.145}$$

Following the assumption of long and small-amplitude waves, we keep only the first-order terms in (3.143):

$$\varphi_0(\xi, \eta, \zeta) \approx \varphi_0(0) + \frac{\partial \varphi_0}{\partial \xi} \xi + \frac{\partial \varphi_0}{\partial \eta} \eta + \frac{\partial \varphi_0}{\partial \zeta} \zeta \tag{3.146}$$

Here, $\varphi_0(0) = \varphi_0(0{,}0{,}0)$ means the potential is calculated at the origin of the coordinate system. We can use this approximate presentation of the potential (3.146) in the formula for amplitude of Froude-Krylov (3.143), keeping in mind that the potential now is a linear function of coordinates:

$$\vec{F}_{Aw}^{FK} = -\rho \nabla \varphi_0(0) \iiint_{V_0} dV + \rho \varphi_0(0) \iint_{S_0} \vec{n} \, dS + \\ + \rho \frac{\partial \varphi_0}{\partial \xi} \iint_{S_0} \xi \vec{n} \, dS + \rho \frac{\partial \varphi_0}{\partial \eta} \iint_{S_0} \eta \, \vec{n} \, dS + \rho \frac{\partial \varphi_0}{\partial \zeta} \iint_{S_0} \zeta \, \vec{n} \, dS \tag{3.147}$$

Using definitions (3.119) and keeping in mind that since ship motions is not included in these derivations, the fixed and global coordinate systems coincide. Also, since the waterplane is actually a horizontal flat plane, where ζ=0, the normal vector $\vec{n}$ coincides with vector $\vec{k}$:

$$\vec{F}_{Aw}^{FK} = -\rho \nabla \varphi_0 V_0 + \rho \vec{k} \cdot \left(\varphi_0 S_0 + \frac{\partial \varphi_0}{\partial \xi} S_1 + \frac{\partial \varphi_0}{\partial \eta} S_2 \right) \tag{3.148}$$

Consider a moment of the Froude-Krylov force:

$$\vec{M}_W^{FK} = -\rho \iint_{S_B} \mathrm{Re}\left[i\omega e^{i\omega t} \zeta_{Aw} \varphi_0 \right] (\vec{r}' \times \vec{n}) dS = \mathrm{Re}\left[i\omega e^{i\omega t} \zeta_{Aw} \vec{M}_{Aw}^{Fk} \right] \tag{3.149}$$

We define amplitude of this moment as

$$\vec{M}_{Aw}^{FK} = -\rho \iint_{S_B} \varphi_0 (\vec{r}' \times \vec{n}) dS \tag{3.150}$$

Following an analogous procedure, the above formula has to be converted into a sum of the volume integral over a submerged part of the hull and a surface integral over the waterplane:

$$\vec{M}_{Aw}^{FK} = -\rho \iiint_{V_0} \nabla \times \varphi_0 \cdot \vec{r}' dV + \rho \iint_{S_0} \varphi_0 (\vec{r}' \times \vec{n}) dS \tag{3.151}$$

Substitution of a Taylor expansion for the incident wave potential (3.146) allows significant simplification of the formula (3.151). We also take into account that S_0 is the horizontal flat surface with ζ=0 and the normal vector $\vec{n}$ coincides with vector $\vec{k}$:

$$\begin{aligned}\vec{M}_{Aw}^{FK} &= -\rho \nabla \varphi_0(0) \times \iiint_{V_0} \vec{r}' dV - \\ &- \vec{i} \rho \varphi_0(0) \iint_{S_0} \eta \, dS + \vec{j} \rho \varphi_0(0) \iint_{S_0} \xi \, dS \\ &- \vec{i} \rho \frac{\partial \varphi_0}{\partial \xi} \iint_{S_0} \xi \eta \, dS + \vec{j} \rho \frac{\partial \varphi_0}{\partial \xi} \iint_{S_0} \xi^2 \, dS \\ &- \vec{i} \rho \frac{\partial \varphi_0}{\partial \eta} \iint_{S_0} \eta^2 \, dS + \vec{j} \rho \frac{\partial \varphi_0}{\partial \eta} \iint_{S_0} \xi \eta \, dS\end{aligned} \tag{3.152}$$

We use definitions (3.119)-(3.122); again, since the ship is considered as an unmovable obstacle and the origin of global coordinate system is placed onto the floatation centre, all the coordinate systems are identical.

$$\begin{aligned}\vec{M}_{Aw}^{FK} &= -\rho V_0 \cdot \nabla \varphi_0(0) \times \vec{r}_B - \\ &- \vec{i} \rho \left[\varphi_0(0) S_2 + \frac{\partial \varphi_0}{\partial \xi} S_{12} + \frac{\partial \varphi_0}{\partial \eta} S_{22} \right] + \\ &+ \vec{j} \rho \left[\varphi_0(0) S_1 + \frac{\partial \varphi_0}{\partial \xi} S_{11} + \frac{\partial \varphi_0}{\partial \eta} S_{12} \right]\end{aligned} \tag{3.153}$$

Here, $\vec{r}_B$ is the radius vector of centre of buoyancy originated from the centre of floatation. If the ship is symmetrical relative to the center-plane and the origin of the coordinate system is still located in the centre of flotation, formulae (3.125-3.127) allow significant simplifying expressions for amplitude of Froude-Krylov force and moment:

$$\vec{F}_{Aw}^{FK} = -\rho V_0 \nabla \varphi_0 + \rho \vec{k} \cdot \varphi_0 S_0 \tag{3.154}$$

$$\vec{M}_{Aw}^{FK} = -\rho V_0 \cdot \nabla \varphi_0(0) \times \vec{r}_B - \vec{i} \rho \frac{\partial \varphi_0}{\partial \eta} S_{22} + \vec{j} \rho \frac{\partial \varphi_0}{\partial \xi} S_{11} \tag{3.155}$$

Assuming the wave being plane and progressive, we can express unit potential φ_0 as

$$\varphi_0 = \frac{\varphi_W}{\zeta_{Aw} \exp(i\omega t)} = \frac{g}{\omega} i \exp\bigl(k\zeta - ik(\xi \cos \beta + \eta \sin \beta)\bigr) \tag{3.156}$$

Where φ_W is the plane progressive wave potential described with formula (3.82).

Substitution of (3.156) into (3.154) and (3.155) allows presenting the complex amplitude of Froude-Krylov forces in a form of vector components. We continue to use numbering for degrees of freedom, extending it to the forces: where numbers 1 to 3 correspond to vector components of the force and numbers 4 to 6 correspond to vector components of the moment. We also are using expression (3.83) for wave number and $i = \sqrt{-1}$.

$$F_{Aw1}^{FK} = \left[\vec{F}_{Aw}^{FK}\right]_{\xi} = \rho V_0 \omega \cos\beta \tag{3.157}$$

$$F_{Aw2}^{FK} = \left[\vec{F}_{Aw}^{FK}\right]_{\eta} = \rho V_0 \omega \sin\beta \tag{3.158}$$

$$F_{Aw3}^{FK} = \left[\vec{F}_{Aw}^{FK}\right]_{\zeta} = -\rho V_0 \omega i + \rho \frac{g}{\omega} S_0 i \tag{3.159}$$

$$F_{Aw4}^{FK} = \left[\vec{M}_{Aw}^{FK}\right]_{\xi} = \rho V_0 \omega \sin\beta \left(z_B + \frac{S_{22}}{V_0} \right) \tag{3.160}$$

$$F_{Aw5}^{FK} = \left[\vec{M}_{Aw}^{FK}\right]_{\eta} = -\rho V_0 \omega \left[x_B i + \cos\beta \left(z_B + \frac{S_{11}}{V_0} \right) \right] \tag{3.161}$$

$$F_{Aw6}^{FK} = \left[\vec{F}_{Aw}^{FK}\right]_{\zeta} = -\rho V_0 \omega \cdot x_B \sin\beta \tag{3.162}$$

Substitution of (3.147)-(3.152) into formulae for Froude-Krylov force and moment (3.141) and (3.149) yields the final expression within limits of small-amplitude and long wave assumptions:

$$F_{W1}^{FK} = -\rho V_0 \zeta_{Aw} \omega^2 \cos\beta \sin\omega t \tag{3.163}$$

$$F_{W2}^{FK} = -\rho V_0 \zeta_{Aw} \omega^2 \sin\beta \sin\omega t \tag{3.164}$$

$$F_{W3}^{FK} = \rho \zeta_{Aw} (V_0 \omega^2 - g S_0) \cos\omega t \tag{3.165}$$

$$F_{W4}^{FK} = \rho V_0 \zeta_{Aw} \omega^2 \left(z_B + \frac{S_{22}}{V_0} \right) \sin\beta \sin\omega t \tag{3.166}$$

$$F_{W5}^{FK} = \rho V_0 \zeta_{Aw} \omega^2 \left[x_B \cos\omega t - \left(z_B + \frac{S_{11}}{V_0} \right) \cos\beta \sin\omega t \right] \tag{3.167}$$

$$F_{W6}^{FK} = -\rho V_0 x_B \omega^2 \zeta_{Aw} \sin\beta \sin\omega t \tag{3.168}$$

3.4.7 Hydrodynamic or Diffraction Wave Forces

Now we consider the hydrodynamic or diffraction component of the wave forces and moments, which describes disturbance to wave pressures caused by the ship's presence. As we pointed out earlier, the disturbance we mean is the wave reflection: we already included the wave generation effect into radiation potentials, so we assume a ship unmovable. As zero forward speed is assumed, Khaskind relations can be used, leading directly to the result in a form of equations (3.183). However, full derivations are carried out here, to demonstrate the technique that can be used for the cases when the forward speed is not zero.

$$\vec{F}_W^H = -\rho \iint_{S_B} \mathrm{Re}\left[i\omega e^{i\omega t}\zeta_{Aw}\varphi_7\right]\vec{n}\,dS \tag{3.169}$$

$$\vec{M}_W^H = -\rho \iint_{S_B} \mathrm{Re}\left[i\omega e^{i\omega t}\zeta_{Aw}\varphi_7\right](\vec{r}'\times\vec{n}\,)dS \tag{3.170}$$

Applying boundary conditions (3.90) for (3.169) and (3.170), we rewrite these formulae as the vector component:

$$F_{Wj}^H = -\rho\,\mathrm{Re}\left(e^{i\omega t}\zeta_{Aw}\iint_{S_B}\varphi_7\frac{\partial\varphi_j}{\partial n}\,dS\right) \tag{3.171}$$

Here, φ_j are the radiation potentials that we have used in subchapter 3.4.4. Following the accepted assumptions of small-amplitudes and long waves, the diffraction potential can be approximated as:

$$\varphi_7 \approx \frac{i}{\omega}\left(\varphi_1\frac{\partial\varphi_0}{\partial\xi}+\varphi_2\frac{\partial\varphi_0}{\partial\eta}+\varphi_3\frac{\partial\varphi_0}{\partial\zeta}\right) \tag{3.172}$$

Here, φ_0 is the unit amplitude potential of incident waves defined for the origin of the coordinate system in the centre of floatation and φ_1, φ_2 and φ_3 are radiation potentials corresponding to three translation motions [Newman, 1977, 1978].

Substitution of this approximation into (3.172) reveals complex coefficients of added masses and wave damping (3.133):

$$F_{Wj}^H = -\mathrm{Re}\left[e^{i\omega t}\zeta_{Aw}\frac{i}{\omega}\left(f_{1j}\frac{\partial\varphi_0}{\partial\xi}+f_{2j}\frac{\partial\varphi_0}{\partial\eta}+f_{3j}\frac{\partial\varphi_0}{\partial\zeta}\right)\right] \tag{3.173}$$

The adopted assumptions allow using formulae (3.146) for complex unit-amplitude potential for incident waves; substitution of expression (3.134) yields the following form for the diffraction component of wave forces:

$$\begin{aligned} F_{Wj}^H = &-\zeta_{Aw}\cos\omega t\left(b_{j1}\omega\cos\beta+b_{j2}\omega\sin\beta-a_{j3}\,\omega^2\right)+ \\ &+\zeta_{Aw}\sin\omega t\left(a_{j1}\omega^2\cos\beta+a_{j2}\omega^2\,\sin\beta+b_{j3}\omega\right) \end{aligned} \tag{3.174}$$

It is possible to simplify formula (3.174) further with time derivatives of wave profile at the origin of the coordinate system. Consider derivatives of wave height first:

$$\begin{aligned} \zeta_w(0,0,t) &= \zeta_{Aw}\cos(\omega t) \\ \dot{\zeta}_w(0,0,t) &= -\zeta_{Aw}\omega\sin(\omega t) \\ \ddot{\zeta}_w(0,0,t) &= -\zeta_{Aw}\omega^2\cos(\omega t) \end{aligned} \tag{3.175}$$

Now, consider motion of the wave profile along the direction of the wave propagation, which is defined as:

$$x_0 = \xi\cos\beta+\eta\sin\beta \tag{3.176}$$

Substitution of the definition (3.176) in the equation for wave profile (3.74) yields the formula for wave height with respect of new axes χ:

$$\zeta_w(x_0,t) = \zeta_{Aw}\cdot\cos(kx_0-\omega t) \tag{3.177}$$

Motion of the wave profile along the axes χ can be derived taking into account orbital motion of the fluid particle in the fluid domain, affected by the wave. This includes fluid particles located on the free surface: their orbital motion actually makes the wave profile move:

$$\chi_w = \sqrt{\zeta_{Aw}^2-\zeta_w^2(t)} = \zeta_{Aw}\cdot\sin(kx_0-\omega t) \tag{3.178}$$

Using (3.176) we can express motion of the wave profile in the coordinate system $\xi\eta\zeta$:

$$\xi_w(\xi,\eta,t) = \zeta_{Aw}\cos\beta\cdot\sin(k\xi\cos\beta + k\eta\sin\beta-\omega t) \tag{3.179}$$

$$\eta_w(\xi,\eta,t) = \zeta_{Aw}\sin\beta\cdot\sin(k\xi\cos\beta + k\eta\sin\beta-\omega t) \tag{3.180}$$

Calculation of the time derivatives of (3.179) and (3.180) is easy:

$$\begin{aligned}\dot\xi_w(0,0,t) &= -\zeta_{Aw}\omega\cos\beta\cdot\cos\omega t\\ \ddot\xi_w(0,0,t) &= \zeta_{Aw}\omega^2\cos\beta\cdot\sin\omega t\end{aligned} \tag{3.181}$$

$$\begin{aligned}\dot\eta_w(0,0,t) &= -\zeta_{Aw}\omega\sin\beta\cdot\cos\omega t\\ \ddot\eta_w(0,0,t) &= \zeta_{Aw}\omega^2\sin\beta\cdot\sin\omega t\end{aligned} \tag{3.182}$$

Substitution of formulae (3.175), (3.181) and (3.182) into (3.174) allows us to re-write the expression for the diffraction component in compact form:

$$F_{Wj}^H = -\sum_{k=1}^{3}\left(a_{jk}\ddot\xi_{Wk} + b_{jk}\dot\xi_{Wk}\right) \tag{3.183}$$

Here, we used the derivatives of generalized coordinates of the wave profile $\ddot\xi_{Wk}$ and $\dot\xi_{Wk}$ in the following sense:

$$\xi_{W1} = \xi_w\ ;\quad \xi_{W2} = \eta_w\ ;\quad \xi_{W3} = \zeta_w \tag{3.184}$$

3.4.8 Body Mass Forces

We completed consideration of the forces acting on the body from a fluid. To write a system of equations for ship motions in waves, we need to add mass forces describing inertia, which are not dependent if the body moves in a fluid or in a vacuum.

Let us consider a body moving in 3-dimensional space. As it is well known from theoretical mechanics, any motion of the body can be presented as translation together with some point that is called the centre of rotation and rotation about this point, and as a result the velocity of any point of the body can be presented with formula (3.51). We copy this formula here with slightly changed nomenclature to avoid possible confusion:

$$\vec V_M = \vec V_C + \vec\Omega\times\vec r_M \tag{3.185}$$

Here $\vec{V}_M$ is velocity of any point in the body, defined with radius vector $\vec{r}_M(x_M, y_M, z_M)$, originating from the centre of rotation *C*, that is involved in translational motion with the velocity $\vec{V}_C$, while the rest of the body rotates about it with the angular velocity $\vec{\Omega}$.

This velocity, however, also can be presented in the following form:

$$\vec{V}_M = \sum_{i=1}^{6} u_i \vec{b}_i \tag{3.186}$$

Here, $\vec{b}_i$ are unit vectors corresponding to each degree of freedom: the first three correspond to translations and the second three to rotations:

$$\begin{array}{lll} \vec{b}_1 = \vec{i} & \vec{b}_2 = \vec{j} & \vec{b}_3 = \vec{k} \\ \vec{b}_4 = \vec{i} \times \vec{r}_M & \vec{b}_5 = \vec{j} \times \vec{r}_M & \vec{b}_6 = \vec{k} \times \vec{r}_M \end{array} \tag{3.187}$$

Values u_i are projections of the vectors of translation and rotation velocities:

$$\begin{array}{lll} u_1 = V_{Mx} & u_2 = V_{My} & u_3 = V_{Mz} \\ u_4 = \Omega_x & u_5 = \Omega_y & u_6 = \Omega_z \end{array} \tag{3.188}$$

Now we consider the amount of kinetic energy of the infinitely small volume within the body:

$$dT = \rho_b \frac{\left|\vec{V}_M\right|^2}{2} dV = \rho_b \frac{\vec{V}_M \cdot \vec{V}_M}{2} dV = \frac{\rho_b}{2}\left(\sum_{i=1}^{6} u_i \vec{b}_i \cdot \sum_{i=1}^{6} u_i \vec{b}_i\right) dV \tag{3.189}$$

The whole amount of kinetic energy is expressed through the integration of the elementary amounts over the body volume:

$$T = \iiint_V \frac{\rho_b}{2}\left(\sum_{i=1}^{6}\sum_{j=1}^{6} u_i u_j \vec{b}_i \cdot \vec{b}_j\right) dV \tag{3.190}$$

Swapping integration and summation reveals the matrix that describes inertial properties of the body and only depends on its geometry:

$$T = \frac{1}{2}\sum_{i=1}^{6}\sum_{j=1}^{6} u_i u_j \iiint_V \rho_b \vec{b}_i \vec{b}_j dV = \frac{1}{2}\sum_{i=1}^{6}\sum_{j=1}^{6} u_i u_j M_{ij} \tag{3.191}$$

$$M_{ij} = \iiint_V \rho_b \vec{b}_i \vec{b}_j dV \tag{3.192}$$

This matrix sized 6 x 6 elements is similar to the added masses matrix (to be exact, the added masses matrix is similar to this one). The units for the terms of this matrix are given with equation (3.63). This matrix is symmetrical because of the corresponding property of the scalar product of vectors. Calculation of the terms of this matrix is not difficult, just bulky, so we give the final formulae only.

The first 9 terms of the matrix have mass units:

$$M_{ij} = \begin{cases} m & i = j \\ 0 & i \neq j \end{cases} \quad i = 1..3; \quad j = 1..3 \tag{3.193}$$

The next 18 terms (i = 4..6, j = 1..3 and i = 1..3, j = 4..6) are static moments and related with the position of centre of mass:

$$M_{15} = M_{51} = -M_{24} = -M_{42} = \iiint_V \rho_B z_M dV = m \cdot z_G \tag{3.194}$$

$$M_{34} = M_{43} = -M_{16} = -M_{61} = \iiint_V \rho_B y_M dV = m \cdot y_G \tag{3.195}$$

$$M_{26} = M_{62} = -M_{35} = -M_{53} = \iiint_V \rho_B z_M dV = m \cdot x_G \tag{3.196}$$

$$M_{14} = M_{41} = M_{25} = M_{52} = M_{36} = M_{36} = 0 \tag{3.197}$$

Finally the last 9 terms (i = 4..6, j = 4..6) are moments of inertia:

$$M_{44} = \iiint_V \rho_B \left(z_M^2 + y_M^2\right) dV = I_{xx} \tag{3.198}$$

$$M_{45} = M_{54} = -\iiint_V \rho_B y_M x_M dV = -I_{xy} = -I_{yx} \tag{3.199}$$

$$M_{46} = M_{64} = -\iiint_V \rho_B z_M x_M dV = -I_{xz} = -I_{zx} \tag{3.200}$$

$$M_{55} = \iiint_V \rho_B \left(x_M^2 + z_M^2\right) dV = I_{yy} \tag{3.201}$$

$$M_{56} = M_{65} = -\iiint_V \rho_B y_M z_M dV = -I_{yz} = -I_{zy} \tag{3.202}$$

$$M_{66} = \iiint_V \rho_B \left(x_M^2 + y_M^2\right) dV = I_{zz} \tag{3.203}$$

All the above formulae are summarized below in the form of the mass matrix:

$$M_{ij} = \begin{vmatrix} m & 0 & 0 & 0 & m \cdot z_G & -m \cdot y_G \\ 0 & m & 0 & -m \cdot z_G & 0 & m \cdot x_G \\ 0 & 0 & m & m \cdot y_G & -m \cdot x_G & 0 \\ 0 & -m \cdot z_G & m \cdot y_G & I_{xx} & -I_{xy} & -I_{xz} \\ m \cdot z_G & 0 & -m \cdot x_G & -I_{yz} & I_{yy} & -I_{yz} \\ -m \cdot y_G & m \cdot x_G & 0 & -I_{zx} & -I_{zy} & I_{zz} \end{vmatrix} \tag{3.204}$$

Besides inertia, mass of the body determines its weight.

$$\vec{W} = -mg \cdot \vec{k} \tag{3.205}$$

Since we put the origin of the coordinate system into the centre of floatation, the weight force creates the moment:

$$\vec{M}_W = \vec{W} \times \vec{r}_G \tag{3.206}$$

Here, $\vec{r}_G$ is the radius vector of the centre of gravity in the global coordinate system.

The following assumption of small motions, the coordinate of the centre of gravity in the global system can be defined with the formulae (3.104)-(3.106)

$$\xi_G = x_G - y_G\psi + z_G\theta \tag{3.207}$$

$$\eta_G = x_G\psi + y_G - z_G\phi \tag{3.208}$$

$$\zeta_G = -x_G\theta + y_G\phi + z_G \tag{3.209}$$

Here, x_G, y_G and z_G are coordinates of the centre of gravity in the ship-fixed coordinate system. Substituting (3.195) and (3.197)-(3.199) into (3.196) yields the following expression for the moment created by the weight force:

$$\vec{M}_W = mg(x_G\psi + y_G - z_G\phi)\vec{i} - mg(x_G - y_G\psi + z_G\theta)\vec{j} \tag{3.210}$$

3.4.9 Linear Equation of Motions

Here, our purpose is put all the forces together and build equations of ship motions in waves. However, before we proceed let us change the coordinate system fixed to a ship. It is both convenient and conventional to place the origin of the ship-fixed coordinate system at the centre of gravity. Also, from now on we shall assume that the ship is symmetrical relative to the center-plane:

$$y_B = 0; \quad y_G = 0 \tag{3.211}$$

It is known from theoretical mechanics that if a body has a plane of symmetry; then the principal axis of inertia is perpendicular to this plane. Since the y –coordinate is perpendicular to the center-plane, it is the principal axis of inertia. . As most vessels of conventional configuration are relatively long bodies (width and depth are smaller than length), it is usually assumed that other two axes relative to the y-axis are also principal axes of inertia and all centrifugal moments of inertia are close to zero.

$$I_{xy} \approx I_{yz} \approx I_{zx} \approx 0 \tag{3.212}$$

Placing the origin of coordinate system at the center of gravity also makes all static moments equal to zero. As a result, the matrix of inertia becomes simpler:

$$M_{ij} = \begin{vmatrix} m & 0 & 0 & 0 & 0 & 0 \\ 0 & m & 0 & 0 & 0 & 0 \\ 0 & 0 & m & 0 & 0 & 0 \\ 0 & 0 & 0 & I_{xx} & 0 & 0 \\ 0 & 0 & 0 & 0 & I_{yy} & 0 \\ 0 & 0 & 0 & 0 & 0 & I_{zz} \end{vmatrix} \tag{3.213}$$

According to d'Alembert's principle, equations of motions can be built by equalizing the vector sum of all the forces and moments acting on the body to principal vector and moment of inertia force. Since the ship-fixed coordinate system uses principal axes of inertia, the motion equations (as was shown by Euler) have the following appearance:

$$\begin{aligned} &mg\ddot{\xi}_G = \sum F_1\,; \quad mg\ddot{\eta}_G = \sum F_2\,; \quad mg\ddot{\zeta}_G = \sum F_3 \\ &I_{xx}\ddot{\phi} - (I_{yy} - I_{zz})\dot{\theta}\dot{\psi} = \sum F_4 \\ &I_{yy}\ddot{\theta} - (I_{zz} - I_{xx})\dot{\phi}\dot{\psi} = \sum F_5 \\ &I_{zz}\ddot{\psi} - (I_{zz} - I_{xx})\dot{\phi}\dot{\theta} = \sum F_6 \end{aligned} \tag{3.214}$$

Following small motion assumption, the components of the principal vector and moment of inertia can be presented as:

$$F_i^{In} = \sum_{j=1}^{6} M_{ij}\ddot{\xi}_j \quad i = 1,2...,6 \tag{3.215}$$

Here, we continue to follow the numbering convention that the first three components correspond to translation motion and the other three to rotation. M_{ij} is the inertia matrix defined with equation (3.213).

Since we have shifted the origin of the ship-fixed coordinate systems to centre of gravity, we have to correct all the moments, because they were written relative to the flotation centre. Let us start with the hydrostatic moment. As is known from theoretical mechanics, the value of this correction to be applied is as follows:

$$\vec{M}_{Hc}^{C} = -\vec{r}_G{}' \times \vec{F}_{Hs} \tag{3.216}$$

Here, $\vec{r}_G$ is the radius vector of the centre of floatation relative to the centre of gravity:

$$\vec{r}_G{}' = (\xi_G - \xi_f)\vec{i} + (\eta_G - \eta_f)\vec{j} + (\zeta_G - \zeta_f)\vec{k} \tag{3.217}$$

Using small-motion assumptions and equations (3.104)-(3.106):

$$\vec{r}_G{}' = (x_G{}' + z_G{}'\theta)\vec{i} + (x_G{}'\psi - z_G{}'\phi)\vec{j} + (-x_G{}'\theta + z_G{}')\vec{k} \tag{3.218}$$

Substitution of (3.218) into (3.216) yields the correction:

$$\vec{M}_C = \rho g V_0\left[(x_G{}'\psi - z_G{}'\phi)\vec{i} - (x_G{}' + z_G{}'\theta)\vec{j}\right] \tag{3.219}$$

Nonlinear terms were excluded from (3.219) to be consistent with the small amplitude motion assumption. Taking into account conditions of static equilibrium in calm water for a freely floating body:

$$x_G = x_B\,; \quad \rho V = m \tag{3.220}$$

The principal moment of hydrostatic forces relative to the centre of gravity is expressed as:

$$\vec{M}'_{Hs} = -\rho g V_0\left[GM_L\theta \cdot \vec{j} + GM \cdot \phi \cdot \vec{i}\right] \tag{3.221}$$

Were GM and GM_L are the transverse and longitudinal metacentric heights respectively – we have used formulae from ship hydrostatics that relate the moment of inertia of the waterplane area, positions of gravity and buoyancy with metacentric height.

Considering the moment created by weight (3.210) we note that it equals zero, since the origin of the coordinate system is located at the point where the weight force is applied – at the centre of gravity.

It is convenient for further derivations to present the sum of the hydrostatic and weight forces and moments in the form of a matrix multiplied with the motions using the same numbering convention as we did for added masses. However, for those derivations with the small amplitude motion assumption, only three elements of this matrix would be different from zero:

$$\begin{aligned} c_{33} &= \rho g S_0 \\ c_{44} &= \rho g V_0 \cdot GM \\ c_{55} &= \rho g V \cdot GM_L \end{aligned} \tag{3.222}$$

Finally, the component of hydrostatic forces is expressed as:

$$F_{Hs\,i} = \sum_{j=1}^{6} c_{ij}\xi_j \tag{3.223}$$

Next, consider forces and moments related with radiation potentials (3.135). This formula contains coefficients a_{ij} and b_{ij} that are respectively real and imaginary parts of the integral (3.133) calculated over the submerged ship surface. This surface usually cannot be expressed in simple analytical functions and, as a result, the integral (3.133) has to be calculated numerically. Then, it is enough to define the surface in the ship-fixed coordinate system with the origin in the centre of gravity, to include these forces in our equation.

Froude-Krylov wave forces can be considered as the principal vector, defined with components (3.163)-(3.165) and the principal moment, components of which are defined with equations (3.166)-(3.168). Correction due to change of the origin from the centre of flotation to the centre of gravity is defined as:

$$\vec{M}_{FK}^{C} = -\vec{r}_G \times \vec{F}_{FK}^{W} \tag{3.224}$$

Here, $\vec{r}_G$ is the radius vector of the centre of floatation relative to the centre of gravity defined as (3.217). The difference with the case of hydrostatic forces is that we do not take into account ship motions while calculating wave forces. As a result, there the radius vector can be presented in the following form:

$$\vec{r}_G' = x_G'\vec{i} + 0\vec{j} + z_G'\vec{k} \tag{3.225}$$

Substitution of (3.225), (3.163)-(3.165) into (3.224) with further addition of the components of the result to corresponding projections (3.166)-(3.168) gives the following formulae for the moment of Froude-Krylov forces relative to the centre of gravity:

$$F_{W4}^{FK} = \rho V_0 \zeta_{Aw}\omega^2 GM \sin\beta \sin\omega t \tag{3.226}$$

$$F_{W5}^{FK} = -\rho\zeta_{Aw}\left[S_0 x_G g\cos\omega t + \omega^2 V_0 GM_L \cos\beta\sin\omega t\right] \tag{3.227}$$

$$F_{W6}^{FK} = 0 \tag{3.228}$$

Here, we have used formulae for ship statics and equilibrium condition for a freely floating body (3.220).

Re-writing of the moments of diffraction wave forces (3.183) into the coordinate system with the origin located at the centre of gravity is automatic. Since this expression uses the same added mass and wave damping coefficients as forces and moments related with the radiation potential, these coefficients should be already calculated in the coordinate system with the origin at the centre of gravity.

Combining formulae (3.215) with (3.223), (3.135), (3.163-166), (3.226-228) and (3.183) we finally write the system of ship motion equation in regular waves:

$$\sum_{j=1}^{6}\left[(M_{ij}+a_{ij})\ddot{\xi}_{Gj} + b_{ij}\dot{\xi}_{Gj} + c_{ij}\xi_{Gj}\right] = F_{Wi}^{FK} - \sum_{k=1}^{3}\left(a_{ik}\ddot{\xi}_{Wk} + b_{ik}\dot{\xi}_{Wk}\right) \qquad i = 1,2...,6 \tag{3.229}$$

This is the system of ordinary differential equations of the second order that will be the subject of further study in the next subchapter.

3.5 Linear Equation of Roll Motions

3.5.1 Adequacy of Linear Equation of Motions

Let us now repeat the set of assumptions made in order to derive the equation of motions in a form of (3.229).

First of all, the fluid is inviscid and there is no rotation motion. This allows conversion of the fluid domain problem in a boundary value problem.

The boundary conditions are linearized: that means that the waves are small, in other words, wave length is large in comparison with wave height.

The ship is small in comparison with waves, or waves are long in comparison with the ship.

And finally, the ship motions are small.

The last assumption is quite clearly incompatible with the ultimate purpose of our study, since motions that might lead to capsizing are definitely not small, so the model (3.229) does not seem to be adequate for a stability study.

Nevertheless, despite the clear inadequacy of the result, all the above derivations and their results are extremely important, because they revealed the structure and nature of forces acting on the ship in waves. The model (3.229) is a foundation for further development. Since the structure of forces is known, we can study them separately and put together a new model for each force that will be adequate for the problem. We shall do so in subchapter 3.5.

There were attempts to derive ship motion equations including nonlinear terms from the very beginning, (such as [Lugovsky, 1980]), but these derivations lead to very complicated equations, which are very difficult to analyze. This is the main reason why we still need a "simplistic", though consistent, linear approach.

3.5.2 Calculation of Forces and Motions

Formulae (3.229) represent a system of six ordinary linear differential equations, which can be solved analytically. Before considering such a solution, let us examine a matrix of added masses in waves again. Since most ships are symmetrical relative to the center-plane, the formula (3.65) can be applied and the matrix for added mass looks as follows (with the matrix of damping coefficients similarly analogous):

$$a_{ij} = \begin{vmatrix} a_{11} & 0 & a_{13} & 0 & a_{15} & 0 \\ 0 & a_{22} & 0 & a_{24} & 0 & a_{26} \\ a_{31} & 0 & a_{33} & 0 & a_{35} & 0 \\ 0 & a_{42} & 0 & a_{44} & 0 & a_{46} \\ a_{51} & 0 & a_{53} & 0 & a_{55} & 0 \\ 0 & a_{62} & 0 & a_{64} & 0 & a_{66} \end{vmatrix} \qquad (3.230)$$

The character of matrix (3.230) points out that the relationships between some kinds of ship motions are stronger than others. This illustrates the known fact [Lewis, 1989] that ship longitudinal motions (surge, heave, and pitch) can be considered separately from transverse motions (sway, roll, and yaw) at least within frames of the linear approach.

Formulae for the Froude-Krylov component of the wave forces (3.163-165), (3.226-228) show that the maximum wave action for surge, and pitch might be expected in longitudinal waves $(\cos\beta = \pm 1)$, while the peak for sway and roll is beam waves $(\sin\beta = \pm 1)$. Heave wave forces do not depend on the direction of the waves' propagation. Since the equations are linear, we can expect the largest ship response when wave forces reach their maximum for this particular kind of motion.

As result, we can carry out calculations separately for sway, heave and roll in beam waves and for heave and pitch in longitudinal waves. Further, we will consider these problems separately, with an emphasis on roll motion.

To make equations (3.219) more practical, some of the assumptions summarized in subchapter 3.5.1 have to be worked around. First of all, it concerns the small size of a ship in comparison with waves.

One (and the oldest) way to solve this problem is introduction of reduction coefficients [Blagoveshchensky, 1962]. These coefficients are used to correct wave forces: the Froude-Krylov component is reduced; the diffraction component might be either reduced or increased. This method is usually accompanied with empirical formulae for added masses and wave damping coefficients.

Another approach is the strip theory where the 3-Dimensional (3D) problem is substituted with a set of 2-Dimensional (2D) problems. A ship is presented as a number of transverse sections. Each of these 2D problems considers an infinite cylinder moving in waves and the transverse section of this cylinder coincides with one taken from the ship hull. The solution of a 2D problem is considered as a force acting in this section of the ship (strip). The result is obtained by the integration of these solutions over the ship length; see [Lewis, 1989] for more information.

A more theoretically consistent approach is offered by slender body theory, where the main assumption is that one dimension of a ship larger than the other, which is true for most of ships. Newman [1977, 1978] covers the fundamentals of this theory.

The panel method allows considering fully 3-D flow without any limitations or assumptions concerning the form of the submerged part of the hull. The idea of the method is to present a body (and sometimes free surface) with a set of flat panels. Then, elementary flow: source or sink is placed in the middle of each panel. Intensity of each source or sink is searched numerically to match initial conditions on the body and (sometimes) on the free surface. Once intensities of the sources and sinks are found, the potential can be found by summation (intensity of source or sinks define the potential of the flow around it) [Chang, 1977; Inglis and Price, 1982].

Introduction of the panel method eliminates any limitations of the geometry of the body and the wave length. However, the motions still are assumed to be small. If this assumption is abandoned, the equation (3.219) becomes nonlinear and their analytical solution is no longer possible. This creates the whole new approach based on numerical integration of differential equations (we consider this later in subchapter 4.2.5) with simultaneous calculations of hydrodynamic forces with the panel method [Scavounos and Nakos, 1988; Lin and Yu, 1990; Shin, *et al*, 1997]. The main advantage of this approach is a possibility to include forces that cannot be obtained with the ideal fluid approach [Belenky, *et al*, 2002]. Comprehensive state-of-the art review of computational methods for ship motions could be found in [Beck and Reed, 2001].

3.5.3 Isolated Linear Equation of Roll Motions

Our main interest lays, of course, in the roll motion equation, since it is where capsizing might happen. Here, we consider the linear equation of roll as a foundation for further nonlinear study. As we have seen from the previous subchapter, there is no influence of pitching and surging within frames of linear theory. Taking into account (3.230), we extract equation number 4 from (3.229) and using (3.183), (3.222) and (3.226):

$$\begin{aligned}&a_{42}\ddot{\eta}_G + \dot{b}_{42}\eta_G + (I_{xx} + a_{44})\ddot{\phi} + b_{44}\dot{\phi} + \rho g V_0 GM \cdot \phi + a_{46}\ddot{\chi} + b_{46}\dot{\chi} = \\ &= \zeta_{Aw}\sin\beta\left[\left(\rho V_0 GM - a_{42}\right)\omega^2 \sin\omega t + b_{42}\omega\cos\omega t\right]\end{aligned} \tag{3.231}$$

To reveal most principal properties of roll motions, we neglect the influence from different degrees of freedom completely: the result is the isolated equation of roll:

$$\begin{aligned}&(I_{xx} + a_{44})\ddot{\phi} + b_{44}\dot{\phi} + \rho g V_0 GM \cdot \phi = \\ &= \zeta_{Aw}\sin\beta\left[\left(\rho V_0 GM - a_{42}\right)\omega^2 \sin\omega t + b_{42}\omega\cos\omega t\right]\end{aligned} \tag{3.232}$$

The isolated roll equation (3.232) is very similar to the equation for a mechanical linear oscillator, for example a pendulum, where wave forces play a role in excitation. We consider this analogy later; at this moment let us present this equation in standard form, we also present the right hand side in a form of a single trigonometric function.

$$\ddot{\phi} + 2\delta\dot{\phi} + \omega_\phi^2 \phi = \alpha_E \sin(\omega t + \varphi_E) \tag{3.233}$$

Here:

$$\omega_\phi = \sqrt{\frac{\rho g V_0 \cdot GM}{I_{xx} + a_{44}}}; \quad 2\delta = \frac{b_{44}}{I_{xx} + a_{44}} \tag{3.234}$$

$$\alpha_E = \frac{\zeta_{Aw}\omega \sin\beta}{I_{xx} + a_{44}} \sqrt{(\rho V_0 GM - a_{42})^2 \omega^2 + b_{42}^2} \tag{3.235}$$

$$\tan(\varphi_E) = \frac{b_{42}}{(\rho V_0 GM - a_{42}) \cdot \omega} \tag{3.236}$$

3.5.4 Other Forms of Linear Equation of Roll Motions

Further simplification of the roll equation can be reached by neglecting the diffraction component of the wave force: it is small in comparison with the Froude-Krylov part. Such equation of roll sometimes is called "shortened":

$$\ddot{\phi} + 2\delta\dot{\phi} + \omega_\phi^2 \phi = \alpha_{ES} \sin \omega t \tag{3.237}$$

The amplitude of the wave forces:

$$\alpha_E = \frac{\zeta_{Aw}\omega^2 \sin\beta}{I_{xx} + a_{44}} \rho V_0 GM = \zeta_{Aw} \frac{\omega^2}{g} \omega_\phi^2 \sin\beta \tag{3.238}$$

As can be seen from the expressions (3.233) and (3.237), the difference between the "full" and shortened equations of roll is only amplitude and initial phase excitation.

It is convenient to express amplitude of wave excitation through the angle of wave slope, which can be found with partial differentiation of the profile equation (3.177). The derivative has to be calculated by the direction of wave propagation x_0:

$$\begin{aligned} \tan(\alpha_W(x_0,t)) &\approx \alpha_W(x_0,t) = \frac{\partial}{\partial x_0} \zeta_{Aw} \cos(kx_0 - \omega t) = \\ &= k\zeta_{Aw} \sin(kx_0 - \omega t) = \alpha_{Aw} \sin(kx_0 - \omega t) \end{aligned} \tag{3.239}$$

Where α_{Aw} is the amplitude of wave slope:

$$\alpha_{Aw} = k\zeta_{Aw} = \frac{\omega^2}{g} \zeta_{Aw} \tag{3.240}$$

The angle of wave slope at the location of a ship:

$$\alpha_W(0,t) = -\alpha_{Aw} \sin \omega t \tag{3.241}$$

The above formulae allow expressing amplitudes of wave excitation for "full" (3.233) and shortened (3.237) equations in terms of amplitudes of wave slope, correspondingly:

$$\alpha_E = \alpha_{Aw} \sin\beta \cdot \sqrt{\left(\omega_\phi^2 - \frac{g a_{42}}{(I_{xx} + a_{44})}\right)^2 + \frac{g^2 b_{42}^2}{\omega^2 (I_{xx} + a_{44})^2}} \tag{3.242}$$

$$\alpha_E = \alpha_{Aw} \omega_\phi^2 \sin\beta \tag{3.243}$$

Some authors [Semenov-Tian-Schansky, *et al*, 1969] consider a slightly different equation of roll:

$$\begin{aligned}(I_{xx} + a_{44})\ddot{\phi} + b_{44}\dot{\phi} + \rho g V_0 GM \cdot \phi = \\ = g\rho V_0 GM \cdot \alpha_W \sin\beta + a_{44}\ddot{\alpha}_W \sin\beta + b_{44}\dot{\alpha}_W \sin\beta\end{aligned} \tag{3.244}$$

The difference between the above expression and the roll equation (3.232) is the way the diffraction forces are written. This form of roll equation allows introduction of so-called relative coordinates:

$$\phi_r = \phi - \alpha_W \sin\beta \tag{3.245}$$

Substituting (3.235) into (3.234) leads to the following:

$$\ddot{\phi}_r + 2\delta\dot{\phi}_r + \omega_\phi^2 \phi_r = \alpha_{Ar}(\omega) \sin\omega t \tag{3.246}$$

Where α_r is the amplitude of excitation in relative coordinates that can be expressed as follows:

$$\alpha_{Ar}(\omega) = \frac{I_{xx}\omega^2 \alpha_{Aw}}{I_{xx} + a_{44}} \tag{3.247}$$

3.5.5 Solution of Linear Equation of Roll Motions

As can be seen clearly from the formula (3.236), the third form of the roll equation has the same structure as the previous two forms written in "absolute" coordinates (this term is used contrary to "relative" coordinates). All these forms of the roll motion presentation are ordinary linear differential equations with constant coefficients and sinusoidal excitation. We briefly review its solution.

The equation (3.233) includes the wave excitation term and therefore is heterogeneous (from Greek *heteros* –different, here it means a differential equation with a term that does not depend on an unknown function). We are interested in all types of roll motions, so we are looking for a general solution.

As it is known from calculus, a general solution of such an equation can be found as a sum of general solutions of a corresponding autonomous equation and partial solution of a heterogeneous equation [Bronshtein and Semendyayev, 1997].

$$\phi(t) = \phi_G(t) + \phi_P(t) \tag{3.248}$$

Here ϕ_G is the general solution of the corresponding autonomous equation and ϕ_p is partial solution of heterogeneous equation.

3.5.6 Linear Roll Motions in Calm Water

The corresponding autonomous equation is obtained from (3.233) by eliminating all the terms that do not depend on an unknown function. In our case it means that the corresponding autonomous equation does not contain wave excitation and therefore it describes roll motion of a ship in calm water (naturally with certain initial roll angle and/or angular velocity):

$$\ddot{\phi}_G + 2\delta\dot{\phi}_G + \omega_\phi^2\phi_G = 0 \qquad (3.249)$$

To solve the ordinary differential equation means to find a function that would turn the equation into true equality. Let us try to search the solution in the form of an exponent:

$$\phi_G(t) = C\exp(\lambda t) \qquad (3.250)$$

Substitution of (3.250) into (3.249) turns the differential equation into an algebraic one. This equation is called the characteristic equation:

$$\lambda^2 + 2\delta\lambda + \omega_\phi^2\lambda = 0 \qquad (3.251)$$

This is the well-known quadratic equation. It has two solutions:

$$\lambda_{1,2} = \frac{-2\delta \pm \sqrt{D}}{2}; \quad D = 4\delta^2 - 4\omega_\phi^2 \qquad (3.252)$$

Solutions of the characteristic equation are called eigenvalues. There are two of them and so there are two solutions of the differential equation (3.251). As it is known, the sum of two solutions of the linear differential equation is also a solution of this equation:

$$\phi_G(t) = C_1\exp(\lambda_1 t) + C_2\exp(\lambda_2 t) \qquad (3.253)$$

Also, if discriminate $D \neq 0$ and $\lambda_1 \neq \lambda_2$, components of the solution for (3.253) are linearly independent (it is impossible to express one through another using linear operations only). Then, as it is proven in mathematics, the solution for (3.253) constitutes a general solution of the differential equation of the second order: it means that any solution of (3.249) can be expressed through (3.253) by changing arbitrary constants C_1 and C_2.

Depending on the sign of discriminate (3.252), eigenvalues may be real or complex. It is quite typical for the roll motion of a ship that:

$$\delta < \omega_\phi \Rightarrow \quad D < 0 \qquad (3.254)$$

So, the eigenvalues are complex:

$$\lambda_{1,2} = -\delta \pm i\sqrt{\omega_\phi^2 - \delta^2}\ ; \quad i = \sqrt{-1} \qquad (3.255)$$

Substitution of eigenvalues (3.255) into general solution (3.253) makes it look as follows:

$$\phi_G(t) = \phi_{0a}\exp(-\delta t)\sin(\omega_0 t + \varepsilon_0) \qquad (3.256)$$

Here, the imaginary part of the eigenvalues makes the frequency:

$$\omega_0 = \sqrt{\omega_\phi^2 - \delta^2} \tag{3.257}$$

Also, we have re-written arbitrary constants in a form of amplitude and phase:

$$\phi_{0a} = \sqrt{C_1^2 + C_2^2}\ ; \quad \varepsilon_0 = \arctan\frac{C_1}{C_2} \tag{3.258}$$

These arbitrary constant can be expressed through initial conditions: to do that we need to set $t = 0$ and require that the initial roll angle and angular velocity would be equal to the specified quantities. Since the roll motion is described with a differential equation of the second order and, as a result we have two arbitrary constants, it is enough to provide two initial conditions for the initial roll angle and angular velocity:

$$\phi_G(t=0) = \phi_0\ ; \quad \dot{\phi}_G(t=0) = \dot{\phi}_0 \tag{3.259}$$

The system of two algebraic equations (3.258) has two unknowns, and therefore may be solved:

$$\phi_{0a} = \frac{1}{\omega_0}\sqrt{\left(\delta\phi_0 + \dot{\phi}_0\right)^2 + \omega_0^2\phi_0^2}\ ; \quad \varepsilon_0 = \arctan\frac{\omega_0\phi_0}{\delta\phi_0 + \dot{\phi}_0} \tag{3.260}$$

Formula (3.260) describes decaying oscillation, shown in fig 3.6. These oscillations however never die out completely; since the term $\exp(-\delta t)$, which is responsible for the decay asymptotically tends to zero:

$$\lim_{t\to\infty}\left[\exp(-\delta t)\right] = 0$$

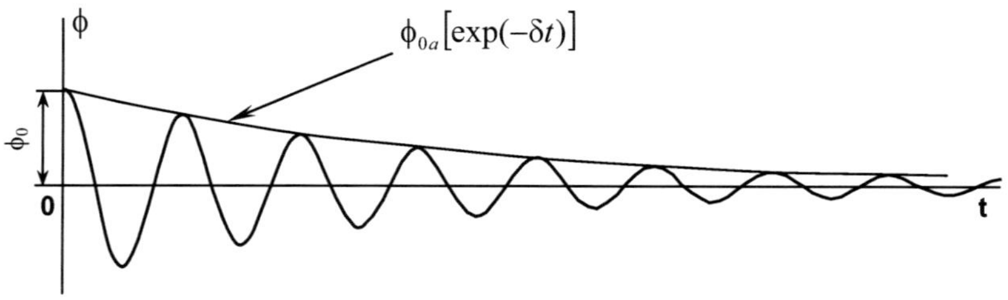

Fig. 3.6 Roll oscillations in calm water

If we imagine that waves are not generated, there will be no damping at all, and the oscillations would be described with the simple sinusoidal curve:

$$\phi(t) = \phi_{0a}\sin(\omega_\phi t + \varepsilon_0) \tag{3.261}$$

This finally clarifies the meaning of coefficient ω_ϕ: it a frequency of small oscillations in assumption that damping is absent or very small. This frequency is called "natural" frequency and contains very important information on a ship as dynamical system. As we have seen from the formula (3.234) above, the natural frequency is dependent on the metacentric height and, therefore, contains important stability information.

Natural frequency does not differ very much from damped roll oscillation frequency ω_0, since small roll motions are generally slightly damped.

3.5.7 Linear Roll in Waves

Partial solution of the heterogeneous equation, needed for (3.238) is searched in a form similar to the form of excitation, e.g. in a form of sine function with the excitation frequency, but with unknown amplitude and phase shift:

$$\phi_P(t) = \phi_a \sin(\omega t + \varepsilon_\phi + \varphi_E) \tag{3.262}$$

Let us substitute the expected solution (3.262) into the linear roll equation (3.233). The first derivative of (3.262) contains a cosine function and therefore is linearly independent on the solution itself. Equalizing sine and cosine function separately allows converting it into a system of two algebraic equations. The solution of this system would define amplitude and phase of the partial solution we are searching for:

$$\phi_a = \frac{\alpha_E}{\sqrt{\left(\omega_\phi^2 - \omega^2\right)^2 + 4\delta^2\omega^2}}; \qquad \varepsilon_\phi = \arctan\frac{2\delta\omega}{\omega_\phi^2 - \omega^2} \tag{3.263}$$

To complete the general solution of the heterogeneous equation, we have to re-define arbitrary constants with the addition of a partial solution:

$$\phi_G(t=0) + \phi_P(t=0) = \phi_0\,; \quad \dot{\phi}_G(t=0) + \dot{\phi}_P(t=0) = \dot{\phi}_0 \tag{3.264}$$

This leads to the following formulae for arbitrary constants:

$$\phi_{0a} = \frac{1}{\omega_0}\sqrt{\left[\delta(\phi_0 - \phi_P(0)) + \left(\dot{\phi}_0 - \dot{\phi}_P(0)\right)\right]^2 + \omega_0^2(\phi_0 - \phi_P(0))^2} \tag{3.265}$$

$$\varepsilon_0 = \arctan\frac{\omega_0\left(\phi_0 - \phi_P(0)\right)}{\delta(\phi_0 - \phi_P(0)) + \left(\dot{\phi}_0 - \dot{\phi}_P(0)\right)} \tag{3.266}$$

The final form for solution of the linear roll equation is

$$\phi(t) = \phi_{0a}\exp(-\delta t)\sin(\omega_0 t + \varepsilon_0) + \phi_a \sin(\omega t + \varepsilon_\phi + \varphi_E) \tag{3.267}$$

It is clear from this formula that the oscillations with frequency ω_0 will decay, and the system will move with the frequency of a wave; see fig. 3.7.

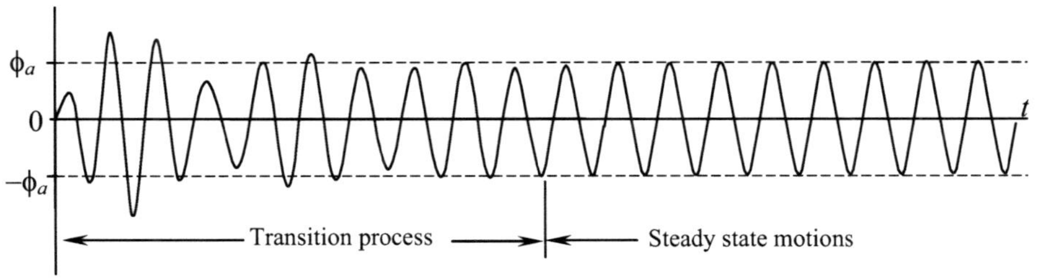

Fig. 3.7 Solution of linear roll equation

As is shown in fig. 3.7, two types of motions can be observed: transition process, during which influence of initial conditions affects the motions; and steady state mode of motions, when the system moves with excitation frequency. This is a manifestation of one of the fundamental properties of a dynamical system: an ability to synchronize itself with external excitation. Linear differential equations have a clear and elegant mechanism for this phenomenon: the general solution of an autonomous equation dies out and leaves the partial solution of the heterogeneous equation "alone", which actually has the excitation frequency.

This also means, since the transition is not something that is found for linear system only, capsizing could happen only during transition. Once roll oscillation in regular waves reaches a steady state, the question on the possibility of capsizing is senseless: it did happen already, or will never happen (with the exception of the quite theoretical situation when capsizing occurs on every period).

3.5.8 Steady State Roll Motions. Memory Effect

A steady state solution of the roll equation defined by partial solution of the heterogeneous equation is capable to reproduce one very important phenomenon: resonance.

If we look at the formula for amplitude of steady state response more closely, we see that it reaches maximum, when frequency of excitation equals frequency of damped oscillation in calm water, (also see fig. 3.8).

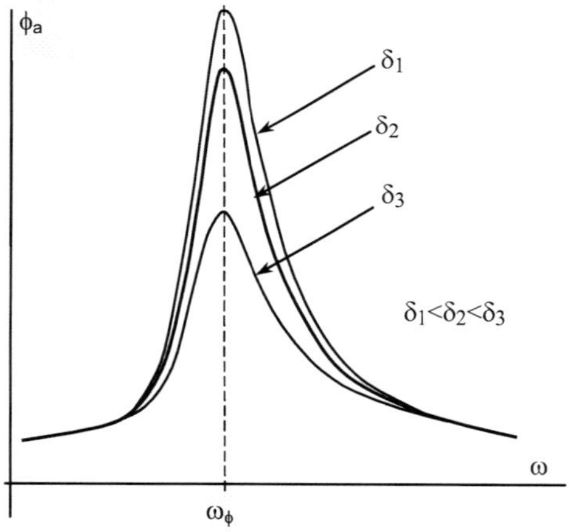

Figure 3.8 Response curves of linear roll motion with different damping.

$$\omega_\phi = \omega \quad \Rightarrow \quad \phi_a = \max \phi_a = \frac{\alpha_E}{2\delta\omega_\phi}$$

The fact that the amplitude of steady state roll has a maximum means that certain wave frequencies lead to larger responses than others. In other words, there is a frequency range where response is amplified. This phenomenon is known as resonance.

It is clear from formulae (3.263) and (3.268) that the amplitude in resonance depends on damping, and less damping means larger response amplitude in resonance regime. The influence of damping is less, or almost none, outside of the resonance region, see fig. 3.8.

Dependence of the response amplitude on excitation frequency, as shown in fig.3.8, is called the "response curve". More information on roll resonance in beam seas is available from Chapter 10 of [Kobylinski and Kastner, 2003].

So far, ship roll in waves looks almost identical to a mechanical oscillator with the exception of the fact that coefficients of the linear roll equation (added mass and wave damping) depend on wave frequency. There is a principal difference between a ship roll and a pendulum, beyond the frequency dependent coefficients. A ship makes waves. These waves depend on ship motions at the moment they were generated. Then, these waves move away, still holding the information of ship motions in the past. Moreover, flow of fluid caused by these waves has an effect on distribution of pressure in entire fluid domain. This means that ship motions in waves cannot be completely determined with the previous time step, as it is true for a mechanical oscillator. This "memory effect" decays with time, since waves move further from the ship. That is why it does not undermine the presentation of a ship as a dynamical system; understanding however, that such a presentation is not exact and should be considered only as an approximation, see [Newman, 1977] for more information.

3.6 Nonlinear Roll Equation

The equation of roll motions considered above is linear. It is the result of the assumptions that the wave is small. As we noted already, it means that the model only describes small ship motions. Transition to large roll angles requires the introduction of nonlinear terms into the ship motion equations. As we pointed out, there is no theory available to derive nonlinear ship motions equations in a consistent manner. So we use a "physics-based" approach, considering forces of a different nature separately and try to combine them into one model of nonlinear roll, suitable for capsizing study.

3.6.1 Classification of Forces

The forces acting on a ship in a seaway can be classified in the following manner:

- Inertial hydrodynamic forces (e.g. added mass),
- Wave damping hydrodynamic forces,
- Viscous damping forces,
- Hydrostatic forces,
- Wave excitation forces,
- Other forces, including:
  - Aerodynamic forces,
  - Forces caused by green water,
  - Appendage forces (including damping from rudders, bilge keels and anti-roll fins),
  - Forces due to fluid motion in internal tanks including anti-roll tanks,
  - Hull "maneuvering" forces, propeller thrust, etc.

There are other ways to classify these forces. This classification assigns forces of a different nature to the same category. However, it reflects a role that force plays in ship motions.

3.6.2 Inertial Hydrodynamic Forces and Moments

Inertial hydrodynamic forces and moments express changing inertial qualities of the body in a fluid as opposed to a vacuum. We considered them in subchapter 3.4.4 within frames of linear assumptions as a part of forces related to radiation potentials.

These linear assumptions suggest that the generated waves have small amplitudes and body motions are themselves small, so changing of underwater geometry was not taken into account. However, when roll angles are large, this influence might be significant. We will return to this matter in Chapter 7, where the problem of deck-in-water is studied.

3.6.3 Hydrodynamic Wave Damping Forces

Generally, damping and resistance forces represent transfer of kinetic energy of the moving body to the fluid. There are several ways to do this, so we are talking about components of a different physical nature. (Here, we use the hydrodynamic rather than general physics meaning of "physical nature". Strictly speaking, all these forces have a gravitational and electromagnetic nature.).

We already identified wave damping while considering forces related with radiation potential in subchapter 3.4.4. These forces describe losses of energy that are taken away by waves generated by the ship. When we examined the linear equation of roll in subchapter 3.5, wave dumping was the only phenomenon included that is "responsible" for roll decay in calm water and synchronization of roll in waves with excitation frequency.

3.6.4 Viscous Damping Forces

Wave generation is not the only way of damping of ship motions.

As it was mentioned in subchapter 3.2.1, viscous forces are significant for roll damping. These forces are the result of an exchange of energy between layers of fluid: we have discussed this briefly in subchapter 3.1.5. So, a body moving in the fluid would transfer some of its energy to the closest layer; then it will be transferred further and further from the body. Part of the kinetic energy of the body finally will be converted into kinetic energy of fluid molecules and will result in a rise of the fluid's temperature. The viscous forces have been profoundly studied by theoretical and applied fluid dynamics. Another phenomenon related to viscosity is generation of vortexes.

As we have seen from subchapter 3.4.4, a linear expression was enough to describe wave damping for small motions. To take into account the viscous and vortex component of roll damping, a second power term is used:

$$M_D = b\dot{\phi} + b_2 \cdot |\dot{\phi}| \cdot \dot{\phi} \tag{3.269}$$

Coefficients b and b_2 have to be determined from the roll decay test: so all three components contribute to both coefficients. Basic information on roll decay test can be found in [Bhattacharyya, 1978].

The absolute value in formula (3.269) might create certain analytical difficulties, since it requires consideration of two domains ($\dot{\phi} < 0$; $\dot{\phi} \geq 0$) separately, so sometimes it is more convenient to present nonlinear damping as:

$$M_D = b\dot{\phi} + b_3\dot{\phi}^3 \qquad (3.270)$$

When the deck enters the water, roll damping changes dramatically: the deck edge generates vortexes, so damping might depend on roll and wave slope angles as well; we will return to this problem in Chapter 7.

At the same time, nonlinearity of damping does not introduce qualitative changes into the roll model. With several exceptions, we can keep damping linear, but it does not mean that only wave forces are taken into account. It would be more accurate to think about the linear coefficient *b* as an equivalent damping coefficient that includes the influence of all three components of roll damping.

The only application where nonlinearity of roll damping may play a qualitative role is parametric resonance, because steady state parametric roll is only possible in a nonlinear system. We will consider parametric resonance in Chapter 6.

3.6.5 Other Forces

Aerodynamic forces and moments are caused by interaction of the non-submerged part of a ship with airflow. The airflow is created by wind for the majority of ships; speed is usually too small to create aerodynamic drag that would be comparable with other forces. The wind heeling moment usually is included in models for stability regulations, see subchapter 3.5 of [Kobylinski and Kastner, 2003].

Numerical values for aerodynamic forces are usually available from model experiments. These values are typically presented in the form of non-dimensional coefficients C_x, C_y, C_z, C_{mx}, C_{my}, C_{mz}. Projection of aerodynamic forces and moments are expressed as:

$$F_x^{Aero} = \frac{1}{2}\rho v_A^2 C_x A_{wx}; \quad F_y^{Aero} = \frac{1}{2}\rho v_A^2 C_y A_{wy}; \quad F_z^{Aero} = \frac{1}{2}\rho v_A^2 C_z A_{wz} \qquad (3.271)$$

$$M_x^{Aero} = \frac{1}{2}\rho v_A^2 C_{mx} A_{wx} l_{wx}; \quad M_y^{Aero} = \frac{1}{2}\rho v_A^2 C_{my} A_{wy} l_{wy};$$
$$M_z^{Aero} = \frac{1}{2}\rho v_A^2 C_{mz} A_{wz} l_{wz} \qquad (3.272)$$

The choice of specific areas A_{wx}, A_{wy}, A_{wz} and specific length l_{wx}, l_{wy}, l_{wz} is arbitrary; this is a matter of convenience of presenting and processing data of the experiment. These values are needed just to make the coefficients non-dimensional. We will return to aerodynamic forces and moments in subchapter 7.1.

Forces caused by greenwater shipping can alter ship behavior in waves. The water on deck causes an additional heeling moment that is also included in the model for stability regulations (see [Kobylinski and Kastner, 2003], subchapter 3.3). That model, however did not consider dynamic effects of the green water behavior. These effects include "sloshing", ingress and outflow through the freeing ports, over the bulwark or deck edge. We discuss these forces and how they affect ship dynamics in subchapter 7.2.

Hull "maneuvering" forces are of a hydrodynamic lifting nature and related with vortex flow. These forces may have a significant influence on the possibility of capsizing due to broaching and will be considered in Chapter 6.

Thrust force, rudder force and other hydrodynamic forces generated by ship equipment might have certain influences on stability. This action will be also addressed when we will be considering specific issues such as stability in following and quartering seas.

Finally, it is important to mention forces created by different anti-roll devices such as fins, and U-tube tanks. This type of equipment creates forces that mitigate roll and therefore, are capable of improving dynamic stability of ships. Their influence, however, is beyond the scope of the present book, besides perhaps, roll damping, which might be included in formulae (3.269) or (3.270).

3.6.6 Wave Excitation Forces

We have examined wave excitation forces and moments earlier in subchapters 3.4.5 through 3.4.7, using assumptions of small-amplitude waves and small ship motions. As we already noted, finite-amplitude waves can be considered using the nonlinear boundary condition on the free surface (3.70) instead of the linearized condition (3.73). The problem of nonlinear wave action on the body is numerically complicated, however there are a number of solutions available [Kim and Yue, 1989; Malenica and Molin, 1995; Ferrant, 1998]. Such an analysis, however, is beyond the scope of this book.

The other linear assumption is that ship motions are small, which allow ignoring changes of the underwater geometry. If this assumption is not made, the complete result can only be reached by time domain simulation only [Lin and Yue, 1990]. The consequences of keeping these assumptions can be mitigated using the roll equation in relative coordinates (3.246) with nonlinear hydrostatic terms; we will examine this option in the next subchapter.

3.6.7 Hydrostatic Forces: Structure of Nonlinear Roll Equation

Hydrostatic forces and moments are the direct outcome from Archimedean law. They are the result of the summation of hydrostatic pressure over the surface of a ship's hull: we have examined them in subchapter 3.4.3 with the assumption that ship motions are small.

This assumption cannot be used anymore. Hydrostatic forces and moments are "responsible" for restoring action and they determine equilibria of the floating body. That is why hydrostatic forces have to be at least qualitatively correctly described in our model.

Dependence of the restoring moment on the heel angle has a nonlinear character. To take into account that roll angles are not small, the term $\rho g V_0 GM \cdot \phi$ in the equation (3.222) has to be substituted with the instantaneous restoring moment: for calm water, it is the well-known *GZ* curve.

Usage of the term "hydrostatic" does not necessarily imply calm water. This term only reflects the fact that the fluid is heavy and the pressures created by its weight are transferred in all directions. If the water is not calm, its surface experiences a wave deformation, which alter hydrostatics just because there is more water in one place and less in the other.

So, strictly speaking, we cannot use the *GZ* curve for the above purpose, since it is calculated for calm water. Theoretically, we have to integrate hydrostatic pressures over the surface of the ship hull, taking into account curvature of the wave surface. Such an approach is used for time domain simulation [Lin and Yue, 1990; Shin, *et al* 1997; Shin *et al* 2003]. However, for the qualitative study, it is not necessary to do it: the restoring moment in the wave is qualitatively the same as that in calm water.

The only exception is when a ship is heading in longitudinal waves: the change in the restoring moment is quite significant there (see [Kobylinski and Kastner, 2003], subchapter 10.10). We will return to this matter later in Chapter 6 where stability in following and quartering seas is discussed.

Another problem is associated with the fact that the buoyancy force is directed perpendicularly to the water surface. If we assume that wave slope $\alpha = \alpha_W \sin\beta$ is an angle of inclination of the water surface, than the restoring moment can be written as:

$$M_R = \rho g V_0 \cdot GZ(\phi - \alpha) = \rho g V_0 \cdot GZ(\phi_r) \tag{3.273}$$

Hence, only the relative roll angle creates the righting arm, which makes the roll equation with relative coordinates (3.246) our first nonlinear model:

$$\ddot{\phi}_r + 2\delta\dot{\phi}_r + \omega_\phi^2 f(\phi_r) = \alpha_{Ar}(\omega)\sin\omega t \tag{3.274}$$

With:

$$f(\phi_r) = \frac{\rho g V_0 \cdot GZ(\phi_r)}{GM(I_{xx} + a_{44})} \tag{3.275}$$

Terms in the roll equation with relative coordinates are differences between moments acting on the ship and corresponding terms of wave excitation. So, the restoring term in (3.246) also is a difference between the roll restoring moment and the Froude-Krylov part of the wave excitation.

In general, it is impossible to separate the Froude-Krylov part of the wave excitation and roll hydrostatic moment: *GZ* does not have an analytical form. To do this, some approximation of the *GZ* curve is required. An odd power polynomial is one of the most popular methods (usually 3rd degree is sufficient):

$$f(\phi_r) = \phi_r - c_3\phi_r^3 \tag{3.276}$$

Let us try to separate Froude-Krylov excitation from hydrostatic restoring by substituting the relative roll angle (3.245) into equation (3.274):

$$\begin{aligned} f(\phi - \alpha) &= \phi - \alpha - c_3(\phi - \alpha)^3 = \\ &= \phi - \alpha - c_3\phi^3 + 3c_3\phi^2\alpha - 3c_3\phi\alpha^2 - c_3\alpha^3 \end{aligned} \tag{3.277}$$

Wave slope angles are much smaller than roll angles, so deleting higher orders of the term α appears to be appropriate:

$$f(\phi-\alpha)=\phi-\alpha-c_3\phi^3+3c_3\phi^2\alpha \quad (3.278)$$

The term $3c_3\phi^2\alpha$ is usually ignored for the sake of simplicity; there are no data indicating this term alters the qualitative behavior of the system. Then, we can separate excitation and restoring terms resulting in an equation with absolute roll angles as follows:

$$\ddot{\phi}+2\delta_\phi\dot{\phi}+\omega_\phi^2(\phi-c_3\phi^3)=\alpha_E\sin(\omega t+\varphi_E) \quad (3.279)$$

Keeping in mind the above simplifications, we can rewrite the equation (3.279) back to a general form:

$$\ddot{\phi}+2\delta_\phi\dot{\phi}+\omega_\phi^2 f(\phi)=\alpha_E\sin(\omega t+\varphi_E) \quad (3.280)$$

An alternative way to introduce restoring nonlinearity into the roll equation is to consider the nonlinear Froude-Krylov wave excitation as:

$$(I_{xx}+a_{44})\ddot{\phi}+b_{44}\dot{\phi}+\rho g V_0 GZ(\phi)=-g\rho V_0 GZ(\alpha)+\frac{1}{\omega^2}(a_{42}\ddot{\alpha}-b_{42}\dot{\alpha}) \quad (3.281)$$

Taking into account that the angle of wave slope is a relatively small value in comparison with roll, nonlinearity of excitation might be dropped and the above equation will be converted into (3.280).

The above derivations demonstrated that the procedure for developing the nonlinear model is artificial and it is not mathematically rigorous. However, it can be accepted so far as the model is qualitatively adequate. Analysis of the adequacy of the mathematical model is closely related with the particular problem to be studied. So, we will be changing the above models further, as necessary, trying to maintain this adequacy.

Finally, the structure of the equations (3.274) (3.280) and (3.281) is the same. It makes all further development equally applicable to all of them[1].

---

[1] With the only exception that for the equation in relative coordinates amplitude of excitation depends on the frequency of excitation

Chapter 4

Nonlinear Roll Motion in Regular Beam Seas[1]

4.1 Free Roll Motion

4.1.1 Free Oscillations of Nonlinear System

Following the methodology used earlier, we will describe free nonlinear roll motion as:

$$\ddot{\phi} + \omega_{\phi}^{2} f(\phi) = 0 \qquad (4.1)$$

We give the initial inclination to the system, say ϕ_a and then let the system move freely. Since there is no damping, the oscillations will never fade: their amplitude is equal to the initial inclination. The period of these oscillations can be found from the equation (4.1) by re-writing it in the first integral form assuming that the initial conditions correspond to an amplitude value for the roll angle and zero for the roll rate:

$$\frac{d\dot{\phi}}{dt} + \omega_{\phi}^{2} f(\phi) = 0 \quad \Leftrightarrow \quad \frac{d\dot{\phi}}{d\phi} \cdot \frac{d\phi}{dt} + \omega_{\phi}^{2} f(\phi) = 0 \quad \Leftrightarrow \quad \dot{\phi} d\dot{\phi} + \omega_{\phi}^{2} f(\phi) d\phi = 0$$

$$-\int_{0}^{\dot{\phi}} \dot{\phi} d\dot{\phi} = \int_{\phi_a}^{\phi} \omega_{\phi}^{2} f(\phi) d\phi \quad \Leftrightarrow \quad \frac{\dot{\phi}^{2}}{2} = \omega_{\phi}^{2} \cdot \left(F(\phi_a) - F(\phi)\right)$$

With $F(\phi) = \int_{0}^{\phi} f(\phi) d\phi$. Further derivations lead to:

$$\frac{d\phi}{dt} = \omega_{\phi} \sqrt{2\left(F(\phi_a) - F(\phi)\right)} \quad \Leftrightarrow \quad dt = \frac{d\phi}{\omega_{\phi} \sqrt{2\left(F(\phi_a) - F(\phi)\right)}}$$

Performing integration from zero to an amplitude value yields a quarter of a period. The entire period is determined as:

$$T(\phi_a) = \frac{2\sqrt{2}}{\omega_{\phi}} \int_{0}^{\phi_a} \frac{d\phi}{\sqrt{F(\phi_a) - F(\phi)}} \qquad (4.2)$$

The integral in formula (4.2) is improper. There are several methods to calculate this integral numerically. The best results however, are achieved with the following substitution (Krein-Sizov formula):

---

[1] The author is grateful to Prof. M.A.S. Neves for his detailed review of this chapter during preparation of the second edition.

$$\phi = \phi_a \cos\varphi$$

$$T(\phi_a) = 2\sqrt{2} \cdot \frac{\phi_a}{\omega_\phi} \cdot \int_0^{\pi/2} \frac{\sin\varphi \, d\varphi}{\sqrt{F(\phi_a) - F(\phi_a \cos\varphi)}} \tag{4.3}$$

A graph indicating the dependence of the free oscillation frequency initial amplitude is given in fig. 4.1. This curve is frequently called the "backbone" curve.

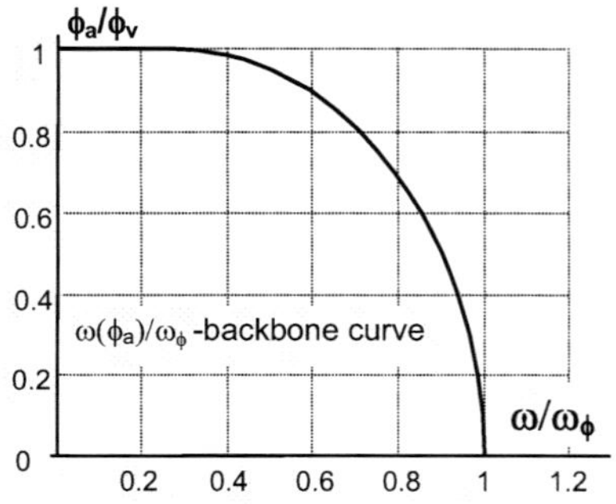

Fig. 4.1 Example of backbone curve

This curve has several important qualities:

The period of free oscillations depends on initial amplitude. This is a nonlinear property: the period of free oscillations of a linear system does not depend on any initial conditions, in other words, free oscillations of a linear system are isochronous. Free oscillations of nonlinear systems are not isochronous.

The period of free oscillations of a nonlinear system is equal to the period of a corresponding linear system, when the initial amplitude is small:

$$\lim_{\phi_a \to 0} T(\phi_a) = \frac{2\pi}{\omega_\phi} \tag{4.4}$$

The period of free oscillation equals infinity when the initial amplitude equals to the angle of vanishing stability. The angle of vanishing stability is an unstable equilibrium (see [Kobylinski and Kastner 2003], subchapter 8.3). If the system is placed exactly at the unstable equilibrium position, it stays there until disturbed. If there is no disturbance (which never happens in the physical world), the system remains there forever.

$$T(\phi_v) = \infty \tag{4.5}$$

Free motions of a nonlinear system is periodic, but may be far from sinusoidal. Nonlinearity influences on the shape of the time history curve are stronger when the initial angle is closer to the angle of vanishing stability, see fig. 4.2.

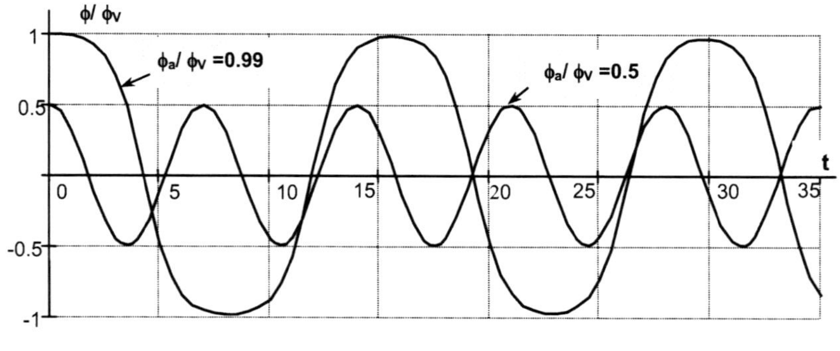

Fig. 4.2 Free nonlinear oscillations

4.1.2 Free Motions of Piecewise Linear System

Capsizing is a transition to oscillation near the upside down equilibrium. It is important, then, to have a model that contains several (at least two) stable equilibria. One of the simplest models that possesses such a capability is a piecewise linear system. Since we will be using this model for further capsizing study, it is worthwhile to show that it has the same qualities as a "conventional" nonlinear system.

The system we are going to use is described by the following differential equation:

$$\ddot{\phi}+2\delta_{\phi}\dot{\phi}+\omega_{\phi}^{2}f_{L}^{*}(\phi)=\alpha_{E}\sin(\omega t+\varphi_{E}) \tag{4.6}$$

The piecewise linear term is responsible for the following equilibria:

$$f_{L}^{*}(\phi)=\begin{cases} k_{f0}\phi & \phi\in[0;\phi_{m0}] \quad \textit{range}\ 0 \\ k_{f1}(\phi_{V}-\phi) & \phi\in]\phi_{m0};\phi_{m1}[\quad \textit{range}\ 1 \\ k_{f2}(\phi-\pi) & \phi\in[\phi_{m1};\phi_{m2}] \quad \textit{range}\ 2 \\ k_{f1}(2\pi-\phi_{V}-\phi) & \phi\in]\phi_{m2};\phi_{m3}[\quad \textit{range}\ 3 \\ k_{f0}(\phi-2\pi) & \phi\in[\phi_{m3};2\pi[\quad \textit{range}\ 4 \end{cases} \tag{4.7}$$

This function is shown in fig. 4.3: we consider positive dissection because of symmetry of the function.

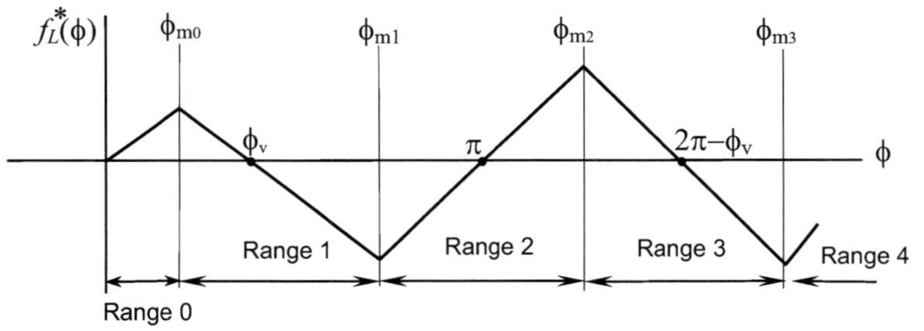

Fig. 4.3 Piecewise linear restoring term

The system in figure 4.2 has two stable equilibria: at the origin of the coordinate system and at π. These two stable equilibria are divided by the unstable equilibrium at the angle of vanishing stability ϕ_v.

It is important to note that the system with the piecewise linear restoring term does not have any discontinuities: the piecewise linear term is actually a force; hence, it is proportional to acceleration. This force is described by a continuous, though, non-smooth function. Acceleration is the second derivative of motion, so it is also continuous. Discontinuity appears only with its derivative, while the third derivative of motion displacement does not have a mechanical sense.

The solution for such a system consists of many pieces. Each piece is a solution of a linear differential equation. The pieces are connected through boundary conditions. When the roll angle is expected to cross the boundary between different linear ranges, the final conditions at the end of the first range are the initial conditions for the second one.

Since the boundaries between the different ranges are known, we have the only one unknown condition, which is angular velocity. It should be the same at the end of one range and at the beginning of the next one.

Here we consider free motion of a system with a piecewise linear term in order to find its backbone curve (period of free oscillations as a function of initial amplitude):

$$\ddot{\phi} + c_i\phi + c_{Fi} = 0; \quad i = 0...n \tag{4.8}$$

The equation (4.8) within the range i is a linear differential equation, its solution is well known:

$$\phi(t_i) = A_i \exp(\lambda_{1i}t_i) + B_i \exp(\lambda_{2i}t_i) + \phi_{Ei} \tag{4.9}$$

The term, ϕ_{Ei}, is the equilibrium position associated with the range i :

$$\phi_{Ei} = -\frac{c_i}{c_{Fi}} \tag{4.10}$$

Time, t_i, is local for range i. It starts from 0 when the system enters the range i. Other variables also have a local meaning:

$$\lambda_{1i,2i} = \pm\sqrt{-c_i} \tag{4.11}$$

Arbitrary constants:

$$A_i = \frac{\dot{\phi}_{0i} + \sqrt{-c_i}\,(\phi_{0i} - \phi_{Ei})}{2\sqrt{-c_i}}; \quad B_i = -\frac{\dot{\phi}_{0i} - \sqrt{-c_i}\,(\phi_{0i} - \phi_{Ei})}{2\sqrt{-c_i}} \tag{4.12}$$

The initial conditions at the boundary of a range are denoted here as ϕ_{0i} , $\dot{\phi}_{0i}$.

In order to find a period of free oscillation, we assume that motion starts from initial conditions (ϕ_a, 0). The initial heel angle, ϕ_a, is located above the maximum ϕ_{m0}, see fig. 4.3.

The solution (4.9) behaves in different ways depending on the sign of coefficient c_i . If it is positive ("*GZ* curve" before maximum), we have the usual linear oscillations. If it is above the maximum, it is no longer periodical. In our particular case $\phi \in \left]\phi_{m0};\phi_v\right[$ the solution (4.9) can be rewritten as follows [Belenky, 1995a]:

$$\phi(t) = \phi_V - (\phi_V - \phi_a)\cosh\left(\omega_\phi\sqrt{k_{f1}} \cdot t\right) \tag{4.13}$$

Since we know the solution of this range, we can find the time necessary to reach the boundary ϕ_{m0} and angular velocity at the time when the system crosses this boundary.

$$t_1 = \frac{1}{\omega_\phi\sqrt{k_{f1}}}\operatorname{arcosh}\frac{\phi_V - \phi_{m0}}{\phi_V - \phi_a} \tag{4.14}$$

$$\dot{\phi}(t_1) = -\omega_\phi\sqrt{k_{f1}}\sqrt{(\phi_V - \phi_{m0})^2 - (\phi_V - \phi_a)^2} \tag{4.15}$$

As soon the boundary ϕ_{m0} is crossed, the motion will be governed by the conventional linear oscillation equation with the well-known solution:

$$\phi(t) = \phi_{a1}\sin\left(\omega_\phi\sqrt{k_{f0}}\cdot t + \varepsilon\right) \tag{4.16}$$

With amplitude and phase depending on initial conditions at the moment of crossing of the boundary ϕ_{m0}:

$$\phi_{a1} = \sqrt{[\phi_{m0}]^2 + \frac{1}{k_{f0}\omega_\phi^2}[\dot\phi(t=t_1)]^2} = \sqrt{\frac{k_{f1}}{k_{f0}}\left[(\phi_V - \phi_{m0})^2 - (\phi_V - \phi_a)^2\right] + \phi_{m0}^2} \tag{4.17}$$

$$\varepsilon = \arctan\frac{\omega_0\phi_{m0}}{\dot\phi(t=t_1)} = \arctan\left(-\sqrt{\frac{k_{f1}}{k_{f0}}}\frac{\phi_{m0}}{\sqrt{(\phi_V-\phi_{m0})^2-(\phi_V-\phi_a)^2}}\right) \tag{4.18}$$

Since the solution at the second range is known, we can easily find the time necessary to reach zero:

$$t_2 = \frac{1}{\omega_\phi\sqrt{k_{f0}}}\arctan\left(\sqrt{\frac{k_{f1}}{k_{f0}}}\frac{\phi_{m0}}{\sqrt{(\phi_V-\phi_{m0})^2-(\phi_V-\phi_a)^2}}\right) \tag{4.19}$$

The sum of times t_1 and t_2 makes one quarter of the period of free motion [Belenky, 2000a]:

$$T(\phi) = \frac{4}{\omega_\phi}\left(\frac{1}{\sqrt{k_{f1}}}\operatorname{arcosh}\frac{\phi_V-\phi_{m0}}{\phi_V-\phi_a} + \right.$$
$$\left. + \frac{1}{\sqrt{k_{f0}}}\arctan\left(\sqrt{\frac{k_{f1}}{k_{f0}}}\frac{\phi_{m0}}{\sqrt{(\phi_V-\phi_{m0})^2-(\phi_V-\phi_a)^2}}\right)\right) \tag{4.20}$$

The backbone curve for the piecewise linear system is shown in fig. 4.4: it is not smooth, since we have quite distinctive boundaries between the ranges. As we shall see from further development, the absence of a smooth backbone curve does not have an affect on the qualitative adequacy of the piecewise linear system. It should be noted that formula (4.20) is correct only for a symmetric piecewise linear term. The solution for the general case is available from [Belenky, 1999].

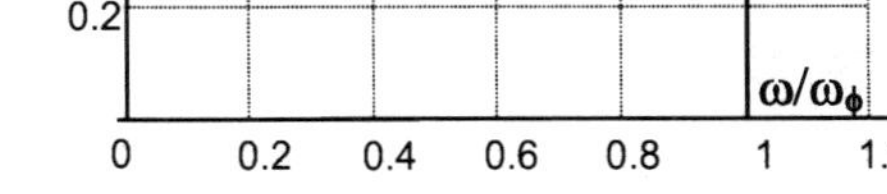

Fig. 4.4 Backbone curve for piecewise linear system

4.2 Steady State of Forced Roll Motions

4.2.1 Equivalent Linearization

The next several subchapters are focused on steady states of forced nonlinear oscillations. Since there is no analytical solution in a closed form for nonlinear systems available, different approximate methods are used. We start from equivalent linearization: the idea of this method is to substitute a nonlinear term with a linear one, but still keeping the value of the period of free oscillations:

$$\ddot{\phi} + 2\delta\dot{\phi} + \omega_\phi^2 f(\phi) = \alpha_E \sin \omega t \tag{4.21}$$

The linearized system is described as follows:

$$\ddot{\phi} + 2\delta\dot{\phi} + \omega_\phi(\phi_a)\phi = \alpha_E \sin \omega t \tag{4.22}$$

The differential equation (4.22) is formally linear, but one of its terms depends on the amplitude of response ϕ_a. We assume here that the amplitude of steady state response can be used for calculation of the natural period by formula (4.3). The steady state solution of the equation (4.22) can be expressed as:

$$\phi(t) = \phi_a \cdot \sin(\omega t + \varepsilon_\phi) \tag{4.23}$$

Where:

$$\phi_a = \frac{\alpha_E}{\sqrt{\left[(\omega_\phi(\phi_a))^2 - \omega^2\right]^2 + 4\delta^2\omega^2}} \tag{4.24}$$

$$\varepsilon_\phi = \arctan \frac{2\delta\omega}{(\omega_\phi(\phi_a))^2 - \omega^2} \tag{4.25}$$

Both these formulae include the natural frequency, $\omega(\phi_a)$, that depends on amplitude and can be calculated using (4.3):

$$\omega(\phi_a) = \frac{2\pi}{T(\phi_a)} \tag{4.26}$$

The formula for the phase angle (4.25) includes the amplitude of steady state motions, so it has to be calculated first. This amplitude depends on natural frequency that also depends on the amplitude. So, to use formula (4.24) it is necessary to express excitation frequency via amplitude of the steady state response:

$$\omega = \sqrt{(\omega_\phi(\phi_a))^2 - 2\delta^2 \mp \sqrt{\left[(\omega_\phi(\phi_a))^2 - 2\delta^2\right]^2 + \left[\alpha_E^2/\phi_a^2 - (\omega_\phi(\phi_a))^4\right]}} \tag{4.27}$$

Formula (4.27) describes the response curve, which has two branches, low and high, according to sign "+" and "-" in front of the internal square root. Fig. 4.5 shows this curve: it can clearly be seen now why the dependence of natural frequency on amplitude is called the backbone curve: it holds the response curve. Fig. 4.5 reveals the difference between linear and nonlinear response curves: the nonlinear one is bent following the backbone curve (Compare fig. 3.8 and fig. 4.5). As a result, it may have a zone where several amplitudes correspond to one excitation frequency. This is a very important nonlinear quality and we will return to it in subchapter 4.5.

Another important feature that is worthwhile to note is limits of the response curve. Its lower limit goes to amplitude excitation. If the frequency is very low in comparison with the natural frequency, action of the excitation is practically static, there is no dynamic amplification and the roll motions just follow the wave. (This is different for roll equation (3.274) as there the amplitude of excitation depends on frequency. The response curve for the equation (3.274) tends to zero for small frequencies. The formula (4.27) needs to be re-written as it does not account that amplitude of excitation depends on frequency, however the methodology of derivation does not differ.)

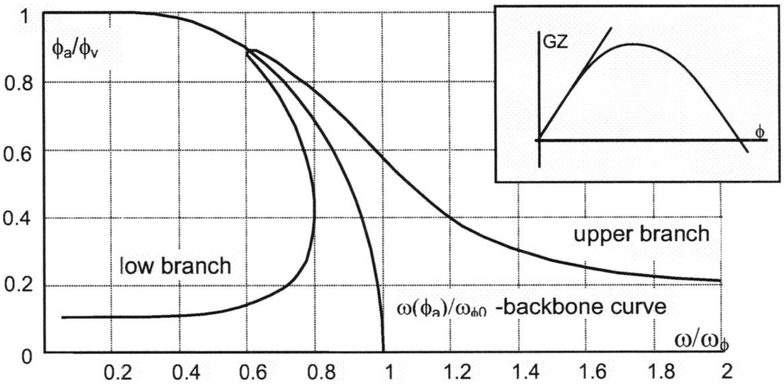

Fig. 4.5 Backbone curve and response curve

If the backbone bends toward the origin of the co-ordinate system, the case is called a "soft" type of nonlinearity. A bend in the opposite direction is called "hard". A "hard" type is partially demonstrated in fig. 4.6, where the response curve for an S-shape *GZ* curve is shown (the model for this GZ curve is taken from [Belenky, 2004]). This sample contains both types of nonlinearities: "hard" for small roll angles and then "soft" for large ones.

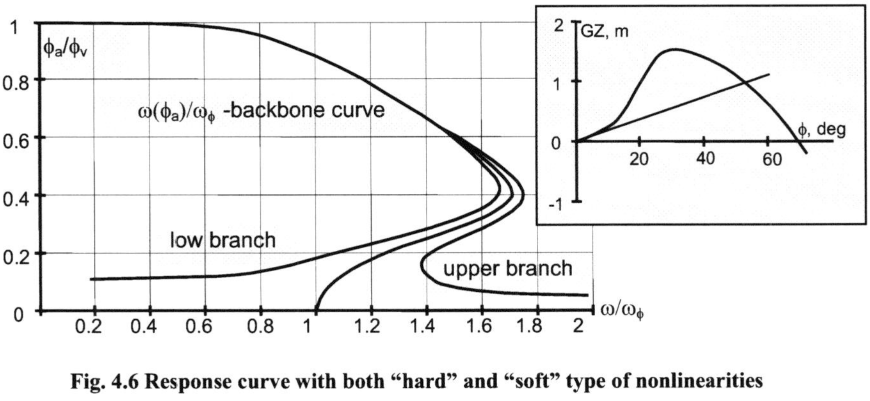

Fig. 4.6 Response curve with both "hard" and "soft" type of nonlinearities

The same procedure can be applied to a piecewise linear system: the result is shown in fig. 4.7. Contrary to a conventional nonlinear system, it is possible to obtain an exact solution for a piecewise linear case and an exact response curve can be plotted (see subchapter 4.2.6). The curve shown in fig. 4.7 is only an approximation. Its shape, however, is very similar to the response curve of the conventional nonlinear system shown in fig. 4.5.

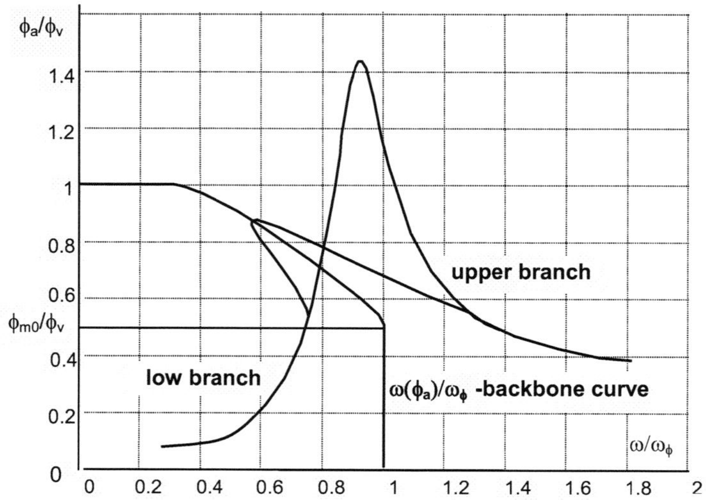

Fig. 4.7 Response curve of equivalently linearized piecewise linear system

4.2.2 Harmonic Balance Method

The next method to be considered here is called the harmonic balance method. The main idea is to search the solution in a form of a truncated Fourier series:

$$\phi(t) = b_1 \sin(\omega_1 t + \varphi_1) + b_2 \sin(\omega_2 t + \varphi_2) + \ldots \tag{4.28}$$

Amplitudes b_i and initial phases φ_i should be found by equalization of the terms that contain sine or cosine functions with the same frequency. The calculation procedure for this method consists of several expansions. The solution of the first expansion is a sine or cosine function with the frequency that is equal to the excitation frequency:

$$\phi(t) = b_1 \sin(\omega t + \varphi_1) \tag{4.29}$$

We consider a nonlinear system with the *GZ* curve presented with the cubic polynomial:

$$\ddot{\phi} + 2\delta\dot{\phi} + \omega_\phi^2 \phi - a_3 \phi^3 = \alpha_E \sin \omega t \tag{4.30}$$

After the substitution of the solution (4.29) into equation (4.30) we shall get:

$$\begin{aligned} &-b_1\omega^2 \sin(\omega t + \varphi_1) + 2\delta b_1 \omega \cos(\omega t + \varphi_1) + \omega_\phi^2 b_1 \sin(\omega t + \varphi_1) - \\ &- a_3 b_1^3 \sin^3(\omega t + \varphi_1) = \alpha_E \sin \omega t \end{aligned} \tag{4.31}$$

The term with the third power sine function can be expanded:

$$\sin^3(\omega t+\varphi_1)=\frac{1}{4}[3\sin(\omega t+\varphi_1)-\sin(3\omega t+3\varphi_1)] \tag{4.32}$$

The term that contains frequency 3ω should be truncated in the first expansion:

$$\begin{aligned}&-b_1\omega^2\sin(\omega t+\varphi_1)+2\delta b_1\omega\cos(\omega t+\varphi_1)+\\&\quad+b_1(\omega_\phi^2-0.75a_3b_1^2)\sin(\omega t+\varphi_1)=\alpha_E\sin\omega t\end{aligned} \tag{4.33}$$

Equation (4.33) can be transformed into a system of two algebraic equations by expanding the sine and cosine functions:

$$\begin{cases}-b_1\omega^2\cos\varphi_1+2\delta b_1\omega\sin\varphi_1+b_1(\omega_\phi^2-0.75a_3b_1^2)\cos\varphi_1=\alpha_E\\-b_1\omega^2\sin\varphi_1-2\delta b_1\omega\cos\varphi_1+b_1(\omega_\phi^2-0.75a_3b_1^2)\sin\varphi_1=0\end{cases} \tag{4.34}$$

The term $(\omega_\phi^2-0.75a_3b_1^2)$ can be considered as natural frequency versus amplitude $(\omega_\phi(b_1))^2$ or the backbone curve and the system (4.34) can be solved:

$$b_1=\frac{\alpha_E}{\sqrt{\left[(\omega_\phi(b_1))^2-\omega^2\right]^2+4\delta^2\omega^2}} \tag{4.35}$$

$$\varphi_1=\arctan\frac{2\delta\omega}{(\omega_\phi(b_1))^2-\omega^2} \tag{4.36}$$

The first expansion of the solution obtained by the harmonic balance method is quite similar to one by equivalent linearization. The difference is in the backbone curve. We express coefficient a_3 via the angle of vanishing stability, this difference becomes clear:

$$a_3=\frac{\omega_\phi^2}{\phi_v^2};\quad \omega_\phi(b_1)=\omega_{\phi 0}\sqrt{1-\frac{3}{4}\cdot\frac{b_1^2}{\phi_v^2}} \tag{4.37}$$

The backbone should reach a zero value for natural frequency when the amplitude is equal to the angle of vanishing stability (see subchapter 4.1), because the system is placed in an unstable equilibrium position and theoretically can stay there for an indefinite time. Thus, the period of free oscillations is equal to infinity and frequency is equal to zero correspondingly. The formula for the backbone curve (4.37) is not exact; it reaches a zero value of frequency when amplitude is equal to $(2/\sqrt{3})\cdot\phi_v$. This error is caused by the truncated terms of the Fourier series. The influence of this error on the response curve is not very large: see fig. 4.8.

Let us consider the second expansion of the harmonic balance method. It is more convenient to use exponential representation of harmonic functions for such analyses because of the significant size of the formulae.

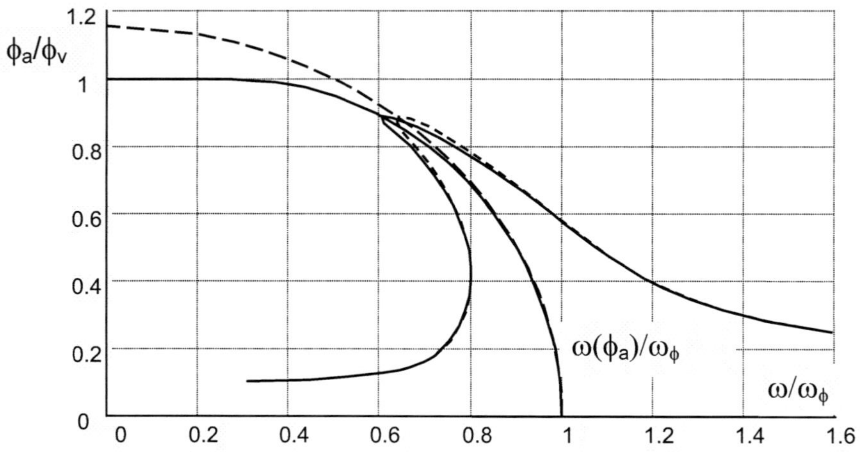

Fig. 4.8 Backbone lines and response curves obtained by equivalent linearization (solid lines) and the first expansion of harmonic balance method (dashed lines)

Using the well-known Euler's formulae:

$$\sin\omega t = \frac{1}{2i}\left(e^{i\omega t} - e^{-i\omega t}\right); \quad \cos\omega t = \frac{1}{2}\left(e^{i\omega t} + e^{-i\omega t}\right) \tag{4.38}$$

We obtain:

$$\begin{aligned}\phi(t) &= b_1 \sin(\omega t + \varphi_1) = C_1 \sin\omega t + C_2 \cos\omega t = \\ &= \frac{1}{2}(C_2 - iC_1)e^{i\omega t} + \frac{1}{2}(C_2 + iC_1)e^{-i\omega t} = b_{C1}e^{i\omega t} + CC\end{aligned} \tag{4.39}$$

Where b_{C1} is the complex amplitude of the first expansion solution and CC is the complex conjugate value. The complex amplitude is related with real amplitude and initial phase angle in the following way:

$$b_{C1} = C_2 - iC_1; \quad b_1 = |b_{C1}| = \sqrt{C_1^2 + C_2^2}; \quad \varphi_1 = \arg(b_{C1}) = \arctan\left(\frac{C_1}{C_2}\right) \tag{4.40}$$

We shall search for the second expansion solution in this form:

$$\phi(t) = b_{C1}e^{i\omega t} + b_{C3}e^{3i\omega t} + CC \tag{4.41}$$

Following the previously described procedure:

$$\begin{aligned}\dot{\phi}(t) &= i\omega b_{C1}e^{i\omega t} + 3i\omega b_{C3}e^{3i\omega t} + CC \\ \ddot{\phi}(t) &= -\omega^2 b_{C1}e^{i\omega t} - 9\omega^2 b_{C3}e^{3i\omega t} + CC \\ \phi^3(t) &= c_1 e^{i\omega t} + c_3 e^{3i\omega t} + c_5 e^{5i\omega t} + c_7 e^{7i\omega t} + c_9 e^{9i\omega t} + CC\end{aligned} \tag{4.42}$$

Where

$$\begin{aligned}&c_1 = 3\left(b_{C1}^2\bar{b}_{C1} + 2b_{C1}b_{C3}\bar{b}_{C3} + b_{C3}\bar{b}_{C1}^2\right); \; c_3 = b_{C1}^3 + 6b_{C1}b_{C3}\bar{b}_{C3} + 3b_{C3}^2\bar{b}_{C3} \\ &c_5 = 3\left(b_{C1}^2 b_{C3} + b_{C3}^2\bar{b}_{C1}\right); \quad c_7 = 3b_{C3}^2 b_{C1}; \quad c_9 = b_{C3}^3\end{aligned} \tag{4.43}$$

All terms that contain frequencies higher than 3ω should be truncated. Equalization of amplitudes of harmonics with the same frequencies allows building-up of the following system of equations:

$$\begin{cases} -b_{C1}\omega^2 + 2i\delta b_{C1}\omega + b_{C1}\omega_\phi^2 - a_3 c_1 = \alpha_E \\ -9b_{C3}\sigma^2 + 2i\delta b_{C3}\omega + b_{C3}\omega_\phi^2 - a_3 c_3 = 0 \end{cases} \quad (4.44)$$

This is the system of two nonlinear algebraic equations, i.e. two complex variables b_{C1} and b_{C3} and it has to be solved numerically.

4.2.3 Perturbation Method

The perturbation method is very popular for analysis of nonlinear systems. Our description of this method will be very brief, as comprehensive literature on this subject is available: [Bogolubov and Mitropolsky, 1961; Nayfeh, 1973, etc]. This method was used for nonlinear roll and ship stability analysis by Wellicome [1975], Cardo, *et al* [1981, 1984] and others.

There are many different versions of the perturbation method. We shall consider the simplest example to illustrate the technique. We continue using roll equation (4.30) with 3rd order power approximation of the restoring term. The solution is represented in the form of a series of successively smaller terms:

$$\phi = \phi_0 + \varepsilon\phi_1 + \varepsilon^2\phi_2 + \varepsilon^3\phi_3 + ... \quad (4.45)$$

Where ε is a small (bookkeeping) parameter. We also assume that:

$$\delta = \varepsilon\nu; \quad \alpha_E = \alpha_{E0}\varepsilon; \quad a_3 = a_{30}\varepsilon \quad (4.46)$$

Frequency of response is related with natural frequency as:

$$\omega_\phi^2 = \omega^2 - \varepsilon\omega_1 - \varepsilon^2\omega_2 - \varepsilon^3\omega_3 - ... \quad (4.47)$$

Substitution of (4.46) and (4.47) into (4.30) yields:

$$\ddot{\phi} + 2\varepsilon\nu\dot{\phi} + \left(\omega^2 - \varepsilon\omega_1 - \varepsilon^2\omega_2 - \varepsilon^3\omega_3 - ...\right)\cdot\phi - a_{30}\varepsilon\phi^3 = \alpha_{E0}\varepsilon\sin\omega t \quad (4.48)$$

We truncate all the series up to the second power of small parameter ε. The third order of the solution for will look like:

$$\phi^3 = \phi_0^3 + 3\varepsilon\phi_1\phi_0^2 + 3\varepsilon^2\left(\phi_1\phi_0^2 + \phi_0\phi_1^2\right) \quad (4.49)$$

Then we substitute (4.45) and (4.49) into (4.48) and equalize the right-hand terms and left-hand terms with the same power of small parameter ε:

$$\varepsilon^0: \quad \ddot{\phi}_0 + \omega^2\phi_0 = 0 \quad (4.50)$$

$$\varepsilon^1: \quad \ddot{\phi}_1 + \omega^2\phi_1 = \alpha_{E0}\sin\omega t - 2\nu\dot{\phi}_0 + \omega_1\phi_0 - a_{30}\phi_0^3 \quad (4.51)$$

$$\varepsilon^2: \quad \ddot{\phi}_2 + \omega^2\phi_2 = -2\nu\dot{\phi}_1 - \omega_1\phi_1 + \omega_2\phi_0 - 3a_{30}\phi_0^2\phi_1 \quad (4.52)$$

A solution of the first equation (4.50) is trivial:

$$\phi_0 = b_0 \sin(\omega t + \varphi) \tag{4.53}$$

After the substitution of the solution (4.53) into the second equation (4.51) we get:

$$\ddot{\phi}_1 + \omega^2 \phi_1 = c_1 \sin(\omega t + \gamma) + 0.25 b_0^3 a_{30} \sin(3\omega t + 3\varphi) \tag{4.54}$$

With:

$$c_1 = \sqrt{c_{S1}^2 + c_{C1}^2}; \quad \gamma = \arctan\left(c_{S1} / c_{C1}\right) \tag{4.55}$$

$$c_{S1} = \alpha_{E0} + 2\nu b_0 \omega \sin\varphi + \omega_x b_0 \cos\varphi \tag{4.56}$$

$$c_{C1} = -2\nu b_0 \omega \cos\varphi + \omega_x b_0 \sin\varphi \tag{4.57}$$

$$\omega_x = \omega_1 - 0.75 a_{30} b_0^2 \tag{4.58}$$

The partial solution of equation (4.54) can be expressed as:

$$\phi_1 = -0.5 c_1 \omega t \cos(\omega t + \gamma) + b_1 \sin(3\omega t + 3\varphi + \beta) \tag{4.59}$$

The equation (4.59) contains the secular term $-0.5 c_1 \omega t \cos(\omega t + \gamma)$, its magnitude increases proportionally to time. If a periodic solution is required, the secular term should be eliminated. It can be realized if $c_1 = 0$. Consequently the following system should be satisfied:

$$\begin{cases} 2\nu b_0 \omega \sin\varphi + b_0 \omega_x \cos\varphi = \alpha_{E0} \\ -2\nu b_0 \omega \cos\varphi + b_0 \omega_x \sin\varphi = 0 \end{cases} \tag{4.60}$$

It is not difficult to see that

$$b_0 = \frac{\alpha_{E0}}{\sqrt{\omega_x^2 + 4\nu^2 \omega^2}}; \quad \varphi = \arctan \frac{2\nu\omega}{\omega_x^2} \tag{4.61}$$

Taking into account (4.46), (4.47) and (4.58):

$$\omega_x = \frac{1}{\varepsilon}\left[\omega^2 - \omega_\phi^2(b_0)\right]; \quad \omega_\phi(b_0) = \sqrt{\omega_{\phi0}^2 - 0.75 a_{30} b_0^2} \tag{4.62}$$

Formulae (4.61) can be rewritten:

$$b_0 = \frac{\alpha_E}{\sqrt{\left[\left(\omega_\phi(b_0)\right)^2 - \omega^2\right]^2 + 4\omega^2 \delta^2}}; \quad \varphi = \arctan \frac{2\delta\omega}{\left(\omega_\phi(b_0)\right)^2 - \omega^2} \tag{4.63}$$

We recognize $\omega_\phi(b_0)$ as the backbone curve. It can be clearly seen that the first expansion of the perturbation method gives the same result as the harmonic balance method. The general form of the formulae for amplitude and initial phase angle are similar to the results of the equivalent linearization method. A reason for such similarity is in assuming a harmonic form in the solution of the steady state oscillation of the nonlinear system.

The second order expansion can be produced by the same sequence. Formula for ϕ_1 (4.59) should be used in formula (4.52) taking into account the condition c_1=0 and the general form of the second expansion solution ϕ_1 can be obtained. Elimination of secular terms yields a system of equations relative to the second order amplitude b_1 and the initial phase angle φ. The second order solution has a harmonic term with frequency 3ω, and consequently this expansion is interesting for high order resonance analysis.

Another important feature regarding the perturbation method is the possibility in obtaining the transient motion. This can be done if the amplitude b_0 and initial phase angle φ will be considered as functions of time. A similar procedure will be shown in the next section during consideration of the method of multiple scales.

4.2.4 Method of Multiple Scales

A method of multiple scales is a version of the perturbation method. A detailed description of this method can be found in [Nayfeh, 1973, 1981]. The technique of multiple scales is illustrated here using roll equation (4.30) with 3rd power approximation of the restoring term. The solution follows Nayfeh and Khdeir [1986, 1986a].

We consider the frequency domain near the main resonance:

$$\omega^2 = \omega_\phi^2 + \varepsilon \cdot \sigma \tag{4.64}$$

Variable σ is called the tuning parameter; we also use our previous assumption to present the coefficients (4.46):

$$\delta = \varepsilon \nu; \quad \alpha_E = \alpha_{E0}\varepsilon; \quad a_3 = a_{30}\varepsilon$$

The solution is searched in the following form:

$$\phi = \phi_0(T_0, T_1, T_2) + \varepsilon\phi_1(T_0, T_1, T_2) + \varepsilon^2\phi_2(T_0, T_1, T_2) \tag{4.65}$$

T_0, T_1 and T_2 are different time scales. They correspond to different orders of expansion:

$$T_0 = t; \quad T_1 = \varepsilon t; \quad T_2 = \varepsilon^2 t \tag{4.66}$$

Now, instead of one independent variable, time t, we have three. Time derivatives can be expressed through partial derivatives by new independent variables:

$$\frac{d}{dt} = D_0 + \varepsilon D_1 + \varepsilon^2 D_2 + \ldots; \quad \frac{d^2}{dt^2} = D_0^2 + 2\varepsilon D_1 D_0 + \varepsilon^2\left[2D_0 D_2 + D_1^2\right] + \ldots \tag{4.67}$$

With $D_i = \dfrac{\partial}{\partial T_i}$ is a partial derivative operator.

After substitution of solution (4.65) into equation (4.30) with taking into account (4.66) and (4.67), we shall get a system of differential equations, each of which corresponds to a certain power of ε:

$$\varepsilon^0: \quad D_0^2\phi_0 + \omega^2\phi_0 = 0 \tag{4.68}$$

$$\varepsilon^1: \quad D_0^2\phi_1 + \omega^2\phi_1 = \alpha_{E0}\sin\omega t - 2D_0D_1\phi_0 - 2\nu D_0\phi + \sigma\phi_0 + a_{30}\phi_0^3 \tag{4.69}$$

$$\begin{aligned}\varepsilon^2: \quad & D_0^2\phi_2 + \omega^2\phi_2 = -2D_0D_1\phi_1 - 2D_0D_1\phi_0 - 2D_1^2\phi_0 + \sigma\phi_1 - \\ & -2\nu D_0\phi_1 - 2\nu D_1\phi_0 + 3a_{30}\phi_0^2\phi_1\end{aligned} \tag{4.70}$$

We search for the first expansion solution in the following form:

$$\phi_0 = A(T_1, T_2)\exp(i\omega T_0) + CC \tag{4.71}$$

Here CC is a complex conjugate term. We assume that amplitude $A(T_1,T_2)$ is a function of time scales of the first and the second order. This assumption is natural: amplitude changes slower than the process itself. The same assumption lies in the background of averaging or the van-der-Poul method, where amplitude is also considered as a slow function of time in comparison with the oscillation process itself, [Andronov, *et al* 1966]. The system of equations for the method of multiple scales (4.68-4.70) has the same principle structure as the system of equations for the perturbation method (4.50-4.52). So, secular terms should also be present in the formula for the second order expansion amplitude and the first order amplitude can be obtained through elimination of the secular terms. Substitution of (4.71) in (4.69) yields:

$$\begin{aligned} & D_0^2\phi_1 + \omega^2\phi_1 = \\ & = \left[0.5\alpha_{E0} - 2i\omega D_1A - 2\nu i\omega A + \sigma A + 3a_{30}A^2\bar{A}\right]\exp(i\omega T_0) + \\ & + a_{30}A^3\exp(3i\omega T_0) + CC \end{aligned} \tag{4.72}$$

It is enough for elimination of secular terms to require:

$$0.5\alpha_{E0} - 2i\omega D_1A - 2\nu i\omega A + \sigma A + 3a_{30}A^2\bar{A} = 0 \tag{4.73}$$

It is a differential equation, i.e. $A(T_1)$, because amplitude A is a function of T_1 only for the first expansion solution. The following form is proposed for solution of the differential equation (4.73):

$$A(T_1) = \frac{1}{2}c(T_1)\exp[i\beta(T_1)] \tag{4.74}$$

Substitution of (4.74) in (4.73) and separation of the real and imagined part gives the following system of differential equations:

$$\begin{cases} \dfrac{dc}{dT_1} = -\nu c - \dfrac{\alpha_{E0}}{2\omega}\sin\beta \\ c\dfrac{d\beta}{dT_1} = -\dfrac{\sigma c}{2\omega} - \dfrac{3}{8\omega}a_{30}c^3 - \dfrac{\alpha_{E0}}{2\omega}\cos\beta \end{cases} \tag{4.75}$$

This is the system of two nonlinear differential equations. It can be solved numerically. Functions $c(T_1)$ and $\beta(T_1)$ are the result of such a solution. Knowledge of these functions allows the transient solution of the first expansion to be built. Initial conditions of roll equation (4.30) $\phi(t=0)$ and $\dot{\phi}(t=0)$ are related with the initial condition of the system (4.75) in the following way:

$$\begin{cases} \phi(t=0) = 0.5c(T_1=0)\exp[i\beta(T_1=0)]\exp(i\omega t) + CC \\ \dot{\phi}(t=0) = D_0\phi + D_1\phi(T_1=0) + CC \end{cases} \tag{4.76}$$

If a steady state solution is required, then derivatives of c and β should be equal to zero; this condition transforms the system of nonlinear differential equations (4.75) into the following system of nonlinear algebraic equations:

$$\begin{cases} \nu c + \dfrac{\alpha_{E0}}{2\omega}\sin\beta = 0 \\ \dfrac{\sigma c}{2\omega} + \dfrac{3}{8\omega}a_{30}c^3 + \dfrac{\alpha_{E0}}{2\omega}\cos\beta = 0 \end{cases} \tag{4.77}$$

The system (4.77) can be resolved with use of formula (4.64). When it is substituted in (4.77) a "backbone" term is revealed:

$$\begin{cases} \nu c + \dfrac{\alpha_{E0}}{2\omega}\sin\beta = 0 \\ \dfrac{c}{2\varepsilon\omega}\left(\omega^2 - \omega_x^2\right) + \dfrac{\alpha_{E0}}{2\omega}\cos\beta = 0 \end{cases} \tag{4.78}$$

With:

$$\omega_x = \sqrt{\omega_\phi^2 - 0.75a_{30}c^2} \tag{4.79}$$

The solution of the system (4.78) can be expressed as:

$$c = \frac{\alpha_E}{\sqrt{\left[\omega_x^2 - \omega^2\right]^2 + 4\delta^2\omega^2}}; \quad \beta = \arctan\frac{2\omega\delta}{\omega_x^2 - \omega^2} \tag{4.80}$$

We end up with the same solution that was obtained by the first expansion of the harmonic balance method and perturbation method. It is logical because we have used harmonic approximation for the first expansion near the principal resonance in all the described methods.

Following Nayfeh and Khdeir [1986, 1986a], we continue with the second expansion of the method of multiple scales. The solution is supposed to be:

$$\phi_1 = kA^3\exp(3i\omega T_0) + CC \quad \text{with} \quad k = -\frac{1}{8}\frac{a_{30}}{\omega^2} \tag{4.81}$$

Substitution of this formula into the third expansion equation (4.70) yields:

$$\begin{aligned} D_0^2\phi_2 + \omega^2\phi_2 = & -18i\omega A^2 D_1\exp(3i\omega T_0) - \\ & -2i\omega D_2 A\exp(i\omega T_0) - D_1^2 A\exp(i\omega T_0) + \\ & +\omega kA^3\exp(3i\omega T_0) - 2\nu kA^3\cdot 3i\omega\exp(3i\omega T_0) \\ & -2\nu D_1 A\exp(i\omega T_0) + 3a_{30}kA^5\exp(5i\omega T_0) \\ & +6a_{30}A^4\bar{A}\exp(3i\omega T_0) + 3a_{30}kA^3\bar{A}^2\exp(i\omega T_0) + CC \end{aligned} \tag{4.82}$$

Elimination of secular terms in the third expansion solution gives an equation for amplitude of the second expansion:

$$-2i\omega D_2 A - D_1^2 A - 2\nu D_1 A + 3a_{30} k\bar{A}^2 A^3 + CC = 0 \tag{4.83}$$

Here, amplitude A should be considered as a function of both scales T_1 and T_2. Derivative $D_1 A$ can be taken from the first expansion solution (4.73). It can be expressed as:

$$D_1 A = \frac{1}{2i\omega}\left\{(-2\nu i\omega + \sigma)A + 3a_{30}A^2\bar{A} + 0.5\alpha_{E0}\right\} + CC \tag{4.84}$$

The second derivative $D_1^2 A$ can be found from (4.84):

$$\begin{aligned} D_1^2 A = &-\frac{\omega_0 \alpha_{E0}}{8\omega^2} + \frac{i\alpha_{E0}\nu}{4\omega^2} + \left[-\frac{\omega_0^2}{4\omega^2} + \frac{i\omega_0 \nu}{\omega} + \nu^2\right]A + \\ &+ \frac{3\alpha_{E0}a_{30}}{8\omega^2}A^2 - \frac{3\alpha_{E0}a_{30}}{8\omega^2}A\bar{A} + \\ &+ \left[-\frac{3a_{30}\sigma}{2\omega^2} + \frac{6a_{30}i\nu}{\omega}\right]A^2\bar{A} - \frac{9a_{30}^2}{4\omega^2}A^3\bar{A}^2 + CC \end{aligned} \tag{4.85}$$

Substitution of (4.84) and (4.85) into (4.83) yields an equation for $D_2 A$. Combining it with equation (4.84), we get a system of differential equations, i.e. $A(T_1,T_2)$:

$$\begin{cases} 2i\omega D_2 A = \left[\frac{\omega_0 \alpha_{E0}}{8\omega^2} + \frac{i\alpha_{E0}\nu}{4\omega}\right] + \left[\frac{\omega_0^2}{4\omega^2} + \nu^2\right]A - \frac{3\alpha_{E0}a_{30}}{8\omega^2}A^2 + \\ + \frac{3\alpha_{E0}a_{30}}{8\omega^2}A\bar{A} + \left[\frac{3a_{30}\omega_0}{2\omega^2} - \frac{3a_{30}i\nu}{\omega}\right]A^2\bar{A} - \frac{15a_{30}^2}{4\omega^2}A^2\bar{A}^2 \\ 2i\omega D_1 A = (-2\nu i\omega + \sigma)A + 3a_{30}A^2\bar{A} + \frac{\alpha_{E0}}{2} \end{cases} \tag{4.86}$$

For the second expansion, the following form of $A(T_1, T_2)$ is expected:

$$A(T_1) = \frac{1}{2}c(T_1,T_2)\exp[i\beta(T_1,T_2)] \tag{4.87}$$

Using the full derivative operator (4.67), we can transform the system (4.86) into one equation in full derivatives:

$$\frac{dA}{dt} = \frac{\partial A}{\partial T_0} + \varepsilon\frac{\partial A}{\partial T_1} + \varepsilon^2\frac{\partial A}{\partial T_2} \tag{4.88}$$

It is clear that:

$$\frac{\partial A}{\partial T_0} = 0$$

After separation of the real and imaginary parts of (4.88), we get a system of differential equations:

$$\begin{cases} \dot{c} = \varepsilon\left(-\nu c - \dfrac{\alpha_{E0}}{2\omega}\sin\beta\right) + \\ \quad + \varepsilon^2\left[-\dfrac{3a_{30}\nu}{2\omega^2}c^3 + \left(-\dfrac{\sigma\alpha_{E0}}{8\omega^3} - \dfrac{9\alpha_{E0}a_{30}}{32\omega^3}c^2\right)\sin\beta + \dfrac{\alpha_{E0}\nu}{4\omega^2}\cos\beta\right] \\ c\dot{\beta} = \varepsilon\left(-\dfrac{\omega_0 c}{2\omega} - \dfrac{3a_{30}c^3}{8\omega} - \dfrac{\alpha_{E0}}{2\omega}\cos\beta\right) + \\ \quad + \varepsilon^2\left[-\left(\dfrac{\omega_0^2}{8\omega^3} + \dfrac{\nu^2}{2\nu}\right)c - \dfrac{3a_{30}\omega_0}{16\omega^3}c^3 + \dfrac{15a_{30}\omega_0}{256\omega^3}c^5 - \right. \\ \quad \left. - \dfrac{\alpha_{E0}\nu}{4\nu^2}\sin\beta + \left(-\dfrac{\sigma\alpha_{E0}}{8\omega^3} - \dfrac{3\alpha_{E0}a_{30}}{32\omega^3}c^2\right)\cos\beta\right] \end{cases} \tag{4.89}$$

If the steady state solution is required then, $\dot{c} = 0$, $\dot{\beta} = 0$ and the system of differential equations (4.89) becomes the system of nonlinear algebraic equations:

$$\begin{cases} \left(-\nu c - \dfrac{\alpha_{E0}}{2\omega}\sin\beta\right) + \\ \quad + \left[-\dfrac{3a_{30}\nu}{2\omega^2}c^3 + \left(-\dfrac{\sigma\alpha_{E0}}{8\omega^3} - \dfrac{9\alpha_{E0}a_{30}}{32\omega^3}c^2\right)\sin\beta + \dfrac{\alpha_{E0}\nu}{4\omega^2}\cos\beta\right] = 0 \\ -\dfrac{\sigma c}{2\omega} - \dfrac{3a_{30}c^3}{8\omega} - \dfrac{\alpha_{E0}}{2\omega}\cos\beta - \left(\dfrac{\sigma^2}{8\omega^3} + \dfrac{\nu^2}{2\nu}\right)\cdot c - \dfrac{3a_{30}\omega_0}{16\omega^3}c^3 + \\ \quad + \dfrac{15a_{30}\sigma}{256\omega^3}c^5 - \dfrac{\alpha_{E0}\nu}{4\nu^2}\sin\beta + \left(-\dfrac{\sigma\alpha_{E0}}{8\omega^3} - \dfrac{3\alpha_{E0}a_{30}}{32\omega^3}c^2\right)\cos\beta = 0 \end{cases} \tag{4.90}$$

The bookkeeping parameter, ε, is assumed as a unit here. A final solution after the second expansion is:

$$\phi(t) = c\cdot\cos(\omega t + \beta) + \frac{a_3}{8\omega^2}c^3\cos(3\omega t + 3\beta) \tag{4.91}$$

4.2.5 Numerical Method

Let us consider numerical methods for the solution of the nonlinear roll equation. These methods are fast and reliable, also they put much less restrictions on the model. We do not need any approximations for the *GZ* curve or damping term. We can include other degrees of freedom as well. Nowadays, numerical methods have become the main tool for research and design, while all other methods play an auxiliary role, helping to understand and interpret the results of the numerical solution.

Numerical methods are intended to solve differential equations and systems of differential equations of the first order. At the same time, all motion equations have second order derivatives, so they have to be transformed to the system of the first order equations.

Consider, for example, a general nonlinear roll equation such as equation (3.270) from chapter 3:

$$\ddot{\phi} + 2\delta_{\phi}\dot{\phi} + \omega_{\phi}^{2} f(\phi) = \alpha_E \sin(\omega t + \varphi_E) \tag{4.92}$$

Let's introduce a new function:

$$\vartheta = \dot{\phi}; \quad \Rightarrow \quad \dot{\vartheta} = \ddot{\phi} \tag{4.93}$$

Its substitution into the original equation (4.92) allows its consideration as a system of differential equations of the first order:

$$\begin{cases} \dot{\vartheta} = \alpha_E \sin(\omega t + \varphi_E) - 2\delta_{\phi}\vartheta - \omega_{\phi}^{2} f(\phi) \\ \dot{\phi} = \vartheta \end{cases} \tag{4.94}$$

It is convenient to present system (4.94) in vector form:

$$\dot{\vec{Y}} = F(\vec{Y}, t) \tag{4.95}$$

$$\dot{\vec{Y}} = \begin{pmatrix} \dot{Y}_1 \\ \dot{Y}_2 \end{pmatrix} = \begin{pmatrix} \dot{\vartheta} \\ \dot{\phi} \end{pmatrix}; \qquad \vec{Y} = \begin{pmatrix} Y_1 \\ Y_2 \end{pmatrix} = \begin{pmatrix} \vartheta \\ \phi \end{pmatrix};$$

$$\vec{F}(\vec{Y}, t) = \begin{pmatrix} \alpha_E \sin(\omega t + \varphi_E) - 2\delta_{\phi} Y_1 - \omega_{\phi}^{2} f(Y_2) \\ Y_1 \end{pmatrix} = \tag{4.96}$$

$$= \begin{pmatrix} \alpha_E \sin(\omega t + \varphi_E) - 2\delta_{\phi}\vartheta - \omega_{\phi}^{2} f(\phi) \\ \vartheta \end{pmatrix}$$

For the sake of simplicity, let us consider first one differential equation of the first order.

$$\dot{y} = F(y, t) \tag{4.97}$$

Suppose, we need to know the solution of the equation at the interval from 0 to T with step Δt. The initial condition is:

$$y(t = 0) = y_0 \tag{4.98}$$

The simplest (but not the best) way to get the solution is the Euler method that is implemented in the following calculating scheme [Bronshtein and Semendyayev, 1997]:

$$\begin{aligned} &t_0 = 0 \cdot \Delta t; \quad \dot{y}_0 = F(y_0, t_0); \quad y_1 = y_0 + \Delta t \cdot \dot{y}_0 \\ &t_1 = 1 \cdot \Delta t; \quad \dot{y}_1 = F(y_1, t_1); \quad y_2 = y_1 + \Delta t \cdot \dot{y}_1 \\ &\ldots. \\ &t_i = i \cdot \Delta t; \quad \dot{y}_i = F(y_i, t_i); \quad y_{i+1} = y_i + \Delta t \cdot \dot{y}_i \\ &\ldots \end{aligned} \tag{4.99}$$

The calculating scheme of the Euler method can be easily interpreted graphically, see fig. 4.9. It has to be noted that the time step does not have to be constant. It just has to be small enough to provide the desired accuracy.

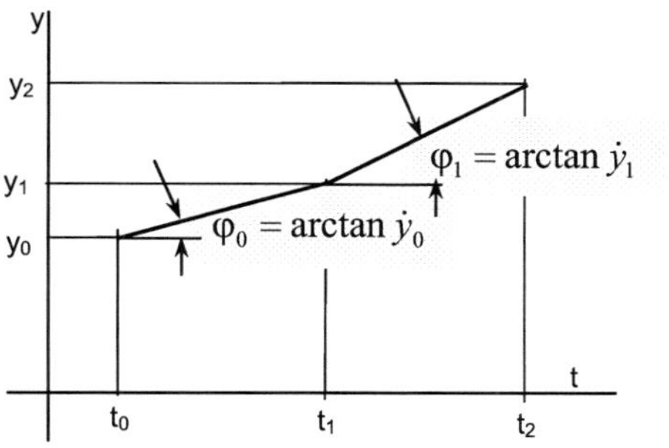

Fig. 4.9 Graphical interpretation of the Euler method

Although we are not discussing problems of accuracy and errors here, the Euler method is not optimal from the point of view of accuracy [Bronshtein and Semendyayev, 1997]. The Runge-Kutta method is one of the most popular ways for a numerical solution of differential equations. It can be considered as generalization of the Euler method: the difference is that function $F(y,t)$ is substituted by the more general expression $\widetilde{F}(y,t)$:

$$\widetilde{F}(y,t) = \frac{1}{6}\left(k_1(y,t) + 2k_2(y,t) + 2k_3(y,t) + k_4(y,t)\right) \tag{4.100}$$

With:

$$\begin{aligned} &k_1(y,t) = F(y,t) \\ &k_2(y,t) = F\left(y + 0.5\Delta t \cdot k_1(y,t)\,,\, t + 0.5\Delta t\right) \\ &k_3(y,t) = F\left(y + 0.5\Delta t \cdot k_2(y,t)\,,\, t + 0.5\Delta t\right) \\ &k_4(y,t) = F\left(y + \Delta t \cdot k_3(y,t)\,,\, t + \Delta t \cdot\right) \end{aligned} \tag{4.101}$$

The calculation scheme is the same as that in the Euler method:

$$\begin{aligned} &t_0 = 0 \cdot \Delta t; \quad \dot{y}_0 = \widetilde{F}(y_0, t_0); \quad y_1 = y_0 + \Delta t \cdot \dot{y}_0 \\ &\ldots. \\ &t_i = i \cdot \Delta t; \quad \dot{y}_i = \widetilde{F}(y_i, t_i); \quad y_{i+1} = y_i + \Delta t \cdot \dot{y}_i \\ &\ldots. \end{aligned} \tag{4.102}$$

Now, let us return to the roll equation in vector form (4.95). All the operations in both methods can be applied to vectors, so the generalization of formulae (4.100)-(4.102) is straightforward:

$$\begin{aligned} &t_0 = 0 \cdot \Delta t; \quad \dot{\vec{Y}}_0 = \vec{\widetilde{F}}(\vec{Y}_0, t_0); \quad \vec{Y}_1 = \vec{Y}_0 + \Delta t \cdot \dot{\vec{Y}}_0 \\ &\ldots. \\ &t_i = i \cdot \Delta t; \quad \dot{\vec{Y}}_i = \vec{\widetilde{F}}(\vec{Y}_i, t_i); \quad \vec{Y}_{i+1} = \vec{Y}_i + \Delta t \cdot \dot{\vec{Y}}_i \\ &\ldots. \end{aligned} \tag{4.103}$$

With:

$$\vec{\widetilde{F}}(\vec{Y},t) = \frac{1}{6}\left(\vec{K}_1(\vec{Y},t) + 2\vec{K}_2(\vec{Y},t) + 2\vec{K}_3(\vec{Y},t) + \vec{K}_4(\vec{Y},t)\right) \tag{4.104}$$

Where:

$$\begin{aligned}
\vec{K}_1(\vec{Y},t) &= \vec{F}(\vec{Y},t) \\
\vec{K}_2(\vec{Y},t) &= \vec{F}\left(\vec{Y}+0.5\Delta t\cdot\vec{K}_1(\vec{Y},t)\,,\,t+0.5\Delta t\right) \\
\vec{K}_3(\vec{Y},t) &= \vec{F}\left(\vec{Y}+0.5\Delta t\cdot\vec{K}_2(\vec{Y},t)\,,\,t+0.5\Delta t\right) \\
\vec{K}_4(\vec{Y},t) &= \vec{F}\left(\vec{Y}+\Delta t\cdot\vec{K}_3(\vec{Y},t)\,,\,t+\Delta t\right)
\end{aligned} \qquad (4.105)$$

Formulae (4.103)-(4.104) complete the calculation scheme for the nonlinear roll equation. It is important to note that there is no limit on the dimensions of vector $\vec{Y}$ and vector valued function $\vec{F}(\vec{Y},t)$, so other degrees of freedom may be easily included if necessary.

4.2.6 Steady State Solution of Piecewise Linear System

We complete our study of the methods for steady state solutions by demonstration that an exact solution is available for a piecewise linear system. Despite the fact that a piecewise linear system consists of two linear solutions, its synchronization behavior (or behavior while reaching steady state) is rather nonlinear. We mean that there is no clear definition when the piecewise linear system reaches a steady state mode since oscillation with the natural frequency is generated each time the boundary between linear ranges is crossed.

Let us consider the general solution of a piecewise linear system. We consider further the case with a symmetric restoring term. General formulae are available from [Belenky, 1999], however they are quite complex. So the forced piecewise linear system is described by the following differential equation:

$$\ddot{\phi}+2\delta\dot{\phi}+\omega_\phi^2 f_L^*(\phi)=\alpha_E\sin(\omega t+\varphi_E) \qquad (4.106)$$

The piecewise linear term is expressed by formula (4.7) and shown in fig.4.3:

$$\phi=\begin{cases}\phi_{0a}e^{-\delta t}\sin(\omega_0 t+\varepsilon)+q_a\sin(\omega t+\beta_q+\varphi_0) & \phi\in\left[-\phi_{m0};\phi_{m0}\right] \\ Ae^{\lambda_1 t}+Be^{\lambda_2 t}+p_a\sin(\omega t+\beta_p+\varphi_1) & \phi\in\left]\phi_{m0};\phi_{m1}\right[\end{cases} \qquad (4.107)$$

With:

Amplitudes and phases of partial solutions:

$$q_a=\frac{\alpha_E}{\sqrt{\left(k_{f0}\omega_\phi^2-\omega^2\right)^2+4\omega^2\delta^2}};\quad \beta_0=\arctan\frac{2\delta\omega}{k_{f0}\omega_\phi^2-\omega^2} \qquad (4.108)$$

$$p_a=\frac{\alpha_E}{\sqrt{\left(k_{f1}\omega_\phi^2+\omega^2\right)^2+4\delta^2\omega^2}};\quad \beta_1=\arctan\frac{2\delta\omega}{k_{f1}\omega_\phi^2+\omega^2} \qquad (4.109)$$

Additional phase angles φ_0 and φ_1 are necessary to keep continues time for the whole solution (time t is assumed to be local for each range).

Arbitrary constants for $\phi \in [-\phi_{m0};\phi_{m0}]$:

$$\phi_{0a} = \frac{1}{\omega_0}\sqrt{\left[\delta(\phi_0 - q_0) + (\dot{\phi}_0 - \dot{q}_0)\right]^2 + \omega_0^2(\phi_0 - q_0)^2} \quad (4.110)$$

$$\varepsilon_0 = \arctan\frac{\omega_0(\phi_0 - q_0)}{\delta(\phi_0 - q_0) + (\dot{\phi}_0 - \dot{q}_0)} \quad (4.111)$$

With:

$$\omega_0 = \sqrt{\omega_\phi^2 k_{f0} - \delta^2} \quad (4.112)$$

Variables $\phi_0, \dot{\phi}_0$ are initial conditions, when $t = 0$.

Variables $q_0, \dot{q}_0$ are values of the partial solution and its derivative, when $t = 0$.

Arbitrary constants for $\phi \in]\phi_{m0};\phi_{m1}[$:

$$A = \frac{(\dot{\phi}_1 - \dot{p}_1) - \lambda_2(\phi_1 - p_1)}{\lambda_1 - \lambda_2}; \quad B = -\frac{(\dot{\phi}_1 - \dot{p}_1) - \lambda_1(\phi_1 - p_1)}{\lambda_1 - \lambda_2} \quad (4.113)$$

With:

$$\lambda_{1,2} = -\delta \pm \sqrt{\omega_\phi^2 k_{f1} + \delta^2} \quad (4.114)$$

Variables $\phi_1, \dot{\phi}_1$ are initial conditions on the border of the range No. 1, when $t = 0$.

Variables $p_1, \dot{p}_1$ are values of partial solution on the border of the range No. 1, when t=0.

The only difference between transition and steady state regimes are values of crossing velocities and periods of time spent in different ranges of the piecewise linear term. So, if we find such figures that provide a periodic solution with excitation frequency, the steady state problem will be solved. These conditions can be formalized as a system of simultaneous algebraic equations. If we look at an unbiased case, it is enough to consider just half of the period [Belenky, 2000a]:

$$\begin{cases} f_0(T_0, \dot{\phi}_0, \varphi_0) = \phi_{m0} \\ \dot{f}_0(T_0, \dot{\phi}_0, \varphi_0) = \dot{\phi}_1 \\ f_1(T_1, \dot{\phi}_1, \varphi_0 + \omega T_0) = \phi_{m0} \\ \dot{f}_1(T_1, \dot{\phi}_1, \varphi_0 + \omega T_0) = -\dot{\phi}_0 \\ T_0 + T_1 = \pi \cdot \omega^{-1} \end{cases} \quad (4.115)$$

Here, functions f_0 and f_1 are solutions for (4.106) at the first and second ranges expressed explicitly through the only unknown initial condition – angular velocity at the boundary crossing, see also fig.4.10:

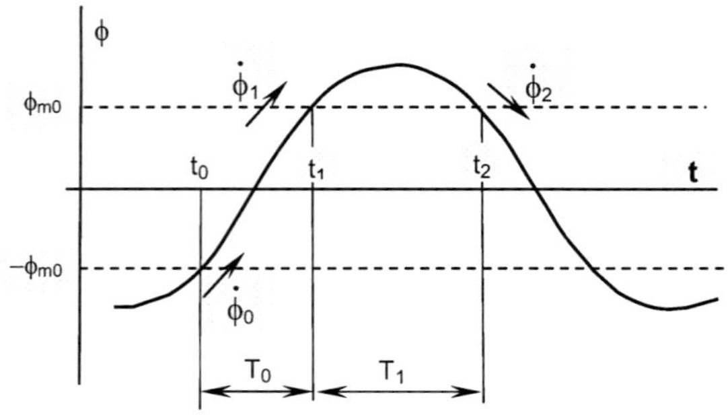

Fig. 4.10 Steady state motion of piecewise linear system

$$f_0(T_0, \dot{\phi}_0, \varphi_0) = \phi_{0a}(\dot{\phi}_0)e^{-\delta t}\sin\left(\omega_0 t + \varepsilon(\dot{\phi}_0)\right) + q_a \sin(\omega t + \beta_q + \varphi_0) \quad (4.116)$$

$$f_1(T_1, \dot{\phi}_1, \varphi_1) = A(\dot{\phi}_1)e^{\lambda_1 t} + B(\dot{\phi}_1)e^{\lambda_2 t} + p_a \sin(\omega t + \beta_p + \varphi_1) + \phi_V \quad (4.117)$$

The system (4.115) can be solved relative to unknown values T_0, T_1, $\dot{\phi}_0, \dot{\phi}_1$, and φ_0 using any appropriate numerical method. The results of the equivalent linearization are used for calculation of initial values of the unknown values, which makes calculations more fast and simple.

The resulting response curve is shown in fig. 4.11. As it could be seen from this figure, it has a quite conventional form, including the area, where three amplitudes correspond to one excitation frequency. We call this steady state solution exact, despite the numerical method was used to calculate crossing characteristics. Accuracy is still controllable: we always can substitute these figures into system (4.115) and check how the solution turns equations into equalities.

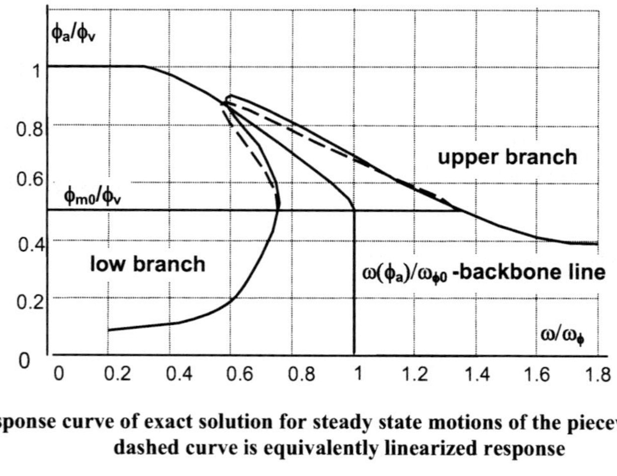

Fig 4.11 Response curve of exact solution for steady state motions of the piecewise linear system, dashed curve is equivalently linearized response

4.3 Stability of Equilibrium

4.3.1 Identification of Equilibria

We already defined stable, unstable and indifferent equilibria in subchapter 8.3 of [Kobylinski and Kastner 2003]. That definition was based on physics. Further, we will need a more formal description of stable and unstable equilibria. We also are going to introduce several mathematical concepts that will be used in the following chapters. At the same time, again, all these figures are just a more formal way to express the same physical contents.

Consider the nonlinear roll equation in general form (4.21):

$$\ddot{\phi}+2\delta\dot{\phi}+\omega_{\phi}^{2}f(\phi)=\alpha_{E}\sin\omega t$$

Suppose that the *GZ* curve presented here by function $f(\phi)$ has three intersections with axes, see fig. 4.12.

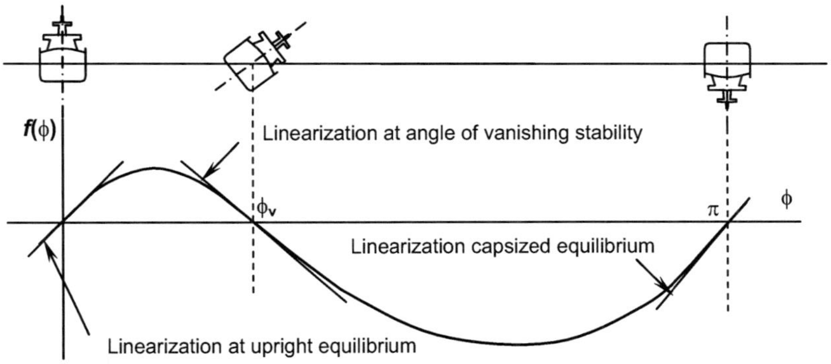

Fig 4.12 Equilibria in roll equation and linearization at vicinity of equilibria

We know that these intersections correspond to three equilibria:

- Stable "normal" equilibrium,
- Unstable equilibrium at angle of vanishing stability,
- Stable "upside down" equilibrium.

We show how this information can be formally deduced from equation (4.21). This procedure will not produce any new physical knowledge, but could be used as an intuitively clear example of a universal procedure that can be (and will be) applied in more complicated cases.

First of all, we delete excitation: it cannot influence the position of equilibria, since it does not depend on roll angle:

$$\ddot{\phi}+2\delta\dot{\phi}+\omega_{\phi}^{2}f(\phi)=0 \tag{4.118}$$

Then we rewrite (4.118) in vector form:

$$\dot{\vec{Y}} = F(\vec{Y}) \tag{4.119}$$

With:

$$\dot{\vec{Y}} = \begin{pmatrix} \dot{Y}_1 \\ \dot{Y}_2 \end{pmatrix} = \begin{pmatrix} \dot{\vartheta} \\ \dot{\phi} \end{pmatrix}; \qquad \vec{Y} = \begin{pmatrix} Y_1 \\ Y_2 \end{pmatrix} = \begin{pmatrix} \vartheta \\ \phi \end{pmatrix};$$
$$\vec{F}(\vec{Y}) = \begin{pmatrix} -2\delta_\phi Y_1 - \omega_\phi^2 f(Y_2) \\ Y_1 \end{pmatrix} = \begin{pmatrix} -2\delta_\phi \vartheta - \omega_\phi^2 f(\phi) \\ \vartheta \end{pmatrix} \tag{4.120}$$

If the system gets into an equilibrium position, it can stay there for an unlimited time (if not disturbed). So there is no motion and, hence, derivatives must be zero:

$$\dot{\vec{Y}} = 0; \quad \Rightarrow \quad \vec{F}(\vec{Y}) = 0 \tag{4.121}$$

This is equivalent to the system:

$$\begin{cases} -2\delta_\phi \vartheta - \omega_\phi^2 f(\phi) = 0 \\ \vartheta = 0 \end{cases} \tag{4.122}$$

Which immediately degenerates to one equation:

$$f(\phi) = 0 \tag{4.123}$$

Equation (4.123) expresses exactly the same what was mentioned above: equilibrium positions are located at intersections of the *GZ* curve with the axes of roll angles. There are three of them (in positive direction of roll angles): "normal" or "mast up", angle of vanishing stability and capsized or "mast down", see fig. 4.12.

As soon as the equilibrium positions are determined, we can evaluate if they are stable or not. To do that, we put the system exactly into an equilibrium position and then give it a small perturbation. If the system returns back, the equilibrium is stable, if not, then this equilibrium is unstable. Since perturbation is meant to be small, nonlinearity of the system cannot be significant and the system (4.118) can be linearized. The linearization has to be done in the vicinity of the corresponding equilibrium position, see fig. 4.12.

4.3.2 Original or "Normal" Equilibrium

Let us look at the "normal" equilibrium first. The linearized system looks like:

$$\ddot{\phi} + 2\delta\dot{\phi} + \omega_\phi^2 \left.\frac{df(\phi)}{d\phi}\right|_{\phi=0} \times (\phi - b) = 0 \tag{4.124}$$

Here, b is a free term that is equal to the heel angle at equilibrium. For equilibrium at $\phi = 0$, it leads to:

$$\ddot{\phi} + 2\delta\dot{\phi} + \omega_\phi^2 \phi = 0 \tag{4.125}$$

The reader is most probably quite familiar with the solution of the equation (4.125). However, it is worthwhile to review it because some intermediate values have crucial importance for stability. Let us search the solution in the form:

$$\phi = \exp(\lambda t) \tag{4.126}$$

Let us find its derivatives and substitute all of them into the original equation:

$$\lambda^2 \exp(\lambda t) + 2\delta\lambda \exp(\lambda t) + \omega_\phi^2 \exp(\lambda t) = 0 \tag{4.127}$$

Both sides of formula (4.127) can be divided by $\exp(\lambda t)$ and we obtain the following algebraic equation:

$$\lambda^2 + 2\delta\lambda + \omega_\phi^2 = 0 \tag{4.128}$$

This equation relates the exponential index, λ, with coefficients of the differential equation (4.125). It is called the "characteristic" equation and its solutions are called "eigenvalues" ("eigen" is a German word for "own"). The equation (4.128) is quadratic and its solutions are:

$$\lambda_{1,2} = -\delta \pm \sqrt{D} = -\delta \pm \sqrt{\delta^2 - \omega_\phi^2} \tag{4.129}$$

Since we have two eigenvalues (solutions of the characteristic equation) we have two independent solutions of the differential equation:

$$\phi_1 = \exp(\lambda_1 t); \quad \phi_2 = \exp(\lambda_2 t) \tag{4.130}$$

These two solutions span the space of all solutions. It means, that any other function that turns equation (4.125) into a true equality can be linearly expressed through these two functions (4.130):

$$\phi_1 = C_1 \exp(\lambda_1 t) + C_2 \exp(\lambda_2 t) \tag{4.131}$$

Where coefficients (arbitrary constants) C_1 and C_2 are to be determined through initial conditions.

Our goal is the determination of stability of equilibrium, so we would like to know where the system does go if displaced. Analysis of the behavior of function (4.131) yields the answer: if it takes the system back or pulls it away from the equilibrium. This behavior is evidently dependent on the eigenvalues (4.129); they also may be complex. Usually, the numerical value of damping, δ, is less than the natural frequency ω_ϕ, so the eigenvalues are complex for roll equation at the original equilibrium "mast up":

$$\begin{aligned}
\phi(t) &= C_1 \exp(\lambda_1 t) + C_2 \exp(\lambda_2 t) = \\
&= C_1 \exp\left(\left(-\delta + \sqrt{\delta^2 - \omega_\phi^2}\right)\cdot t\right) + C_2 \exp\left(\left(-\delta - \sqrt{\delta^2 - \omega_\phi^2}\right)\cdot t\right) = \\
&= C_1 \exp\left(\left(-\delta + i\sqrt{\omega_\phi^2 - \delta^2}\right)\cdot t\right) + C_2 \exp\left(\left(-\delta - i\sqrt{\omega_\phi^2 - \delta^2}\right)\cdot t\right) = \\
&= \exp(-\delta t)\left(C_1 \exp\left(i\sqrt{\omega_\phi^2 - \delta^2}\cdot t\right) + C_2 \exp\left(-i\sqrt{\omega_\phi^2 - \delta^2}\cdot t\right)\right) = \\
&= \phi_A e^{-\delta t} \sin\left(\sqrt{\omega_\phi^2 - \delta^2}\cdot t + \gamma\right)
\end{aligned} \tag{4.132}$$

Where ϕ_A and γ are a new form of arbitrary constants determined by initial conditions:

$$\phi_A = \sqrt{C_1^2 + C_2^2}; \quad \gamma = \arctan\left(-\frac{C_1}{C_2}\right) \tag{4.133}$$

Formula (4.132) describes a dying oscillation, which will return the system back to equilibrium after infinite time (see fig. 4.13):

$$\lim_{t\to\infty}\phi(t) = \lim_{t\to\infty}\left\{\phi_A e^{-\delta t}\sin\left(\sqrt{\omega_\phi^2 - \delta^2}\cdot t + \gamma\right)\right\} = 0 \tag{4.134}$$

That means that the equilibrium is stable: being displaced, the system returns back (the fact that it takes infinite time is usually expressed in the term "asymptotically stable").

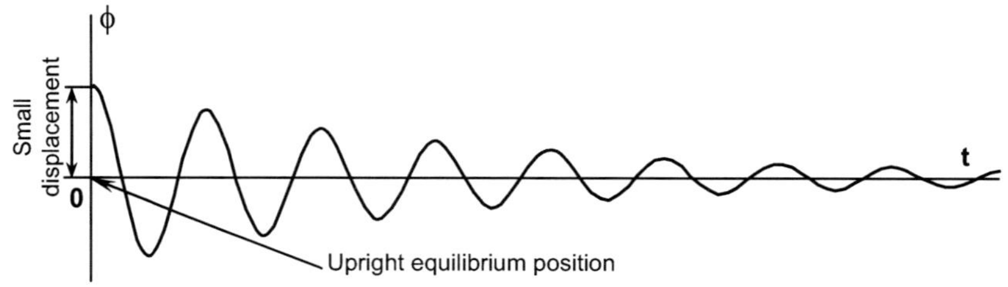

Fig 4.13 Asymptotically stable upright equilibrium

4.3.3 Equilibrium at Angle of Vanishing Stability

Let's use the above technique to analyze the equilibrium at the angle of vanishing stability. The linearized system looks like:

$$\ddot{\phi} + 2\delta\dot{\phi} + \omega_\phi^2 \left.\frac{df(\phi)}{d\phi}\right|_{\phi=\phi_v} \times (\phi - b) = 0 \tag{4.135}$$

Here b is a free term that equals the heel angle at equilibrium. For equilibrium at $\phi = \phi_v$, the derivative of the restoring term gives:

$$\left.\frac{df(\phi)}{d\phi}\right|_{\phi=\phi_v} = -k_1 \tag{4.136}$$

The linearized differential equation of roll is:

$$\ddot{\phi} + 2\delta\dot{\phi} - \omega_\phi^2 k_1\phi + \omega_\phi^2 k_1\phi_v = 0 \tag{4.137}$$

It is almost the same as in the first case of "normal" or "mast up" equilibrium (equation 4.125); the difference is the negative restoring term and the presence of a constant. We search for the solution in the following form:

$$\phi = \exp(\lambda t) + C_0 \tag{4.138}$$

Substitution of the prospective solution (4.138) into the equation (4.137) leads to the following formula:

$$\lambda^2 \exp(\lambda t) + 2\delta\lambda \exp(\lambda t) - \omega_\phi^2 k_1 \exp(\lambda t) - \omega_\phi^2 k_1 C_0 + \omega_\phi^2 k_1 \phi_v = 0 \quad (4.139)$$

The equation (4.139) contains both constant and time dependent terms. In order to satisfy this equation, sums of both types of terms have to be equal to zero independently:

$$\begin{cases} \lambda^2 \exp(\lambda t) + 2\delta\lambda \exp(\lambda t) - \omega_\phi^2 k_1 \exp(\lambda t) = 0 \\ -\omega_\phi^2 k_1 C_0 + \omega_\phi^2 k_1 \phi_v = 0 \end{cases} \quad (4.140)$$

The second equation yields a value for the constant C_0:

$$C_0 = \phi_v \quad (4.141)$$

The first equation yields the characteristic equation:

$$\lambda^2 + 2\delta\lambda - \omega_\phi^2 k_1 = 0 \quad (4.142)$$

Eigenvalues are:

$$\lambda_{1,2} = -\delta \pm \sqrt{D} = -\delta \pm \sqrt{\delta^2 + \omega_\phi^2 k_1} \quad (4.143)$$

We see here a very important change: under no circumstances is the discriminate negative, so the eigenvalues are always real. Then any solution of the equation (4.137) is of the form:

$$\phi(t) = C_1 \exp(\lambda_1 t) + C_2 \exp(\lambda_2 t) + \phi_v \quad (4.144)$$

With arbitrary constants C_1 and C_2 to be determined from initial conditions.

It is very important to note that one of the eigenvalues is always positive and the other one is always negative. Since one of the eigenvalues is positive, the solution is not bounded and it always tends to infinity in the positive or negative direction, depending on whether the sign of C_1 if λ_1 is positive:

$$\lim_{t\to\infty} \phi(t) = \lim_{t\to\infty} \{ C_1 \exp(\lambda_1 t) + C_2 \exp(\lambda_2 t) + \phi_v \} = \pm\infty \quad (4.145)$$

That means, that the equilibrium in not stable; being displaced, the system never returns back, see fig. 4.14.

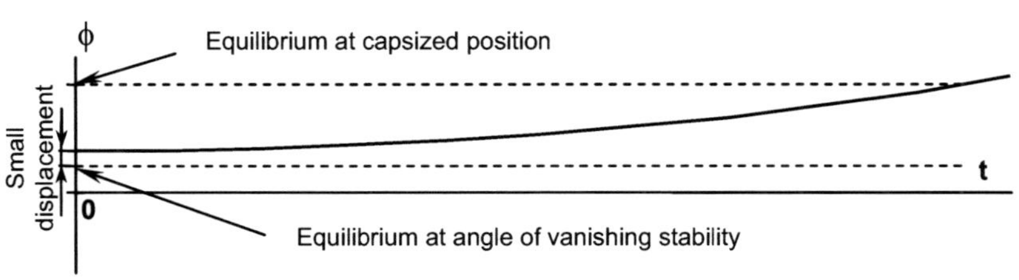

Fig. 4.14 Unstable equilibrium at angle of vanishing stability

4.3.4 Equilibrium at Capsized Position

The last equilibrium we are going to analyze, is capsized or "mast down" position. The sequence of the derivation is the same as that in the previous two cases:

$$\ddot{\phi} + 2\delta\dot{\phi} + \omega_\phi^2 \left.\frac{df(\phi)}{d\phi}\right|_{\phi=\pi} \times (\phi - b) = 0 \tag{4.146}$$

Here b is a free term that is equal to the heel angle at equilibrium. For equilibrium at $\phi = \pi$, the derivative of the restoring term gives:

$$\left.\frac{df(\phi)}{d\phi}\right|_{\phi=\pi} = k_2 \tag{4.147}$$

The linearized differential equation of roll is:

$$\ddot{\phi} + 2\delta\dot{\phi} + \omega_\phi^2 k_2 \phi + \omega_\phi^2 k_2 \phi_v = 0 \tag{4.148}$$

The equation (4.148) is similar to both previous cases: it has a positive restoring term as in (4.125) and a constant as in (4.137). Following the same procedure, we can write a solution as:

$$\phi(t) = C_1 \exp(\lambda_1 t) + C_2 \exp(\lambda_2 t) + \pi \tag{4.149}$$

With the eigenvalues:

$$\lambda_{1,2} = -\delta \pm \sqrt{D} = -\delta \pm \sqrt{\delta^2 - \omega_\phi^2 k_2} \tag{4.150}$$

Formally, eigenvalues are going to be complex here, if we are not taking into account changing of damping in the capsized position. Let us consider the case with unchanged damping first: the solution is dying oscillation (analogous to equation (4.132)):

$$\phi(t) = \phi_A e^{-\delta t} \sin\left(\sqrt{\omega_\phi^2 - \delta^2} \cdot t + \gamma\right) + \pi \tag{4.151}$$

and

$$\lim_{t\to\infty} \phi(t) = \lim_{t\to\infty} \left\{ \phi_A e^{-\delta t} \sin\left(\sqrt{\omega_\phi^2 - \delta^2} \cdot t + \gamma\right) + \pi \right\} = \pi \tag{4.152}$$

The equilibrium is stable, since the system being displaced returns back to the equilibrium, see fig. 4.15.

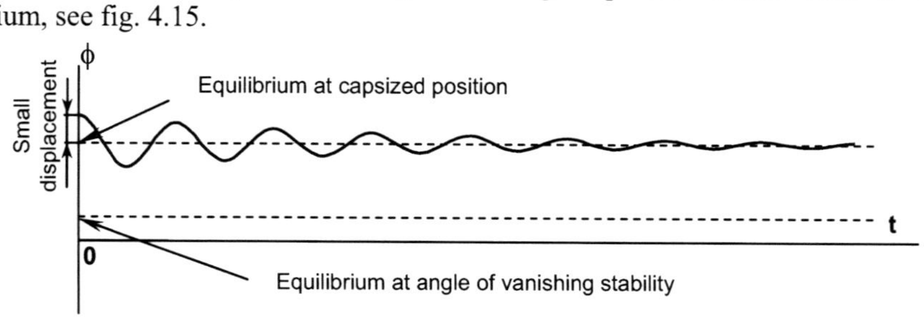

Fig. 4.15 Asymptotically stable equilibrium at capsized position

If we take into account a changing damping coefficient in the capsized position, the eigenvalues may be complex or real. They are real if the damping is very high (larger than critical damping): $\delta > \omega_\phi \sqrt{k_2}$. Let us consider this more closely. The solution also has an exponential form as in the case with equilibrium at vanishing stability, but both eigenvalues are negative. Really, for any positive damping coefficient δ.

$$\text{if} \quad \delta > \omega_\phi \sqrt{k_2} \quad \Rightarrow \quad \sqrt{\delta^2 - \omega_\phi^2 k_2} < \delta \quad \Rightarrow \quad \lambda_{1,2} < 0 \tag{4.153}$$

Since eigenvalues are negative, the solution is bounded, and the system returns back to the equilibrium:

$$\lim_{t\to\infty} \phi(t) = \lim_{t\to\infty} \left\{ C_1 \exp(\lambda_1 t) + C_2 \exp(\lambda_2 t) + \pi \right\} = \pi \tag{4.154}$$

Again, the equilibrium is stable disregarding the damping coefficient value.

All of the above analyses are convenient to present in the form of a table: (see table 4.1)

Table 4.1 Summary of equilibria and their stability

Equilibrium	Eigenvalues	Presence of positive real eigenvalue or positive real part of complex eigenvalue	Stable or unstable
"Normal" or "Mast-up"	Complex	No	Stable
Angle of vanishing stability	Real	Yes	Unstable
Capsized or "Mast-down"	Complex	No	Stable
	Real	No	Stable

Finally, the formal indicator of stability or instability of equilibrium is the presence of a positive real eigenvalue or positive real part of a complex eigenvalue.

4.3.5 Phase Plane in Vicinity of Equilibria

In the previous subchapter we have used time history to show how the solution behaves. However, the time history does not show the entire picture, since the dynamic state of the system without forcing at any moment of time is fully characterized by two values: displacement and its derivative, or roll angle and angular velocity. There is a way to show both these values in one plot: to choose them to be co-ordinate axes. It is called the phase plane.

Every moment of time is mapped on the phase plane with a particular roll angle and angular velocity. Motion of the system leaves a trace on the phase plane that is called "phase trajectory". The entire set of phase trajectories is called the "phase picture" or "phase portrait".

If the solution is known, building the phase picture is easy; let us take the solution (4.132) as an example:

$$\begin{aligned} \phi(t) &= \phi_A e^{-\delta t} \sin\left(\sqrt{\omega_\phi^2 - \delta^2} \cdot t + \gamma\right) \\ \dot{\phi}(t) &= \phi_A \sqrt{\omega_\phi^2 - \delta^2} \cdot e^{-\delta t} \cos\left(\sqrt{\omega_\phi^2 - \delta^2} \cdot t + \gamma\right) - \phi_A \delta e^{-\delta t} \sin\left(\sqrt{\omega_\phi^2 - \delta^2} \cdot t + \gamma\right) \end{aligned} \tag{4.155}$$

Assuming particular initial conditions and using time as a parameter, we obtain two columns of numbers corresponding to the roll angle and angular velocity. This is the phase trajectory. Then, we change initial conditions and repeat the procedure. The resulting phase picture is shown in fig. 4.16.

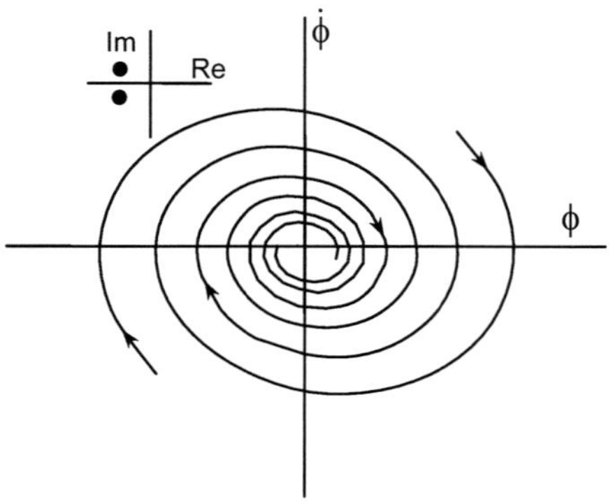

Fig. 4.16 Phase picture at upright equilibrium position: stable focus

Equilibrium is presented by a point with zero angular speed; the phase picture consists of a number of spirals. Direction of movement along these spirals clearly indicates stability of the equilibrium. The phase picture on fig. 4.16 is supplemented by a graph showing the position of the eigenvalues on the complex plane. Usually phase pictures have names, this one is called "stable focus".

The other two cases considered in the above subchapter: angle of vanishing stability and position "mast down" with large damping are shown in figures 4.17 and 4.18 correspondingly. (Position "mast down" with small damping is not much different from the one shown on fig. 4.16). These phase pictures are called "saddle" (fig.4.17) and "stable node" (fig. 4.18).

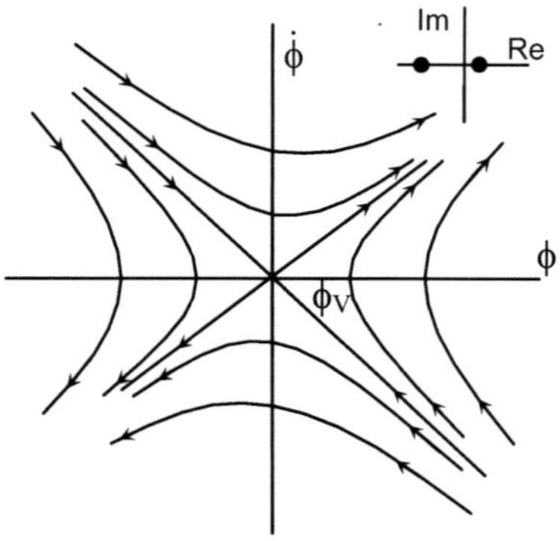

Fig. 4.17 Saddle

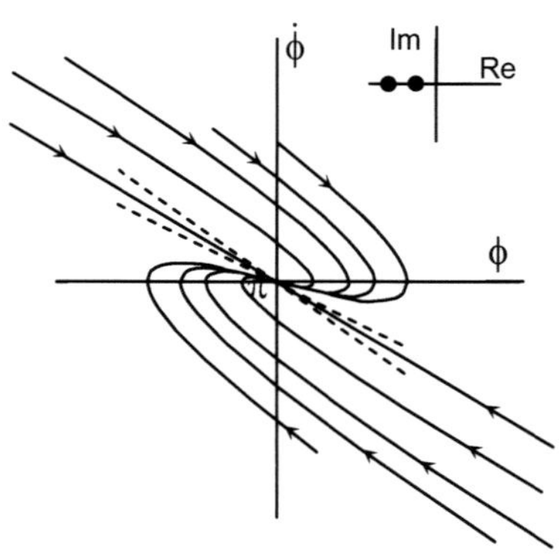

Fig. 4.18 Stable node

These phase pictures can be encountered while studying the roll equation. To complete this consideration, we show a complete list of all possible phase pictures along with its names and characteristic positions of eigenvalues on the complex plane. Since a linear dynamic system describing roll contains only two coefficients: natural frequency and damping, every possible phase picture is associated with a pair of these coefficients, see fig.4.19.

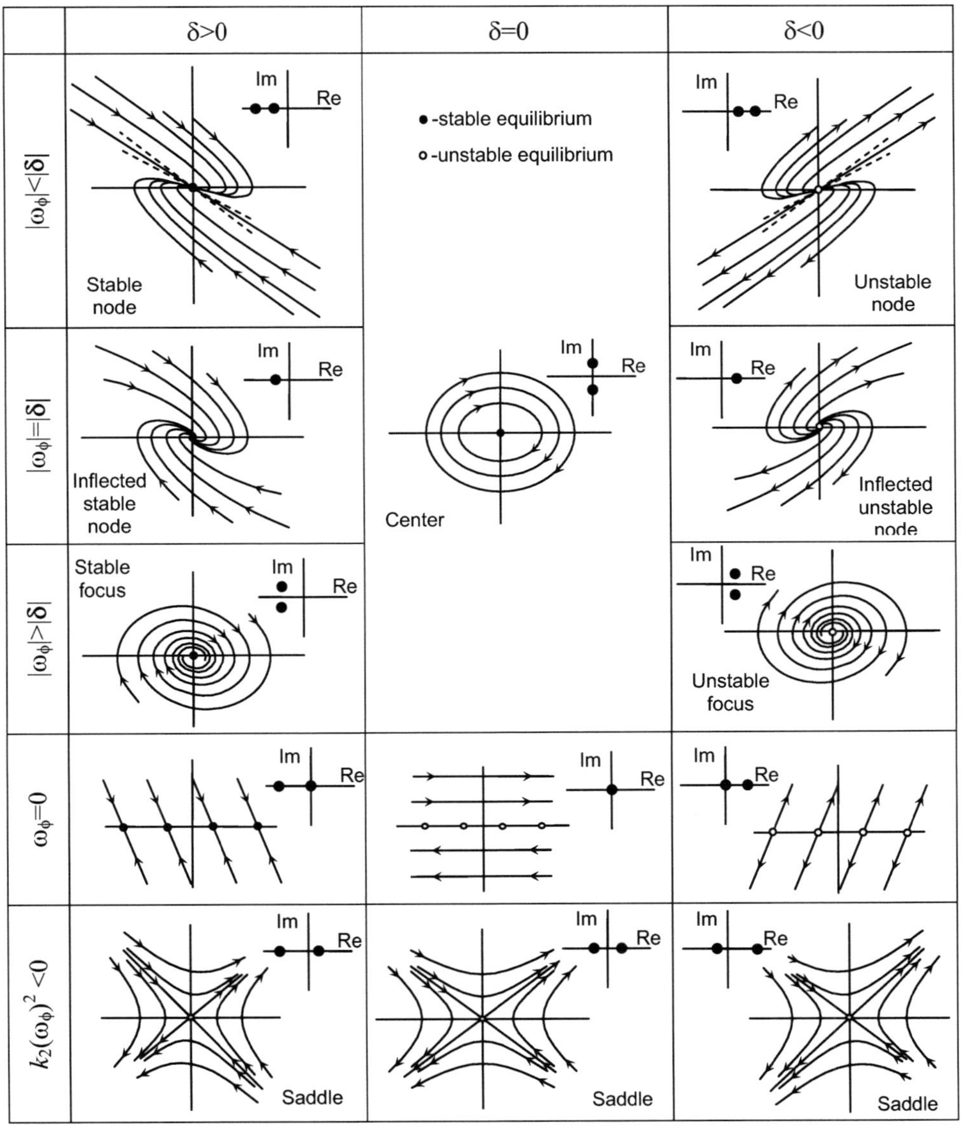

Fig. 4.19 Phase pictures of linear system along with eigenvalues

4.4 Stability of Roll Motion

Steady state periodic motion or limit cycle (an alternative term) is similar to equilibrium in a certain sense. Once the system achieves either of these states, it holds there for an indefinite time if not perturbed. We have seen already what happens, if the system is disturbed from equilibrium: if it is stable, the system will sooner or later return there. If the equilibrium is unstable, the system "tries" to find a stable state. The same applies to steady state regimes of motions. They may be either stable or unstable. If steady state periodic motion is stable, a small perturbation of its motion would not lead to drastic changes: the system would return to the original regime after a finite or infinite time. (The last case defines "asymptotic" stability – similarity with stability of equilibria holds completely).

This subchapter is focused on different methods for studying motion stability of roll in beam seas.

4.4.1 Lyapunov Direct Method

We begin our consideration from a very brief review of the direct Lyapunov method. At the end of the 19th century, Lyapunov [1954] applied the method for the motion of a rigid body floating in fluid.

What do we mean when we are speaking about motion stability? Let us imagine steady state motion of some dynamic system that is describing ship roll. We apply some small perturbation to the system and observe its behavior. If the response (difference between perturbed and non-perturbed motions) is also small, and the system returns to its initial regime after a finite or even infinite time, we recognize this steady state regime as stable, (if return to initial regime requires infinite time, the regime is called asymptotic stable). If the response of the system to the small perturbation is not small and the system reaches some other regime of motion, such a steady state regime should be recognized as unstable. An unstable steady state regime cannot exist in physical system during a long time, analogously to unstable equilibrium.

To make judgments about stability, we need an auxiliary function that is called the 'Lyapunov function'. This function should be defined in phase coordinates. If we consider a dynamic system with one degree of freedom, the Lyapunov function depends on roll angle and roll velocity: $F(\phi,\dot{\phi})$. It will look like the surface in fig. 4.20.

Let us intersect this surface by some plane $F = const$. A trace of this intersection is the boundary of stability. If the phase trajectory of perturbed motion crosses this boundary only in the inward direction, such a motion is recognized as stable. In other words, a substantial derivative of the Lyapunov function that is calculated along the phase trajectory should be negative all the time.

Such usage of the Lyapunov function forces us to stipulate that $F(\phi,\dot{\phi})$ should be a positive definite function. The condition of positive definite is the following:

$$F(0,0) = 0 \quad \text{and} \quad F(\phi,\dot{\phi}) > 0 \qquad (4.156)$$

Now, we can give the formulation of the Lyapunov stability theorem that is essential in the Lyapunov direct method (English formulation is taken from Kuo and Odabasi [1975]): "If the differential equations of perturbed motion are such that it is possible to find a positive definite function $F(\phi,\dot{\phi})$ for which the substantial derivative, taken along the motion trajectory is negative definite, then the non-perturbed motion is stable". Negative definite function should be understood analogously with the positive definite one. So, we have two conditions of stability:

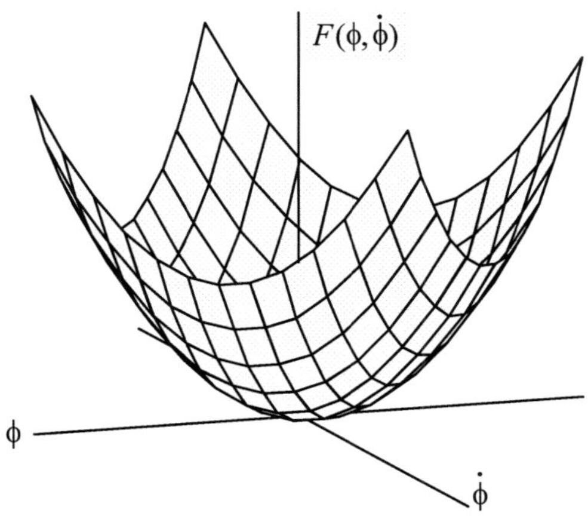

Fig. 4.20 Appearance of Lyapunov function

$$\frac{dF(\phi,\dot{\phi})}{dt}=\frac{\partial F}{\partial t}+\frac{\partial F}{\partial \phi}\dot{\phi}+\frac{\partial F}{\partial \dot{\phi}}\ddot{\phi}<0 \quad \text{and} \quad \frac{dF(0,0)}{dt}=0 \tag{4.157}$$

To provide an example of this procedure, we consider an example taken from Kuo and Odabasi [1975]: a nonlinear roll equation with a cubic restoring term:

$$\ddot{\phi}+2\delta\dot{\phi}+\omega_{\phi}^{2}\phi-a_{3}\phi^{3}=0 \tag{4.158}$$

The Lyapunov function is taken in the following form:

$$F(\phi,\dot{\phi})=\frac{1}{2}(\omega_{\phi}^{2}\phi^{2}+\dot{\phi}^{2}) \tag{4.159}$$

It is clear that function (4.162) is positive definite, and its substantial derivative is:

$$\frac{dF}{dt}=\omega_{\phi}^{2}\phi\dot{\phi}+\dot{\phi}\ddot{\phi}=\omega_{\phi}^{2}\phi\dot{\phi}+\dot{\phi}(-2\delta\dot{\phi}-\omega_{\phi}^{2}\phi+a_{3}\phi^{3})=a_{3}\dot{\phi}\phi-2\delta\dot{\phi}^{2} \tag{4.160}$$

To provide a negative definite of the substantial derivative of Lyapunov function, we should require:

$$\frac{dF}{dt}=a_{3}\dot{\phi}\phi-2\delta\dot{\phi}^{2}<0 \tag{4.161}$$

Roll motion stability analysis where the direct Lyapunov method was used can be found in Martin, *et al* [1982], Phillips [1986, 1986a], Caldeira-Saraiva [1986, 1986a]. This approach to roll motion stability analysis mainly encounters the problem of finding the Lyapunov function.

4.4.2 Floquett Theory

We continue our consideration of roll equation (4.158) including harmonic excitation [Stoker, 1950, Wellicome, 1975]:

$$\ddot{\phi}+2\delta\dot{\phi}+\omega_{\phi}^{2}\phi-a_{3}\phi^{3}=\alpha_{E}\sin\omega t \tag{4.162}$$

To study stability of the steady state solution, we introduce some small perturbation $\xi(t)$:

$$\tilde{\phi}(t) = \phi(t) + \xi(t) \tag{4.163}$$

and substitute (4.163) in (4.162):

$$\ddot{\tilde{\phi}} + 2\delta\dot{\tilde{\phi}} + \omega_\phi^2\tilde{\phi} - a_3\tilde{\phi}^3 = \alpha_E \sin\omega t \tag{4.164}$$

or

$$(\ddot{\phi} + \ddot{\xi}) + 2\delta(\dot{\phi} + \dot{\xi}) + \omega_\phi^2(\phi + \xi) - a_3(\phi + \xi)^3 = \alpha_E \sin\omega t \tag{4.165}$$

Taking into account that $\phi(t)$ is a steady state solution of differential equation (4.162) and its substitution into the equation transforms it into a true equality, we get an equation relative to the small perturbation $\xi(t)$. Such an equation is called the "variation equation". After high powers of small quantity $\xi(t)$ are dropped, the variation equation looks like:

$$\ddot{\xi} + 2\delta\dot{\xi} + (\omega_\phi^2 + 3a_3\phi^2)\xi = 0 \tag{4.166}$$

Expression (4.166) is an ordinary linear differential equation of the second order with a time varying coefficient; $\phi(t)$ is the steady state solution, so it should be periodical because of harmonic excitation, so equation (4.166) has a periodical coefficient.

The general solution of (4.166) can be expressed as:

$$\xi(t) = c_1\xi_1(t) + c_2\xi_2(t) \tag{4.167}$$

Here, $c_{1,2}$ are arbitrary constants and $\xi_{1,2}$ are linear independent solutions. There are two of them, since the differential equation (4.166) is of the second order. Linear independence of the solutions $\xi_{1,2}$ means that they cannot be expressed one through another using linear operations (such as the sine function cannot be expressed through the cosine function of the same argument with linear operation only). These solutions are also called "fundamental" and the following determinant never equals zero:

$$W(t) = \det\begin{vmatrix} \xi_1(t) & \xi_2(t) \\ \dot{\xi}_1(t) & \dot{\xi}_2(t) \end{vmatrix} \neq 0 \tag{4.168}$$

Such a determinant has a special name. It is called the "Wronsky determinant".

As we have mentioned, the equation (4.166) contains the periodic coefficient $(\omega_\phi^2 + 3a_3\phi^2)$ with period T. (If we are using harmonic approximation for the steady state solution, the period is equal to double the excitation period, because of the second power.) The periodicity of the coefficient is very important, because it makes $\xi_1(t+T)$ and $\xi_2(t+T)$ fundamental solutions if $\xi_1(t)$ and $\xi_2(t)$ are fundamental ones as well, [Stoker, 1950]. At the same time $\xi_1(t+T)$ and $\xi_2(t+T)$ are solutions of differential equation (4.166) and can be expressed via a linear combination of fundamental solutions at the time t:

$$\begin{cases} \xi_1(t+T) = b_{11}\xi_1(t) + b_{12}\xi_2(t) \\ \xi_2(t+T) = b_{21}\xi_1(t) + b_{22}\xi_2(t) \end{cases} \tag{4.169}$$

The Wronsky determinant of ξ_1(t+T) and ξ_2 (t+T) can be obtained from (4.169):

$$W(t+T) = \det\begin{vmatrix} b_{11} & b_{12} \\ b_{21} & b_{22} \end{vmatrix} \cdot W(t), \tag{4.170}$$

Since $\xi_1(t+T)$ and $\xi_2(t+T)$ are also fundamental solutions:

$$\det\begin{vmatrix} b_{11} & b_{12} \\ b_{21} & b_{22} \end{vmatrix} \neq 0 \tag{4.171}$$

Here is the essence of the Floquett theory: the solution, we search for is not necessarily periodic; it changes after a period by a constant:

$$\xi(t+T) = \lambda \cdot \xi(t); \qquad \lambda = \text{Const} \tag{4.172}$$

If the constant λ is more than one, the solution of variation equation (4.166) increases, the difference between the original and perturbed process becomes larger and our process is clearly unstable.

The solution (4.172) as any other can be expressed via fundamental solutions in accordance with formulae (4.167) and (4.169):

$$\begin{aligned} \xi(t+T) &= c_1\xi_1(t+T) + c_2\xi_2(t+T) = \\ &= c_1[b_{11}\xi_1(t) + b_{12}\xi_2(t)] + c_2[b_{21}\xi_1(t) + b_{22}\xi_2(t)] \end{aligned} \tag{4.173}$$

Using formulae (4.172) and (4.167):

$$\xi(t+T) = \lambda c_1\xi_1(t) + \lambda c_2\xi_2(t) \tag{4.174}$$

Equalizing (4.173) and (4.174) we can get an equation relatively λ:

$$[c_1(b_{11}-\lambda) + c_2 b_{12}]\xi_1(t) + [c_1 b_{21} + c_2(b_{22}-\lambda)]\xi_2(t) = 0; \tag{4.175}$$

Let's transform it into the system of linear equations using the linear independence of $\xi_1(t)$ and $\xi_2(t)$:

$$\begin{cases} c_1(b_{11}-\lambda) + c_2 b_{12} = 0 \\ c_1 b_{21} + c_2(b_{22}-\lambda) = 0 \end{cases} \tag{4.176}$$

Adding the equations we shall get:

$$(c_1 - c_2)[(b_{11}-\lambda)(b_{22}-\lambda) - b_{12}b_{21}] = 0 \tag{4.177}$$

and

$$(b_{11}-\lambda)(b_{22}-\lambda) - b_{12}b_{21} = 0 \tag{4.178}$$

Analogously to linear differential equations, expression (4.178) is called the "characteristic equation" and its roots are known as "eigenvalues".

A matter of particular interest is the case when $\lambda_{1,2} = \pm 1$. If $\lambda_1 = 1$, the solution will have the period T, or the same period that periodic coefficient $(\omega_\phi^2 + 3a_3\phi^2)$ has. For $\lambda_2 = -1$, the period of solution equals $2T$. Really:

$$\xi_2(t+2T) = -1\cdot\xi_2(t+T) = (-1)^2\xi_2(t) = \xi_2(t) \tag{4.179}$$

Following Nayfeh [1986a, 1990], here we solve the variation equation numerically by the Runge-Kutta method. To receive a pair of fundamental solutions, we set up initial conditions $\xi(t=0)=1$, $\dot{\xi}(t=0)=0$ for the first solution and $\xi(t=0)=0$, $\dot{\xi}(t=0)=1$ for the second one. The solution and its derivatives are shown in fig. 4.21 and 4.22.

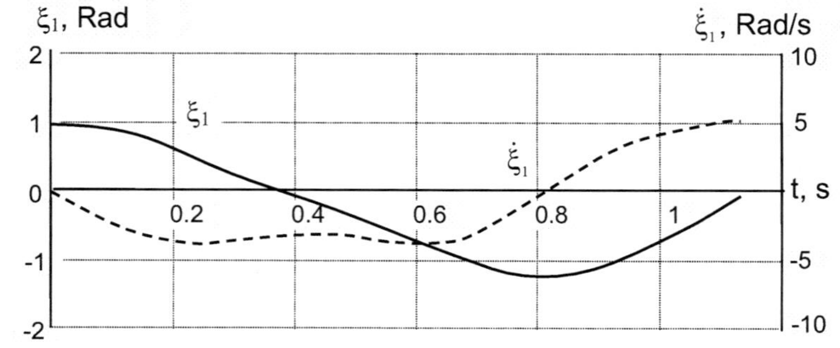

Fig. 4.21 First fundamental solution of variation equation ($\omega/\omega_\phi = 1.05$)

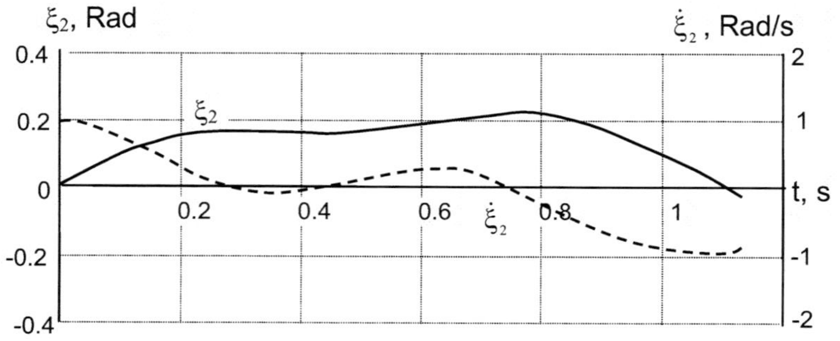

Fig. 4.22 Second fundamental solution of variation equation ($\omega/\omega_\phi = 1.05$)

Using formula (4.170) for solution after time $t=T$ passes:

$$\begin{cases}\xi_1(T) = b_{11}\xi_1(t=0) + b_{12}\xi_2(t=0)\\ \xi_2(T) = b_{21}\xi_1(t=0) + b_{22}\xi_2(t=0)\end{cases} \tag{4.180}$$

We can write:

$$\begin{aligned} b_{11} &= \xi_1(T) \quad b_{12} = \dot{\xi}_1(T)\\ b_{21} &= \xi_2(T) \quad b_{22} = \dot{\xi}_2(T)\end{aligned} \tag{4.181}$$

Eigenvalues of the differential equation (4.166) are the same as the eigenvalues of matrix $\mathbf{b}_{i,j}$. The eigenvalues can be found from the following quadratic equation:

$$\lambda^2 - (b_{11}+b_{22})\lambda + [b_{11}b_{22} - b_{12}b_{21}] = 0 \tag{4.182}$$

Coefficients of this equation have special names or nominations, see Thompson [1986]:

Trace:

$$Tr = b_{11} + b_{22} \tag{4.183}$$

Determinant:

$$Dt = b_{11}b_{22} - b_{12}b_{21} = \det(b_{ij}) \tag{4.184}$$

The eigenvalues can be expressed as:

$$\lambda_{1,2} = 0.5 \cdot \left(Tr \pm \sqrt{Tr^2 - 4 \cdot Dt}\right) \tag{4.185}$$

We already mentioned that our process is stable if both eigenvalues are less than one. However eigenvalues (4.185) may be complex. To generalize stability conditions we require that an absolute value of both eigenvalues should be less then unity, or the eigenvalues must stay in the unit circle in a complex plane. These eigenvalues are the main result of the motion stability study at this stage. Subchapter 4.5 shows how they are used and what information may be extracted from their evaluation.

4.4.3 Poincare Map and Numerical Method for Motion Stability

This subchapter is focused on a description of a numerical approach to motion stability. Its main advantage is no limitations for nonlinear terms: they can even be defined numerically. Also, no limitations for degrees of freedom are imposed.

First, we introduce the concept of the Poincare map. It is a very convenient tool to work with periodic forced motions, because it allows visualizing complex transition processes. The idea is to keep just one point of phase trajectory per expected period of motion (it is usually period of exciting force). Then, instead of a messy phase picture of transition, we get a series of points that clearly indicate the tendency, see figure 4.23. The steady state periodic regime is shown by just one point.

Now, let's return to the problem of motion stability. Since it is meant to get a solution numerically, our dynamic system describing roll (coupled with other ship motions, if necessary) is presented in the vector form (4.95):

$$\dot{\vec{Y}} = F(\vec{Y}, t)$$

Assume we know initial conditions corresponding to the steady state solution (or limit cycle), which is also called the "fixed point", because it looks like a point on the Poincare map. Where might we obtain these initial conditions? One of the possible answers – from the approximate solution received by any method described in subchapter 4.2. Another option is "Global analysis". It is when we subdivide the entire phase plane with a mesh and then check all the initial conditions defined by nodes of this mesh. Those nodes are the ones who lead to the solution that repeats itself after a given period are ones we are looking for. We will return to this method later in Chapter 6.

Another possible question here is: do we need motion stability analysis for a numerical solution at all? It might seem unnecessary, since we can track the process itself and see if the system can keep this particular steady state regime for a while. In fact, the stability analysis is still necessary. One reason is that instability may not be very strong and the development of a significant deviation may take a long time. A stability analysis shows the "fate" of the regime within one period. Also, we really need to know where the system goes if this regime is unstable. This is discussed in subchapter 4.5, but to be prepared for it we need to know a value of certain criterion, like eigenvalues such as in the Floquett theory described in the previous subchapter 4.4.2.

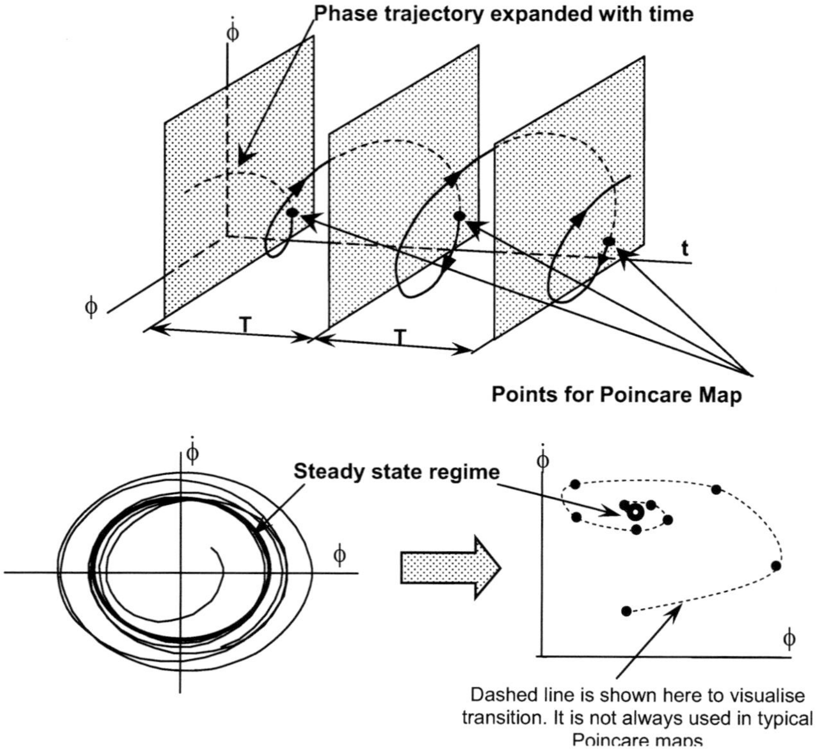

Fig. 4.23 On Poincare map definition

Let's introduce a small perturbation by steady state solution (also known as a "limit cycle" and "fixed point") to the system (4.95) at the moment $t = t_0$, when initial conditions were $\vec{Y} = \vec{Y}_S$. As a result, after the perturbation was applied, the system state becomes:

$$\vec{Y}\Big|_{t=t_0} = \vec{Y}_0 = \vec{Y}_S + \vec{\xi}_0 \tag{4.186}$$

$$\dot{\vec{Y}}_0 = F(\vec{Y}_s + \vec{\xi}_0, t_0) = F(\vec{Y}_0, t_0) \tag{4.187}$$

With

$$\dot{\vec{Y}}_0 = \begin{pmatrix} \dot{Y}_{01} \\ \dot{Y}_{02} \end{pmatrix} = \begin{pmatrix} \dot{\vartheta}_0 \\ \dot{\phi}_0 \end{pmatrix}; \quad \vec{Y} = \begin{pmatrix} Y_{01} \\ Y_{02} \end{pmatrix} = \begin{pmatrix} \vartheta_0 \\ \phi_0 \end{pmatrix}; \quad \vec{\xi}_0 = \begin{pmatrix} \xi_{01} \\ \xi_{02} \end{pmatrix}$$
$$\vec{F}(\vec{Y}_0, t_0) = \begin{pmatrix} -2\delta_\phi(Y_{01}) - \omega_\phi^2 f(Y_{02}) \\ Y_{01} \end{pmatrix} = \begin{pmatrix} -2\delta_\phi(\vartheta_0) - \omega_\phi^2 f(\phi_0) \\ \vartheta_0 \end{pmatrix} \tag{4.188}$$

Here ξ_{01} is a small perturbation by roll angle and ξ_{02} is a small perturbation by angular velocity at the moment t_0 when they were introduced. (One of them may be zero). After period T, the system status is described as:

$$\dot{\vec{Y}}_1 = \vec{F}(\vec{Y}_1, t_0 + T) \tag{4.189}$$

Consider the Poincare map. It has two points corresponding to the state at t_0 and t_0+T. Point $\vec{Y}_0$ after time T becomes point $\vec{Y}_1$. So, we can consider a vector-valued function $\vec{G}$ that maps the first point to the second one. Actually, it is the value of the solution taken after a given period of time. The solution in general depends on initial conditions and time. The Poincare map is dependent on the previous point (which plays role of initial conditions in this case) only:

$$\vec{Y}_1 = \vec{Y}_1(\vec{Y}_0, t = T) = \begin{pmatrix} Y_{11}(Y_{01}, Y_{02}, t = T) \\ Y_{12}(Y_{01}, Y_{02}, t = T) \end{pmatrix} = \vec{G}(\vec{Y}_0) = \begin{pmatrix} G_1(Y_{01}, Y_{02}) \\ G_2(Y_{01}, Y_{02}) \end{pmatrix} \tag{4.190}$$

It is important to note that the steady state periodic solution is mapped to itself:

$$\vec{Y}_s = \vec{G}(\vec{Y}_s) \tag{4.191}$$

Using (4.186):

$$\vec{Y}_1 = \vec{Y}_s + \vec{\xi}_1 = \begin{pmatrix} Y_{s1} + \xi_{11} \\ Y_{s2} + \xi_{12} \end{pmatrix} = \vec{G}(\vec{Y}_0) = \begin{pmatrix} G_1(Y_{s1} + \xi_{01}, Y_{s2} + \xi_{02}) \\ G_2(Y_{s1} + \xi_{01}, Y_{s2} + \xi_{02}) \end{pmatrix} \tag{4.192}$$

Here, ξ_{11} is a small perturbation by roll angle and ξ_{12} is a small perturbation by angular velocity at the moment t_0+T. The equation (4.192) in scalar form is:

$$\begin{cases} Y_{s1} + \xi_{11} = G_1(Y_{s1} + \xi_{01}, Y_{s2} + \xi_{02}) \\ Y_{s2} + \xi_{12} = G_2(Y_{s1} + \xi_{01}, Y_{s2} + \xi_{02}) \end{cases} \tag{4.193}$$

Since the perturbation is small, equations (4.192) or (4.193) can be expanded in a Taylor series in the vicinity of the fixed point (with a point on the Poincare map corresponding to steady state regime or limit cycle):

$$\begin{cases} Y_{s1} + \xi_{11} = G_1(Y_{s1}) + \dfrac{\partial G_1}{\partial Y_{s1}} \cdot \xi_{01} + \dfrac{\partial G_1}{\partial Y_{s2}} \xi_{02} + \dots \\ Y_{s2} + \xi_{12} = G_2(Y_{s2}) + \dfrac{\partial G_2}{\partial Y_{s1}} \cdot \xi_{01} + \dfrac{\partial G_2}{\partial Y_{s2}} \xi_{02} + \dots \end{cases} \tag{4.194}$$

Taking into account (4.191) and limiting ourselves with the terms of the first order:

$$\begin{cases} \xi_{11} = \dfrac{\partial G_1}{\partial Y_{s1}} \cdot \xi_{01} + \dfrac{\partial G_1}{\partial Y_{s2}} \xi_{02} \\ \xi_{12} = \dfrac{\partial G_2}{\partial Y_{s1}} \cdot \xi_{01} + \dfrac{\partial G_2}{\partial Y_{s2}} \xi_{02} \end{cases} \qquad (4.195)$$

or

$$\vec{\xi}_1 = \begin{pmatrix} \xi_{11} \\ \xi_{12} \end{pmatrix} = \begin{bmatrix} \dfrac{\partial G_1}{\partial Y_{s1}} & \dfrac{\partial G_1}{\partial Y_{s2}} \\ \dfrac{\partial G_2}{\partial Y_{s1}} & \dfrac{\partial G_2}{\partial Y_{s2}} \end{bmatrix} \times \begin{pmatrix} \xi_{01} \\ \xi_{02} \end{pmatrix} = \mathrm{J}(\vec{G}) \times \vec{\xi}_0 \qquad (4.196)$$

The matrix that consists of partial derivatives of a vector valued function, as it appears in equation (4.196), has a special name. It is called a Jacobean matrix and it has a meaning of the generalized derivative for a vector-valued function. It is usually denoted as $\mathrm{J}(\vec{G})$.

Eigenvalues of the Jacobean matrix contain the most important information. If eigenvalues are inside the unit circle at the complex plane (modulus of the eigenvalues does not exceed one), the steady state regime (fixed point or limit cycle) is stable. The derivation is not difficult and a detailed description was made by Thompson and Stewart [1986], so it is not necessary to copy it here. However, it is interesting to note certain similarities with the Floquett theory described in subchapter 4.4.2.

Also, it is convenient to track behavior of the steady state solution (fixed point or limit cycle) using Trace Determinant plane. These quantities are defined through the Jacobean matrix as:

$$Dt = \frac{\partial G_1}{\partial Y_{s1}} \cdot \frac{\partial G_2}{\partial Y_{s2}} - \frac{\partial G_1}{\partial Y_{s2}} \cdot \frac{\partial G_2}{\partial Y_{s1}} \qquad (4.197)$$

$$Tr = \frac{\partial G_1}{\partial Y_{s1}} + \frac{\partial G_2}{\partial Y_{s2}} \qquad (4.198)$$

These figures will be used later in subchapter 4.5 when we will consider bifurcations.

The last question to answer is how to calculate the Jacobean matrix having the system in the form (4.95)? Actually, the simplest way is by direct numerical differentiation. As an alternative, Virgin [1987] proposed to use one more point:

$$\vec{Y}_2 = \vec{Y}_2(\vec{Y}_1, t = T) = \vec{Y}_1(\vec{Y}_0, t = 2T) = \vec{Y}_s + \vec{\xi}_2 \qquad (4.199)$$

Considering the perturbation small, we can assume that the same function $\vec{G}$ (4.190) defines mapping of point $\vec{Y}_1$ to point $\vec{Y}_2$. This allows the addition of two more equations into the system (4.195):

$$\begin{cases} \xi_{11} = \dfrac{\partial G_1}{\partial Y_{s1}} \cdot \xi_{01} + \dfrac{\partial G_1}{\partial Y_{s2}} \xi_{02} \\ \xi_{12} = \dfrac{\partial G_2}{\partial Y_{s1}} \cdot \xi_{01} + \dfrac{\partial G_2}{\partial Y_{s2}} \xi_{02} \\ \xi_{21} = \dfrac{\partial G_1}{\partial Y_{s1}} \cdot \xi_{11} + \dfrac{\partial G_1}{\partial Y_{s2}} \xi_{12} \\ \xi_{22} = \dfrac{\partial G_2}{\partial Y_{s1}} \cdot \xi_{11} + \dfrac{\partial G_2}{\partial Y_{s2}} \xi_{12} \end{cases} \qquad (4.200)$$

There are four unknown members of a Jacobean matrix and four linear equations in the system (4.200), A value of initial perturbation $\vec{\xi}_0(\xi_{01}, \xi_{02})$ is given. The values of variation $\vec{\xi}_1(\xi_{11}, \xi_{12})$ after period T and $\vec{\xi}_2(\xi_{21}, \xi_{22})$ after two periods - $2T$ are known from numerical calculations. Then, the solution of the system (4.200) does not encounter any difficulties.

4.4.4 Motion Stability of Piecewise Linear System

We complete our treatment of stability of steady state motions with a similar consideration of the piecewise linear system, using solution (4.115)-(4.117). The main problem here is that the steady state solution consists of four functions (for the symmetric case) smoothly attached to each other like a chain. We could, of course, treat this solution as numerical and use the method described in subchapter 4.4.3, but there is a more elegant way.

An analogous problem was solved by Murashige, et al [1998] for a piecewise nonlinear system. Following the principle described there, we calculate the Jacobean matrix for each range first. The resulting Jacobean matrix is their product.

Here, we use the same functions f_0 and f_1 as they were defined in subchapter 4.2.6 with the only difference that we have to add the initial conditions for the roll angle at the moment of crossing the boundary between two linear ranges. Since we have to consider the full period of motion (instead of the half-period as in subchapter 4.2.6) both positive and negative boundaries are to be crossed, see fig. 4.24.

In order to take into account this circumstance, the formulae (4.116) and (4.117) are used. However, the solution for the first range has to be re-written as:

$$f_0(T_0, \dot{\phi}_0, \varphi_0) = \phi_{0a}(\phi_{n0,} \dot{\phi}_0) e^{-\delta t} \sin\left(\omega_0 t + \varepsilon(\phi_{n0,} \dot{\phi}_0)\right) + q_a \sin(\omega t + \beta_q + \varphi_0) \qquad (4.201)$$

With $\phi_{n0} = -\phi_{m0}$ the Jacobean matrix is expressed as:

$$J_0 = \begin{vmatrix} \dfrac{\partial f_0(T_0, \dot{\phi}_0, \varphi_0)}{\partial \phi_{n0}} & \dfrac{\partial \dot{f}_0(T_0, \dot{\phi}_0, \varphi_0)}{\partial \phi_{n0}} \\ \dfrac{\partial f_0(T_0, \dot{\phi}_0, \varphi_0)}{\partial \dot{\phi}_0} & \dfrac{\partial \dot{f}_0(T_0, \dot{\phi}_0, \varphi_0)}{\partial \dot{\phi}_0} \end{vmatrix} \qquad (4.202)$$

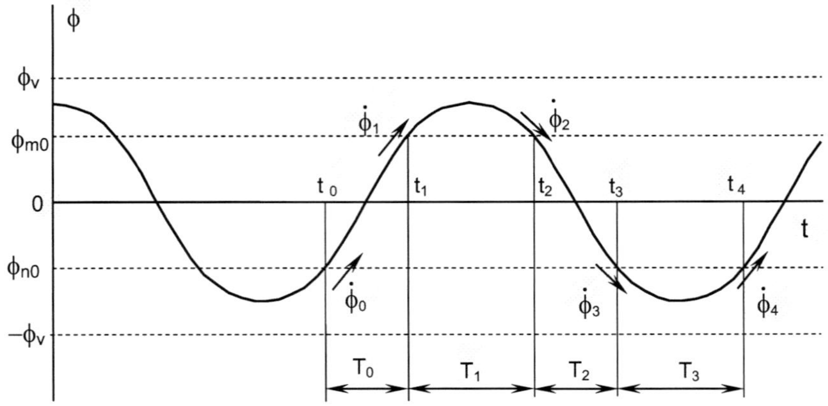

Fig. 4.24 On motion stability of piecewise linear system

The steady state solution for the second range is defined by formula (4.117). We have rewritten it to emphasize that the initial roll angle here is the positive boundary:

$$f_1(T_1, \dot{\phi}_1, \varphi_1) = A(\phi_{m0}, \dot{\phi}_1)e^{\lambda_1 t} + B(\phi_{m0}, \dot{\phi}_1)e^{\lambda_2 t} + p_a \sin(\omega t + \beta_p + \varphi_1) + \phi_V \qquad (4.203)$$

This formula leads to the following expression for the Jacobean matrix:

$$J_1 = \begin{vmatrix} \dfrac{\partial f_1(T_1, \phi_{m0}, \dot{\phi}_1,, \varphi_0 + \omega T_0)}{\partial \phi_{m0}} & \dfrac{\partial \dot{f}_1(T_1, \phi_{m0}, \dot{\phi}_1, \varphi_0 + \omega T_0)}{\partial \phi_{m0}} \\ \dfrac{\partial f_1(T_1, \phi_{m0}, \dot{\phi}_1,, \varphi_0 + \omega T_0)}{\partial \dot{\phi}_1} & \dfrac{\partial \dot{f}_1(T_1, \phi_{m0}, \dot{\phi}_1, \varphi_0 + \omega T_0)}{\partial \dot{\phi}_1} \end{vmatrix} \qquad (4.204)$$

The third range is again controlled by the solution (4.116), but we have to rewrite it with respect to the positive boundary:

$$f_2(T_0, \dot{\phi}_0, \varphi_0) = \phi_{0a}(\phi_{m0,} \dot{\phi}_0)e^{-\delta t} \sin\big(\omega_0 t + \varepsilon(\phi_{m0,} \dot{\phi}_0)\big) + q_a \sin(\omega t + \beta_q + \varphi_0) \qquad (4.205)$$

The Jacobean matrix is:

$$J_2 = \begin{vmatrix} \dfrac{\partial f_2(T_0, \phi_{m0}, \dot{\phi}_2, \varphi_0 + \omega(T_0 + T_1))}{\partial \phi_{m0}} & \dfrac{\partial \dot{f}_2(T_0, \phi_{m0}, \dot{\phi}_2, \varphi_0 + \omega(T_0 + T_1))}{\partial \phi_{m0}} \\ \dfrac{\partial f_2(T_0, \phi_{m0}, \dot{\phi}_2, \varphi_0 + \omega(T_0 + T_1))}{\partial \dot{\phi}_2} & \dfrac{\partial \dot{f}_2(T_0, \phi_{m0}, \dot{\phi}_2, \varphi_0 + \omega(T_0 + T_1))}{\partial \dot{\phi}_2} \end{vmatrix} \qquad (4.204)$$

Motions on the fourth range are governed by the solution (4.117) by taking into account that both the boundary and angle of vanishing stability are negative:

$$f_3(T_1, \dot{\phi}_1, \varphi_1) = A(\phi_{n0}, \dot{\phi}_1)e^{\lambda_1 t} + B(\phi_{n0}, \dot{\phi}_1)e^{\lambda_2 t} + p_a \sin(\omega t + \beta_p + \varphi_1) - \phi_V \qquad (4.205)$$

With the Jacobean matrix:

$$J_3 = \begin{vmatrix} \dfrac{\partial f_3(T_1, \phi_{n0}, \dot{\phi}_3, \varphi_0 + \omega(2T_0 + T_1))}{\partial \phi_{n0}} & \dfrac{\partial \dot{f}_3(T_1, \phi_{n0}, \dot{\phi}_3, \varphi_0 + \omega(2T_0 + T_1))}{\partial \phi_{n0}} \\ \dfrac{\partial f_3(T_3, \phi_{n0}, \dot{\phi}_3, \varphi_0 + \omega(2T_0 + T_1))}{\partial \dot{\phi}_3} & \dfrac{\partial \dot{f}_3(T_1, \phi_{n0}, \dot{\phi}_3, \varphi_0 + \omega(2T_0 + T_1))}{\partial \dot{\phi}_3} \end{vmatrix} \tag{4.206}$$

The resulting Jacobean matrix can be expressed as a product:

$$J(\vec{G}) = J_3 \cdot J_2 \cdot J_1 \cdot J_0 \tag{4.207}$$

Calculation of the partial derivatives is not so easy, so we suggest numerical differentiation for practical calculations. It is possible to do more analytical work, trying to get formulae without numerical procedures, using the chain formula for differentiation of multivariable functions f_0, f_1, f_2, f_3. However, application of this formula would require working with inverse functions, that cannot be expressed via elementary functions, so a numerical procedure seems to be more rational than any other approximate solution involving, for example, a multivariable series expansion.

A more general case of motion stability of a piecewise linear system, including initial heel, is considered in [Belenky, 1999].

4.5 Bifurcation Analysis

4.5.1 General

We have just completed a review of different methods of determining whether the steady state mode of motion is stable or not. The next question to be answered is what is going to happen when the steady state motion loses its stability? To address this issue, we use a standard approach of nonlinear dynamics: we pick a parameter, change it systematically and see what happens to the system. This parameter has a special name: it is called a "control parameter".

In general, the choice of control parameter strongly depends on the context of the problem. We are going to use excitation frequency as a control parameter for the roll equation. This choice is especially convenient because we usually present the solution for steady state motions as a response curve, where amplitude and phase are plotted against the excitation frequency.

The analysis consists of the following steps:

- Assign the frequency;
- Calculate amplitude and phase for the steady state solution using any of the approximate methods described in subchapter 4.2;
- Calculate initial conditions, corresponding to steady state mode of motions;
- Evaluate stability of the steady state solution: the result is presented as a pair of eigenvalues and as a point on the Trace-Determinant plane;

- Visualize the motion by numerical solution with the above initial conditions;
- Plot phase trajectory and Poincare map;
- Plot the series of eigenvalues makes a trace on the complex plane. Such a plot usually is called a "locus".

Following Nayfeh and Sanchez [1990] we consider an example for biased ship; roll motions are described as:

$$\ddot{\phi}+2\delta\dot{\phi}+\delta_3\dot{\phi}^3+\omega_\phi^2\phi+a_3\phi^3+a_5\phi^5=\alpha_E\cos\omega t \qquad (4.208)$$

The bias makes it easy to see all the instabilities. (However, it is possible to find the same phenomena of non-biased systems as well [Belenky, 1999].) The following implements the bias for the system (4.208), see Nayfeh and Sanchez [1990] for more details:

$$\phi(t)=\phi_S+u(t) \qquad (4.209)$$

Numerical values for the coefficients are given in table 4.2.

Table 4.2 Numerical values

Static bias, ϕ_s, degree	6	Restoring coefficient a_3/ω_ϕ^2	-1.3
Damping coefficient δ, 1/s	0.086	Restoring coefficient a_5/ω_ϕ^2	0.3
Damping coefficient δ_3, s/rad^2	0.108	Amplitude of effective wave slope, Rad	0.23
Natural frequency ω_ϕ, 1/s	5.278	Inertial coefficient a_{44}/I_{xx}	0.25

4.5.2 Fold Bifurcation

We are moving from the origin of the coordinate system. The solution is trivial before point A at the response curve on fig. 4.25.

There are three solutions after point A. Let us look at the low amplitude first. Nothing special happens while moving from point A to point B in fig. 4.25. The locus for the eigenvalues is shown in fig. 4.26.

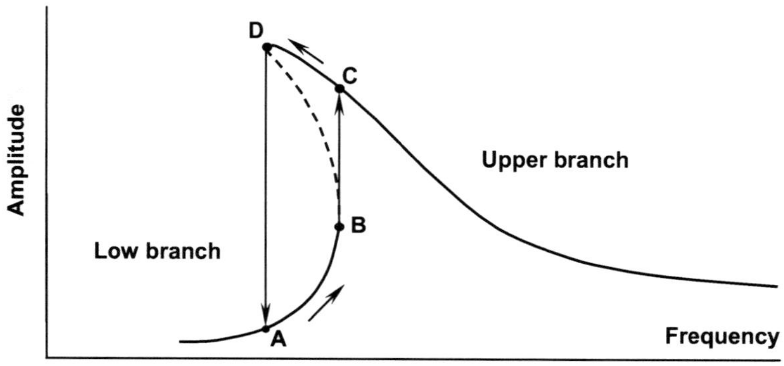

Fig 4.25 "Jump" phenomenon or fold bifurcation

We observe that the eigenvalues remain complex for a while. They become real at $\omega/\omega_\phi = 0.7293$ and leave the unit circle in a positive direction at $\omega/\omega_\phi = 0.7298$, which corresponds to point B in fig. 4.25. It means that the steady state regime middle amplitude is unstable. What does it mean?

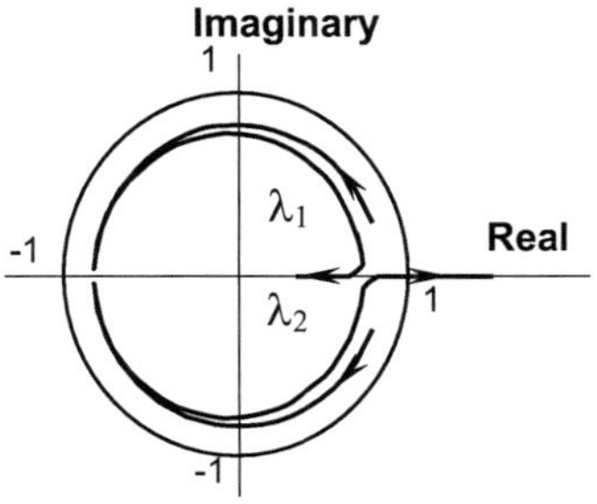

Fig. 4.26 Locus: track of eigenvalues

An unstable steady state regime cannot be realized physically or even numerically for a long time. Small perturbations that always exist in the real world will rise and eventually take the system away from the unstable regime to the stable one. If we are at the point B, the most probable alternative is a high amplitude regime at the point C on fig 4.25. (The other alternative is the stable regime near another stable equilibrium, which means capsizing; these kinds of transitions are considered in Chapter 5.)

A numerical solution of the equation (4.209) is similar to a physical experiment in the sense that an unstable regime cannot be reproduced as well. The computer presents values with a finite number of digits. This means that all the physical values in the computer are approximate, and these associated inaccuracies play the role of a small perturbations, which will increase and take the system to a stable state. So the numerical method cannot provide us with the unstable solution lasting for a significant time. (It is possible, however, to get an unstable steady state solution for a short time, like one or two periods, but this requires very accurate initial conditions, which could be found only with a really fine mesh. Such a procedure requires a significant amount of computing.)

A numerical calculation shows that the system makes the "jump" to higher amplitude being started with the initial conditions corresponding to point B. The phase trajectory of such a jump is shown in fig. 4.27.

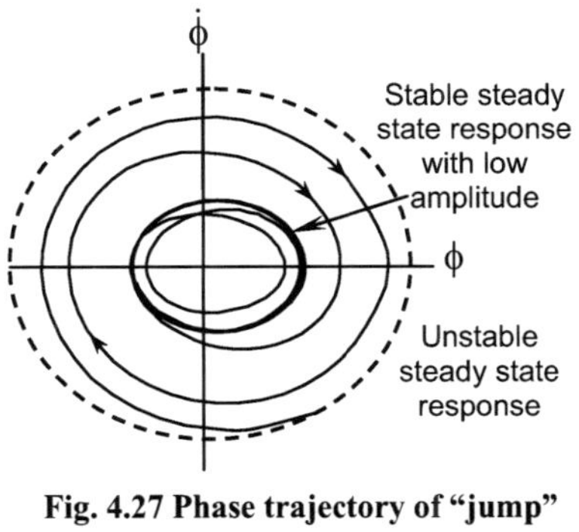

Fig. 4.27 Phase trajectory of "jump"

Let's continue moving from point B to point D (in fig. 4.25) along the dashed line. To do that, we decrease excitation frequency and assign initial conditions for the middle amplitude case. The numerical solution cannot be obtained for a long time: the system "jumps" low or to high amplitude mode. The choice where to jump depends on initial conditions. We will address the transition problem later in Chapter 5 when we will be considering capsizing, which is also a transition to another stable state of the system.

When we reach point D in fig. 4.25, the eigenvalues return back to the unit circle and the steady state regime becomes stable again, fig 4.28.

In order to move further along the response curve (fig. 4.25), we have to increase the excitation frequency again. After we reach point C, there are no longer several solutions available.

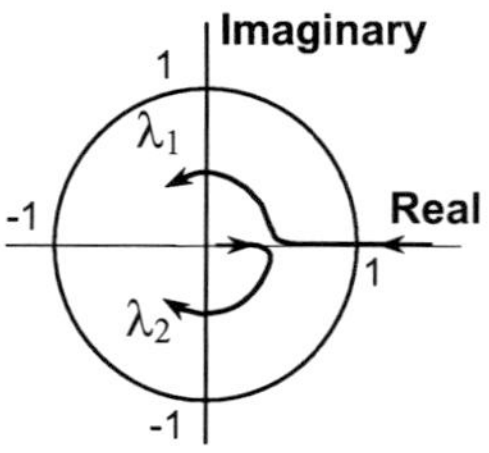

Fig. 4.28 Eigenvalues (through point D)

Now, let us decrease the excitation frequency from point C and move back along the response curve in fig. 4.25. The amplitudes increase until we reach point D, where only one solution is available. Here, the system experiences "jump back" to point A with a dramatic decrease of amplitude.

Since there are two stable steady state solutions (they are always separated by the unstable one), we see the phenomena of hysteresis here: with "jump up" and "jump down" happening at different frequencies.

Generally, the phenomenon we just observed is called "bifurcation". It is an instant qualitative change in the behavior of the system. In the absence of bifurcation, a small change of the control parameter leads to small changes in the response. Bifurcation breaks this continuity.

There are many different bifurcations in a nonlinear dynamical system and some of them were found for large-amplitude ship roll. Classification of bifurcation exists [Thompson and Stewart 1986], but standard terminology is not yet established. The bifurcation we have just seen happens when eigenvalues leave the unit circle in a positive direction and is known under names "fold bifurcation", or "tangent instability".

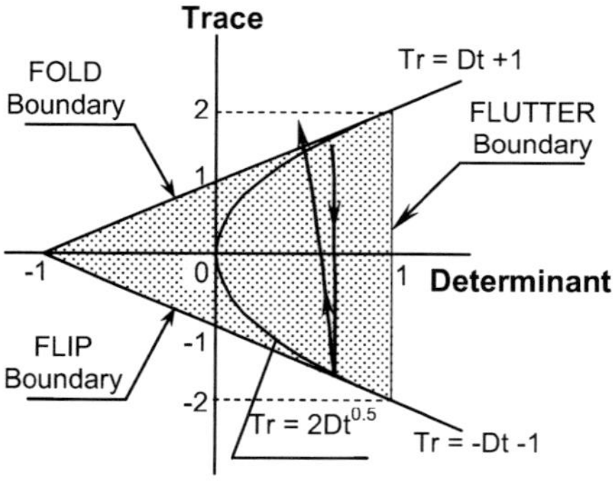

Fig. 4.29 Trace –Determinant plane

Another convenient way to observe bifurcation is by use of the Trace-Determinant plane, see fig. 4.29. It is not so difficult to see from formula (4.185) that stability boundaries in the Trace-Determinant plane by three straight lines: $Tr = Dt + 1$, $Tr = -Dt - 1$ and $Dt = 1$. Curve $Tr = 2\sqrt{Dt}$ is an area of points where the discriminate of the quadratic equation (4.182) is zero: eigenvalues are complex in the inner area of this curve.

Analogously to the locus of eigenvalues, the type of bifurcation can be seen by the boundary that is crossed by the image point. The "upper" boundary corresponds to fold bifurcation, the lower boundary corresponds to flip or period doubling bifurcation (we will be looking at this bifurcation in subchapter 4.5.3) and the left boundary corresponds to flutter bifurcation, which is impossible in the roll equation we study.

4.5.3 Period Doubling and Deterministic Chaos

To observe flip or period doubling bifurcation, we start from high frequencies and decrease excitation frequency - our control parameter. The picture of behavior of the eigenvalues – the locus, is shown in fig. 4.30. The eigenvalues stay complex, become real (ω/ω_ϕ=1.053) and leave the unit circle in a negative direction (ω/ω_ϕ=1.000).

First the phase trajectory becomes clearly asymmetric (ω/ω_ϕ=0.93), fig. 4.31a. Further decreasing of the excitation frequency leads to the first doubling of the response period

(ω/ω_ϕ=0.89), fig. 4.31b. The next steps are $4T$ response (ω/ω_ϕ=0.885), fig. 4.31c and $8T$ response (ω/ω_ϕ=0.88) fig. 4.31d. Then further doubling of the period leads the system to a chaotic response (ω/ω_ϕ=0.87) fig. 4.31e. Poincare maps and fragments of time histories are shown along with phase trajectories.

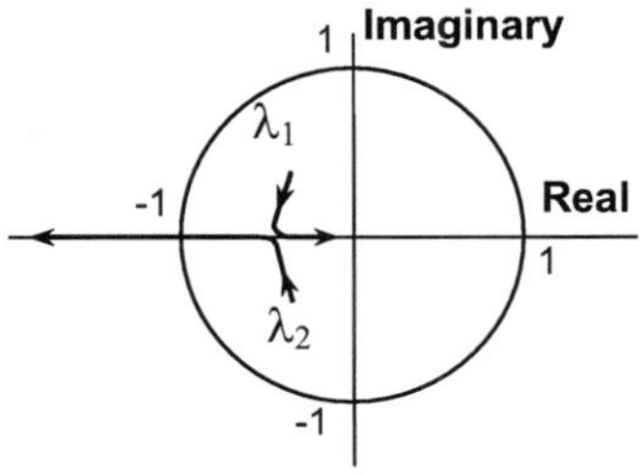

Fig 4.30 Locus for eigenvalues indicating flip bifurcation

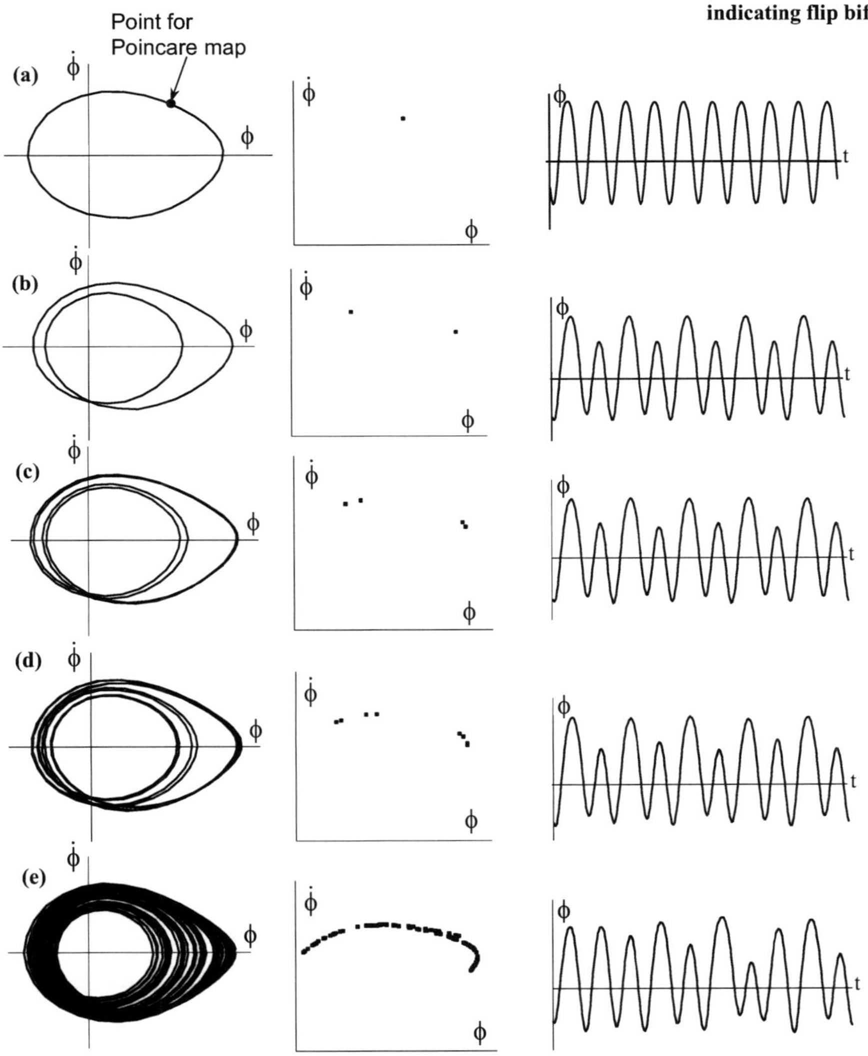

Fig. 4.31 Flip bifurcation: phase planes, Poincare maps and time histories

Deterministic chaos is a typical phenomenon for a nonlinear system. As we have just seen, nonlinear roll is not an exception. There are several ways for a general nonlinear system to develop a chaotic response, for more information, see Thompson and Stewart [1986]. Further decreasing of excitation frequency takes the system out of the chaotic state and we observe inverse development of period doubling or flip bifurcation. It is convenient to show the whole picture of period doubling and chaos in the bifurcation diagram shown in fig. 4.32. The Trace - Determinant plane is shown in fig. 4.33.

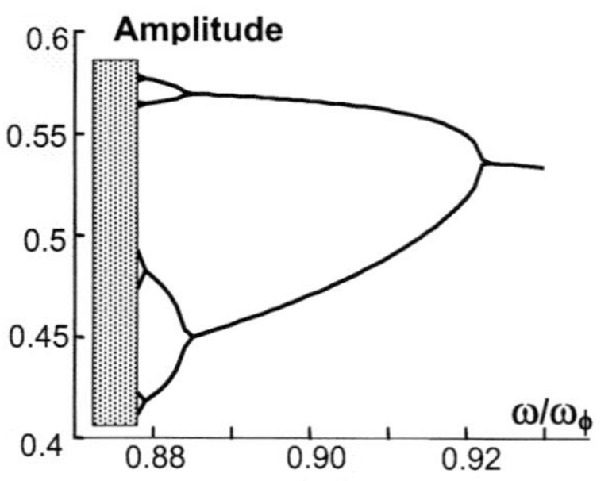

Fig 4.32 Flip bifurcation diagram

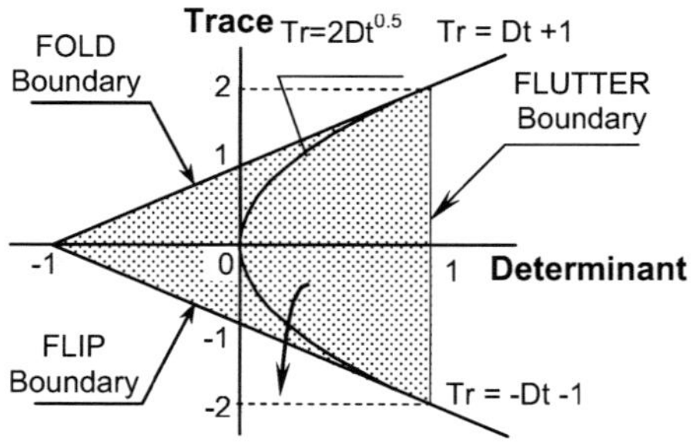

Fig. 4.33 Trace-Determinant plane for flip bifurcation

4.5.4 Bifurcations of Piecewise Linear System

So far, we were able to show that a piecewise linear system (4.6) has the same properties as a "conventional" nonlinear equation of roll (4.30). We obtained the backbone line in subchapter 4.1.2, the exact steady state solution in subchapter 4.2.6 and showed that stability of its steady state solution can also be checked easily (subchapter 4.4.4). Now, we are going to look at its bifurcation behavior.

As can be seen from fig.4.34, the motion stability analysis indicates the presence of flip and fold bifurcations. Existence of fold bifurcation is clearly seen from fig.4.11 where we have a range with three amplitudes corresponding to the same frequency. As we have just seen, the conventional nonlinear system has three responses in the fold bifurcation zone: two stable ones and one unstable between them.

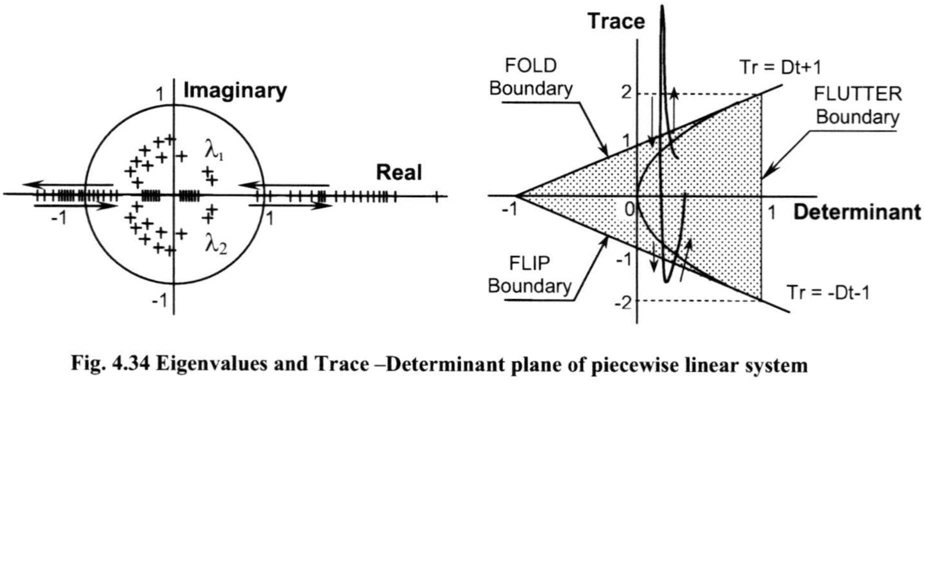

Fig. 4.34 Eigenvalues and Trace –Determinant plane of piecewise linear system

Working with the piecewise linear system, we also get three responses in this area, one of them is pure linear or trivial, so it is definitely stable. Two piecewise linear responses were obtained from the same system of equation (4.115) using two different initial points. One of these initial points corresponds to a high amplitude response of the equivalently linearized solution; another one is from the middle one. The middle solution is expected to be unstable, the high one - stable.

These expectations are correct. Fold bifurcation is illustrated in fig. 4.35, where both "jumps" are shown. The system was started from initial conditions corresponding to the unstable steady state regime (with middle amplitude). After a certain time, an increasing perturbation finally took the system towards one of the stable steady states.

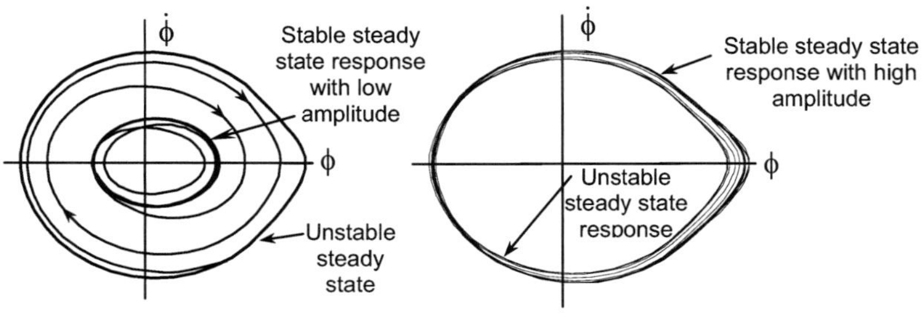

Fig. 4.35 Fold bifurcation in piecewise linear system: phase trajectory of transitions from unstable steady state regime (with middle amplitude) towards stable regimes with high or low amplitude

The general appearance of eigenvalues behavior shown in fig. 4.34 indicates the possibility of flip bifurcation. Our task is simply to show that flip bifurcation and consequent deterministic chaos can be found in a piecewise linear system. Figures 4.36 and 4.37 show phase planes, time histories and Poincare maps. Table 4.3 contains the numerical data for this example.

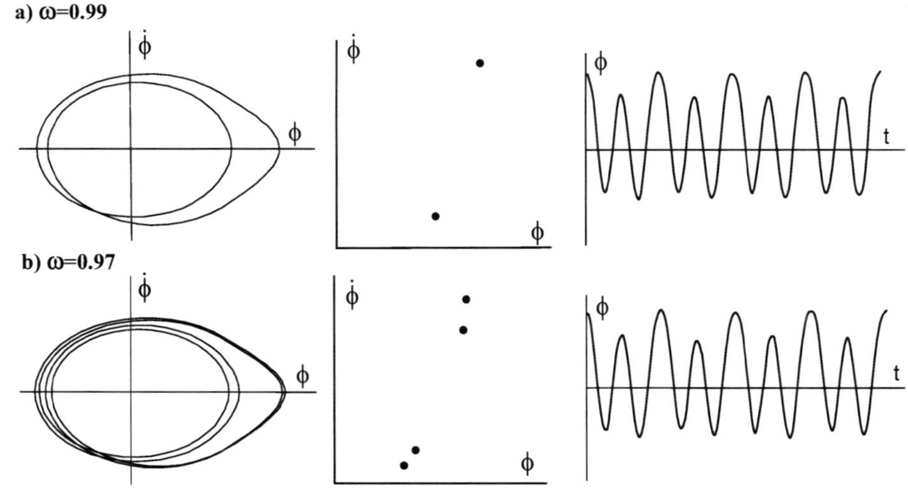

Fig 4.36 Development of flip bifurcation in piecewise linear system

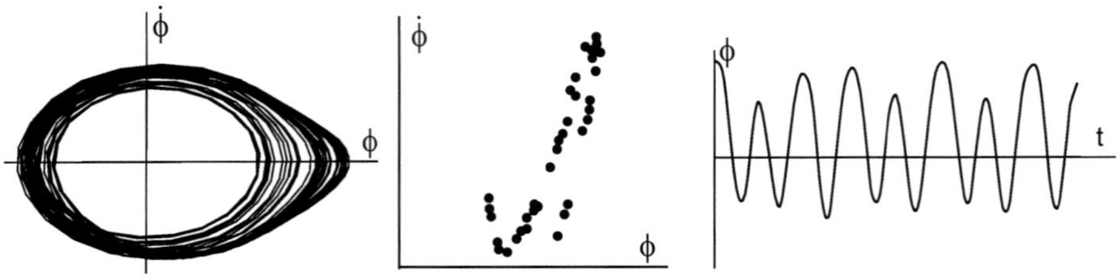

Fig. 4.37 Deterministic chaos in piecewise linear system (ω=0.92439)

Table 4.3 Numerical values for the considered example of piecewise linear system

Damping coefficient	0.1	Angle of vanishing stability	1
Excitation amplitude	0.2	Bias	0.05
Excitation frequency	0.99-0.92	Number of point per period	50

As we have seen from our review of bifurcations, the piecewise linear system qualitatively makes no difference with a conventional nonlinear system. Belenky [1999] contains more details.

4.6 High Order Resonances[1]

4.6.1 General

We continue our analysis of the nonlinear qualities of the ship roll equation with ultra and sub-harmonic resonance phenomena. The linear system has only one resonance: when the excitation frequency is close to the natural frequency, we observe a dramatic increase in response amplitude. The nonlinear dynamical system also shows an increase of oscillation amplitude when it is excited near the natural frequency, but, as it is well known, resonance phenomena are possible at frequencies that are a multiple of the natural one.

If the excitation frequency $\omega = k \cdot \omega_\phi$ where k is an integer, then sub-harmonic resonance takes place.

If excitation frequency $\omega = \omega_\phi / m$ where m is an integer, then ultra-harmonic resonance takes place.

A nonlinear system possesses an infinite number of high order (ultra and sub-harmonic) resonances. Here, we shall study the simplest example of these resonance cases using the nonlinear roll equation with linear damping and a cubic presentation of the restoring term:

$$\ddot{\phi} + 2\delta\dot{\phi} + \omega_\phi^2\phi - a_3\phi^3 = \alpha_E \cdot \sin \omega t \qquad (4.210)$$

[1] The author is grateful to Prof. Francescutto for fruitful discussion of the materials of this subchapter.

4.6.2 Ultra-harmonic Resonance

Following Cardo, *et al* [1981] we shall search for an ultra-harmonic solution of the first expansion in the following form:

$$\phi_0(t) = c \cdot \sin(3\omega t + \varphi_3) + b \cdot \sin \omega t \quad (4.211)$$

Where b is the main harmonic amplitude and c is the ultra-harmonic amplitude.

The steady state solution (4.211) of the nonlinear differential equation (4.211) can be obtained by the any appropriate method, some of which were considered in subchapter 4.2. Following Cardo, *et al* [1981] we use the perturbation method here. If ε is a small value, it is chosen as a bookkeeping parameter:

$$\delta = \varepsilon \nu; \quad a_3 = a_{30}\varepsilon \quad (4.212)$$

And:

$$\omega_\phi^2 = 9\omega^2 - \varepsilon\omega_1 - \varepsilon^2\omega_2 - \varepsilon^3\omega_3 - \ldots \quad (4.213)$$

The following form of the whole stable state solution is assumed:

$$\phi = \phi_0 + \varepsilon\phi_1 + \varepsilon^2\phi_2 + \varepsilon^3\phi_3 + \ldots. \quad (4.214)$$

Substitution of (4.212), (4.213) and (4.214) into (4.210) yields:

$$\ddot{\phi} + 2\varepsilon\nu\dot{\phi} + \left(9\omega^2 - \varepsilon\omega_1 - \varepsilon^2\omega_2 - \varepsilon^3\omega_3 - \ldots\right)\phi - a_{30}\varepsilon\phi^3 = \alpha_E \varepsilon \sin \omega t \quad (4.215)$$

Further, we shall truncate all the series up to the second degree of bookkeeping parameter ε; our third order solution looks like:

$$\phi^3 = \phi_0^3 + 3\varepsilon\phi_1\phi_0^2 + 3\varepsilon^2\left(\phi_1\phi_0^2 + \phi_0\phi_1^2\right) \quad (4.216)$$

Then we substitute (4.216) into (4.215) and equalize the right hand and left hand terms with the same power of small parameter ε:

$$\varepsilon^0: \quad \ddot{\phi}_0 + 9\omega^2\phi_0 = \alpha_{E0} \sin \omega t \quad (4.217)$$

$$\varepsilon^1: \quad \ddot{\phi}_1 + 9\omega^2\phi_1 = -2\nu\dot{\phi}_0 + \omega_1\phi_0 - a_{30}\phi_0 \quad (4.218)$$

$$\varepsilon^2: \quad \ddot{\phi}_2 + 9\omega^2\phi_2 = -2\nu\dot{\phi}_1 - \omega_1\phi_1 + \omega_2\phi_0 - 3a_{30}\phi_0\phi_1 \quad (4.219)$$

The first expansion equation (4.217) is heterogeneous here. Its solution assumed in the form of (4.211) can be interpreted as consisting of a general solution of the autonomous equation $c \cdot \sin(3\omega t + \varphi_3)$ and partial solution of heterogeneous equation $b \cdot \sin \omega t$. The amplitude of the last one can be found from the equation (4.217) directly, taking into account (4.214). It can be expressed as:

$$b = \frac{\alpha_{E0}}{9\omega^2 - \omega^2} = \frac{\alpha_{E0}}{\omega_\phi^2 - \omega^2} \quad (4.220)$$

Amplitude c and initial phase angle φ_3 of the general solution of the autonomous equation can be found by the condition of elimination of the secular terms in the second order equation (4.218) using with the perturbation technique considered in subchapter 4.2.3. This condition is expressed in the following system of equations:

$$\begin{cases} c\cdot\left(\frac{3}{2}a_{30}b^2+\omega_1+\frac{3}{4}a_{30}c^2\right)\cdot\cos\varphi_3+6\omega\nu c\sin\varphi_3=-\frac{1}{4}a_{30}b^3 \\ c\cdot\left(\frac{3}{2}a_{30}b^2+\omega_1+\frac{3}{4}a_{30}c^2\right)\cdot\sin\varphi_3-6\omega\nu c\cos\varphi_3=0 \end{cases} \quad (4.221)$$

The system (4.221) can be reduced to one nonlinear equation, i.e. ultra-harmonic amplitude c by summing the second power of both equations:

$$A\cdot c^6+B\cdot c^4+C\cdot c^2+D=0 \quad (4.222)$$

Where:

$$A=\frac{9}{16}a_{30}^2;\quad B=\frac{9}{4}a_{30}^2b^2+\frac{3}{2}a_{30}(9\omega^2-\omega_\phi^2);\quad D=-\frac{1}{16}a_{30}^2b^6$$
$$C=36\omega^2\nu^2+3a_{30}b^2(9\omega^2-\omega_\phi^2)+(9\omega^2-\omega_\phi^2)^2+\frac{9}{4}a_{30}^2b^4 \quad (4.223)$$

Here, we have taken into account: $\omega_1 = 9\omega^2-\omega_\phi^2$, that was derived from (4.213).

Equation (4.222) can be solved analytically. There are 6 roots: 1 real, 1 imaginary and 4 complex or 3 real and 3 imaginary. The first case corresponds to a single stable state ultra-harmonic solution, the second one reflects the possibility of three stable state ultra-harmonic solutions, one of which is unstable, see fig. 4.38. A time history of the stable mode of the ultra-harmonic oscillation is shown in fig. 4.39

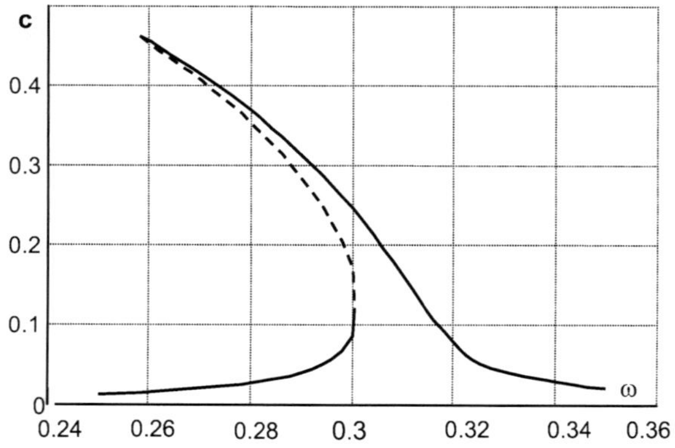

Fig. 4.38 Ultra-harmonic response curve ($\delta=\nu=0.006$, $\alpha_E=0.2$, $a_3=a_{30}=1.75$, $\omega_\phi=1$). Dashed line means unstable regime

The initial phase angle can be found from the second equation (4.221) when the ultra-harmonic amplitude has been found:

$$\varphi_3=\arctan\left(-\frac{6\omega\nu}{\frac{3}{2}a_{30}b^2+(9\omega^2-\omega_\phi^2)+\frac{3}{4}a_{30}c^2}\right) \quad (4.224)$$

According to Cardo, *et al* [1981], the ultra-harmonic resonance phenomenon is very sensitive to roll damping. Because of that, we are forced to choose a very small roll damping value coefficient for our numerical example presented in the figures above.

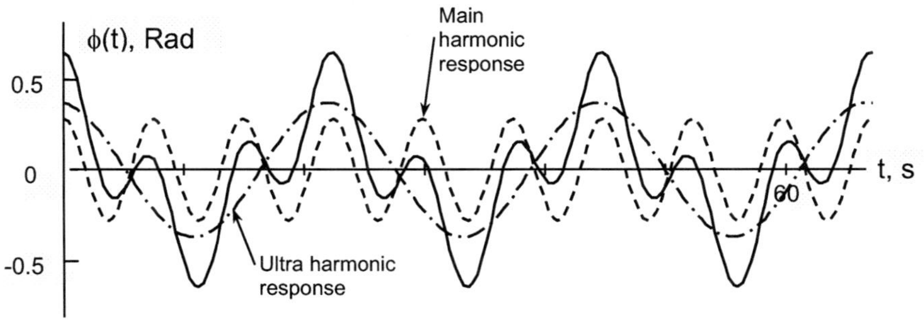

Fig. 4.39 Time history of ultra-harmonic oscillation, high amplitude mode (ω=0.28)

4.6.3 Sub-harmonic Resonance

Following Cardo, *et al* [1981] we continue our study of high order resonance by consideration of the sub-harmonic response with the frequency three times greater than the natural one. The first order expansion of the steady state solution should be taken in the following form:

$$\phi_0(t) = c \cdot \sin\left(\frac{1}{3}\omega t + \varphi_{1/3}\right) + b \cdot \sin \omega t \tag{4.225}$$

Using the perturbation technique for finding the elements is practically the same as in the previous case with ultra-harmonic response:

$$\omega_\phi^2 = \frac{1}{9}\omega^2 - \varepsilon\omega_1 - \varepsilon^2\omega_2 - \varepsilon^3\omega_3 - ... \tag{4.226}$$

The system of consequent linear differential equations, each of which corresponds to a certain power of bookkeeping parameter ε will look like:

$$\varepsilon^0: \qquad \ddot{\phi}_0 + \frac{1}{9}\omega^2\phi_0 = \alpha_{E0} \sin \omega t \tag{4.227}$$

$$\varepsilon^0: \qquad \ddot{\phi}_1 + \frac{1}{9}\omega^2\phi_1 = -2\nu\dot{\phi}_0 + \omega_1\phi_0 - a_{30}\phi_0 \tag{4.228}$$

$$\varepsilon^0: \qquad \ddot{\phi}_2 + \frac{1}{9}\omega^2\phi_2 = -2\nu\dot{\phi}_1 - \omega_1\phi_1 + \omega_2\phi_0 - 3a_{30}\phi_0\phi_1 \tag{4.229}$$

The first expansion equation (4.227) is heterogeneous here. Its solution assumed in the form of (4.225) can be interpreted as consisting of the general solution of the autonomous equation and a partial solution of the heterogeneous equation. The amplitude of the latter can be found from the equation (4.227), analogous to the previous case:

$$b = \frac{\alpha_{E0}}{\frac{1}{9}\omega^2 - \omega^2} = \frac{\alpha_{E0}}{\omega_\phi^2 - \omega^2} \tag{4.230}$$

Amplitude c and initial phase angle φ_3 of the general solution of the autonomous equation can be found by elimination of the secular terms in the second order equation (4.228) that can be expressed in the following system of equations:

$$\begin{cases} c\cdot\left(\frac{3}{2}a_{30}b^2 + \omega_1 + \frac{3}{4}a_{30}c^2\right)\cos\varphi_{1/3} - \frac{2}{3}\omega\nu c\sin\varphi_{1/3} = \frac{3}{4}a_{30}bc^2\cos(2\varphi_{1/3}) \\ c\cdot\left(\frac{3}{2}a_{30}b^2 + \omega_1 + \frac{3}{4}a_{30}c^2\right)\sin\varphi_{1/3} + \frac{2}{3}\omega\nu c\cos\varphi_{1/3} = -\frac{3}{4}a_{30}bc^2\cos(2\varphi_{1/3}) \end{cases} \tag{4.231}$$

The system (4.231) can be reduced to one nonlinear equation, i.e. sub-harmonic amplitude c by summing the second power of both equations:

$$A\cdot c^4 + B\cdot c^2 + C = 0 \tag{4.232}$$

Where:

$$\begin{aligned} &A = \frac{9}{16}a_{30}^2;\quad B = \frac{27}{16}a_{30}^2 b^2 + \frac{3}{2}a_{30}\left(\frac{1}{9}\omega^2 - \omega_\phi^2\right)^2 \\ &C = \frac{4}{9}\omega^2\nu^2 + \left(\frac{1}{9}\omega^2 - \omega_\phi^2\right)^2 + 3a_{30}b^2\left(\frac{1}{9}\omega^2 - \omega_\phi^2\right) + \frac{9}{4}a_{30}^2 b^4 \end{aligned} \tag{4.233}$$

Equation (4.232) is bi-quadratic; it has 4 roots:

$$c_{1,2,3,4} = \pm\sqrt{\frac{-B \pm \sqrt{B^2 - 4AC}}{2A}} \tag{4.234}$$

We are interested in the positive real roots only. Formally, it is possible to obtain one or two of them, one to stable steady state. The discriminate of the equation is the threshold of the sub-harmonic response; see Cardo, *et al* [1981]. The response curve is shown in fig. 4.40 and a sample of the time history in fig 4.41.

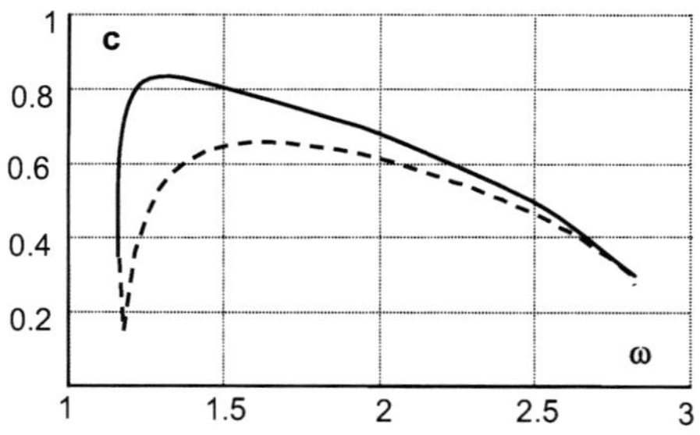

Fig. 4.40 Sub-harmonic response curve (δ=ν=0.006 α_E=0.2, a_3=a_{30}=1.75, ω_ϕ=1). Dashed line means unstable regime

The initial phase angle can be found from any equation of (4.231) once the sub-harmonic amplitude is known:

$$\delta_{1/3} = \arctan\left(-\frac{6\omega\nu}{\frac{3}{2}a_{30}b^2 + \left(\frac{1}{9}\omega^2 - \omega_\phi^2\right) + \frac{3}{4}a_{30}c^2} \right) \tag{4.235}$$

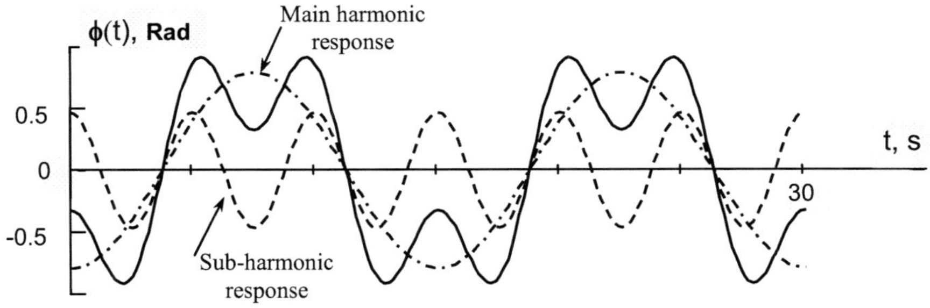

Fig. 4.41 Time history of sub-harmonic oscillation (ω=1.252)

Now, it is possible to plot the whole response curve including main, ultra- and sub-harmonic resonance features. To do so, we need to calculate the maximum roll amplitude versus excitation frequency. The approximate formula for the maximum roll amplitude near the ultra-harmonic resonance is given in Cardo, *et al* [1982]:

$$\phi_{max}(\omega) = c_{ultra}(\omega) + \frac{9\alpha_E}{8\omega_\phi^2} \tag{4.236}$$

The maximum roll amplitude near sub-harmonic resonance has been calculated numerically. The whole response curve is given in fig. 4.42. The main resonance peak has one feature: it is not closed as it was in our previous example. Such a result can be explained by the error caused by the first expansion, see Nayfeh and Khdeir [1986] that is especially seen in conditions of light damping. A similar picture can be found in Francescutto [1991].

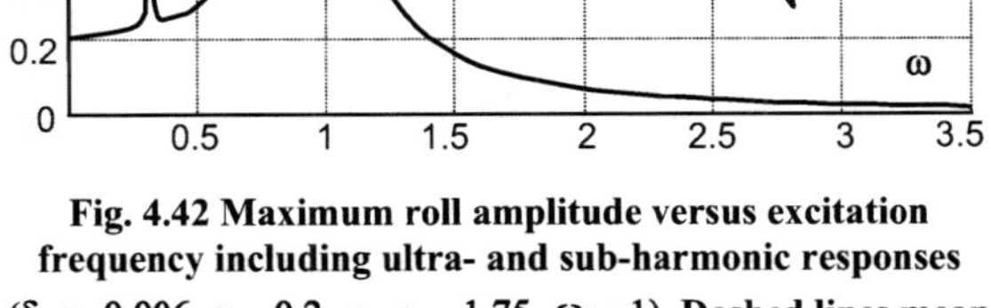

Fig. 4.42 Maximum roll amplitude versus excitation frequency including ultra- and sub-harmonic responses (δ=ν=0.006, α_E=0.2, a_3=a_{30}=1.75, ω_ϕ =1). Dashed lines mean unstable regimes.

Chapter 5

Capsizing in Regular Beam Seas

If we want to study a phenomenon, we should define it first. The most obvious definition of capsizing says, “capsizing is a transition to motion at another stable equilibrium that is dangerous from practical point of view” [Sevastianov and Pham, 1979]. However, not all the methods of stability evaluation use this definition directly. This chapter reviews these methods based on what definition of capsizing (or ship stability) they use.

5.1 Classical Definition of Stability

We start from the definition of ship stability. It is quite natural to start with the classical definition, the contemporary formulation of which was given by Krylov [1958]. It says: “we call stability the ability of a ship to float in an upright position and, if inclined under the action of an external cause, to return to the above said position after the external cause ceased acting”.

Let us look at this definition from the point of view of the theory of oscillations. It means that the external force vanishes at the moment of stability determination and consequent motion is free. The definition seems a bit artificial, because wind and wave action in a real sea could not be stopped at the moment when we are interested to see if our ship is stable or not. However, as it will be shown further, in subchapter 5.3, this classical definition works pretty well for practical purposes.

5.1.1 Concept of Separatrix

So we consider an autonomous dynamic system describing nonlinear ship roll motion in calm water:

$$\ddot{\phi}+\delta_{\phi}\dot{\phi}+\omega_{\phi}^{2}f(\phi)=0 \tag{5.1}$$

We do not consider nonlinearity of roll damping for simplicity’s sake: we keep nonlinearity for the restoring moment only. Function f that represents the GZ curve should be considered until the upside down position while we are examining capsizing, fig.5.1.

As we have seen in subchapter 4.3, the dynamic system (5.1) has three equilibria. Two of them are stable: the upright position and upside down (capsized) position. This means that the ship can stay there indefinitely if external forces are absent, and if some small excitation does occur. There is unstable equilibrium between two stable ones: it is the angle of vanishing stability. Theoretically, the system also can stay at this position for an infinitely long term, but in real life any small excitation can disturb the system and it will move to one of the stable equilibria positions.

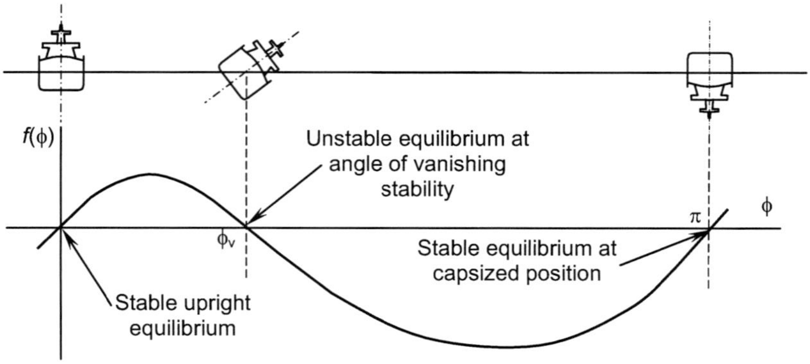

Fig. 5.1 *GZ* curve and positions of equilibria

We already implemented the phase plane concept in subchapter 4.3.5. The main advantage of the use of the phase plane in comparison with time domain is an ability to see all possible motions simultaneously. Again, the phase plane for a dynamic system (5.1) is defined as a rectangular co-ordinate axis system $(\phi, \dot{\phi})$. The state of the system at any time is reflected in the phase plane by a point with corresponding co-ordinates that is usually called the "image point". When the system is in motion, the image point is creating a trace on the phase plane. Such a trace is called "phase trajectory" and the whole aggregate of phase trajectories is called the "phase portrait". Equilibrium of the dynamic system corresponds to the so-called "singular point". So we first consider our dynamic system of the phase plane from the analysis of phase trajectories near a singular point, e.g. the above mentioned equilibria.

Let us consider the system without damping first:

$$\ddot{\phi} + \omega_\phi^2 f(\phi) = 0 \tag{5.2}$$

As it is easy to see from equation (4.132) with $\delta = 0$, its solution in the vicinity of the original or "normal" equilibrium looks like:

$$\phi = \phi_A e^{-\delta t} \sin\left(\sqrt{\omega_\phi^2 - \delta^2} \cdot t + \gamma\right) = \phi_A \sin\left(\omega_\phi \cdot t + \gamma\right) \tag{5.3}$$

With the derivative:

$$\dot{\phi} = \phi_A \omega_\phi \cos\left(\omega_\phi \cdot t + \gamma\right) \tag{5.4}$$

The phase portrait near the original equilibrium now looks like a family of ellipses. The same can be stated on the phase portrait in the vicinity of capsized equilibrium. Such a singular point is called the "centre", and we have seen it already in fig. 4.18 of subchapter 4.3.5 that contains a summary of all linear phase portraits (singular points). As can be seen from that figure, there are no visible changes for unstable equilibria, it is a saddle.

The general view of phase portrait is defined by all three singular points, see fig. 5.2. New elements, which should be particularly pointed out, are phase trajectories that separate areas, where the system is attracted by upright or capsized equilibria or experience rotation motion. Such a phase trajectory is called "separatrix".

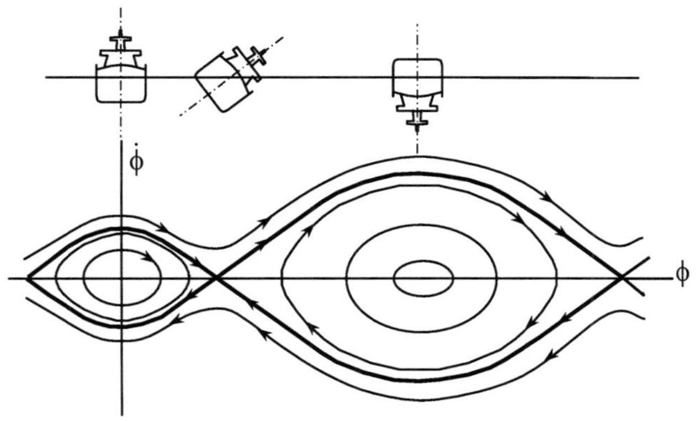

Fig. 5.2 Phase plane of free nonlinear roll motion

Separatorices (plural for separatrix) start from the unstable equilibrium as straight lines at the saddle point. These straight lines are defined by the following formulae [Andronov, *et al*, 1966]:

$$\dot{\phi} = \pm\omega_{\phi}\sqrt{-k_1}\cdot\phi \tag{5.5}$$

They are the only phase trajectories, which lead to the angle of vanishing stability. It is quite clear, however, that moving along these trajectories, the system would achieve the unstable equilibria only after infinite time. The separatorices are the only phase trajectories that are inclined to the axis ϕ with the angle different from $\pi/2$.

If we use a cubic approximation for the GZ curve:

$$\ddot{\phi} + \omega_{\phi}^2(\phi - a_3\phi^3) = 0 \tag{5.6}$$

After we introduce the non-dimensional time $\tau = \omega_{\phi}t$, the system changes to:

$$\frac{d^2\phi}{d\tau^2} + \phi - a_3\phi^3 = 0 \tag{5.7}$$

There is a close form for the separatorices [Vishnubhota, *et al*, 2000]:

$$\begin{cases} \phi(\tau) = \dfrac{1}{\sqrt{a_3}}\tanh\left(\dfrac{\tau-\tau_0}{\sqrt{2}} + \dfrac{1}{2}\right) \\ \dot{\phi}(\tau) = \dfrac{1}{\sqrt{2a_3}}\operatorname{sech}^2\left(\dfrac{\tau-\tau_0}{\sqrt{2}} + \dfrac{1}{2}\right) \end{cases} \tag{5.8}$$

Where τ_0 is initial time.

Now, let's consider the influence of damping. The phase picture in the vicinity of stable equilibria is now a stable focus, see fig. 5.3. Now separatorices are not closed. There is no rotation regime possible and there are only two attraction areas to the stable equilibria. All phase trajectories lead the system to one of the stable equilibria: to the normal or capsized one. Separatorices in the vicinity of the angle of vanishing stability are still straight lines defined by the equations:

$$\dot{\phi} = \left(-\delta \pm \sqrt{\delta^2 + \omega_\phi^2 k_1}\right) \cdot \phi \tag{5.9}$$

The separatorices subdivide the phase plane in two parts. One of them is called the "safe basin" and includes phase trajectories that lead to the upright equilibrium position. Another one contains phase trajectories that lead the system to the upside down equilibrium position, which means capsizing.

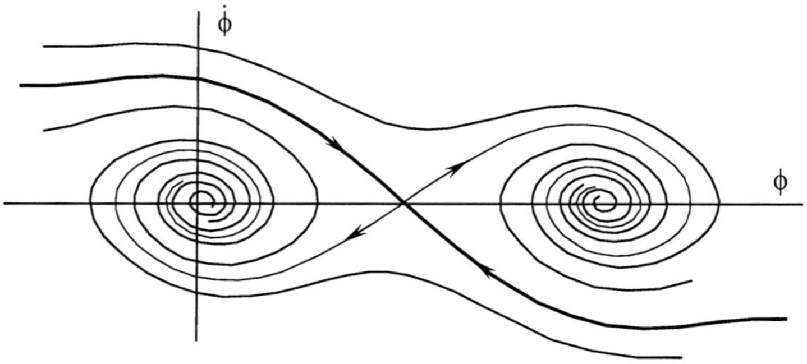

Fig. 5.3 Phase plane for free roll motion with damping

Now we can relate the classical stability definition with capsizing phenomenon. Assume the external forces created a certain angle of heel and angular velocity. It means, they took the system to a point in the phase plane and then let it go. In other words, the external forces created initial conditions.

If external forces stop their action when the angle of heel and angular velocity correspond to an image point which is situated inside the safe basin, so capsizing is impossible and the ship should be considered as stable according to the classical definition of stability.

This analysis can be generalized for an example of ship motion with all six degrees of freedom. We should stop the action of external forces at the moment we would like to assess the ship's stability: if the ship will return to her initial position, then she is stable. We can proceed with this analysis until the time the ship will not return to the normal equilibrium position. Such an initial condition can be considered as critical. In other words, this point belongs to a separatrix hyper-surface in a multidimensional phase space. Similar to the case with only one degree of freedom, crossing of the separatrix hyper-surface is related with capsizing.

Summarizing all of the above, we can state that according to the classical definition of stability, crossing of the separatrix (separatrix surface) in the phase plane (space) by the forced phase trajectory leads to capsizing.

5.1.2 Calculation of Separatrix

The next problem to consider is a method of separatrix calculation. There is no general analytical way, as there is no analytical method for solving a nonlinear differential equation of the general kind. Therefore, we have to use numerical methods. Because the separatrix divides domains of two types of solution of the differential equation (5.1), we can integrate this equation, changing initial conditions systematically. We are looking for the point when the system will not return to its initial equilibrium position. Then the iteration method can be applied for a precise location of a point that belongs to the separatrix. Repeating this procedure, we can find co-ordinates of the separatrix as accurately as needed.

There is an alternative method. If we take into account that the separatrix is also a phase trajectory and we know at least one point of this phase trajectory, it is the angle of vanishing stability. We can begin numerical integration of equation (5.1) in inverse time. The problem here is the correct choice of initial conditions. The separatrix has two branches. If we start exactly from the angle of vanishing stability, we get just one branch of the separatrix, and it is difficult to predict which one. Why does this happen?

The angle of vanishing stability is the unstable equilibrium (see subchapter 4.3). The system cannot stay there for a long time: a very small perturbation will take the system to one of the stable equilibria. As we noted in subchapter 4.5.2, the computer presents floating point values with certain accuracy, so the last digit always contains a rounding error. This error plays a role of a small perturbation during numerical integration. Since we do not have control over this error, the direction of motion will be chosen randomly.

To gain control, we have to introduce perturbation in a required direction to ensure we get both branches. These initial perturbations can be calculated using formula (5.9):

$$\begin{cases} \xi_1 = \Delta \\ \dot{\xi}_1 = \left(-\delta - \sqrt{\delta^2 + \omega_\phi^2 k_1}\right) \cdot \xi_1 \end{cases} \qquad \begin{cases} \xi_2 = -\Delta \\ \dot{\xi}_2 = \left(-\delta - \sqrt{\delta^2 + \omega_\phi^2 k_1}\right) \cdot \xi_2 \end{cases} \tag{5.10}$$

The numerical value for Δ has to be defined depending on the number of digits in the engineering value format. For example, if there are six digits and the maximum roll angle is in units of radians, it might be recommended to set Δ:

$$\Delta = 5 \cdot 10^{-6} ... 1 \cdot 10^{-5} \tag{5.11}$$

The initial condition for numerical integration of the equation (5.1):

$$\text{Branch 1:} \begin{cases} \phi_0 = \phi_v + \xi_1 \\ \dot{\phi}_0 = \dot{\xi}_1 \end{cases} \qquad \text{Branch 2:} \begin{cases} \phi_0 = \phi_v + \xi_2 \\ \dot{\phi}_0 = \dot{\xi}_2 \end{cases} \tag{5.12}$$

The saddle has two asymptotes; we have used only one. The other asymptote does not belong to the separatrix in the case of a damped system (5.1), but if there is no damping, the second asymptote is also a part of the separatrix; compare fig.5.2 and 5.3.

This method was used by Sevastianov and Pham [1979], and Umeda, *et al* [1990] for calculation of the separatrix.

5.1.3 Separatrix, Eigenvalues and Eigenvectors

There is a very important relationship between eigenvalues at the unstable equilibrium and the separatrix. The reader probably already noted the geometrical meaning of the eigenvalues for this type of singular point: they are angle coefficients for asymptotes in the phase picture.

The concept of asymptotes works fine for the phase plane. However, what are we going to do if there are more degrees of freedom? It is very possible we will have to deal with phase space beyond three dimensions that we can neither plot nor imagine. In fact, the relationship between separatrix and eigenvalues is much more general and does not depend on how many degrees of freedom are involved.

This relationship is supported through eigenvectors. Eigenvectors (as the eigenvalues) are a concept of linear algebra, see, Apostol [1997] for an example. Before we give a formal definition of the eigenvectors, let's review the roll equation linearized in the vicinity of the angle of vanishing stability (4.137):

$$\ddot{\phi} + 2\delta\dot{\phi} - \omega_\phi^2 k_1 \phi + \omega_\phi^2 k_1 \phi_v = 0$$

For the sake of simplicity we transfer the origin of the co-ordinate system into the angle of vanishing stability:

$$\varphi = \phi - \phi_v \tag{5.13}$$

Therefore:

$$\ddot{\varphi} + 2\delta\dot{\varphi} - \omega_\phi^2 k_1 \varphi = 0 \tag{5.14}$$

Also, we rewrite this equation in vector form:

$$\dot{\vec{Y}} = F(\vec{Y}, t) \tag{5.15}$$

With:

$$\dot{\vec{Y}} = \begin{pmatrix} \dot{Y}_1 \\ \dot{Y}_2 \end{pmatrix} = \begin{pmatrix} \dot{\vartheta} \\ \dot{\varphi} \end{pmatrix}; \quad \vec{Y} = \begin{pmatrix} Y_1 \\ Y_2 \end{pmatrix} = \begin{pmatrix} \vartheta \\ \varphi \end{pmatrix};$$
$$\vec{F}(\vec{Y}, t) = \begin{pmatrix} -2\delta_\phi Y_1 + \omega_\phi^2 k_1 Y_2 \\ Y_1 \end{pmatrix} = \begin{pmatrix} -2\delta_\phi \vartheta + \omega_\phi^2 k_1 \varphi \\ \vartheta \end{pmatrix} \tag{5.16}$$

The vector valued function $\vec{F}(\vec{Y}, t)$ contains only a linear operation and can be presented as:

$$\vec{F}(\vec{Y}, t) = \begin{pmatrix} -2\delta_\phi & \omega_\phi^2 k_1 \\ 1 & 0 \end{pmatrix} \cdot \begin{pmatrix} \vartheta \\ \varphi \end{pmatrix} = \mathbf{A} \cdot \vec{X} \tag{5.17}$$

Eigenvalues of the matrix **A** are the same as that for the differential equation (5.14). Using the definition for eigenvalues [Bronshtein and Semendyayev, 1997]:

$$\det(\mathbf{A} - \mathbf{I} \cdot \lambda) = 0 \tag{5.18}$$

Where **I** is the identity matrix:

$$\mathbf{I} = \begin{pmatrix} 1 & 0 \\ 0 & 1 \end{pmatrix} \tag{5.19}$$

The characteristic equation of matrix **A** (5.18) is exactly the same characteristic equation of roll at the angle of vanishing stability (4.142):

$$\lambda^2 + 2\delta\lambda - \omega_\phi^2 k_1 = 0 \tag{5.20}$$

Naturally, the eigenvalues of matrix **A** are identical to the eigenvalues of the roll equation linearized at the angle of vanishing stability defined by formula (4.143):

$$\lambda_{1,2} = -\delta \pm \sqrt{\delta^2 + \omega_\phi^2 k_1} \tag{5.21}$$

Now, the eigenvectors of matrix **A** are defined to satisfy the following equation [Bronshtein and Semendyayev, 1997]:

$$(\mathbf{A} - \lambda \cdot \mathbf{I}) \cdot \vec{X} = 0 \tag{5.22}$$

Equation (5.22) is a homogenous system of linear equations. It has non-trivial solutions because the determinant of its coefficient matrix is zero due to (5.18). Let us solve it for the eigenvalue λ_1:

$$\lambda_1 = -\delta + \sqrt{\delta^2 + \omega_\phi^2 k_1} \tag{5.23}$$

The system of linear equations looks like:

$$\begin{cases} (-2\delta - \lambda_1)X_1 + \omega_\phi^2 k_1 X_2 = 0 \\ X_1 + (-\lambda_1)X_2 = 0 \end{cases} \tag{5.24}$$

We can express the relationship between X_1 and X_2 from the second equation of the system (5.24) immediately:

$$X_1 = \lambda_1 X_2 \tag{5.25}$$

After substitution of formula (5.25) into the system (5.24), the first equation of the system is:

$$(-2\delta - \lambda_1)\lambda_1 X_2 + \omega_\phi^2 k_1 X_2 = 0 \tag{5.26}$$

Then we substitute (5.23) and see that:

$$\left(\left(-\delta - \sqrt{\delta^2 + \omega_\phi^2 k_1}\right) \cdot \left(-\delta + \sqrt{\delta^2 + \omega_\phi^2 k_1}\right) + \omega_\phi^2 k_1\right) X_2 = 0$$

Trying to simplify the equation (5.25) we obtain:

$$\left(\delta^2 - \left(\delta^2 + \omega_\phi^2 k_1\right) + \omega_\phi^2 k_1\right) \cdot X_2 = 0$$

We find that the expression in the external parentheses equals zero:

$$0 \cdot X_2 = 0 \tag{5.27}$$

Equation (5.27) does not contain any particular information about X_2. Any value satisfies this equation. It means that the system (5.24) has an infinite number of solutions as soon as its co-ordinates comply with (5.25). Provided that a is any real number, it can be expressed as:

$$\vec{X} = a \cdot \begin{pmatrix} \lambda_1 \\ 1 \end{pmatrix} \quad (5.28)$$

Analogously, we can find the second set of eigenvectors corresponding to the second eigenvalue λ_2:

$$\vec{Z} = a \cdot \begin{pmatrix} \lambda_2 \\ 1 \end{pmatrix} \quad (5.29)$$

Now, it is easy to see that these two sets of eigenvectors actually define directions of asymptotes (5.9). It means that when we want to get the separatrix of the system in the position of unstable equilibria, the system has to be perturbed in the direction of the eigenvectors, see fig 5.4.

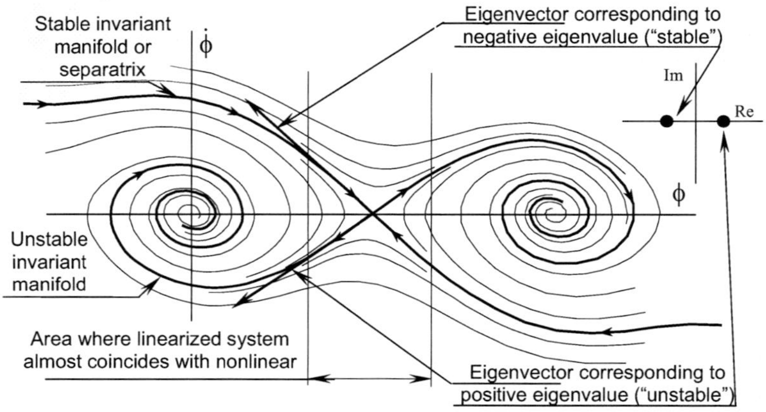

Fig 5.4 Eigenvectors and separatorices

This is a very important outcome. As it was stated earlier, it is applicable to any number of degrees of freedom: we did not assume any specific size of matrixes and vectors while making derivations (5.15-29), so it is applicable for any size.

The calculation of eigenvectors is a standard numerical procedure; corresponding subroutines or functions can be found in almost any math package. These packages, however, usually calculate eigenvectors in the normalized form (provided their length is 1), so if we would use them, the answer would be in the following form:

$$\vec{X} = \begin{pmatrix} \lambda_1 / \sqrt{\lambda_1^2 + 1} \\ 1/\sqrt{\lambda_1^2 + 1} \end{pmatrix} \quad \text{and} \quad \vec{Z} = \begin{pmatrix} \lambda_2 / \sqrt{\lambda_2^2 + 1} \\ 1/\sqrt{\lambda_2^2 + 1} \end{pmatrix} \quad (5.30)$$

5.1.4 Numerical Validation of Classical Definition of Stability

Now we have the mathematical equivalent of the classical definition of ship stability: capsizing is crossing of the separatrix. However, we are still in doubt [Sevastianov, 1978]: how can we deal with the changing excitation which can not only push the motion out of the separatrix, but to take the image point back "by force" as well? Really, there are some artificial elements in the classical definition of stability: wind and wave excitation never disappear in real seas at the moment when we would like to determine if the ship is stable or not. To erase these doubts, the classical definition of stability has to be validated.

Such a validation answer to this question was given by computer simulation, which was carried out at the Naval Architecture department of Kaliningrad Institute of Technology (Russia) by one of the authors and Pham Ngock Hoeh [Sevastianov, 1977]. The following mathematical model was studied:

$$\ddot{\phi}_r + 2\omega_\phi \mu \dot{\phi}_r + \omega_\phi^2 \phi_V f(\phi_r / \phi_V) = -\frac{I_x}{I_x + a_{44}} \ddot{\alpha}_m \qquad (5.31)$$

Here ϕ_r means relative roll angle (see Chapter 3). The non-dimensional damping coefficient was taken as a fraction of critical damping:

$$\mu = \delta / \omega_\phi \qquad (5.32)$$

The restoring term was approximated with a polynomial of the 5th order:

$$f(\phi_r / \phi_V) = a_1 \cdot (\phi_r / \phi_V) + a_3 \cdot (\phi_r / \phi_V)^3 + a_5 \cdot (\phi_r / \phi_V)^5 \qquad (5.33)$$

1. Wave excitation was asymptotically regular and was defined as follows:

$$\alpha = \alpha_0 (1 - \exp(-p^2 t^2)) \sin \omega t \qquad (5.34)$$

Where p is the parameter of the speed of increasing waves.

Such a form was adopted to avoid the influence of initial conditions and to make transition motion softer. There were 32 runs. From the first two series, that included 10 runs, it was found that the most dangerous regime can be obtained when $\omega / \omega_\phi = 0.8$ and $p / \omega = 0.1$. Capsizing was observed in three runs. Crossing of the separatrix was observed only in these three runs when capsizing also took place. One of these runs is described here: numerical data is collected in Table 5.1 and phase trajectory is shown in fig. 5.5.

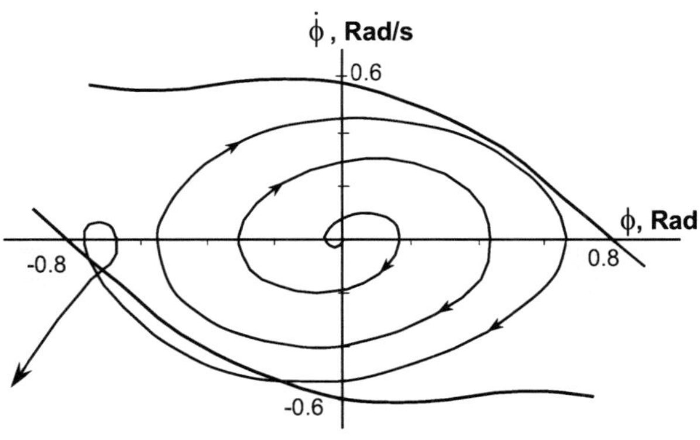

Fig. 5.5 Results of numerical simulations of ship roll and capsizing on phase plane

Table 5.1 Characteristics of the model

Displacement W, metric tons	100
Transverse moment of inertia I_{xx}, metric tons m²	300
Added moment of inertia A_{44} , metric tons m²	100
Initial metacentric height GM_0, m	0.64
Additional changing of metacentric height ΔGM, m	-0.15
Non-dimensional damping coefficient, μ	0.05
Angle of vanishing stability, ϕ_V, Rad	0.8
Amplitude of wave slope angle, α_0, degree	12
GZ approximating coefficient a_1	1
GZ approximating coefficient a_3	-11/7
GZ approximating coefficient a_5	4/7

It can be seen clearly from fig. 5.5 that crossing of the separatrix by forced motion phase trajectory leads to capsizing, perhaps not immediately, but in a short time. The other numerical simulations that consisted of 15 runs confirmed this conclusion.

At least, the validity of the classical definition of stability was not disproved by the described numerical simulations. A theoretical explanation of these observations was found using the piecewise linear model and it is considered in subchapter 5.2.

5.2 Piecewise Linear Model of Capsizing

5.2.1 General

We have discussed above (subchapter 5.1) the classical definition of stability: "the ability of a ship to float in an upright position and, if inclined under action of an external cause, to return to the above said position after the external cause ceased acting". We have noted that this definition sounds artificial because external forces are supposed to disappear at the moment we would like to assess stability. However, as we discussed in subchapter 5.1.4, the numerical simulation test has shown that the classical definition of stability is adequate.

In this subchapter we are going to use a piecewise linear system to implement a direct definition of capsizing " as a transition to oscillation near a stable position of equilibrium that is dangerous from a practical point of view" [Sevastianov and Pham, 1979].

We introduced the piecewise linear system in Chapter 4, showing that it is the simplest system that describes capsizing directly as a transition to another equilibrium. We have seen that we can treat a piecewise linear system as conventional nonlinear. We can find the natural period of large amplitude oscillations (subchapter 4.1.2) and the steady state regime of forced motions (subchapter 4.2.6). We were able to check the stability of steady state motion (subchapter 4.4.4) and to see the development of bifurcations (subchapter 4.5.4). Again, all these studies were done to convince the reader that despite the "strange" approximation of the *GZ* curve, the piecewise system is a qualitatively adequate model of nonlinear roll motions of ship.

Generally speaking, the piecewise linear model is a particular case of a nonlinear dynamic system. Andronov, *et al* [1966] applied this approach for clock theory and the description of dry friction. Komuro [1988, 1988a, 1992, 1994] carried out comprehensive theoretical studies of the piecewise linear vector field that is a general form of the model we use. The piecewise linear system has been applied to a wide range of engineering problems [Smolnikov and Byohkov, 1972; Choi and Noah, 1988; Kim and Noah, 1991; Troesch, *et al*, 1992; Karr, *et al*, 1995; Zuo and Hjelmstad, 1998].

5.2.2 Capsizing in Piecewise Linear System

The main question we are interested in is how capsizing occurs in the piecewise linear system. Let us consider it in a form (4.106):

$$\ddot{\phi} + 2\delta\dot{\phi} + \omega_\phi^2 f_L^*(\phi) = \alpha_E \sin(\omega t + \varphi_E)$$

Piecewise linear term $f_L^*(\phi)$ is described in formula (4.7) and shown in fig.4.3. The system has a solution that "switches" at the moment of crossing boundary, see formula (4.107):

$$\phi = \begin{cases} \phi_a e^{-\delta t} \sin(\omega_0 t + \varepsilon) + q_a \sin(\omega t + \beta_q + \varphi_0) & \phi \in [-\phi_{m0}; \phi_{m0}] \\ Ae^{\lambda_1 t} + Be^{\lambda_2 t} + p_a \sin(\omega t + \beta_p + \varphi_1) + \phi_V & \phi \in]\phi_{m0}; \phi_{m1}[\end{cases}$$

With eigenvalues and arbitrary constants defined by formulae (4.108-114), we look at the system when it crosses the boundary ϕ_{m0}, rolling in a positive direction, i.e. coming from range 0 to range 1. According to the general solution (4.107), its second equation is about to take control over the motion:

$$\phi = Ae^{\lambda_1 t} + Be^{\lambda_2 t} + p_a \sin(\omega t + \beta_p + \varphi_1) + \phi_v \tag{5.35}$$

It clearly can be seen from the formula (4.114) that in the pair of the eigenvalues, one is always positive and another is always negative:

$$\begin{aligned} \lambda_1 &= -\delta + \sqrt{\omega_\phi^2 k_{f1} + \delta^2} > 0 \\ \lambda_2 &= -\delta - \sqrt{\omega_\phi^2 k_{f1} + \delta^2} < 0 \end{aligned} \tag{5.36}$$

Evidently the term $A\exp(\lambda_1 t)$ with positive index λ_1 is unbounded and terms with negative index λ_2 and harmonic term $p(t) = p_a \sin(\omega t + \beta_1 + \varphi_1) + \phi_V$ are bounded.

As soon as a ship reaches angle ϕ_{m0} and enters the range 1, her further motion will be defined by the unbounded term, which will lead the motion out of the range 1. This can be done in two ways: back to range 0 or further to range 2. The sign of the coefficient *A* defines the direction of this term's action.

If $A>0$ and the ship has reached ϕ_{m0}, her further motion will lead her up to range 2, where the stable oscillation regime around the stable equilibrium "mast down" exists. So this is capsizing.

If $A<0$ and the ship has reached ϕ_{m0}, her further motion takes her back to the range 0, where the stable oscillation regime around "mast up" equilibrium exists. So the ship will not capsize during the next semi-period of roll.

Consequently, capsizing in a piecewise linear system can be defined as the event of crossing of the boundary ϕ_{m0} in the positive direction and a positive value of coefficient A [Belenky, 1989, 1993]. The example of the time history of the roll process with capsizing is given in fig. 5.6 and corresponding phase trajectory in fig. 5.7.

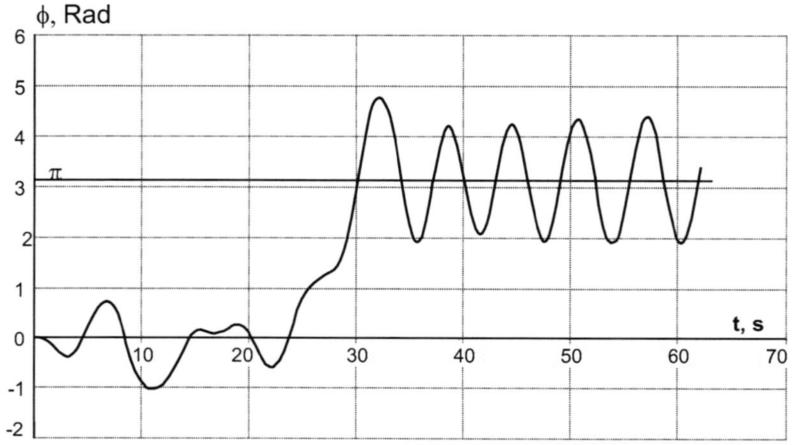

Figure 5.6 Time history of capsizing for piecewise linear system

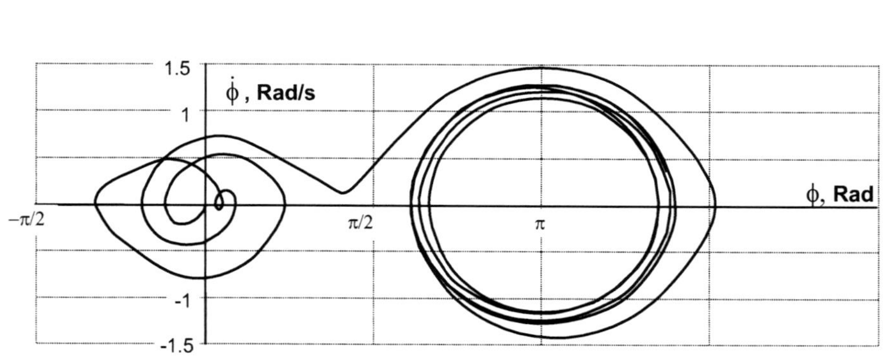

Figure 5.7 Phase trajectory of capsizing for piecewise linear system

5.2.3 Piecewise linear System and Classical Definition of Stability

We have seen that a possibility of capsizing is determined by the exponential term with a positive argument, which belongs to the general solution of the autonomous equation. Let us try to determine the influence of the partial solution:

$$p(t) = p_a \sin(\omega t + \beta_p + \varphi_1) + \phi_V \tag{5.37}$$

With amplitude and phase given by formulae (4.109).

This question is important because the partial solution reflects the action of external forces and their contribution towards the possibility of capsizing.

This influence can be estimated using the formula for the arbitrary constant A, which includes the value of the partial solution p_1 and its derivative $\dot{p}_1$ at the moment it enters into range 1:

$$A = \frac{(\dot{\phi}_1 - \dot{p}_1) - \lambda_2(\phi_1 - p_1)}{\lambda_1 - \lambda_2} \tag{5.38}$$

The partial solution (5.37) consists of a constant value and harmonic function. The constant value is equal to the angle of vanishing stability and reflects the fact that we are moving towards an unstable position of equilibrium that is not located in the origin of the coordinate system. So, all the action of external forces is included in the harmonic term. Its maximal value is equal to the amplitude of the partial solution that is defined by formulae (4.109). Let's rewrite it here:

$$p_a = \frac{\alpha_E}{\sqrt{(k_{f1}\omega_\phi^2 + \omega^2)^2 + 4\delta^2\omega^2}} \tag{5.39}$$

Looking at formula (5.39), we can see its similarity with the well-known formula (4.108) that expresses the amplitude of response of a common linear oscillator. Let's rewrite it here too:

$$q_a = \frac{\alpha_E}{\sqrt{(k_{f0}\omega_\phi^2 - \omega^2)^2 + 4\delta^2\omega^2}} \tag{5.40}$$

The difference between the formulae (5.39) and (5.40) is that the subtraction in (5.34) has been changed to sum in formula (5.40). This is a very significant difference, because this subtraction is "responsible" for resonance phenomenon. When excitation and natural frequencies are equal, amplitude (5.40) reaches its maximal value. Formula (5.39) does not have such a property and numerical values of amplitude p_a are significantly less than the values of amplitude q_a, see fig. 5.8.

The physical reason for the different behavior of partial solutions in ranges 0 and 1 is the following: we have instant negative GM in range 1 and the natural frequency becomes an imaginary value, so resonance and a large amplitude value is impossible there.

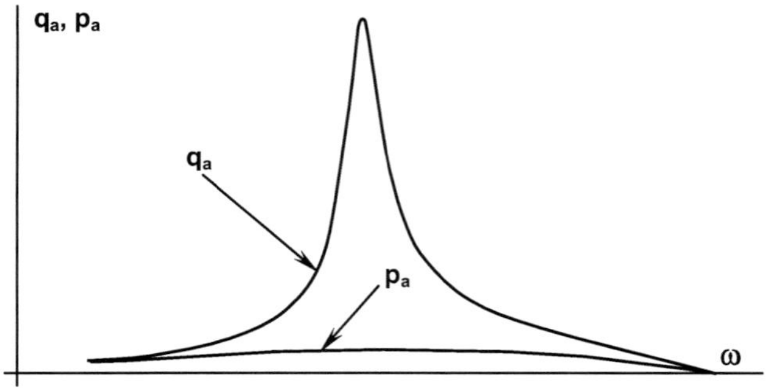

Fig. 5.8 Response curves of partial solutions on range 0 (q_a) and range 1 (p_a).

The same could be applied to the real *GZ* curve. Instant *GM* is positive before the maximum of the *GZ* curve, is equal to zero at the exact location of the maximum and is negative after the maximum. When the roll angle reaches the maximum of the *GZ* curve, the ability of the ship to accept periodical external loads dramatically decreases.

The above means that action of external forces has a small influence on stability in large roll angles. That is why we really can take external forces out of consideration, when we are defining stability [Belenky, 1991, 1993]. This is the theoretical explanation of the numerical results of Sevastianov and Pham [1979] that we have described in subchapter 5.1.4.

5.2.4 Shapes of Capsizing Trajectories

Capsizing is governed by the solution (5.36) that either takes the system to another equilibrium or returns it to its previous equilibrium. Let us see how these two options make the shape of the time history and phase trajectory.

Let us rewrite formula (5.36), converting exponents into hyperbolic sine and cosine functions. We substitute formulae for eigenvalues (5.36) and arbitrary constants (4.113) into the solution (5.36):

$$\phi(t) = e^{-\delta t}\left[C_1 \sinh \omega_1 t + C_2 \cosh \omega_1 \tau_1\right] + p_a \sin(\omega t + \beta_1 + \varphi_1) + \phi_V \tag{5.41}$$

$$C_1 = \frac{\delta(\phi_1 - p_1) + (\dot{\phi}_1 - \dot{p}_1)}{\omega_1}; \quad \text{and} \quad C_2 = \phi_1 - p_1 \tag{5.42}$$

$$\omega_1 = \sqrt{\omega_\phi^2 k_{f1} - \delta^2} \tag{5.43}$$

The expression (5.41) can be presented as a sum of functions. It can be done in two ways: expressing the general solution via hyperbolic sine or via hyperbolic cosine:

$$C_1 = H \cdot \sinh \varepsilon_1 \quad \text{and} \quad C_2 = H \cdot \cosh \varepsilon_1 \tag{5.44}$$

Hyperbolic cosine form:

$$\phi(\tau_1) = H \cdot e^{\delta t} \cosh(\omega_1 t + \varepsilon_1) + p_a \sin(\omega t + \beta_1 + \varphi_1) + \phi_V \tag{5.45}$$

As a result, the solution at the range 1 looks almost like the solution for range 0 in the system (4.107), but there is a hyperbolic function instead of a trigonometric function.

New arbitrary constants H and ε can be obtained from the equations (5.42) using the well-known formulae for hyperbolic functions:

$$\cosh^2 \varepsilon_1 - \sinh^2 \varepsilon_1 = 1 \quad \text{and} \quad \tanh \varepsilon_1 = \frac{\sinh \varepsilon_1}{\cosh \varepsilon_1} \tag{5.46}$$

$$H = \frac{1}{\omega_1}\sqrt{\omega_1^2(\phi_1 - p_1)^2 - \left[\delta(\phi_1 - p_1) + (\dot{\phi}_1 - \dot{p}_1)\right]^2} \tag{5.47}$$

$$\varepsilon_1 = \operatorname{arctanh} \frac{\delta \cdot (\phi_1 - p_1) + (\dot{\phi}_1 - \dot{p}_1)}{\omega_1(\phi_1 - p_1)} \tag{5.48}$$

To express the solution (5.41) via hyperbolic sine the following substitution should be used:

$$C_1 = H \cdot \cosh \varepsilon_1 \quad \text{and} \quad C_2 = H \cdot \sinh \varepsilon_1 \tag{5.49}$$

Substitution (5.49) allows expression of solution (5.41) in a form of the hyperbolic sine:

$$\phi(t) = H \cdot e^{-\delta t} \sinh(\omega_1 t + \varepsilon_1) + p_a \sin(\omega t + \beta_1 + \varphi_1) + \phi_V \tag{5.50}$$

New arbitrary constants H and ε can be obtained from the equations (5.49) using formulae (5.46):

$$H = \frac{1}{\omega_1}\sqrt{\left[\delta(\phi_1 - p_1) + (\dot{\phi}_1 - \dot{p}_1)\right]^2 - \omega_1^2(\phi_1 - p_1)^2} \tag{5.51}$$

$$\varepsilon_1 = \operatorname{arctanh} \frac{\omega_1(\phi_1 - p_1)}{\delta \cdot (\phi_1 - p_1) + (\dot{\phi}_1 - \dot{p}_1)} \tag{5.52}$$

Formulae for arbitrary "amplitudes" H (5.47) and (5.52) differ only by the sign of the expression under the square root. Consequently, the hyperbolic cosine and sine forms of the presentation of the solution (5.41) cannot exist simultaneously:

$$\text{Hyperbolic sine}: \quad \left|\delta(\phi_1 - p_1) + (\dot{\phi}_1 - \dot{p}_1)\right| \geq |\omega_1(\phi_1 - p_1)| \tag{5.53}$$

$$\text{Hyperbolic cosine}: \quad \left|\delta(\phi_1 - p_1) + (\dot{\phi}_1 - \dot{p}_1)\right| \leq |\omega_1(\phi_1 - p_1)| \tag{5.54}$$

Taking into account the above consideration about the small contribution of the partial solution of range 1, and the fact that the initial condition for the range 1, $\phi_1 = \phi_{m0}$, these formulae can be rewritten as:

$$\text{Hyperbolic sine}: \quad \left|\delta(\phi_{m0} - \phi_v) + \dot{\phi}_1\right| \geq |\omega_1(\phi_{m0} - \phi_v)| \tag{5.55}$$

$$\text{Hyperbolic cosine}: \quad \left|\delta(\phi_{m0} - \phi_v) + \dot{\phi}_1\right| \leq |\omega_1(\phi_{m0} - \phi_v)| \tag{5.56}$$

Let us show that inequality (5.55) corresponds to the condition of capsizing and inequality (5.56) to return to previous equilibrium. Taking into account that value ω_1 is positive and using formula (5.38) the condition of transition can be rewritten as:

$$\dot{\phi}_1 + \delta(\phi_{m0} - \phi_v) > -\omega_1(\phi_{m0} - \phi_v) \tag{5.57}$$

For positive roll angles $\phi_V > \phi_{m0}$ the right hand term in (5.57) is positive:

$$-\omega_1(\phi_{m0} - \phi_v) = \left|\omega_1(\phi_{m0} - \phi_v)\right| \tag{5.58}$$

Inequality (5.57) provides that the left hand term is also positive:

$$\dot{\phi}_1 + \delta(\phi_{m0} - \phi_v) = \left|\dot{\phi}_1 + \delta(\phi_{m0} - \phi_v)\right| \tag{5.59}$$

The substitution of (5.58) and (5.59) into (5.57) yields (5.53). So the time history of oscillation during capsizing might be described by a hyperbolic sine. If there is no capsizing during this semi-period of roll, the system returns to its previous equilibrium, its return trajectory may be described by a hyperbolic cosine function [Belenky, 1999], see fig. 5.9.

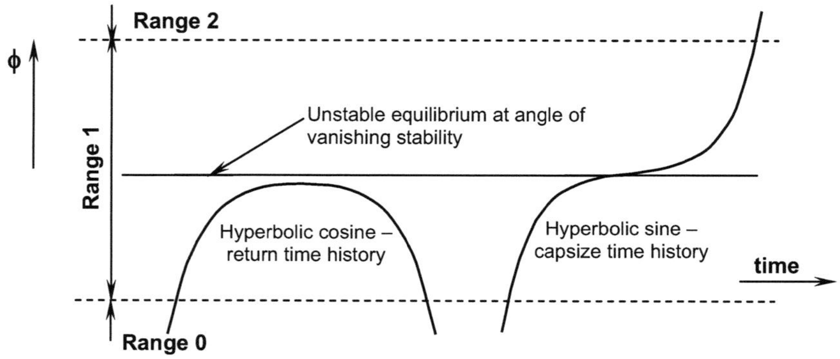

Fig. 5.9 Time histories at the vicinity of unstable equilibrium

The above statement, however, is not strict: there is a "grey area", when the arbitrary constant is very close to zero. In the case of $A = 0$, equation (5.36) indicates that the ship experiences periodic motions at the angle of vanishing stability. These periodic motions are unstable: small perturbations would break condition $A = 0$, which would result in escape from the range 1.

An example of phase trajectories when evolution when A is nearly equal to 0 is shown on fig. 5.10, where some "strange shapes" of trajectories can be observed. The origin of the "strange shapes" near the angle of vanishing stability is that usually the shape of the trajectory is defined by the term $A\exp(\lambda_1 t)$; a small value of A that allows the other terms of equation (5.36) to put their contribution into the motion [Belenky and Umeda, 1997]. That is why we can sometimes see loops at the angle of vanishing stability, as in fig. 5.4.

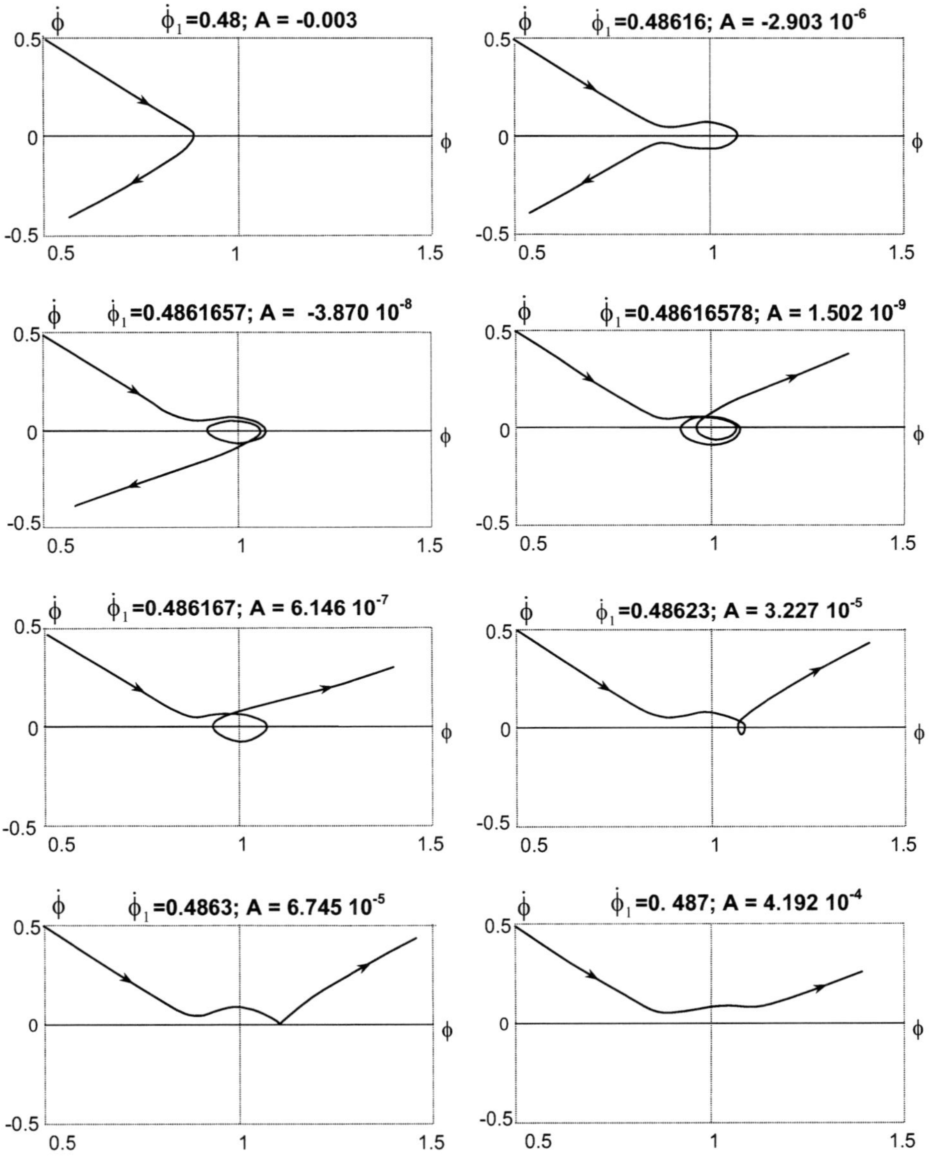

Fig. 5.10 Shapes of phase trajectories in vicinity of unstable equilibrium

5.3. Nonlinear Dynamics and Capsizing

5.3.1 General

Nonlinear dynamics is the most recent development in the theory of oscillators. Nowadays, instead of the term "oscillator" we use the term "dynamical system". This change reflects a widening of the applications from mechanical and electrical oscillation towards complex processes in physics, chemistry, ecology and many other fundamental and applied sciences. Also, this terminological change emphasizes that we no longer limit ourselves to periodical processes.

The linear theory of dynamic systems found its most comprehensive presentation in the famous work of Rayleigh [1896]. Works of Poincare, Lyapunov and Birkhoff developed the theory for nonlinear systems. The well-known book by Andronov, *et al* [1966] was the milestone in the theory for one-degree-of-freedom systems.

Further development was accelerated with increasing computation capabilities and allowed the consideration of dynamic systems with many degrees of freedom, but, as usual, interpretation and presentation of computation results was a problem. Geometrical methods were the answer. The monograph by Guckenheimer and Holmes [1983] was the third milestone and formed Nonlinear Dynamics in its contemporary form.

Among the many excellent books on this subject, [Thompson and Stewart, 1986] can be especially recommended for engineers. It gives simple and visual reviews of major nonlinear phenomena and methods of their study.

Nonlinear dynamics offers the most completed view on any nonlinear dynamic system and any nonlinear phenomenon, including nonlinear roll and capsizing.

5.3.2 Sensitivity to Initial Conditions: Safe Basin

We have already seen that behavior of a nonlinear system depends strongly on initial conditions. Depending on initial conditions, a nonlinear system may choose a high- or low-amplitude response if the excitation frequency is in the range where three responses are possible.

A piecewise linear system (4.106) could be used to see the qualitative relationship between initial conditions and capsizing. Let us assume that roll motions start within range 0 with initial conditions $\phi_0, \dot{\phi}_0$, initial phase φ_0 and reach the boundary between ranges at the moment t_1. We know that capsizing in a piecewise linear system is associated with a positive sign for the arbitrary constant (5.36):

$$A = \frac{(\dot{\phi}_1 - \dot{p}_1) - \lambda_2(\phi_1 - p_1)}{\lambda_1 - \lambda_2}$$

Here $\phi_1, \dot{\phi}_1$ are values of the roll angle and angular velocity at the moment of crossing. The value of the roll angle at crossing is fixed: it is equal to the boundary between ranges:

$$\phi_1 = \phi_{m0} \tag{5.60}$$

Roll angular velocity is defined by the "previous history" of roll motion at the range 0:

$$\begin{aligned}\phi_1 &= \phi_a(\phi_1, \dot{\phi}_1) \cdot e^{-\delta t_1}\left\{\omega_0 \cos(\omega_0 t_1 + \varepsilon(\phi_1, \dot{\phi}_1)) - \delta(\omega_0 t_1 + \varepsilon(\phi_1, \dot{\phi}_1))\right\} + \\ &+ q_a \omega \cos(\omega t_1 + \beta_q + \varphi_0)\end{aligned} \tag{5.61}$$

Arbitrary constants: $\phi_a(\phi_0, \dot{\phi}_0)$ and $\varepsilon(\phi_0, \dot{\phi}_0)$ depend on initial conditions ϕ_0, $\dot{\phi}_0$, so the value of A and hence, possibility of capsizing depends on these initial conditions.

The technique for the numerical study of influence of initial conditions is simple. The phase plane has to be covered with a grid of initial conditions, each of which are used for numerical simulation of forced roll oscillations. Simulations have to last long enough to clarify the outcome: capsizing or steady state regime of roll motions. So there are only two answers to consider: capsizing or non-capsizing. The result is quite easy to visualize; marking in black the point on the phase plane corresponding to initial conditions that do not lead to capsizing. If there is a capsizing, the point remains white.

The area that contains the initial conditions that do not lead to capsizing is called the safe basin. Numerical studies of the safe basin were done by Rainey, *et al* [1990], Nayfeh and Sanchez [1990], Rainey and Thompson [1991], Kan and Taguchi [1991, 1992, 1993], Kan, *et al* [1992], Kan [1992], and others.

The most important fact found is as soon as excitation becomes large enough, the safe basin experiences erosion: safe and unsafe conditions are located close to each other, forming a very complex picture, Fig. 5.11.

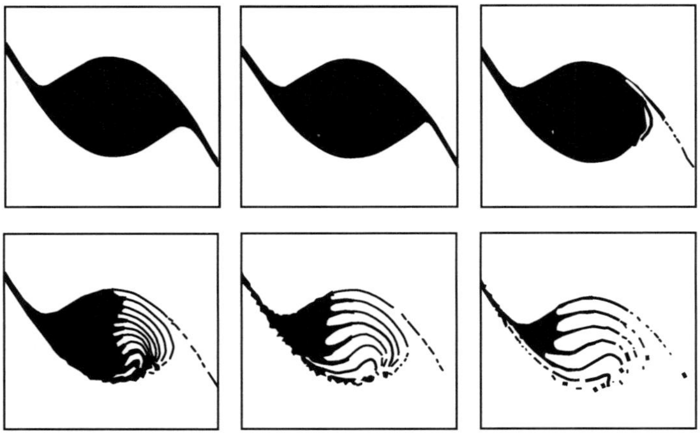

Fig. 5.11 Fractal erosion of safe basin (after Kan, et al [1992], black area corresponds to safe initial conditions, white area corresponds to capsizing)

Close examination of this picture reveals fractal structure of erosion of the safe basin. Fractals are geometrical figures recurrently repeating themselves, see a sample fractal in fig. 5.12.

The piecewise linear system also demonstrates erosion of the safe basin at high levels of excitation [Belenky, 2000a].

It is obvious that erosion of the safe basin increases changes to capsizing. Rainey and Thompson [1991] proposed to use the safe basin as a stability criterion for given wave parameters, see also Rainey, *et al* [1990], Soliman [1990]. Rainey and Thompson [1991] plotted the area of the safe basin versus wave height and found the threshold where the safe basin suddenly erodes, fig 5.13.

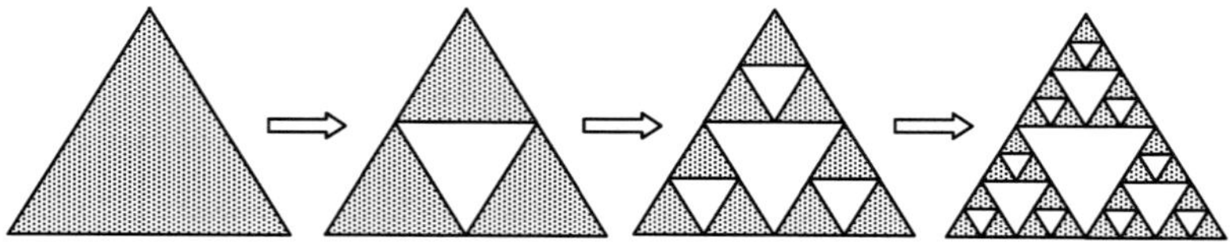

Fig. 5.12 Sample fractal

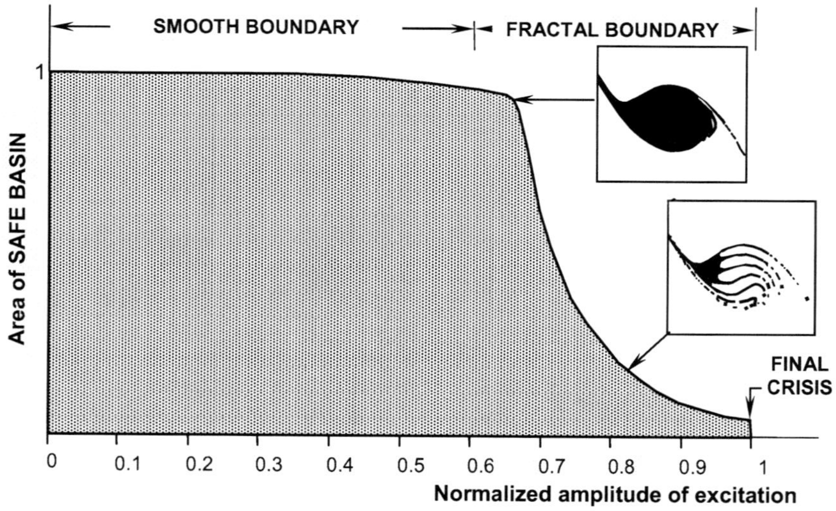

Fig. 5.13 Normalized area of the safe basin versus normalized excitation amplitude

This threshold or critical wave height is also called capsizial resistance. Repeating this calculation for different situations, for example changing position of the ship relative to waves, initial heel, period of wave, etc., Rainey and Thompson [1991] built a transient capsize diagram.

The transient capsizing diagram has the capability to take into account many important factors and might be a good background for new stability criterion [Vassalos, 1994].

5.3.3 Concept of Invariant Manifold

The phenomenon of fractal erosion of the safe basin requires a physical explanation. In order to present it, we have to introduce the concept of invariant manifold.

We introduced the concept of separatrix in subchapter 5.1.1 as a special phase trajectory that separates different areas of attraction in the roll equation (5.1). That equation described roll without excitation, so it was a homogeneous dynamic system. Invariant manifold is a generalization of separatrix for the case with excitation (heterogeneous dynamic system). So the separatrix is a particular case of the invariant manifold when there is no excitation.

First let us see what the words "invariant manifold" mean. "Manifold" generally means a subspace with dimensions less than n. A line is a manifold for a plane, a plane a manifold for 3-D space, etc. However, it is not necessary that the manifold of n dimensional space has n-1 dimensions, it is enough that it just has less dimensions than the original space. A line is a manifold for a 3-D space too.

Why is it "invariant"? Actually, the phase plane does not give complete information of the state of the system with excitation. To describe such a system state completely, we need one more coordinate: current phase of excitation. Phase plane for the excited system is just a projection, so almost all phase trajectories depend on phase of the excitation. The only exception is the boundary between different areas of attraction – phase trajectory that leads towards (or in opposite direction from) unstable equilibrium. That is why this boundary is called the "invariant manifold", for more information see the monograph by Guckenheimer and Holmes [1983].

Generally, it is more convenient to use for an excited system a Poincare map rather than a phase plane, because we can delete one dimension. As we have seen from subchapter 4.4.3, the steady state regime (limit cycle) maps itself into one point (fixed point).

Similar to the separatrix, the invariant manifold has two branches: one leads toward unstable equilibrium another one leads in opposite direction. The branch that tries to take the system to unstable equilibria is called "stable invariant manifold", the other branch is named "unstable invariant manifold". Similar to the separatrix, only the stable invariant manifold is a boundary between two areas of attraction. Fig. 5.14 shows a sample of a Poincare map with two fixed points (steady state motion regimes) around normal and capsized equilibria positions, unstable equilibrium and both branches of the invariant manifold.

Similar to the separatrix, the invariant manifold is tangent to the eigenvectors at unstable equilibrium, which is a saddle point in our case. (The proof can be found in a monograph by Guckenheimer and Holmes [1983].)

Calculation of the invariant manifold for the roll equation only does not differ much from the procedure for the separatrix. First, we calculate eigenvectors for the saddle point at the angle of vanishing stability. These eigenvectors will provide the directions of initial displacement, similar to (5.10-12). Using these initial points, we integrate the equation in inverse time for the stable invariant manifold and in direct time for the unstable one. Since we are working with a Poincare map, only one point per excitation period is used. As a result, the invariant manifold leaves a trace of points on the Poincare map. If these points are not sufficient to clarify the character of the curve, the initial phase of excitation has to be changed and all the calculations have to be repeated starting from the same initial conditions.

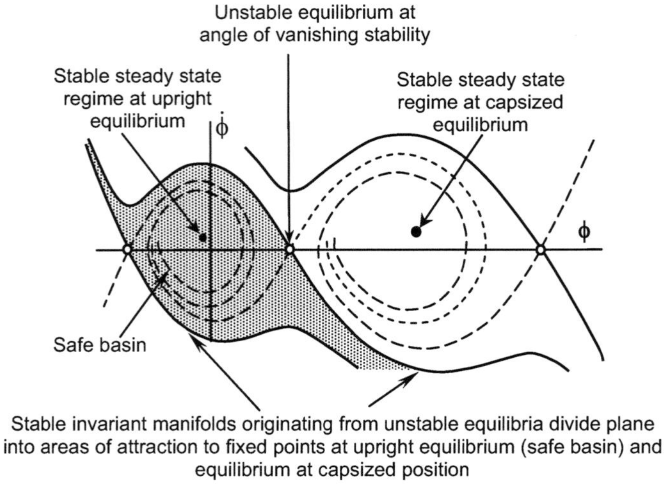

Fig. 5.14 Invariant manifolds on Poincare map. (Solid line – stable invariant manifold, dashed line – unstable invariant manifold)

The sample of the invariant manifold in fig. 5.14 has just one fixed point (steady state motion regime) at safe basin. As we have seen in subchapter 4.5.2 there is a frequency range where three steady state regimes exist. This would make three fixed points on a Poincare map. From the previous study we know that two of them are stable and one, which is in the middle, is unstable. We also mentioned that the choice of the stable steady state mode depends on initial conditions and we promised to address the problem of how to separate these initial conditions.

There is quite an evident similarity between equilibrium and steady state mode (fixed point) on a Poincare map. Both these figures attract motion if they are stable and repel it if they are unstable. We have seen from Chapter 4 that their stability is defined in a very similar way: through the eigenvalues. This similarity extends for invariant manifolds.

The procedure for calculating the invariant manifold for an unstable fixed point (unstable steady state regime) is the same as that for the unstable equilibrium. First, the eigenvectors corresponding to eigenvalues for the Jacobean matrix in formula (4.196) have to be calculated. Then, initial conditions have to be evaluated using equations (5.10-12). Finally, the stable invariant manifold can be calculated by numerical integration of the roll equation in inverse time and the unstable invariant manifold might be obtained by integration in "normal" time. The stable invariant manifold divides the safe basin into two areas of attraction for each stable steady state regime of motion. The sample of the Poincare map with three fixed points is shown in fig. 5.15.

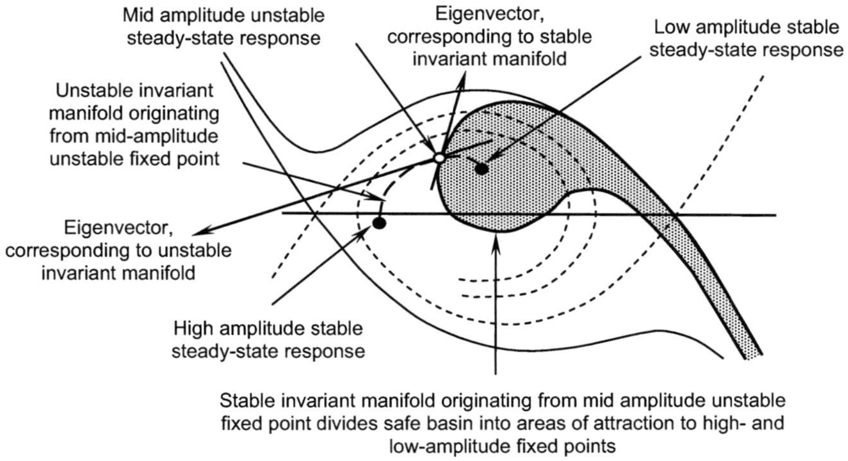

Fig. 5.15 Invariant manifolds on Poincare map for three-amplitude response. (Solid line – stable invariant manifold, dashed line – unstable invariant manifold)

Now we can observe the fold bifurcation on a Poincare map [Falzarano, *et al*, 1995]. The procedure is almost the same as that described in subchapter 4.5.2. Excitation frequency is our control parameter. We change it from zero and observe changes in behavior, see fig. 5.16.

Before we reach the frequency corresponding to points A and D on the response curve at fig. 5.16, we have only one steady state solution and our Poincare map has only one fixed point in the vicinity of upright equilibrium (as in fig. 5.14).

As soon as we reach the frequency that corresponds to points A and D, we have the second fixed point. Further increasing of frequency makes this new point to bifurcate into two fixed points. One of these points corresponds to high amplitude response (which may be stable), another to middle amplitude response (which is always unstable). Together with low amplitude response (which is stable) we have three fixed points and our Poincare map now is similar to fig. 5.15.

However, the possibility to observe only two fixed points is theoretical, because there is just one frequency that provides such a picture. Since all our calculations are approximate, we cannot obtain exactly this result, however we can be quite close. It means that we get three points anyway, but two of them are very close.

A further change of frequency leads to movement of the unstable fixed point (which corresponds to unstable mid-amplitude response) towards the stable fixed point, which corresponds to stable low amplitude response. When the frequency reaches the values corresponding to points B and C on the response curve, unstable and low amplitudes merge and disappear. As a result, only one fixed point, which corresponds to high amplitude response, can be observed.

We already mentioned about the similarity between fixed points on a Poincare map and equilibria in the phase plane. Here we see one more aspect of this similarity: the structure of motions around an unstable fixed point is almost identical to unstable equilibrium: that is why we could consider it as a saddle point on a Poincare map. Analogously, a stable fixed point can be considered as a stable focus on a Poincare map.

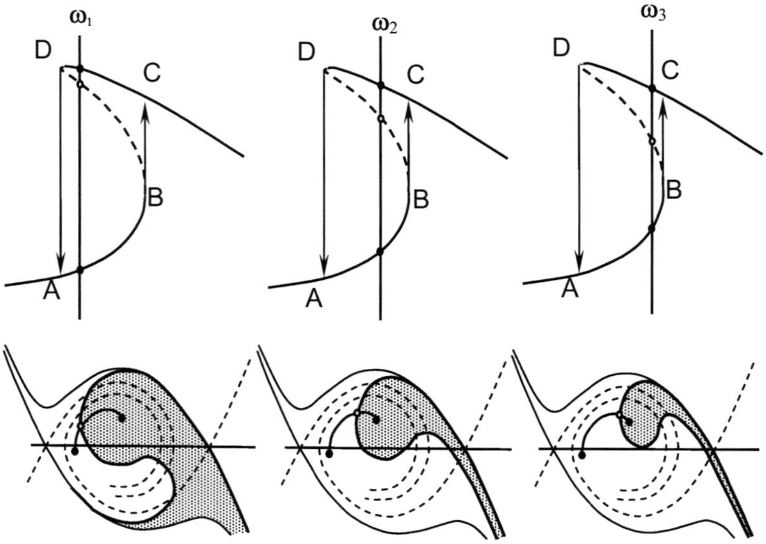

Fig. 5.16 Evolution of areas of attraction in three-amplitudes response zone

5.3.4 Invariant Manifold and Erosion of Safe Basin. Melnikov Function[1]

Now, let's choose another control parameter – amplitude of the excitation and check its influence on the invariant manifold calculated for unstable equilibrium at angle of vanishing stability [Falzarano, *et al*, 1992]. For small amplitude excitation, both stable and unstable invariant manifolds are quite similar to the separatrix for damped non-forced roll, see fig. 5.17a.

An increasing in the amplitude leads to a drastic change: the stable invariant manifold originating from positive angle of vanishing stability intersects with the unstable invariant manifold originating from negative angle of vanishing stability, fig. 5.17b. Once, at least one intersection occurs, there will be an infinite number of intersections [Falzarano, *et al*, 1992], [Moon, 1987]. Baker and Gollub [1996] offered a quite simple and visual explanation of this phenomenon.

An infinite number of intersections of manifolds might make two neighboring points belong to two different areas of attraction. So, we observe fractal erosion of the safe

---

[1] The author is grateful to Prof. Falzarano for fruitful discussion of the materials of this subchapter.

basin. Such a picture has a specific name: it is called a "heteroclinic tangle". Heteroclinic means that manifolds come from different saddle points (from the Greek word ετεροσ heteros – different, distinct).

Besides the explanation why the safe basin experiences erosion, the above provides us with a powerful numerical tool. It is not necessary to perform complex calculations as was described in subchapter 5.3.2, it is enough to calculate the stable invariant manifolds from positive angle of vanishing stability and the unstable invariant manifold from negative angle of vanishing stability (or vice versa). If they have intersected at least once, we observe erosion of the safe basin. In other words, this amplitude of excitation is too high for this particular *GZ* curve i.e. the ship is unsafe for this particular wave [Falzarano, *et al*, 1992].

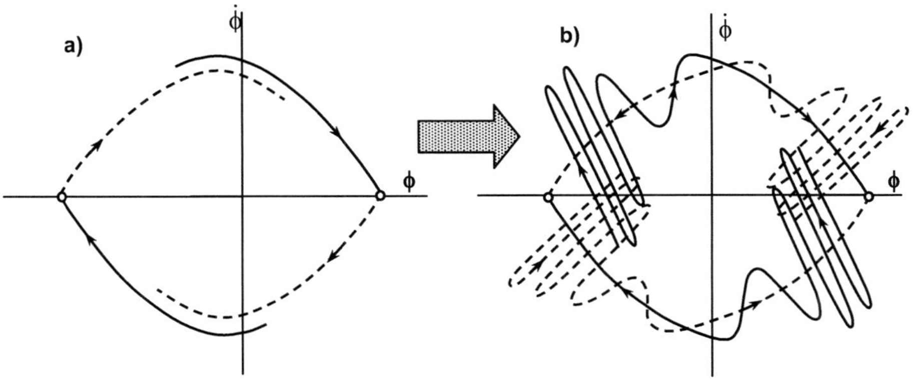

Fig. 5.17 Change of shape of invariant manifolds with increasing amplitude of excitation

Moreover, there is an approximate analytical solution for the distance between stable and unstable invariant manifolds. It is called the "Melnikov function" and the technique of the derivation is usually called the "Melnikov method".

The Melnikov method can be applied to any dynamic system that can be presented as:

$$\dot{\vec{Y}} = \vec{F}(\vec{Y}) + \varepsilon\vec{G}(\vec{Y},t) \qquad (5.62)$$

With $\varepsilon << 1$ is a small perturbation. The vector-valued function $\vec{F}(\vec{X})$ here does not depend on time, so the unperturbed system (when $\varepsilon = 0$) is homogeneous:

$$\dot{\vec{Y}} = F(\vec{Y}) \qquad (5.63)$$

Perturbation $\varepsilon\vec{G}(\vec{Y},t)$ is periodic. Falzarano, *et al* [1992] considered damping and excitation to be small quantities in comparison with other terms; a similar assumption is known to work quite well for perturbation methods (see subchapter 4.2). Also, following the above reference we rewrite our general system (3.270) in non-dimensional time $\tau = \omega_\phi t$ (we have used a similar procedure in subchapter 5.1.1):

$$\ddot{\phi} + 2\delta\dot{\phi} + \omega_\phi^2 f(\phi) = \alpha_E \sin(\omega t + \varphi_E)$$

$$\frac{d^2\phi}{d\tau^2} + 2\mu\frac{d\phi}{d\tau} + f(\phi) = \alpha_{E\tau}\sin(\omega_\tau t + \varphi_E) \tag{5.64}$$

With the following dimensionless quantities:

$$\mu = \frac{\delta}{\omega_\phi};\quad \alpha_{E\tau} = \frac{\alpha_E}{\omega_\phi^2};\quad \omega_\tau = \frac{\omega}{\omega_\phi} \tag{5.65}$$

Converting the system (5.64) into vector form (5.62):

$$\dot{\vec{Y}} = \begin{pmatrix} \dot{Y}_1 \\ \dot{Y}_2 \end{pmatrix} = \begin{pmatrix} \frac{d\vartheta}{d\tau} \\ \frac{d\phi}{d\tau} \end{pmatrix};\qquad \vec{Y} = \begin{pmatrix} Y_1 \\ Y_2 \end{pmatrix} = \begin{pmatrix} \vartheta \\ \phi \end{pmatrix}$$

$$\vec{F}(\vec{Y}) = \begin{pmatrix} -f(Y_2) \\ Y_1 \end{pmatrix} = \begin{pmatrix} -f(\phi) \\ \vartheta \end{pmatrix} \tag{5.66}$$

$$\vec{G}(\vec{Y},t) = \begin{pmatrix} \alpha_{E0\tau}\sin(\omega_\tau\tau + \varphi_E) - 2\mu_0 Y_1 \\ 0 \end{pmatrix} =$$

$$= \begin{pmatrix} \alpha_{E0\tau}\sin(\omega_\tau\tau + \varphi_E) - 2\mu_0\vartheta \\ 0 \end{pmatrix}$$

With:

$$\alpha_{E0\tau} = \varepsilon\alpha_{E\tau};\quad \mu_0 = \varepsilon\mu \tag{5.67}$$

As we did before, we use a cubic approximation for the *GZ* curve:

$$f(\phi) = \phi - a_3\phi^3 \tag{5.68}$$

Note that coefficient a_3 does not have to be small here. The Melnikov function for the system (5.62) is defined by the following integral:

$$M(\tau_0) = \int_{-\infty}^{\infty} \vec{P}(\vec{\Gamma}(\tau)) \cdot \vec{G}(\vec{\Gamma}(\tau), \tau + \tau_0)\, d\tau \tag{5.69}$$

Where $\vec{\Gamma}(\tau)$ is the invariant manifold defined by formulae (5.8) (positive value of τ provides the unstable invariant manifold and negative values of τ yield the stable invariant manifold). The vector-valued function $\vec{P}(\vec{\Gamma})$ is perpendicular to the vector valued function $\vec{F}(\vec{\Gamma})$:

$$\vec{P}(\vec{\Gamma}) \cdot \vec{F}(\vec{\Gamma}) = 0 \tag{5.70}$$

Operation of multiplication implies a scalar product of vectors in equations (5.69) and (5.70). The integral (5.69) can be evaluated analytically [Falzarano, *et al*, 1992]:

$$M(\tau_0) = -\frac{\alpha_{E0\tau}\pi\omega_\tau\sqrt{2}}{\sqrt{a_3}}\operatorname{csch}\left(\frac{\omega_\tau\pi}{\sqrt{2}}\right)\cos(\omega_\tau\tau_0) + \frac{2\sqrt{2}\mu_0}{3a_3} \tag{5.71}$$

The approximate distance between the stable and unstable invariant manifolds is related to the Melnikov function (5.71) as [Falzarano, *et al*, 1992; Guckenheimer and Holmes, 1983] (with the accuracy of ε^2):

$$d(\tau_0) = \frac{\varepsilon M(\tau_0)}{\left|\vec{F}\left(\vec{\Gamma}(\tau_0)\right)\right|} \tag{5.72}$$

With $\left|\vec{F}\left(\vec{\Gamma}(\tau_0)\right)\right|$ means length of the vector valued function $\vec{F}\left(\vec{\Gamma}(\tau_0)\right)$ defined at the moment τ_0.

We are actually interested in whether these invariant manifolds have intersected or not, which means we are interested if the Melnikov function ever crosses zero. The value of excitation amplitude that satisfies this condition is given by Falzarano, *et al* [1992]:

$$\alpha_{Cr} = \frac{2\mu_0}{3\sqrt{a_3}\,\pi\omega_\tau \cosh\left(\dfrac{\omega_\tau \pi}{\sqrt{2}}\right)} \tag{5.73}$$

Formula (5.73) provides us with the critical excitation amplitude, when erosion of the safe basin begins.

Another phenomenon associated with erosion of the safe basin is chaotic response that we have observed in subchapter 4.5.3. Further exploration about this relationship can be found in [Baker and Gollub, 1996].

5.3.5 Loss of Motion Stability and Capsizing

We already considered stability of steady state roll motion in subchapter 4.4 and reviewed several major methods to define the motion stability. We were mainly focused on the methods that were able to provide us with eigenvalues, since eigenvalues were later used as the indicator of what is going to happen to roll motion when the motion stability is lost. We have seen two different types of behavior: jump to high amplitude response (fold bifurcation) and period doubling (flip bifurcation). The last one is the way to deterministic chaos (subchapter 4.5.3). As we have mentioned in the subchapter 5.3.4, chaotic response corresponds to erosion of the safe basin, which increases chances of capsizing.

There are several works aimed on applications of motion stability theory for the study of capsizing. Their ultimate goal was developing ship stability criterion based on roll motion stability. Let us give them a brief review.

We already mentioned works by Martin, *et al* [1982], Phillips [1986, 1986a], Caldeira-Saraiva [1986, 1986a], which were focused on the Lyapunov direct method (see subchapter 4.4.1).

Wellicome [1975], Wright and Marshfield [1980], Nayfeh and Khdeir [1986, 1986a], Nayfeh and Sanchez [1990] applied the Floquett theory to capsizing analysis; we have considered it in detail in subchapter 4.4.2.

Application of the Floquett theory allows us to solve a variation equation for a given particular case and to judge the motion stability. However, the solution is not always necessary, because some of the practical cases lead to the well-known Mathieu equation (see also subchapter 11.1.2 of [Kobylinski and Kastner 2003]) or to the more complex Hill equation.

The Mathieu equation is linear ordinary differential equation with periodic coefficients:

$$\ddot{\xi} + (p + q\cos t)\xi = 0 \tag{5.74}$$

The character of the solution of the Mathieu equation (5.74) depends on parameters p and q. A diagram, where areas of bounded and unbounded solutions of the Mathieu equation are shown, is called an Ince-Strutt diagram, see fig. 5.18. This diagram allows us to check if the solution is bounded or not, and, consequently if our steady state motion is stable or not.

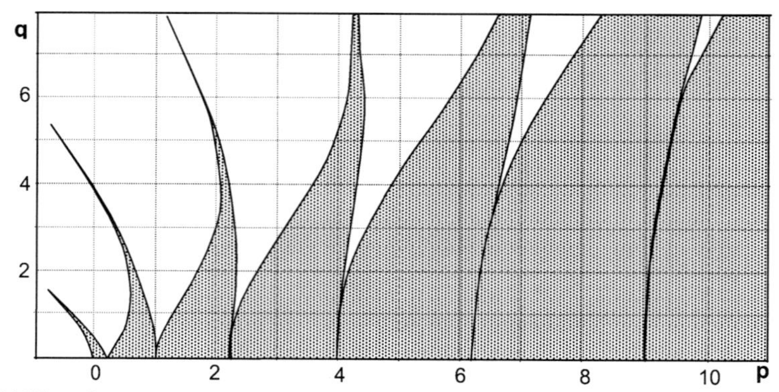

Fig. 5.18 Ince-Strutt diagram. (The diagram is symmetric relative to the abscissa, filled areas correspond to bounded solutions)

Ananiev [1981] used a more complicated equation, which is called the Hill equation. The Mathieu equation is a particular case of the Hill equation. A general view of the Hill equation is the following:

$$\ddot{\xi} + (c + \Psi(t))\xi = 0 \tag{5.75}$$

Here: $\Psi(t)$ is a periodic function.

Analogously to the Mathieu equation, the Hill equation has several areas where the solution is unbounded. Ananiev [1981] considered the first area of instability, because it is located in the region of main resonance, fig. 5.19. If amplitude of the excitation value is larger, it is possible to observe the area of instability of the zero order. The appearance of the response curve with both areas is shown in fig. 5.20.[1]

It is important to note that there are some frequency gaps where steady state oscillation is unstable. As we already know these gaps are due to period doubling or flip bifurcation. As we previously discussed, it does not necessarily mean immediate capsizing, but it is quite clear an indication of advancing danger.

[1] Response curves in fig. 5.19 and 5.20 were evaluated for roll equation in relative coordinates (3.274).

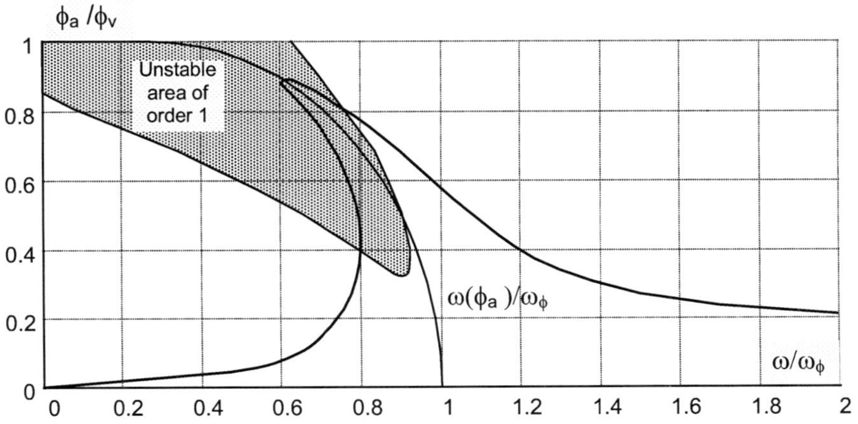

Fig. 5.19 Response curve with area of instability. $\alpha_E=\pi/20$, $\delta/\omega_\phi=0.05$

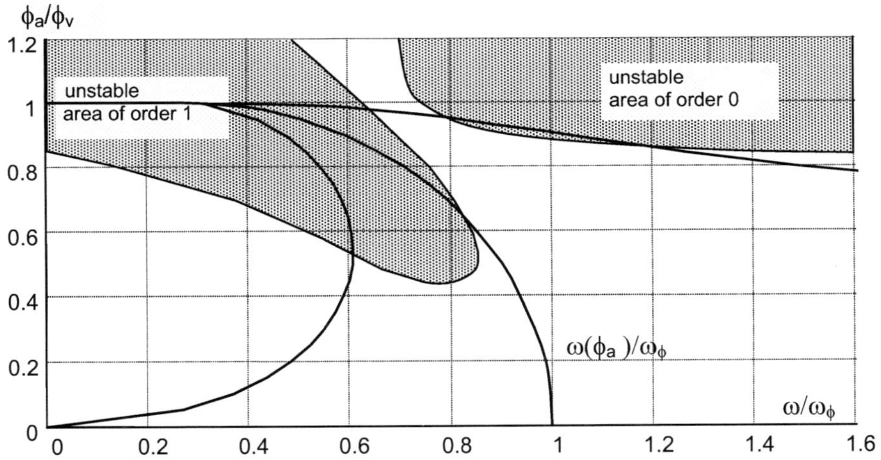

Fig. 5.20 Response curve with areas of instability. $\alpha_E=\pi/5$, $\delta/\omega_\phi=0.10$

Chapter 6

Capsizing in Regular Following and Quartering Seas

6.1 Variation of the *GZ* Curve in Longitudinal Waves. Pure Loss of Stability

6.1.1 Description of Phenomenon

Most vessels designed to sail with medium speed and above have a relatively narrow hull below the waterline to decrease drag and improve flow around the propeller. If the vessel is intended to transport light cargo, host people or equipment, additional space can be added by using such elements as a flared bow and stern overhang. Flared bow also may be useful to decrease greenwater shipping and protect fore part of the deck from spray.

When a crest of a relatively long wave (more precisely when wave length is comparable with the length of the ship) is located near the midship section, troughs of this wave may be near the fore and aft parts of the vessel, where the hull is narrow. This situation results in the waterline that is narrower that it would be in calm water (see fig. 6.1).

When a trough of a long wave is located near the midship section, the flared bow and stern overhang may be submerged as the crests of this wave may be located there, see fig. 6.1.

As a direct consequence of this combination of wave location and hull geometry, the waterline is wider when the wave trough is located near amidships and narrower when the wave crest is about amidships, see fig. 6.1.

The righting moment is dependent on the width of the waterline. As a result, stability changes when the ship is located in longitudinal waves comparable with ship length: it is better when the midship section is located near the wave trough and it is worse when the midship section passes the wave crest (see also [Kobylinski and Kastner 2003]).

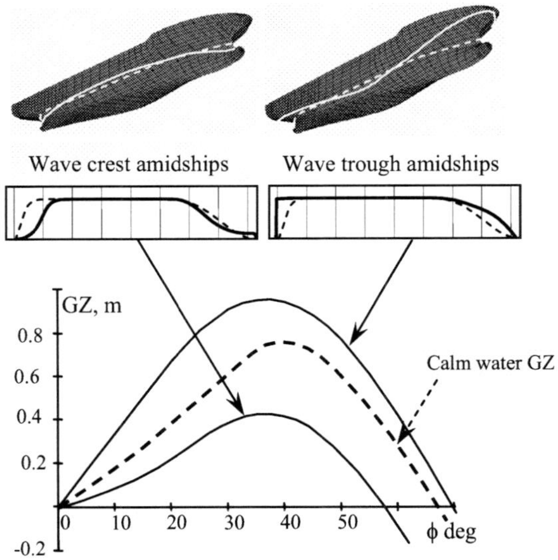

Fig. 6.1 Variation of the GZ curve in longitudinal waves

Changing stability from wave trough to wave crest is also known as the wave pass effect.

6.1.2 Methods of Calculations

The GZ curve is a lever of the restoring moment that counteracts another moment created by an external force acting in a transverse direction. The restoring moment is created by pressures applied all over the submerged part of the hull and there is a result of the surface integration of these pressures.

Hydrodynamic forces and moments as a result of pressure integration were considered in subchapters 3.4 and 3.5. Here, components of pressures are principally the same, but assumptions of small motions and zero speed are no longer applicable:

- Hydrostatic pressures;
- Pressures in a wave without accounting for presence of the vessel – Froude-Krylov hypothesis;
- Pressures caused by wave diffraction on a ship, since the ship is an obstacle for wave propagation;
- Pressures caused by waves generated by periodic ship motions: ship motions are affected by wave damping;
- Pressures caused by waves generated by forward heading: it is a wave component of the resistance;
- Pressures caused by non-potential flow, including vortices and boundary layer.

Pure hydrostatic calculation assumes that *GZ* curve variations are only due to changes of submersed volume: it means that we only take into account the fact that the waterline is no longer a plane [ABS 2004, Shin, *et al* 2004]. Calculation of *GM* is relatively easy and can be preformed in a spreadsheet:

$$
\begin{aligned}
&d_C(x_i, x_{Cj}) = d_m - 0.5h_w \cos\left(\frac{2\pi(x_i - x_{Cj})}{\lambda}\right); \quad i = 1,.., N_{St}; \quad j = 0,.., N_C \\
&I_X(x_{Cj}) = \frac{2}{3}\int_{-0.5L}^{0,5L}[y(x, x_{Cj})]^3\,dx; \quad \Omega(x_i, x_{Ci}) = \int_0^{d_C(x_i, x_{Cj})} y(x_i, x_{Cj}, z)dz \\
&\nabla(x_{Cj}) = 2\int_{-0.5L}^{0.5L}\Omega_C(x, x_{Cj})dx; \quad M_\Omega(x_i, x_{Cj}) = \int_0^{d_C(x_i)} z\cdot y(x_i, x_{Cj}, z)dz \\
&VCB(x_{Cj}) = \frac{2}{\nabla}\int_{-0.5L}^{0.5L} M_\Omega(x, x_{Cj})dx; \quad BM(x_{Cj}) = \frac{I_X(x_{Cj})}{\nabla(x_{Cj})} \\
&GM(x_{Cj}) = BM(x_{Cj}) - KG + VCB(x_{Cj})
\end{aligned}
\tag{6.1}
$$

The formulae (6.1) are written for the wave length λ and wave height h_w; d_C is the draft at each station when the wave crest is located at a distance x_C from the origin of the coordinate system, N_C is the number of wave to be considered, x is abscissa of a station, N_{St} is the number of stations included in the calculations, d_m is a draft amidships, $y(x, x_C)$ is the half beam at each station for the wave crest position at x_C.

$I_x(x_C)$ is the moment of inertia of the area of the waterplane for the wave crest position at x_C; Ω is the submerged area of each station while ∇ and M_Ω are volumetric displacement and vertical static moment. Other values in the formulae (6.1) follows conventional ship hydrostatic notation: *VCB* - the vertical position of the center of buoyancy and *BM* – metacentric radius. All the above values are calculated for wave crest position at x_C.

Despite *GM* value does not provide complete information on stability in waves, formulae (6.1) could be used for initial assessment on how strong wave pass effect may be for this particular vessel.

Complete calculation of GZ curve was implemented by program EUREKA. A description of the method is available in [Paulling 1961]. The current version of EUREKA has an option to include a correction for "Smith effect" – the changes of pressures in a wave using Froude-Krylov hypothesis.

More elaborate calculations of wave pass effect are available from time domain potential solvers. One of them is FREDYN, a potential time-domain code based on strip theory [de Kat and Paulling, 1989] that includes other types of forces based on systematic model test, developed by CRNAV[1] (see also see de Kat *et al* [1994], de Kat and Thomas [1999, 2000]). GL SIMBEL is another time-domain potential code based on strip theory, [Brunswig, *et al* 2006]. LAMP (Large Amplitude Motion Program[2]) is potential time domain code based on panel method [Lin and Yue, 1990, 1993] with capability to include forces of other physical natures.

Solution of a potential flow problem through a time-domain code allows including pressures caused by diffraction, radiation and ship generated waves in addition to hydrostatic and Froude-Krylov-related pressures into calculation of restoring moment in waves. Including the forces of a viscous and a vortex nature would require use of CFD or model-test based data.

In conclusion, we give a brief review of theoretical and experimental works that did not get a proper circulation in English language literature.

First of all, we have to reference two special monographs by Nechaev [1978] and by Boroday and Netsvetaev [1982] that contain profound reviews of this kind of research.

Boroday [1967], Boroday and Netsvetaev [1969, 1982] studied hydrodynamic and hydrostatic components of the *GZ* curve changes. Lugovsky [1966] tackled this problem from the point of view of hydrodynamic theory of ship motion and used parabolic approximations of ship lines.

Nechaev [1978, 1989] developed the method based on model tests in a hydrochannel at the Kaliningrad Institute of Technology. Since inverse motion was used, no limitation on towing tank length existed. An obstacle on the bottom of the hydrochannel created the wave. Results of these experiments were presented in a form of regression polynomials. Using these polynomials, it is possible to estimate *GZ* curves on the wave crest and at the

---

[1]The Cooperative Research Navies consortium (www.crnav.org).

[2] LAMP was developed by SAIC (Science Application International Corporation) under sponsorship from US Navy, US Coast Guard, ABS and SAIC.

wave trough, if the length of wave is not very different from the length of the ship. The method is described in Appendix.

6.1.3 Pure Loss of Stability

As the restoring moment may be significantly decreased while the wave is located about the midship section, a vessel may suffer from large roll angles or even capsizing, if she spends enough time in the situation of decreased stability. Capsizing caused by pure loss of stability in wave crest in following was observed by Paulling *et al* [1972, 1974, 1975] during experiments in San Francisco Bay, see also Kobylinski and Kastner [2003].

Timing may be the critical factor for pure loss of stability. Therefore consideration of surging and surf-riding may be an important part of analysis for pure loss of stability. We shall return to this problem in subchapter 9.3.

Pure loss of stability caused by variation of righting arm in waves is considered as one of the modes of stability failures in near longitudinal seas along with parametric resonance that will be addressed in subchapter 6.2.

6.1.4 Equation of Roll Motions

For the purpose of further study, we assume that all elements of the *GZ* curve are changing according to the cosine function while a wave is passing by. This is not completely accurate, but such an approximation can simulate essential ship behavior caused by variations in the *GZ* curve in following and quartering seas:

$$GZ(\phi,t) = GZ_m(\phi) + GZ_A(\phi)\cos\omega_E t \qquad (6.2)$$

Here, ω_E is the encounter frequency caused by the ship heading:

$$\omega_E = \omega - \frac{\omega^2}{g} v\cos\beta \qquad (6.3)$$

Where β is the course of the ship relative to the waves.

The mean value of the *GZ* curve, $GZ_m(\phi)$ does not necessarily coincide with the *GZ* curve calculated for calm water. The dependence of the *GZ* curve on wave position is not symmetric; see fig. 6.1. The formula for the variation of the *GZ* curve in waves (6.2) gives the following equation of roll in following seas:

$$\ddot{\phi} + 2\delta\dot{\phi} + \frac{mg}{I_{xx} + a_{44}}(GZ_m(\phi) + GZ_A(\phi)\cos\omega_E t) = 0 \qquad (6.4)$$

Where m is mass of the ship. Assuming GZ_A to be constant and introducing a cubic approximation for the *GZ* curve we obtain the following roll equation :

$$\ddot{\phi} + 2\delta\dot{\phi} + \omega_m^2(1 + a_p\cos\omega_E t)\phi - a_3\phi^3 = 0 \qquad (6.5)$$

Here a_p is an amplitude of parametric excitation, determined from the variation of the *GZ* curve, frequency ω_m corresponds to initial part of mean *GZ* curve - GZ_m. Expression (6.5) is a nonlinear equation with a periodically changing coefficient. It is the subject for the next subchapter.

6.2 Parametric Resonance

6.2.1 Description of Phenomenon

The parametric resonance is caused by parametric excitation that is periodic changes of the coefficients in the roll equation (see also [Kobylinski and Kastner, 2003]). One possible reason of such changes, we have just discussed in subchapter 6.1, is *GZ* curve variation in longitudinal seas, see equation (6.5). Parametric resonance caused by this variation is also called low cycle resonance. It has been studied beginning in the 1950's: [Grim, 1952; Basin, 1953; Kerwin, 1955; Paulling and Rosenberg, 1959; Paulling, 1961].

Interest in parametric roll resonance caused by wave pass effect was renewed after the case of heavy roll of a post-Panamax containership in near head seas [France, *et al* 2003]. We will return to this case in subchapter 8.4.4.

It can be seen from the investigation described by France *et al* [2003] that wave pass effect is capable of generating large roll angles.

Assume that a vessel is sailing exactly in longitudinal regular waves. Once she reaches wave crest, a transverse heeling moment is applied causing an initial roll angle and/or roll velocity. When a vessel is at the position "wave crest amidships", her stability is decreased. As the restoring moment is less, a vessel will roll further than she would in calm water. When the amplitude angle is reached the wave trough reaches the midship section and stability increases. The larger restoring moment means a larger acceleration and the vessel returns to zero-angle with the velocity larger than she would in calm water. Meanwhile a new wave crest reaches the midship section, stability is decreased and the vessel rolls even further.

As a result, energy is added every quarter of a period, when stability is decreased, additional potential energy is collected; while stability is increased additional kinetic energy is being picked up. This simple description shows why a frequency ratio 2:1 is the most favorable for development of parametric resonance (principal parametric resonance). It should be also clear that the parametric resonance is possible for other frequency ratios. For the frequency ratio 1:1 energy is being added every half a period (fundamental parametric resonance).

Other causes for parametric excitation are coupling between roll and pitch and between roll and heave. Despite the evident difference between these sources of parametric excitation, the outcome is the same: a significant increase in roll motion in longitudinal or quartering seas where wave excitation is small. The phenomenon is described by the same roll equation with a periodic coefficient, such as (6.5). The only difference is how to determine the amplitude of the parametric excitation from variation of the *GZ* curve in longitudinal seas as in subchapter 6.1 or from coupling between roll and heave or pitch.

6.2.2 Parametric Resonance in Linear System. Mathieu equation

We start from the linear equation of roll with parametric excitation, simply by dropping the nonlinear term in (6.5). Let us consider the system without damping first, $\delta = 0$:

$$\ddot{\phi} + \omega_m^2 \left(1 + a_p \cos \omega t\right) \cdot \phi = 0 \qquad (6.6)$$

Here a_p is an amplitude of parametric excitation, determined from the variation of the *GZ* curve or from coupling of roll with heave or pitch. ω is a frequency of parametric excitation that has a practical meaning of encounter frequency defined by equation (6.3). The following substitutions transform (6.6) into a more simple form:

$$\tau = \omega t; \quad p = \frac{\omega_m^2}{\omega^2}; \quad q = \frac{\omega_m^2 a_p}{\omega^2} \qquad (6.7)$$

$$\frac{d^2\phi}{d\tau^2} + (p + q \cos \tau) \cdot \phi = 0 \qquad (6.8)$$

This is known as the Mathieu equation. We already mentioned in subchapter 5.3.5. There are two possible behaviour types of the solution: the bounded and unbounded type. The unbounded solution of equation (6.8) corresponds to a case of parametric resonance. Examples of bounded and unbounded solutions are shown in fig. 6.2.

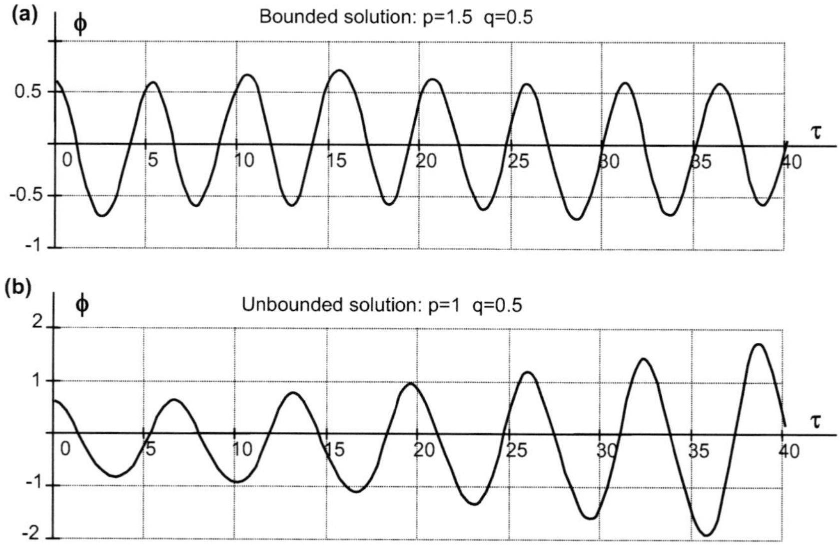

Fig. 6.2 Bounded (a) and unbounded (b) solution of Mathieu equation

The Mathieu equation has only two parameters, so it is convenient to show conditions corresponding to bounded (stable) or unbounded (unstable) solutions on a plane with coordinates p and q. It is also called the Ince-Strutt diagram; see fig .5.4.

Let us start the review of this diagram for the case when the amplitude of parametric excitation equals zero, $q=0$ [Magnus, 1976]. In this case equation (6.8) no longer has a periodic coefficient:

$$\frac{d^2\phi}{d\tau^2}+p\cdot\phi=0 \tag{6.9}$$

The solution of this equation is well known. (We discussed solutions of this equation with damping in subchapters 3.5.6 and 4.3):

$$\phi=\begin{cases}C_1\cdot\exp(\sqrt{-p}\cdot\tau)+C_2\cdot\exp(-\sqrt{-p}\cdot\tau) & p<0\\ A\cdot\cos(\sqrt{p}\cdot\tau+\varphi) & p>0\end{cases} \tag{6.10}$$

Where C_1, C_2, A and φ are arbitrary constants. It is evident that the first solution in (6.10) is unbounded and the second one is bounded and periodic. This is actually shown in the diagram on fig. 5.18. All the solutions for negative p when $q=0$ are unbounded. All the solutions for positive p are bounded.

Now let us start increasing q. We see more zones for unbounded solutions starting at the following values of p:

$$p=0.25,\ 1,\ 2.25,\ 4,\ldots\left(\frac{i}{2}\right)^2;\qquad i=1,2,3\ldots. \tag{6.11}$$

Widths of these zones of instability decrease with increasing p, so only the first and the second zones have practical importance. Oscillations in the first zone usually are called "Principal Parametric Resonance"; oscillations in the second zone are called "Fundamental Parametric Resonance".

The boundaries of these zones can be approximated for small values of q. Linear formulae for the first 5 boundaries are below [Magnus, 1976]:

$$\begin{aligned}&p_0=-\frac{1}{2}q^2\\ &p_1^{left}=\frac{1}{4}-\frac{q}{2};\quad p_1^{right}=\frac{1}{4}+\frac{q}{2}\\ &p_2^{left}=1-\frac{1}{12}q^2;\quad p_2^{right}=1+\frac{1}{12}q^2\end{aligned} \tag{6.12}$$

Let us consider the case with damping:

$$\ddot{\phi}+2\delta\dot{\phi}+\omega_\phi^2\left(1+a_p\cos\omega t\right)\cdot\phi=0 \tag{6.13}$$

Introduction of non-dimensional time $\tau=\omega t$, non dimensional damping coefficient $\mu=\delta/\omega$ and substitution $\phi(\tau)=x(\tau)\cdot\exp(-\mu\tau)$ lead to Mathieu equation (6.8):

$$\frac{d^2x}{d\tau^2}+(p+q\cos\tau)\cdot x=0$$

Definitions for frequency parameter is slightly different here:

$$p = \left(\frac{\omega_m^2}{\omega^2} - \mu^2 \right)$$

The detailed derivation is described by Shin, *et al* [2004]. Damping does not have a large influence on stability or instability of solutions. The only difference is that instability zones no longer touch the horizontal axis [Hayashi 1953]. This creates a certain threshold for amplitude of parametric excitation. If the amplitude is less than this threshold, unbounded parametric oscillations would not be generated, as the energy from parametric excitation cannot accumulate in the dynamical system: it is being dispersed by damping.

The bounded or stable solution of the Mathieu equation asymptotically tends to zero now. The unbounded solution of the Mathieu equation still tends to infinity. So the solution with p and q exactly corresponding to the boundary between zones has to be periodic.

Also, we can consider the case for quartering seas by introducing the excitation into the equation (6.13):

$$\ddot{\phi} + 2\delta\dot{\phi} + \omega_\phi^2 \left(1 + a_p \cos\omega t\right) \cdot \phi = \alpha_E \cos\omega t \qquad (6.14)$$

It has been shown that the equation (6.14) has the same domains of stability as the equation (6.12) [Rosenberg, 1954; Roberts, 1980a; Skomedal, 1982].

It means that the Ince–Strutt diagram may be enough to predict the rise of parametric roll motions; it is actually used for ABS susceptibility criteria [ABS 2004] that allow determination if a vessel with a particular hull geometry and speed capability may be vulnerable for parametric roll resonance in head or following waves. Derivation and validation of the ABS susceptibility criteria is described by Shin, *et al* [2004].

6.2.3 Parametric Resonance in Nonlinear System

Linear theory does not allow one to find the amplitude of roll motion in parametric resonance mode: there is no stabilization of parametric oscillation in the linear system. Once the nonlinearity of restoring is introduced, the dynamical system can be stabilized at a certain amplitude. The nonlinear restoring term makes the instantaneous natural frequency to change, so sooner or later it will take the system out of range of parametric resonance and, therefore, limit flow of energy from parametric excitation into the dynamical system. Once energy balance is reached the steady state mode of motion is established.

Consider equation (6.5):

$$\ddot{\phi} + 2\delta\dot{\phi} + \omega_\phi^2 (1 + a_p \cos\omega t)\phi - a_3\phi^3 = 0$$

Here symbol ω_ϕ is used in a generic sense; it does necessarily mean natural frequency in calm water, rather a frequency corresponding to a mean restoring moment in waves. Equally, symbol ω could have a meaning of encounter frequency.

Following Sanchez and Nayfeh [1990] we solve this equation using the method of multiple scales (see subchapter 4.2.4 for a detailed explanation of the method).

First, we look into the principal parametric resonance (which corresponds to the first zone of instability on the Ince-Strutt diagram at fig. 5.18). The encounter frequency is supposed to be close to twice the natural frequency:

$$\frac{1}{4}\omega^2 = \omega_\phi^2 + \varepsilon \cdot \sigma \tag{6.15}$$

Where ε is a bookkeeping parameter; it is small. Other coefficients of the equation (6.5) are presented as:

$$\delta = \varepsilon \nu; \quad \alpha_E = \alpha_{E0}\varepsilon; \quad a_3 = a_{30}\varepsilon; \quad a_p = a_{p0}\varepsilon \tag{6.16}$$

The solution is searched in the following form:

$$\phi = \phi_0(T_0, T_1, T_2) + \varepsilon\phi_1(T_0, T_1, T_2) + \varepsilon^2\phi_2(T_0, T_1, T_2) \tag{6.17}$$

T_0, T_1 and T_2 are different time scales. They correspond to a different order of expansion:

$$T_0 = t; \quad T_1 = \varepsilon t; \quad T_2 = \varepsilon^2 t \tag{6.18}$$

Now, instead of one independent variable time t, we have three. Time derivatives can be expressed through partial derivatives by new independent variables:

$$\frac{d}{dt} = D_0 + \varepsilon D_1 + \varepsilon^2 D_2 + ...; \quad \frac{d^2}{dt^2} = D_0^2 + 2\varepsilon D_1 D_0 + \varepsilon^2\left[2D_0 D_2 + D_1^2\right] + ... \tag{6.19}$$

With $D_i = \dfrac{\partial}{\partial T_i}$ is a partial derivative operator.

After substitution of solution (6.17) into equation (6.5) and taking into account (6.18) and (4.19), we obtain a system of differential equations, each of which corresponds to certain power of ε. We limit ourselves to the first expansion, a more accurate solution is available from [Sanchez and Nayfeh, 1990]:

$$\varepsilon^0: \quad D_0^2\phi_0 + \frac{1}{4}\omega^2\phi_0 = 0 \tag{6.20}$$

$$\begin{aligned} \varepsilon^1: \quad D_0^2\phi_1 + \frac{1}{4}\omega^2\phi_1 &= -2D_0 D_1\phi_0 - 2\nu D_0\phi_0 + \sigma\phi_0 - \\ &- \frac{1}{4}\omega^2 a_{p0}\cos(\omega T_0) - a_{30}\phi_0^3 \end{aligned} \tag{6.21}$$

We search for the first expansion solution in the following form:

$$\phi_0 = A(T_1)\exp\left(\frac{1}{2}i\omega T_0\right) + CC \tag{6.22}$$

Here CC is a complex conjugate term. We assume that amplitude $A(T_1)$ is a function of the first order of time scale. Substitution of (6.22) into (6.21) yields:

$$D_0^2\phi_1 + \frac{1}{4}\omega^2\phi_1 =$$

$$= \left[-2i\omega D_1 A - 2\nu i\omega A + \sigma A - \frac{1}{8}\omega^2 a_{p0}\overline{A} - 3a_{30}A^2\overline{A}\right]\exp(i\omega T_0) + \qquad (6.23)$$

$$+\left[-\frac{1}{8}\omega^2 a_{p0}A - a_{30}A^3\right]\exp(3i\omega T_0) + CC$$

It is enough for elimination of the secular terms to require:

$$-2i\omega D_1 A - 2\nu i\omega A + \sigma A - \frac{1}{8}\omega^2 a_{p0}\overline{A} - 3a_{30}A^2\overline{A} = 0 \qquad (6.24)$$

It is a differential equation, with unknown function $A(T_1)$. The following form is used for the solution of the differential equation (6.24):

$$A(T_1) = \frac{1}{2}c(T_1)\exp[i\beta(T_1)] \qquad (6.25)$$

Substitution of (6.25) in (6.24) and separation of the real and imaginary parts gives the following system of differential equations:

$$\begin{cases} \dfrac{dc}{dT_1} = -\nu c + \dfrac{1}{8}\omega c a_{p0}\sin 2\beta \\ c\dfrac{d\beta}{dT_1} = -\dfrac{\sigma c}{\omega} - \dfrac{3}{8\omega}a_{30}c^3 + \dfrac{1}{8}\omega c a_{p0}\cos 2\beta \end{cases} \qquad (6.26)$$

Since a steady state solution is required (subchapter 4.2.4), then derivatives of c and β should be equal to zero; this condition transforms the system of nonlinear differential equations (6.26) into the following system of nonlinear algebraic equations:

$$\begin{cases} -\nu c + \dfrac{1}{8}\omega c a_{p0}\sin 2\beta = 0 \\ -\dfrac{\sigma c}{\omega} - \dfrac{3}{8\omega}a_{30}c^3 + \dfrac{1}{8}\omega c a_{p0}\cos 2\beta = 0 \end{cases} \qquad (6.27)$$

The first equation of the system (6.27) obtains the formula for phase β:

$$\beta = \frac{1}{2}\arcsin\left(\frac{8\nu}{\omega a_{p0}}\right) \qquad (6.28)$$

Using formula (6.28), it is easy to express $\cos 2\beta$, since:

$$\cos 2\beta = \sqrt{1-\sin^2 2\beta} = \sqrt{1 - \frac{64\nu^2}{\omega^2 a_{p0}^2}} \qquad (6.29)$$

Substitution of expression (6.29) into the second equation of the system (6.27) yields the formula for amplitude:

$$c = \sqrt{\frac{4\sigma}{3a_{30}} - \frac{a_{p0}\omega^2}{6a_{30}}\sqrt{1 - \frac{64\nu^2}{\omega^2 a_{p0}^2}}}\;; \quad c = 0 \tag{6.30}$$

Formally, we have three solutions for the amplitude, since the second equation is of the third order. Since the amplitude is a positive value, we ignore the negative solution. So, only two solutions remain, one of them equals zero. This is also a legitimate steady state solution, but it is called "trivial".

In order to get the final formula for non-trivial amplitude, we substitute equation (6.15) into (6.30), assuming $\varepsilon = 1$:

$$c = \frac{\sqrt{\omega^2 - 4\omega_\phi^2 - \frac{\omega}{2}\sqrt{\omega^2 a_p^2 - 64\delta^2}}}{\sqrt{3a_3}} \tag{6.31}$$

There are two important outcomes from the formula (6.31). First, it is very clear now that nonlinearity keeps the amplitude limited, decreasing nonlinear coefficient a_3 leads to an increase of the amplitude, while the linear system has an infinite amplitude:

$$\lim_{a_3 \to 0} c = \infty \tag{6.32}$$

Secondly, formula (6.31) gives a threshold for rise parametric oscillations: it is derived from the condition:

$$\omega^2 a_p^2 - 64\delta^2 > 0 \tag{6.33}$$

The threshold amplitude of the parametric excitation can be expressed as:

$$a_p^{Tr} = \frac{8\delta}{\omega} \tag{6.34}$$

Also, it can be clearly seen that, if damping equals zero, there is no threshold for the parametric excitation.

However, all these formulae (6.28), (6.31) and (6.34) are only approximations and the exact values may be slightly different. For more precise calculations, use a solution of the second expansion [Sanchez and Nayfeh, 1990] or use a numerical method. An approximate solution is still useful as in the last case, since it would provide the nearly steady state initial conditions.

Sanchez and Nayfeh [1990] considered fundamental parametric resonance looking for the solution in the vicinity of the natural frequency:

$$\omega^2 = \omega_\phi^2 + \varepsilon \cdot \sigma, \tag{6.35}$$

The procedure for the derivation of the solution is similar to the one above. The only difference is that here we cannot limit ourselves by the first expansion, since the secular term does not contain the amplitude of parametric excitation, so the second expansion is necessary (the technique of the second expansion is described in subchapter 4.2.4).

The resulting system of nonlinear algebraic equations is given below (compare the terms with an analogous system for external excitation (4.90)):

$$\begin{cases} -\nu c-\dfrac{3a_{30}\nu}{2\omega^2}c^3-\dfrac{a_p^2}{8\omega^3}c^4\sin 2\beta=0 \\ -\dfrac{\sigma c}{2\omega}-\dfrac{3a_{30}c^3}{8\omega}-\left(\dfrac{\sigma^2}{8\omega^3}+\dfrac{\nu^2}{2\omega}+\dfrac{a_p^2}{12\omega^3}+\dfrac{a_p^2}{8\omega^3}\cos 2\beta\right)\cdot c- \\ \quad -\dfrac{3a_{30}\sigma}{16\omega^3}c^3+\dfrac{15a_{30}}{256\omega^3}c^5=0 \end{cases} \qquad (6.36)$$

Values of amplitude c and phase β have to be calculated numerically from the system (6.36). The steady state solution is expressed as [Sanchez and Nayfeh, 1990]:

$$\phi(t)=c\cdot\cos(\omega t+\beta)+\frac{a_p}{6\omega^2}c\cos(2\omega t+\beta)+$$
$$+\frac{a_3}{8\omega^2}c^3\cos(3\omega t+3\beta)-\frac{a_p}{2\omega^2}\cos\beta \qquad (6.37)$$

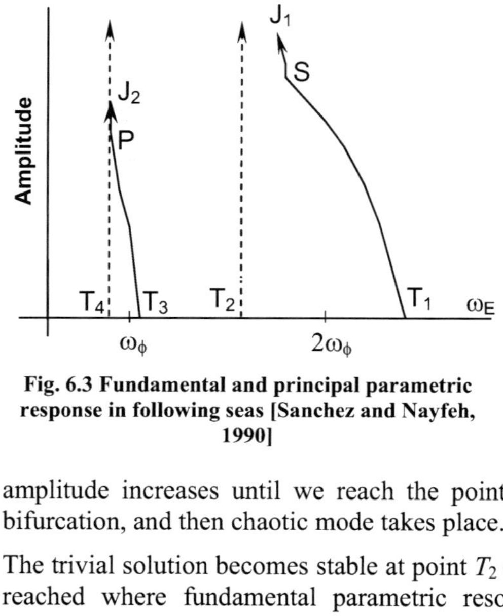

Fig. 6.3 Fundamental and principal parametric response in following seas [Sanchez and Nayfeh, 1990]

Both responses are shown in fig. 6.3. Further analysis of nonlinear equations does not differ from what we considered for the system with external excitation in Chapter 4. After a steady state solution is obtained and its stability evaluated, bifurcation analysis follows.

One can see from fig. 6.3 that the amplitude of roll response is equal to zero (trivial solution) outside of fundamental and principal parametric domains. Following Sanchez and Nayfeh [1990] we decrease frequency and reach the point T_1 where the trivial solution loses its stability and principal parametric resonance occurs. The amplitude increases until we reach the point S. Here we meet flip (period doubling) bifurcation, and then chaotic mode takes place. Finally, capsizing occurs at point J_1.

The trivial solution becomes stable at point T_2 and then keeps its stability until point T_3 is reached where fundamental parametric resonance is encountered. The fundamental resonance response remains stable up to the point P where period doubling bifurcation takes place, that leads to chaos and then to capsizing at point J_2. The trivial solution becomes stable again at point T_4.

A nonlinear system with parametric excitation has basically the same properties that a nonlinear system with external excitation. Chaotic response is associated with fractal erosion of the safe basin and an infinite number of crossings between stable and unstable invariant manifolds, similar to what we have seen in subchapter 5.3, see Sanchez and Nayfeh [1990], Kan, *et al* [1992], Kan [1992], Esparza and Falzarano [1993]. Comprehensive study of parametric roll with analytical methods was performed by Neves and Rodríguez [2006].

6.3 Surf-Riding in Following Seas

6.3.1 General

Another capsizing mode, associated with following and quartering seas is surf-riding and consequent broaching. Large following waves acting on the ship can force her to move with the same speed - the ship begins to move with the wave simultaneously. This phenomenon is called surf-riding. The majority of ships are directionally unstable during surf-riding so the ship may experience an unsteered turn to a beam position relative to the waves. Such uncontrolled turning is known as broaching. Broaching is dangerous because of the occurrence of significant heel angles caused by circulation and wave heeling moment usually acting in the same direction.

The first systematic research of surf-riding and broaching was carried out in the 1960's by Du Cane and Goodrich [1962], Wahab and Swaan [1964], Ananiev [1966] considered surf-riding as loss of motion stability of surging. Makov [1969] (see also [IMO, 1969a]) considered changing topology of the surging phase plane while gradually increasing the nominal Froude number. These two works appeared to be ahead of their time and actually receive less circulation than they deserved. It is especially true about the last work: contemporary consideration of surf-riding employs almost the same technique, however nowadays it is based on a solid foundation of nonlinear dynamics.

6.3.2 Forces and Equation of Motions

Let us consider the surging equation in pure following seas crest:

$$(m+a_{11})\ddot{\xi}_G = X_p - R + F_{XE}(t) \tag{6.38}$$

Here X_p - propeller thrust, R - resistance in calm water, $F_{XE}(t)$ - wave excitation force. The physical reason for the wave excitation is pressure pulsation caused by wave motion. We consider it in detail, in order to reveal its dependence on time: the wave elevation in fixed coordinate system is expressed as follows:

$$\zeta_W(t) = -\zeta_{Aw}\cos(k\xi - \omega t) \tag{6.39}$$

Here ζ_{Aw} is the wave amplitude and $k=2\pi/\lambda_w=\omega^2/g$ is the wave number. Dynamic pressures induced by waves are as follows:

$$p(t) = -\zeta_{Aw} g\rho \exp(k\zeta)\cos(k\xi - \omega t) \tag{6.40}$$

In order to get the force acting on a ship we need to integrate these pressures over the surface of the ship's hull. Since the ship is moving along axis ξ with heading speed v_s, it would be more convenient to do the integration in the coordinate system moving with a ship with the same speed. (To be exact, the integration has to be done in the coordinate system fixed to a ship, but we assume surging motion to be small in comparison with sailing: this allows us to use the same coordinate system, both for evaluation of the wave forces and consideration of motion behavior.) These two coordinate systems are connected by the following evident expressions:

$$x = \xi - v_s t; \quad y = \eta; \quad z = \zeta \tag{6.41}$$

The pressure is presented as:

$$\begin{aligned} p(t,x,y,z) &= -\zeta_{Aw} g\rho \exp(kz)\cos(\omega t - k(x + v_s t)) \\ &= -\zeta_{Aw} g\rho \exp(kz)\cos(\omega_e t - kx) \end{aligned} \tag{6.42}$$

Formula (6.42) does not take into account any disturbance of the pressure field that is caused by the presence of a moving ship. So integration of the above only produces the Froude-Krylov force, it could be considered enough for qualitative study.

$$\begin{aligned} F_{XE}(t) &= \int_S p(t,x,y,z)\cos(\hat{n,x})dS = \int_V \frac{\partial}{\partial x}[p(t,x,y,z)]dV = \\ &= \rho g k \zeta_A [A_S \sin(\omega_E t - kx) - A_C \cos(\omega_E t - kx)] \end{aligned} \tag{6.43}$$

Here we use the well-known technique of the transition from a surface to a volume integral (Gauss theorem). Coefficients A_S and A_C are defined by the following expressions:

$$A_S = \int_V \exp(kz)\cos(kx)dV \tag{6.44}$$

$$A_C = \int_V \exp(kz)\sin(kx)dV \tag{6.45}$$

Another possible assumption – that the ship is symmetric relative to her midship section – allows us to re-write (6.43) in more compact form:

$$F_{XE}(t) = \rho g \kappa \zeta_A A_S \sin(\omega_E t - kx) \tag{6.46}$$

$$A_S = \int_{-L/2}^{L/2} \int_0^T \exp(kz)\cos(kx)dzdx \tag{6.47}$$

The other term of the surging equation (6.38) is the difference between propeller thrust and resistance in calm water. This difference can be presented in the form of a function dependent on calm water speed v_s, number of propeller revolutions n, and surging velocity $\dot{x}_G$:

$$X_p(v) - R(v) = -\delta_1(\dot{x}_G, n, v_s) \tag{6.48}$$

Finally, the surging equation can be expressed as follows:

$$\ddot{x}_G + \delta_x(\dot{x}_G, n, v_s) = f_x \sin(\omega_E t - kx_G) \tag{6.49}$$

Where: $\delta_x = \delta_1 / (m + a_{11})$ and $f_x = \rho g \kappa \zeta_A A_S / (m + a_{11})$.

6.3.3 Equilibria

Once the model is formulated, its behavior can be studied following procedures of nonlinear dynamics (see Chapter 4). The equation (6.49) has to be presented in vector form, decreasing the order of the differential equation:

$$\dot{\vec{y}} = \begin{pmatrix} \dot{y}_1 \\ \dot{y}_2 \end{pmatrix} = \begin{pmatrix} f_x \sin(\omega_E t - ky_2) - \delta_x(y_1, n, v_s) \\ y_1 \end{pmatrix} = \vec{F}(\vec{y}) \tag{6.50}$$

The next step is a choice of control parameter – the value that we are going to change in order to observe changes in the behavior of the system. Since we do not consider maneuvers, the number of revolutions is the only value we can change. It is convenient to express it in a form of nominal Froude number that has to be calculated using the heading speed that would be achieved, if such a number of revolutions would be set in calm water. Then, the nominal Froude number has to be included as a parameter into the terms defining thrust and resistance:

$$(m + m_{11})\ddot{x}_G = X_p(Fn) - R(Fn) + F_{XE}(t) \tag{6.51}$$

The motion equation in vector form also can be rewritten:

$$\dot{\vec{y}} = \begin{pmatrix} \dot{y}_1 \\ \dot{y}_2 \end{pmatrix} = \begin{pmatrix} f_x \sin(\omega_E t - ky_2) - \delta_x(y_1, Fn) \\ y_1 \end{pmatrix} = \vec{F}(\vec{y}) \tag{6.52}$$

Now, let us find positions of equilibria. To do this, simply let $\dot{\vec{y}} = 0$ and consider the rest as a system of nonlinear algebraic equations. Dependence on time from now on may be ignored since the value of the encounter frequency is close to zero:

$$\begin{pmatrix} f_x \sin(-ky_2) - \delta_x(y_1, Fn) \\ y_1 \end{pmatrix} = \vec{F}(\vec{y}, Fn) = 0 \tag{6.53}$$

This system of equations (6.53) degenerates to one equation since is clear that surging velocity is zero, when the system is in equilibrium. So we have only one nonlinear algebraic equation:

$$f_x \sin(-ky_2) = \delta_x(Fn) \tag{6.54}$$

It is evident that (6.54) does not have the solution for all Froude numbers. The solution is only possible when $\delta_x(Fn)$ is equal or less than the amplitude of a sine function. Physically it means that surf-riding is only possible when the Froude number crosses a certain threshold, or in other words, heading speed is close to wave celerity. When the threshold is crossed, we immediately get an infinite number of solutions as these equilibria are possible on each wave, see fig. 6.4.

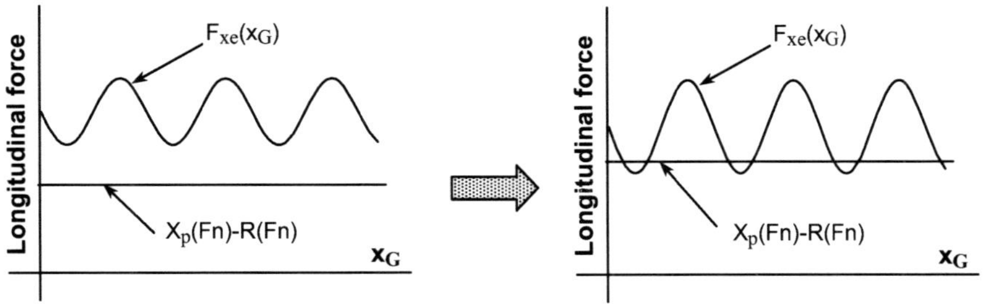

Fig. 6.4 On threshold for surf-riding equilibria

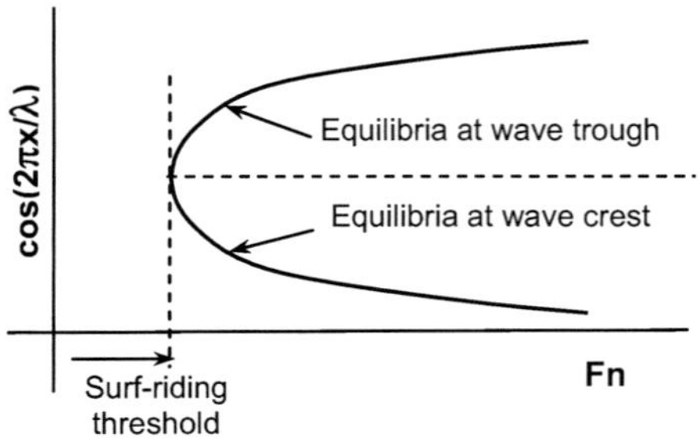

Fig. 6.5 Positions of surf-riding equilibrium vs. Froude number

In order to get rid of infinity, wave only could be considered, or following Spyrou [1995] introduce surging displacement in the form of $\cos(2\pi x/\lambda)$ (where λ is length of the wave). That would enable us to stay within one wave period all the time. Now we get only two distinct solutions for equilibrium: one is located near the wave trough and another – near wave crest. Fig. 6.5 shows their dependence on Froude number.

6.3.4 Stability of Equilibria

The fact that the system of equations (6.53) is satisfied means that there is a combination of variables and control parameters that makes all forces compensate each other, which is equilibrium. It does not mean, however that the system can stay in this position for a considerable amount of time. Equilibrium may be unstable, so any small disturbance, which is always possible, will take the system away from it. So the next logical step would be looking for stability of these equilibria.

As the disturbance is small, the system can be linearized at the vicinity of equilibrium.

The linearization procedure is the substitution of a nonlinear function with a straight line that is tangent to a curve at a given point. As it is well known, a derivative yields this tangent, see fig. 6.6.

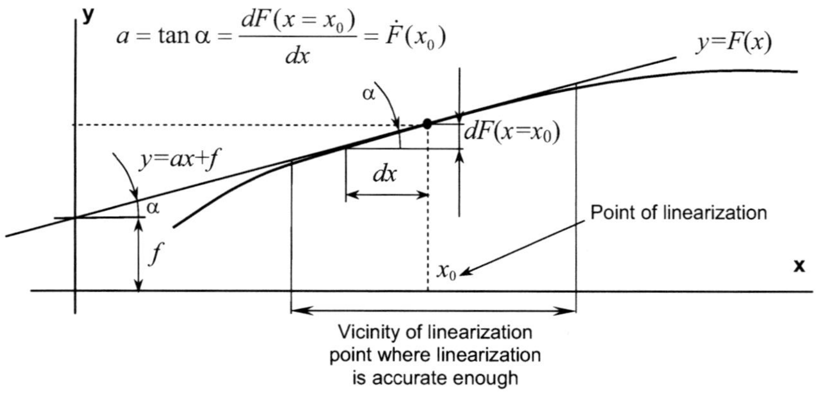

Fig. 6.6 Linearization and geometrical meaning of the derivative

Linearization of the vector valued function $\vec{F}(\vec{y})$ makes no substantial difference with the above. It is proven in functional analysis that the Jacobean matrix plays a role of the derivative for a vector valued function, which can be expressed as:

$$D\left[\vec{F}(\vec{y},Fn)\right]=D\left[\begin{pmatrix} F_1(\vec{y},Fn) \\ F_2(\vec{y},Fn) \end{pmatrix}\right]=\begin{pmatrix} \dfrac{\partial F_1(\vec{y},Fn)}{\partial y_1} & \dfrac{\partial F_2(\vec{y},Fn)}{\partial y_1} \\ \dfrac{\partial F_1(\vec{y},Fn)}{\partial y_2} & \dfrac{\partial F_2(\vec{y},Fn)}{\partial y_2} \end{pmatrix} \tag{6.55}$$

Finally, vector valued function linearized at equilibrium point $\vec{y}_e$ can be expressed as:

$$\vec{y}=D\left[\vec{F}(\vec{y},Fn)\right]\cdot\vec{y}+\vec{f}=\mathbf{A}\cdot\vec{y}+\vec{f} \tag{6.56}$$

Where $\vec{f}$ is a vector of free terms, it can be found by standard methods of linear algebra, but actually it is not needed for further consideration.

So, what we have now is a mathematical model of ship surging (6.52) linearised at the point of equilibrium, the stability of which is studied:

$$\dot{\vec{y}}=F(\vec{y},Fn) \quad \text{linearized at} \quad \vec{y}=\vec{y}_e \quad \rightarrow \quad \dot{\vec{y}}=\mathbf{A}\cdot\vec{y}+\vec{f} \tag{6.57}$$

The linearized model (6.32) is a system of ordinary linear differential equations of the first order:

$$\dot{\vec{y}}=\mathbf{A}\cdot\vec{y}+\vec{f}=\begin{cases} \dot{y}_1=A_{11}y_1+A_{12}y_2+f_1 \\ \dot{y}_2=A_{21}y_1+A_{22}y_2+f_2 \end{cases} \tag{6.58}$$

As it is well known the solution of the system (6.58) looks like:

$$\vec{y}=\begin{cases} y_1=C_1\exp(\lambda_1 t)+C_2\exp(\lambda_2 t)+C_{f1} \\ y_2=C_1\exp(\lambda_2 t)+C_2\exp(\lambda_1 t)+C_{f2} \end{cases} \tag{6.59}$$

Where coefficients C_i and C_{fi} have to be determined from the initial conditions:

$$t=0; \quad \dot{\vec{y}}=0; \quad \vec{y}=\vec{y}_e; \tag{6.60}$$

However, again, we will not need these figures for further analysis. Values λ_i are eigenvalues of the matrix **A**: eigenvalues are defined to satisfy the following equation (for any vector $\vec{z}\neq 0$), there are 2 eigenvalues for a 2x2 matrix:

$$\left\{\begin{pmatrix} A_{11} & A_{12} \\ A_{21} & A_{22} \end{pmatrix}-\lambda_i\right\}\cdot\begin{pmatrix} z_1 \\ z_2 \end{pmatrix}=0 \tag{6.61}$$

Calculation of the eigenvalues is a standard procedure of linear algebra, corresponding software is a part of any math package. Eigenvalues may be real or complex (in the latter case they come in conjugate pairs.)

It is most important to note about the eigenvalues results the presence of any positive number (if real) or positive real part (if complex). If this happens, then the solution (6.59) is unbounded and the system will leave the equilibrium position, being displaced even by a very small disturbance (subchapter 4.3).

Also, as we have seen from subchapter 4.3, every linear phase portrait (singular point) for a 2D system, has a typical arrangement of a pair of eigenvalues. The saddle point is associated with two real eigenvalues, one of which is positive and the other, negative. Applying these considerations to the surf-riding case, we find that the equilibrium near the wave crest is a saddle and it is unstable, the equilibrium near the wave trough is stable (stable focus).

6.3.5 Bifurcation Analysis[1]

Besides equilibrium, the system also has periodical motions – surging.

As can be clearly seen from fig. 6.7 there are two critical values for the control parameter. Before the first critical Froude number Fn_{cr1} is reached, periodical surging is the only steady state of the system. There are two steady states in between two critical Froude numbers: periodical surging and equilibria. Then, only equilibria exist above the second critical Froude number Fn_{cr2}.

While crossing the Fn_{cr1}, the system experiences a global bifurcation: the topology of the phase plane changes dramatically and it's impossible to map points before and after bifurcation to each other. Above Fn_{cr1} the system has alternatives: it may be periodic surging or stable surf-riding at the equilibrium point near the wave trough.

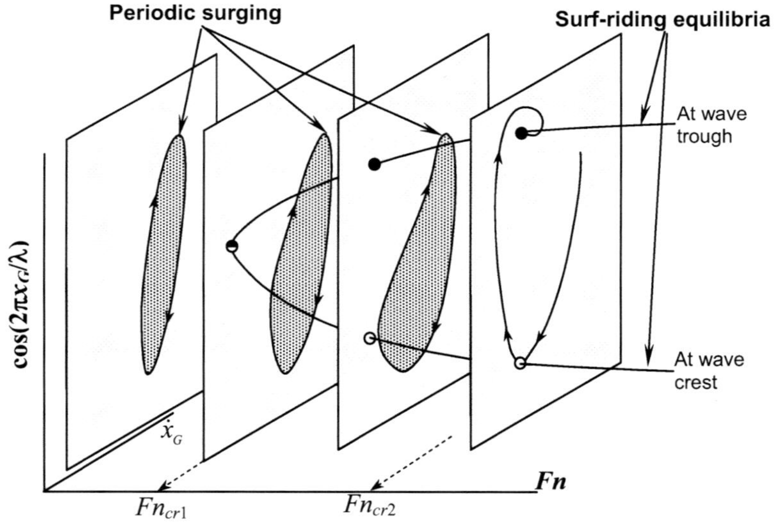

Fig 6.7 Changing of surging/surf-riding behavior with increasing nominal Froude number

[1] The author is grateful to Prof. K. Spyrou for fruitful discussion of the materials of this subchapter

Further increasing of the nominal Froude number makes the trajectory of the periodic motion to move close to unstable equilibrium near the wave crest, see fig. 6.8. The trajectory that leads to unstable equilibrium (saddle inset) reaches it in indefinite time. So being close to such an equilibrium means a significant slowing down of surging, giving the impression that a vessel is caught by the wave near the wave crest [Spyrou, 1996]. At the same time, as we have seen, this equilibrium is unstable and the ship cannot be held there for a significant amount of time.

As soon as a periodic orbit touches the inset (the trajectory that leads to the saddle – unstable equilibrium near the wave crest) the periodic surging ceases to exist and the system experiences another global bifurcation that is called "homoclinic connection".

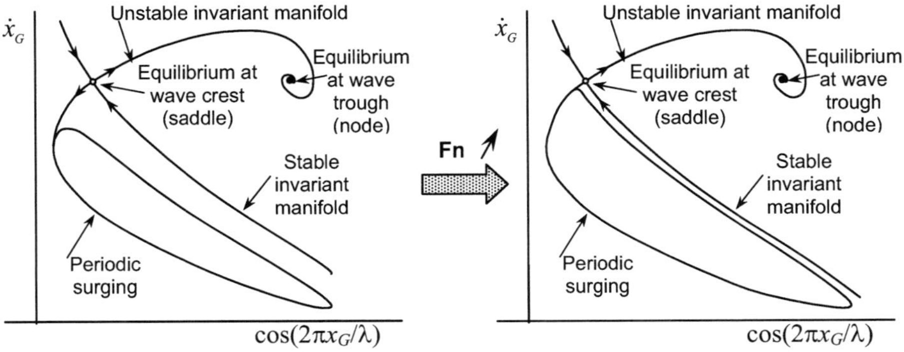

Fig. 6.8 Structure of surging /surf-riding phase picture and its changing when Fn_{cr2} is approached

Identification of the bifurcation has a direct practical meaning. It is known that a hysteresis phenomenon is associated with homoclinic connection. So, an attempt to escape from surf-riding just by decreasing the number of revolutions may not be smooth. See [Kan, 1990] and [Spyrou, 1996] for more. Fig. 6.9 contains samples of the surging phase plane for below Fn_{cr1}, (a), between Fn_{cr1} and Fn_{cr2} (b) and above Fn_{cr2}, (c) taken from [Makov, 1969]. It is noteworthy that these results were obtained without the use of nonlinear dynamics.

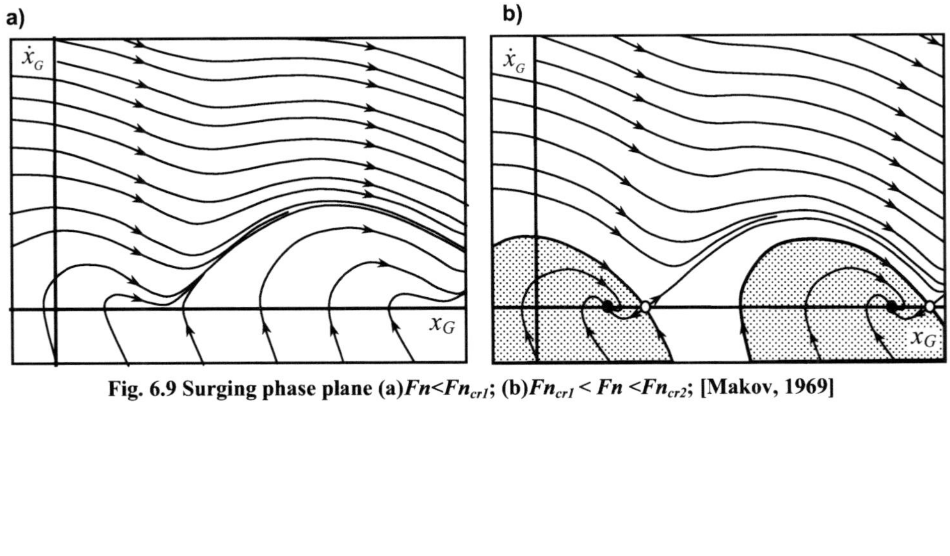

Fig. 6.9 Surging phase plane (a)$Fn<Fn_{cr1}$; (b)$Fn_{cr1} < Fn < Fn_{cr2}$; [Makov, 1969]

Ananiev [1994] associated surf-riding with loss of motion stability of the steady state solution of the surging equation, which allowed getting quite practical results, again without the use of nonlinear dynamics. However, only nonlinear dynamics provides a tool that allows observation of the entire picture.

c)
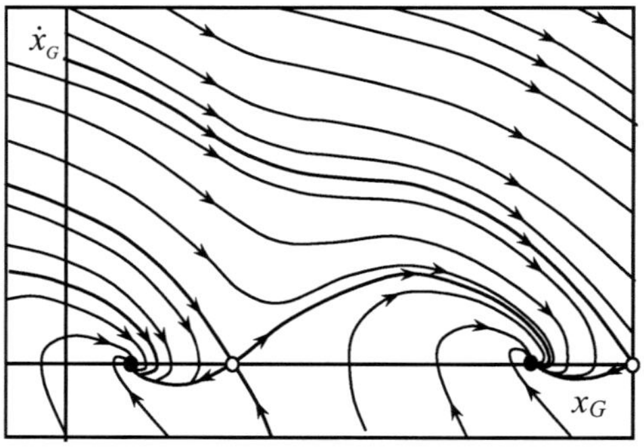

Fig. 6.9 (continued) Surging phase plane (c) $Fn > Fn_{cr2}$ [Makov, 1969]

6.4 Model of Ship Motion in Quartering Seas[1]

This and the following two subchapters are focused on broaching behavior and the capsizing danger associated with it. The problem has to be considered in terms of quartering seas: as soon as the turn occurs, the following seas assumption no longer holds. Also, the danger of surf-riding and consequent broaching also exists for quartering seas.

Broaching is a phenomenon that results in a ship being unable to maintain her course despite application of the maximum steering effort [Umeda and Renilson, 1992]. Usually we consider the interaction of environment and a ship as a solid body. Here, we have to take into account the third factor: actions of the human operator or autopilot. Otherwise, the forthcoming model would be inadequate, since it is difficult to imagine a ship significantly deviating from its desired course and no actions being taken. Moreover, the majority of currently operated ships are equipped with autopilot devices, which will try to keep the course without human intervention.

This actually means that to build an adequate broaching model we need to take controllability and control theory equations and consider them together with the roll equation. So, broaching is a phenomenon that lies at the border between stability and the two above-mentioned sciences.

6.4.1 General

While surf-riding problem could be solved just by consideration of the surging equation (e.g. the dynamical system with one degree of freedom), sailing in quartering seas generates swaying motion, which causes us to consider the system with several degrees of freedom. The first work to examine such a system with regard to broaching was Davidson [1948]. He studied a linear sway-yaw system and found that a ship that is directionally stable in calm water is not necessary directionally stable in following seas.

[1] Written in co-authorship with Prof. Nayoa Umeda of Osaka University

Further development was carried out in the 1960's and 1970's and involved models sometimes with three degrees of freedom: surge-sway-yaw. See Wahab and Swan [1964], Ananiev [1964], Eda [1972]. The approach adopted there was based on motion stability analysis of a linear or linearized system. However, it cannot describe the broaching picture completely since it is a nonlinear phenomenon.

Ananiev [1968] considered three degrees of freedom: surge-yaw-roll, the system was fully nonlinear and treated numerically. The results of this simulation yielded the first qualitative picture of broaching in following seas based on a nonlinear model that included Froude – Krylov and hydrodynamic force components. Unfortunately, this work was published only in Russian and did not get the circulation it deserved.

The work of Motora, *et al* [1982] changed the direction of broaching research. Combining computer simulation of a fully nonlinear system (surge-sway-yaw-roll) with a model and full-scale experiment, it was shown that the Froude-Krylov force component is insufficient for an adequate description of broaching. This attracted attention to the force problem, see Terao [1980], Okhusu [1986], Umeda and Renilson [1992a], Ananiev [1995], Umeda, *et al* [1995]. Motora, *et al* [1982] and Renilson [1982] found the broaching occurs when the wave induced yaw moment exceeds the maximal moment being produced by the rudder deflection.

The increasing power of computers made time domain simulation much easier than before, but also raised another challenge: initial conditions. As any other nonlinear phenomenon, broaching phenomenon significantly depends on initial conditions and this makes it difficult to identify.

The answer was in the application of nonlinear dynamics that considers behavior of sets of solutions of multi degree-of-freedom systems. Umeda and Renilson [1992, 1994] considered a 4 degrees of freedom dynamical system: surge-sway-yaw-rudder and identified its equilibria.

Umeda and Vassalos [1996] studied stable state periodic motions of the above system using the averaging method. Such modes may be unstable when the heading speed of a ship is close to the wave celerity and encounter frequency is small. It was shown that such instability actually means broaching.

Spyrou [1995, 1995a, 1996, 1996a, 1997, 2000] considered nonlinear surge-sway-yaw-roll and studied topology of its phase space. Four distinct broaching scenarios were found. Analysis of nonlinear yawing motion revealed flip and fold bifurcations at lower speed (also see subchapter 6.5).

Umeda, *et al* [1997] proposed an approximate method to predict conditions when broaching becomes inevitable. The method does not require time domain simulation; also it was validated by model experiments. Umeda (1999) studied behavior of unstable invariant manifold and identified capsizing due to broaching with heteroclinic bifurcation (also see subchapter 6.6).

6.4.2 Equations of Horizontal Ship Motions

Consider a ship heading in regular quartering seas with speed V_S that is close to wave celerity – so surf-riding is possible. The wave has frequency ω and runs with angle χ

relative to the course of the ship. We place the origin of the coordinate system at the wave trough, axis ξ coincides with the direction of the wave proliferation and axis ζ is directed downwards. We also introduce another coordinate system that is fixed to the ship. We shall need this system later when we will be considering forces. Since we have to consider rolling, let us allow this fixed system to yaw, but not to roll and to pitch. Such a system was proposed in Hamamoto, *et al* [1994] and is known as the "Horizontal Body Axes Coordinate System", see Fig. 6.10.

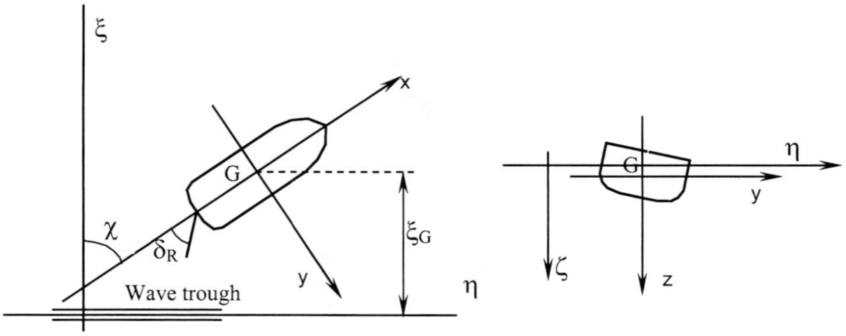

Fig. 6.10 Horizontal body axes coordinate system

As we mentioned above, we need to compose the broaching model from three models: rolling and capsizing, autopilot and controllability. Let us start from the last one and review briefly the procedure of derivation, using mainly Lewis [1989] notation (see also the detailed derivation of maneuvering equations in Appendix 3 of [ABS 2006]):

$$\begin{cases} m\cdot\ddot{\xi}_G = F_\xi \\ m\cdot\ddot{\eta}_G = F_\eta \\ I_{zz}\cdot\ddot{\psi} = N \\ I_{xx}\cdot\ddot{\phi} = K \end{cases} \tag{6.62}$$

System (6.62) takes into account only four degrees of freedom. This can be considered sufficient for an adequate description of broaching phenomenon. The heading speed is assumed to be close to wave celerity, because surf-riding is the pre-requisite for broaching. So encounter frequency is small. The natural frequency for heaving and pitching usually is comparatively high, so these types of motions may not be significant for these frequencies. Surging, swaying and yawing have no restoring terms, so their reaction to low frequency excitation may be significant. The roll restoring term is much smaller than the pitching one: longitudinal stability is always better than transverse, see [Umeda, *et al*, 1997] or [Umeda, 1999].

For simplicity's sake, we start with the surge-sway-yaw model and then add roll and autopilot equations. We also consider the linear model and add nonlinear terms later as we did for the roll equation.

Forces and moments in (6.62) usually are presented in a coordinate system fixed to a ship, see Fig. 6.10. Both coordinate systems are related by the following expressions:

$$\begin{cases} x = (\xi - \xi_G)\cos\psi + (\eta - \eta_G)\sin\psi \\ y = (\eta - \eta_G)\cos\psi - (\xi - \xi_G)\sin\psi \end{cases} \tag{6.63}$$

At the same time, we need the original coordinate system $\xi\eta\zeta$ to study broaching behavior. To be able to use existing calculation methods for the forces, we need first to apply the coordinate transformation (6.63) to them:

$$\begin{cases} F_x = F_\xi \cos\psi + F_\eta \sin\psi \\ F_y = F_\eta \cos\psi - F_\xi \sin\psi \end{cases} \tag{6.64}$$

Then, we need the accelerations $\ddot{\xi}_G$ and $\ddot{\eta}_G$ via projection of the speed vector of a ship $\vec{V}(\dot{\xi}_G, \dot{\eta}_G)$ in projections on axes *x* and *y*, because hydrodynamic forces are usually measured in a coordinate system that is fixed to the body:

$$\begin{cases} \dot{\xi}_G = u(t)\cos[\psi(t)] - v(t)\sin[\psi(t)] - c \\ \dot{\eta}_G = u(t)\sin[\psi(t)] + v(t)\cos[\psi(t)] \end{cases} \tag{6.65}$$

Here, *u* and *v* are just nomenclature for the longitudinal and transverse projections of ship speed, as it is conventional for controllability. Differentiation of both parts of (6.65) with respect to time yields the necessary accelerations $\ddot{\xi}_G$ and $\ddot{\eta}_G$:

$$\begin{cases} \ddot{\xi}_G = \dot{u}\cos\psi - \dot{v}\sin\psi - u\dot{\psi}\sin\psi - v\dot{\psi}\cos\psi \\ \ddot{\eta}_G = \dot{u}\sin\psi + \dot{v}\cos\psi + u\dot{\psi}\cos\psi - v\dot{\psi}\sin\psi \end{cases} \tag{6.66}$$

Substitution of (6.63) and (6.64) into (6.62) yields:

$$\begin{cases} m(\dot{u} - v\dot{\psi}) = F_x(u, v, \dot{u}, \dot{v}, \dot{\psi}, \ddot{\psi}) \\ m(\dot{v} + u\dot{\psi}) = F_y(u, v, \dot{u}, \dot{v}, \dot{\psi}, \ddot{\psi}) \\ I_{zz} \cdot \ddot{\chi} = N(u, v, \dot{u}, \dot{v}, \psi, \dot{\psi}, \ddot{\psi}) \end{cases} \tag{6.67}$$

Hydrodynamic forces here are considered as functions of the speed components and their derivatives. The linearized controllability model can be derived from (6.68) by expanding the terms into a Taylor series:

$$F(x_1, \ldots x_n) = \sum_{i1} \ldots \sum_{in} \frac{1}{i1! \ldots in!} \frac{\partial^{i1+\ldots+in} F}{\partial x^{i1} \cdot \ldots \cdot \partial x^{in}} (x_1 - x_{10})^{i1} \cdot \ldots (x_n - x_{n0})^{in} \tag{6.68}$$

Where $\{x_{01} \ldots x_{0n}\}$ represents an initial point. Limiting ourselves to the linear terms, we shall obtain expression as follows:

$$F_x(u, v, \dot{u}, \dot{v}, \dot{\psi}, \ddot{\psi}) = \frac{\partial F_x}{\partial u}(u - u_0) + \frac{\partial F_x}{\partial v}(v - v_0) + \ldots + \frac{\partial F_x}{\partial \ddot{\chi}}(\ddot{\psi} - \ddot{\psi}_0) \tag{6.69}$$

If we choose the initial point at the time when the ship was sailing on a straight course, then, only u_0 is not equal to zero, the rest of the coordinates are zeroes:

$$u_0 \neq 0; \quad v_0 = \dot{u}_0 = \dot{v}_0 = \dot{\psi}_0 = \ddot{\psi}_0 = 0 \tag{6.70}$$

Let us also take a close look at some of the partial derivatives in (6.68). Some of them may be zeroes, just because a ship usually can be treated as a symmetric body. Symmetry relative to the centerline plane makes the longitudinal component of the hydrodynamic force independent from transverse velocity, yawing angular velocity and angular acceleration:

$$\frac{\partial F_x}{dv} = 0; \quad \frac{\partial F_x}{d\dot{\psi}} = 0; \quad \frac{\partial F_x}{d\ddot{\psi}} = 0 \tag{6.71}$$

The same reasoning makes the transverse component of the hydrodynamic force independent from the longitudinal velocities. In reality, we cannot produce a lifting force by changing heading speed and acceleration:

$$\frac{\partial F_y}{du} = 0; \quad \frac{\partial F_y}{d\dot{u}} = 0 \tag{6.72}$$

The same can be applied to the moments:

$$\frac{\partial N}{du} = 0; \quad \frac{\partial N}{d\dot{u}} = 0 \tag{6.73}$$

Following a conventional derivation procedure for the controllability model, we introduce a new nomenclature for the rest of the derivatives:

$$X_u = \frac{\partial F_x}{du}; \quad X_{\dot{u}} = \frac{\partial F_x}{d\dot{u}} \tag{6.74}$$

$$Y_v = \frac{\partial F_y}{dv}; \quad Y_{\dot{v}} = \frac{\partial F_y}{d\dot{v}}; \quad Y_r = \frac{\partial F_y}{\partial\dot{\psi}}; \quad Y_{\dot{r}} = \frac{\partial F_y}{\partial\ddot{\psi}} \tag{6.75}$$

$$N_v = \frac{\partial N_y}{dv}; \quad N_{\dot{v}} = \frac{\partial N}{d\dot{v}}; \quad N_r = \frac{\partial N}{\partial\dot{\psi}}; \quad N_{\dot{r}} = \frac{\partial N}{\partial\ddot{\psi}} \tag{6.76}$$

The values defined by (6.74-76) are usually called "hydrodynamic derivatives". Also:

$$r = \dot{\psi}; \quad \dot{r} = \ddot{\psi} \tag{6.77}$$

The only item left is rudder action. It can be expressed in the first expansion (so far we take just rudder action; interaction with the hull and propeller will be taken into account later) as:

$$F_X^{Rd} = 0; \quad F_Y^{Rd} = Y_\delta \delta_R; \quad N^{Rd} = N_\delta \delta_R \tag{6.78}$$

Here δ_R is the angle of rudder deflection. Now, we are ready to write a set of controllability equations, we just expand the right hand side terms of (6.67) into a Taylor series up to the first order derivatives and take into account (6.71-73), using the nomenclature of (6.74-76). To get it in a conventional form, all terms expressing hull forces (including inertia) should be gathered on the left side and the "rudder" terms (6.78) should be on the right side:

$$\begin{cases} m(\dot{u} - vr) - X_u(u - u_0) - X_{\dot{u}}\dot{u} = 0 \\ m(\dot{v} + ur) - Y_v v - Y_{\dot{v}}\dot{v} - Y_r r - Y_{\dot{r}}\dot{r} = Y_\delta \delta_R \\ I_{zz}\dot{r} - N_v v - N_{\dot{v}}\dot{v} - N_r r - N_{\dot{r}}\dot{r} = N_\delta \delta_R \end{cases} \quad (6.79)$$

To complete the procedure, we neglect the nonlinear term *mvr* (both *v* and *r* are small) and regroup the terms so the linear controllability model looks like:

$$\begin{cases} -X_u(u - u_0) + \dot{u}(m - X_{\dot{u}}) = 0 \\ -Y_v v + \dot{v}(m - Y_{\dot{v}}) - (Y_r - mu)r - Y_{\dot{r}}\dot{r} = Y_\delta \delta_R \\ -N_v v - N_{\dot{v}}\dot{v} - N_r r + \dot{r}(I_{zz} - N_{\dot{r}}) = N_\delta \delta_R \end{cases} \quad (6.80)$$

System (6.80) could be simplified even more, however it is not necessary here as we proceed further with our study of broaching.

6.4.3 Surging and Surge Wave Force

Let us start from the first equation of (6.80); it describes surging. We have already considered it when we were talking about surf-riding, and there is not too much to add. The only item we can mention is that the coefficient $X_{\dot{u}}$ represents the longitudinal added mass a_{11} in conventional seakeeping nomenclature. The only difference is that conventional controllability does not consider dependence of added mass on wave frequency, since conventional controllability deals with calm water. Here, we also have to consider added mass at excitation frequency is zero since the encounter frequency is very small, so we use symbol m_x instead of a_{11}:

$$(m + m_x)\ddot{\xi}_G + R(c + \dot{\xi}_G) - T(c + \dot{\xi}_G) = X_W(\xi_G / \lambda, \chi) \quad (6.81)$$

6.4.4 Swaying and Sway Wave Force

As can be clearly seen from fig. 6.11, [Umeda, *et al*, 1995]. The Froude-Krylov hypothesis is insufficient to describe the swaying wave force. So we have to take into account the interaction between ship induced fluid motion with incident waves. The theoretical solution of Umeda, *et al* [1995] is based on slender body theory. The general solution [Mauro, 1967] was simplified for the case of zero encounter frequency. Basic application of slender body theory can be found in [Newman, 1977].

Since the derivation is too lengthy, we have to limit ourselves to a discussion of the final results. The final formulae are given below:

$$Y_w(\xi_G / \lambda, u, \psi) = Y_{FK}(\xi_G / \lambda, \psi) + Y_H(\xi_G / \lambda, u, \psi) \quad (6.82)$$

The Froude-Krylov part is just a result of integration of wave pressure on the ship surface:

$$Y_{FK}(\xi_G / \lambda, \psi) = \rho g \zeta_{Aw} k \sin\psi \int_{AP}^{FP} C_1(x) S(x) e^{-0.5kd(x)} \sin[k(\xi_G + x\cos\psi)dx \quad (6.83)$$

Here:

$$C_1(x) = \frac{\sin[0.5B(x) \cdot k \sin\psi]}{0.5B(x) \cdot k \sin\psi} \tag{6.84}$$

B(x) is the waterplane breadth of the ship section at abscissa *x*; *d(x)* is the draft abscissa *x;* *S(x)* is the area of the submersed part of the ship section at abscissa *x*. *AP* means "aft perpendicular" and *FP* means "fore perpendicular".

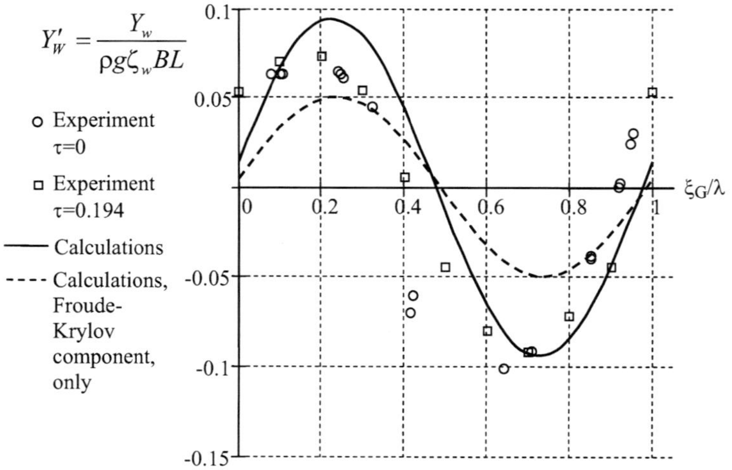

Fig. 6.11 Sway wave force [Umeda, *et al*, 1995]

The hydrodynamic part here is mainly contributed by free vortex layers. The diffraction influence may be small because of the small value of the encounter frequency:

$$Y_H(\xi_G/\lambda, u, \psi) = -\zeta_{Aw}\omega \cdot u \sin\psi \left[\rho S_y(x) e^{-0.5kd(x)} \cos[k(\xi_G + x\cos\psi)\right]_{AP}^{FP} \tag{6.85}$$

$S_y(x)$ is the two-dimensional added mass for the ship section at abscissa *x* for zero frequency. Theoretically, it can be estimated using any appropriate method. A comparison of the results of different theoretical methods with model tests indicates a high sensitivity of the final result to accuracy of its calculation. (See [Hamamoto, 1973] and compare with [Umeda, *et al*, 1995] or fig. 6.11) Then we obtain the following formula based on (6.80) :

$$\dot{v}(m - Y_{\dot{v}}) + (mu - Y_r)r - Y_v v - Y_{\dot{r}}\dot{r} - Y_\phi \phi = Y_\delta \delta_R + Y_w(\xi_G/\lambda, u, \psi) \tag{6.86}$$

We also have added term $Y_\phi \phi$ that expresses influence of the roll angle on swaying motion. Coefficient $Y_{\dot{v}}$ is added mass 22 taken with a "minus" sign. Since the encounter frequency is very low and vortexes are "responsible" for the generation of hull and rudder forces, the best way to get their numerical values is by conventional captive tests in calm water, see [Umeda and Vassalos, 1996]. Experiments by Fujino, *et al* [1983] showed that wave influence on these coefficients is small.

6.4.5 Yaw Motions and Yaw Wave Moment

The third equation of (6.80) describes yaw motions :

$$(I_{zz}+J_{zz})\dot{r}-N_v v-N_r r-N_\phi\phi=N_\delta\delta_R+N_w(\xi_G/\lambda,u,\psi) \qquad (6.87)$$

Here, N_w is the wave yawing moment. Analogously to the swaying wave force, this hydrodynamic component is significant; see fig. 6.12 [Umeda, *et al* 1995]. As in previous cases the term $-N_{\dot{r}}$ has the meaning of added moment of inertia J_{zz} (or m_{66} in seakeeping notation).

We consider the yawing wave moment as a sum of the Froude-Krylov and hydrodynamic components. The free vortex layer contributes to the latter one:

$$N_w(\xi_G/\lambda,u,\psi)=N_{FK}(\xi_G/\lambda,\chi)+N_H(\xi_G/\lambda,u,\psi) \qquad (6.88)$$

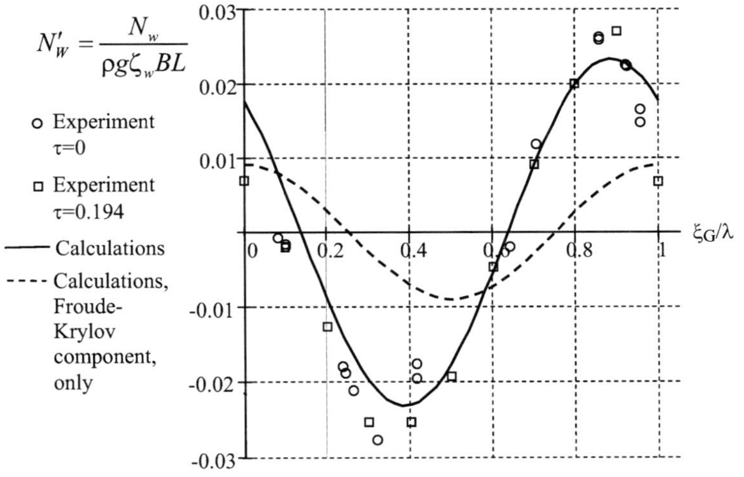

Fig. 6.12 Yaw wave moment [Umeda, *et al*, 1995]

Froude-Krylov component:

$$N_{FK}(\xi_G/\lambda,\psi)=\rho g\zeta_{Aw}k\sin\psi\int_{AP}^{FP}C_1(x)S(x)e^{-0.5kd(x)}x\sin[k(\xi_G+x\cos\psi)dx \qquad (6.89)$$

Hydrodynamic component :

$$N_H(\xi_G/\lambda,u,\psi)=\zeta_{Aw}\omega u\sin\psi\int_{AP}^{FP}\rho S_y(x)e^{-0.5kd(x)}\sin[k(\xi_G+x\cos\psi)dx$$
$$-\zeta_{Aw}\omega\cdot u\sin\psi\left[\rho S_y(x)e^{-0.5kd(x)}x\cos[k(\xi_G+x\cos\psi)\right]_{AP}^{FP} \qquad (6.90)$$

We also have dropped the term $N_{\dot{v}}\dot{v}$ because it is small for the majority of ships, see [Lewis, 1989]. The term $N_\phi\phi$ was added to take into account the roll influence on yaw motion. Numerical values for hydrodynamic derivatives are suggested to be found by a conventional captive model test in calm water, see [Umeda and Vassalos, 1996].

6.4.6 Roll Equation for Broaching Study

The next step is to add the roll equation. First, consider the linear equation:

$$(I_x + a_{44})\ddot{\phi} + b_{44}\dot{\phi} + mg \cdot GM \cdot \phi = M_W(t) \tag{6.91}$$

First of all, if we study capsizing we need to consider the nonlinear restoring term: $mgGZ(\phi)$. Then, since the broaching situation assumes a very low encounter frequency, there will be no time dependence for the excitation moment:

$$\begin{aligned} M_W(t) &= K_w(\xi_G/\lambda, u, \psi) = \\ &= K_{FK}(\xi_G/\lambda, \psi) + K_H(\xi_G/\lambda, u, \psi) + Y_w(\xi_G/\lambda, u, \psi) \cdot \overline{OG} \end{aligned} \tag{6.92}$$

Model tests show that the hydrodynamic component is significant for the roll excitation moment as well, see fig. 6.13 [Umeda, *et al*, 1995].

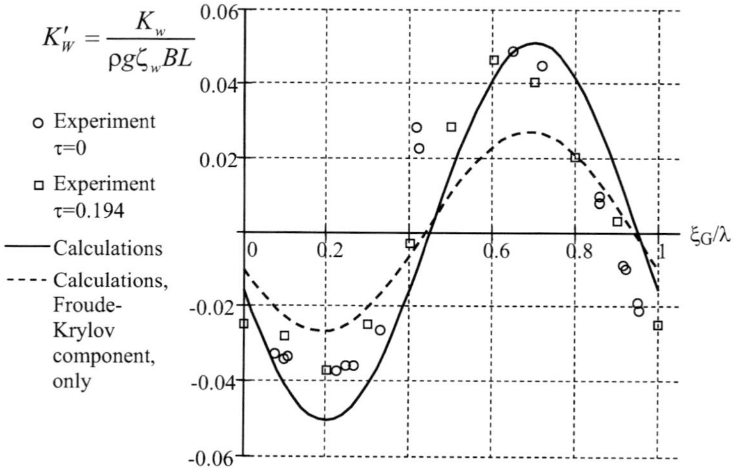

Fig. 6.13 Roll excitation moment [Umeda, *et al*, 1995]

The following expressions for these components were deduced by Umeda, *et al* [1995] :

$$\begin{aligned} &K_{FK}(\xi_G/\lambda, \psi) = \\ &= -\rho g \zeta_{Aw} k \sin\psi \int_{AP}^{FP} C_1(x)[B(x)/2][d(x)]^2 e^{-0.5kd(x)} \sin[k(\xi_G + x\cos\psi)]dx - \\ &- \rho g \zeta_{Aw} k^2 \sin\psi \int_{AP}^{FP} C_4(x)[B(x)/2]^3 d(x) e^{-0.5kd(x)} \sin[k(\xi_G + x\cos\psi)]dx \end{aligned} \tag{6.93}$$

Here, C_1 is defined by formula (6.84) and

$$C_4(x) = \frac{2\sin[k\sin\psi \cdot B(x)/2] - k\sin\psi \cdot B(x)\cos[k\sin\chi \cdot B(x)/2]}{[k\sin\psi \cdot B(x)/2]^3} \tag{6.94}$$

A comparison of the calculation and the experimental results has shown that linear expansion of the hydrodynamic component is sufficient for an adequate description of the roll excitation:

$$K_H(\xi_G/\lambda, u, \psi) = \zeta_w \omega \cdot u \sin\psi \cdot \left[\rho I_{Sy}(x) e^{-0.5kd(x)} \cos[k(\xi_G + x\cos\psi)\right]_{AP}^{FP} \tag{6.95}$$

Hydrodynamic excitation here is created by a free vortex layer, the same as used for swaying and yawing. $I_{Sy}(x)$ is a two-dimensional added roll moment due to sway acceleration. The third component is just a heeling moment, caused by the sway excitation force: here $\overline{OG}$ is the distance between the waterline (point O) and the centre of gravity of the ship. The force Y_w is applied at the waterplane level, because added masses for ship sections $I_{Sy}(x)$ and $S_y(x)$ usually are obtained from the coordinate system with the origin at the waterplane.

Roll damping consists here mainly of the linear component because of relatively large forward velocity, so it will considered linear, in controllability nomenclature:

$$b_{44}\dot{\phi} = -K_p p; \qquad p \equiv \dot{\phi} \tag{6.96}$$

It is easy to get the rest of the terms; they should be just moments of corresponding forces in swaying equations:

$$\begin{array}{llllll} Y_v v & \rightarrow & K_v v & \text{or} & b_{22}\dot{\eta}_G & \rightarrow & b_{24}\dot{\eta}_G \\ Y_r r & \rightarrow & K_r r & & b_{26}\dot{\psi} & \rightarrow & b_{46}\dot{\psi} \\ Y_\phi \phi & \rightarrow & K_\phi \phi & & c_{24}\phi & \rightarrow & c_{24} z_H \phi \\ Y_\delta \delta_R & \rightarrow & K_\delta \delta_R & & & & \end{array} \tag{6.97}$$

Again, the reader should not be misled by usage of symbol b; it does not mean that the forces have a wave nature – vice versa: they are vortex forces and moments! Here z_H is the height of the centre of lateral force.

Now, we can write the roll equation in both notations: "seakeeping" and "controllability" :

$$\begin{aligned} &(I_{xx} + a_{44})\ddot{\phi} + b_{44}\dot{\phi} + mg \cdot GZ(\phi) + b_{24}\dot{\eta}_G + b_{46}\dot{\psi} + c_{24} z_H \phi = \\ &\quad = K_w(\xi_G/\lambda, u, \psi) + K_\delta \delta_R \end{aligned} \tag{6.98}$$

$$\begin{aligned} &(I_{xx} + J_{xx})\dot{p} - m_x ur \cdot z_H - K_v v - K_r r - K_p p - K_\phi \phi + mg \cdot GZ(\phi) = \\ &\quad = K_\delta \delta_R + K_w(\xi_G/\lambda, u, \psi) \end{aligned} \tag{6.99}$$

J_{xx} is added mass 44 in controllability notation. It meant to be defined for zero wave frequency. Hydrodynamic derivatives can be found in the same way as in the previous cases. More details can be found in [Umeda, *et al*, 1995], [Umeda and Vassalos, 1996].

6.4.7 Equation of Autopilot

The last equation should describe action of the autopilot. Generally, this kind of differential equation is a subject of control theory. Here, a brief description is given.

The simplest course-keeping procedure can be described by the following linear equation:

$$\delta_R = -K_R(\psi - \psi_C) \tag{6.100}$$

Here χ_C is desired course. Action is very simple – rudder deflection is just proportional to deviation from the desired course. K_R is a proportional coefficient that is called "rudder gain constant" or just "rudder gain" (sometimes it is called "yaw gain"). This is just simple feedback, too simple to be applied practically: one of the reasons, such an "autopilot" would always be late because it would react on course deviation only. To make it react in advance we can add some reaction on yaw rate too:

$$\delta_R = -K_R(\psi - \psi_C) - K_R T_D r \tag{6.101}$$

Here, T_D is a correction for rudder gain, to keep proportionality to yaw rate. Such control is called "differential" because we have added reaction on the derivative and T_D is called "time constant for differential control".

The last step to get an appropriate equation of autopilot is to take into account that the steering gear cannot react instantly – rudder deflection takes some time – so we can add deflection velocity into (6.101):

$$T_E\dot{\delta}_R + \delta_R = -K_R(\psi - \psi_C) - K_R T_D r \tag{6.102}$$

Here T_E is another proportionality coefficient that is called "time constant for steering gear". All coefficients K_R, T_D and T_E are subjects of autopilot tuning. For more information, see [Lewis, 1989].

6.4.8 Model for Broaching

Summarizing, the system of equations was obtained that may be suitable for the study of ship behavior in quartering seas. It consists of conventional controllability equations with the addition of roll and autopilot equations:

$$\begin{cases} (m+m_x)\dot{u} + R(u) - T(u) = X_w(\xi_G/\lambda, \psi) \\ (m+m_y)\dot{v} + (mu - Y_r)r - Y_v v - Y_\phi \phi = Y_\delta \delta_R + Y_w(\xi_G/\lambda, u, \psi) \\ (I_{zz} + J_{zz})\dot{r} - N_v v - N_r r - N_\phi \phi = N_\delta \delta_R + N_w(\xi_G/\lambda, u, \psi) \\ (I_{xx} + J_{xx})\dot{p} - m_x ur \cdot z_H - K_v v - K_r r - K_p p - K_\phi \phi + mg \cdot GZ(\phi) = \\ \qquad = K_\delta \delta_R + K_w(\xi_G/\lambda, u, \psi) \\ T_E \dot{\delta}_R + \delta_R = -K_R(\psi - \psi_C) - K_R T_D r \end{cases} \tag{6.103}$$

To proceed with the dynamics of broaching phenomenon, we need first to present (6.103) as a system of differential equations of the first order:

$$\begin{cases} \dot{\xi}_G = u\cos - v\sin\psi - c \\ \dot{u} = \{X_w(\xi_G,\chi) - R(u) + T(u)\}/(m+m_x) \\ \dot{v} = \{-(M+m_x)ur + Y_v v + Y_r r + Y_\phi \phi + Y_\delta \delta_R + Y_w(\xi_G,u,\psi)\}/(m+m_y) \\ \dot{\psi} = r \\ \dot{r} = \{N_v v + N_r r + N_\phi \phi + N_\delta \delta_R + N_w(\xi_G,u,\psi)\}/(I_{zz}+J_{zz}) \\ \dot{\phi} = p \\ \dot{p} = \{m_x ur \cdot z_H + K_v v + K_r r + K_p p + K_\phi \phi - \\ \qquad - W\cdot GZ(\phi) + K_\delta \delta_R + K_w(\xi_G,u,\psi)\}/(I_{xx}+J_{xx}) \\ \dot{\delta}_R = \{-\delta_R - K_R(\chi-\chi_C) - K_R T_D r\}/T_E \end{cases} \tag{6.104}$$

It would be more convenient to present system (6.104) in vector form:

$$\dot{\vec{x}} = F(\vec{x},\vec{b}) \tag{6.105}$$

Where $\vec{x}$ is called the state vector:

$$\vec{x} = \{\xi_G, u, v, \psi, r, \phi, p, \delta_R\}^T \tag{6.106}$$

Sign "T" here means transposition that converts vector –row into vector –column. The vector of parameters $\vec{b}$ consist here just from two parameters:

$$\vec{b} = \{\psi_C, n\} \tag{6.107}$$

Here, n is number of propeller revolutions. It has to be included in all terms connected with thrust and rudder actions. Function $F(\vec{x},\vec{b})$ is vector valued and forms the right hand side of the system (6.105):

$$F(\vec{x},\vec{b}) = \begin{cases} x_2\cos x_4 - x_3 \sin x_4 - c \\ \{X_w(x_1,x_4) - R(x_2) + T(x_2)\}/(M+m_x) \\ \{-(m+m_x)x_2x_5 + Y_v x_3 + Y_r x_5 + Y_\phi x_6 + \\ \qquad + Y_\delta x_8 + Y_w(x_1,x_2,x_4)\}/(m+m_y) \\ x_5 \\ \{N_v x_3 + N_r x_5 + N_\phi x_6 + N_\delta x_8 + \\ \qquad + N_w(x_1,x_2,x_4)\}/(I_{zz}+J_{zz}) \\ x_7 \\ \{m_x x_2 x_5 \cdot z_H + K_v x_3 + K_r x_5 + K_p x_7 + K_\phi x_6 - \\ \qquad - mg\cdot GZ(x_6) + K_\delta x_8 + K_w(x_1,x_2,x_4)\}/(I_{xx}+J_{xx}) \\ \{-x_8 - K_R(x_4 - b_1) - K_R T_D x_5\}/T_E \end{cases} \tag{6.108}$$

The next step is application of a nonlinear dynamics procedure to the dynamical system (6.105).

6.5 Ship Behavior in Quartering Seas[1]

Following an established nonlinear dynamics procedure, positions of equilibrium are searched in dynamical system (6.105). First, we will look at an unsteered vessel. It is the same model, just without the autopilot equation: the rudder deflection angle has to be considered as a control parameter. Then, a steered vessel is considered, having a rudder deflection angle as one of the state variables and desired course as a control parameter.

A comparison between steered and unsteered cases has practical importance: it shows how broaching behavior can be altered by the action of the autopilot or helmsman. This comparison can answer a question: what can we do to avoid broaching?

Another control parameter is commanded speed that can be expressed in a form of number of revolutions of main engine or in a form of nominal Froude number, expected for these engine settings in calm water.

6.5.1 Equilibria of Unsteered Vessel

Consider an unsteered vessel. If she is in the equilibrium position, then a sum of all the forces and moments acting on her is zero:

$$\vec{F}_u(\vec{x}_u, \vec{b}_u) = 0 \qquad (6.109)$$

Here index "u" reminds us that the vessel is unsteered.

The procedure is similar to the surf-riding analysis in following seas. We just have more equations in the system. As all derivatives disappear, expression (6.109) degrades to a system of nonlinear algebraic equations that can be solved numerically.

Before we proceed, let us once again consider the physical meaning of the equilibrium for the model adopted here. Assuming the derivatives are zero, we however, did not take away wave action. All the wave forces that depend on position (like the Froude-Krylov wave forces) are still included into (6.109). It means that despite wave action, the ship does not move, so she may be captured by a wave. It means that the equilibrium position in this model is a surf-riding mode in quartering seas.

As we have seen from subchapter 6.3, the surf-riding mode can exist for a certain range of nominal Froude numbers, when ship speed is close to wave celerity. Similar limitations exist for rudder deflection; surf-riding is a phenomenon of following and quartering seas.

The system (6.109) consists of 7 equations, so the equilibrium is a point in a space of 7 dimensions and several solutions are possible since the system (6.109) is nonlinear. Changing rudder deflection angle will cause these points to move in eight-dimensional space (we have added one dimension – rudder deflection angle). We can only study projections of this trace on a plane. Fig. 6.14 [Spyrou, 1996] shows such projections on a coordinate plane position on wave (b) and heading angles (a) vs. rudder deflection angle.

As can be seen from fig. 6.14, there may be none, two or four equilibria, depending on the value of the rudder deflection angle. Following established terminology, we

[1] Written in co-authorship with Prof. Kostas J. Spyrou of the National Technical University of Athens

distinguish equilibrium positions located near the wave crest and wave trough. They can be easily identified from figure 6.14b, but not that evident in figure 6.14a.

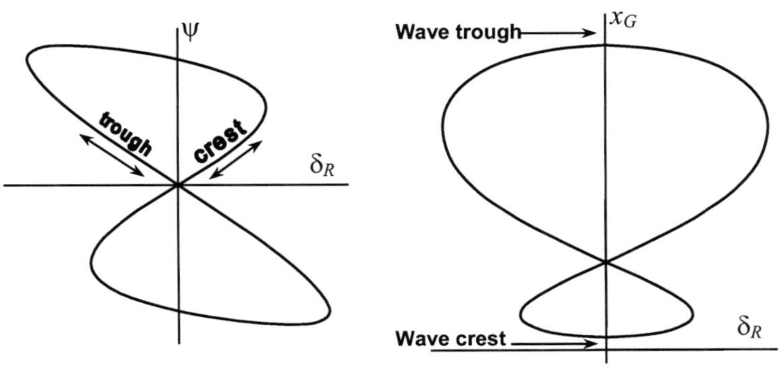

Fig. 6.14 Surf-riding equilibria in quartering seas [Spyrou, 1995]

The second control parameter, the nominal speed (in the form of nominal Froude number), is capable of altering the shape of both curves, although not too significantly in a qualitative sense, see fig. 6.15. So, it is possible to limit further consideration just by one curve, assuming, however, that the nominal Froude number remains within the limits making surf-riding possible.

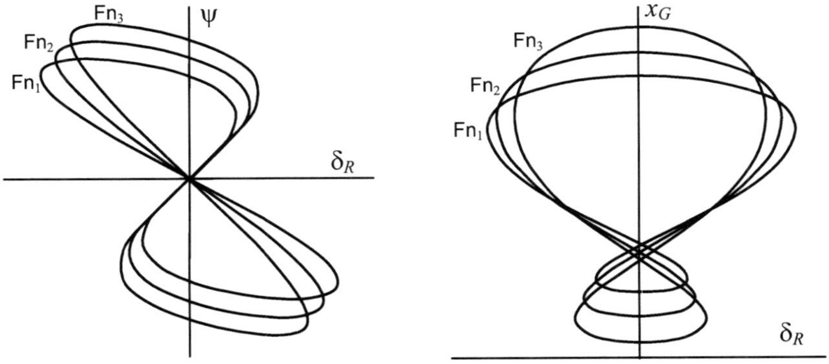

Fig. 6.15 Surf-riding equilibria for different Froude numbers ($Fn_1 < Fn_2 < Fn_3$) [Spyrou, 1995]

6.5.2 Stability of Equilibria of Unsteered Ship

The procedure of stability analysis of equilibria is already described for surf-riding in pure following seas in a quite general way, see subchapter 6.3. We have to point out, however, on some aspects of eigenvalues behavior that are especially important in a multidimensional case, where visualization is a problem.

The number of eigenvalues corresponds to the dimension of the model; it is the number of equations in (6.105). That means, each eigenvalue can be associated with a variable from our system. For the unsteered vessel it looks like:

$$\begin{aligned}
&\lambda_1 \rightarrow x_1 = \xi_G \\
&\lambda_2 \rightarrow x_2 = u \\
&\lambda_3 \rightarrow x_3 = v \\
&\lambda_4 \rightarrow x_4 = \psi \\
&\lambda_5 \rightarrow x_5 = p = \dot{\phi} \\
&\lambda_6 \rightarrow x_6 = \phi \\
&\lambda_7 \rightarrow x_7 = r = \dot{\psi}
\end{aligned} \qquad (6.110)$$

A positive eigenvalue for a certain variable means that the system will escape in the particular direction from an unstable equilibrium position.

Actually eigenvalues provide even more valuable information; they not only constitute the stability or instability of the equilibrium and direction of possible escape. Behavior of the eigenvalues (i.e. how do they move when we change the control parameter) gives a complete qualitative picture of motion in the vicinity of equilibrium (subchapter 4.3). This information may be very useful as a first step towards the whole picture of possible motions.

We are looking at how the phase portrait is changing while we are moving along the curve of heading vs. rudder angle. The only problem here is the number of dimensions. When we are looking at roll, we had just one degree of freedom and everything was taking place on the phase plane, which is a two-dimensional phase space. Here, we have seven variables resulting in 7-D phase space, so the complete visualization is impossible! We can plot only 2-D and 3-D intersections or projections.

It is not necessary to consider all the eigenvalues all the time. We will be looking only on those that indicate instability that are positive or show a tendency to become positive.

It has been shown that there are no stable equilibria for the unsteered vessel Spyrou [1996]. All cases have shown one or two positive real eigenvalues or a positive real part of complex eigenvalues. See fig. 6.16 where eigenvalues and corresponding phase pictures are shown.

Let us start from zero and track the eigenvalues' behavior, moving first along the curve corresponding to the wave crest.

The positive eigenvalue corresponds essentially to surging motion. It is real and its "pair-mate" is real and negative. This makes a saddle type of the phase plane that governs motion of the system in the vicinity of the equilibrium. Moreover, we can see that the saddle repels the system in the surging direction.

Now, let us move towards point D in fig. 6.16. While moving, the negative real eigenvalue moves towards zero and reaches it at the point D, see figure 6.17.

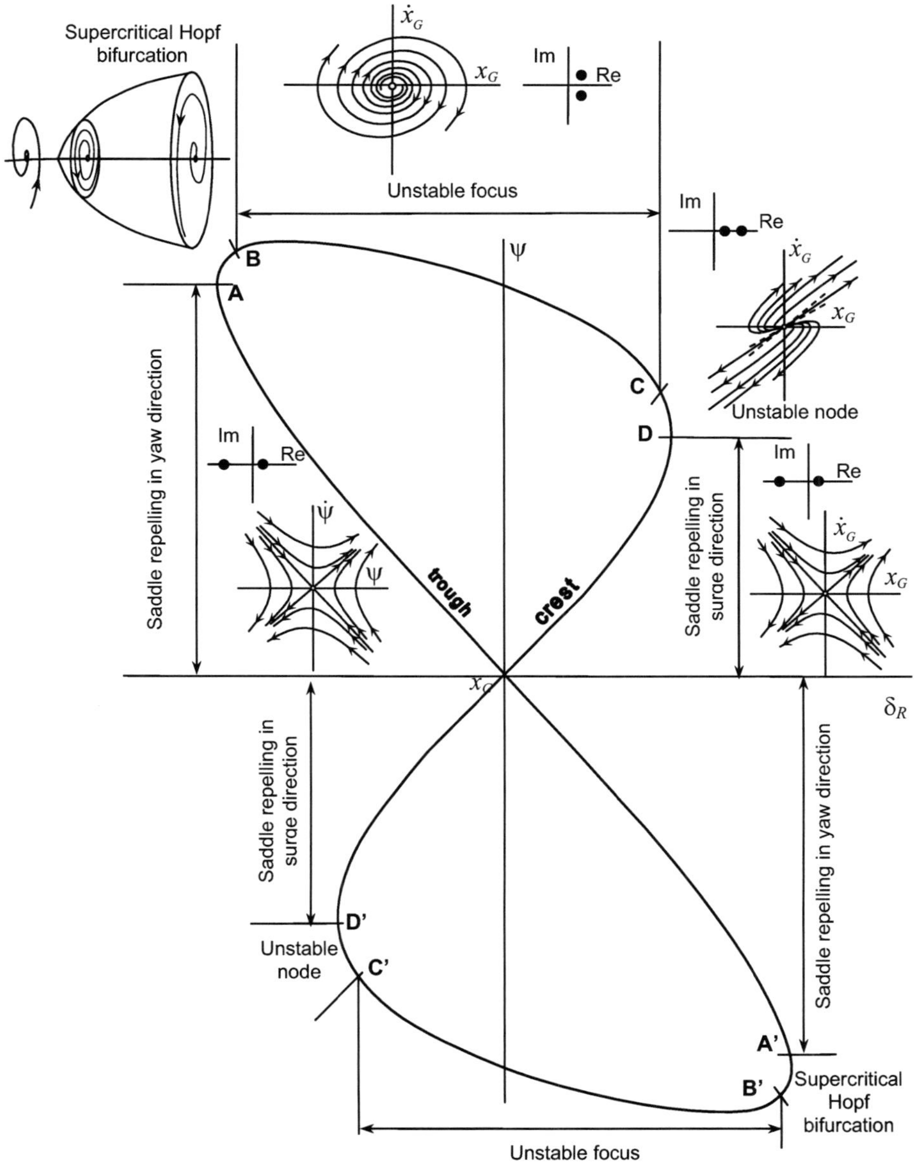

Fig. 6.16 Stability of surf-riding equilibria in quartering seas

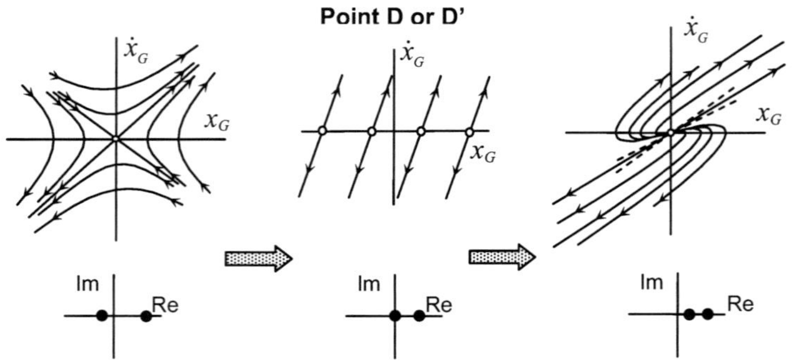

Fig. 6.17 Saddle-node bifurcation for an unsteered vessel in quartering seas

When it crosses zero, we have a situation with two real positive eigenvalues. As it is known, locally this constitutes a nodal type of phase plane (see also subchapter 4.3). We see here a bifurcation, which is known under the names "saddle-node bifurcation", "global fold bifurcation", or "limit point bifurcation". More information about bifurcations in visual form can be found in [Thompson and Stewart, 1986].

Further developments of the situation make the second eigenvalue (that one, which was negative in the first place) to reach its mate at the point C on fig. 6.16. It is separately shown on fig. 6.18.

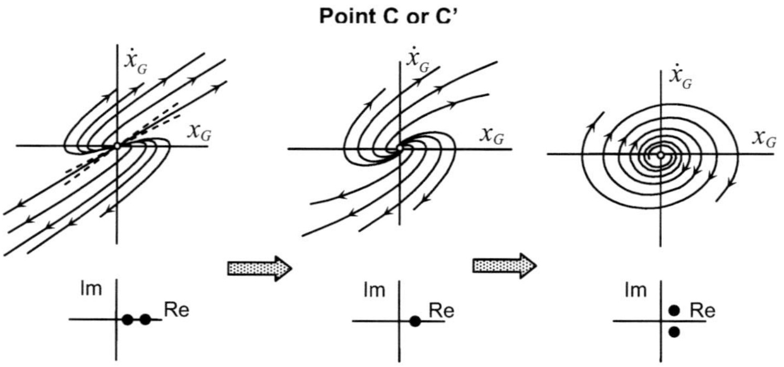

Fig. 6.18 Transition from focal to nodal instability for an unsteered vessel in quartering sea

Then, the eigenvalues become complex conjugates and the imaginary part increases. Now, we have two complex eigenvalues with two positive real parts, which makes an unstable focus type of phase plane. We observe the transition from nodal to focal instability here.

The next transition is more complex. Spyrou [1996] found that there may be several scenarios. Here, we show only one, another one is described in the above reference.

The real parts of complex eigenvalues cross the zero level, see fig. 6.19. Nonlinear numerical calculations show the existence of a limit cycle (periodic surge motions) caused by nonlinear properties of the model. While complex eigenvalues move towards zero, the amplitude of these periodic motions decreases and the motion itself disappears as soon as the complex eigenvalues reach zero level. After it has been crossed, no positive eigenvalues are left. This means the equilibrium becomes stable and the phase space is transformed into a stable focus, see fig.6.19. Such a transition is called "supercritical Hopf bifurcation", see [Thompson and Stewart, 1986]. (Traditionally it is considered vice versa stability towards instability).

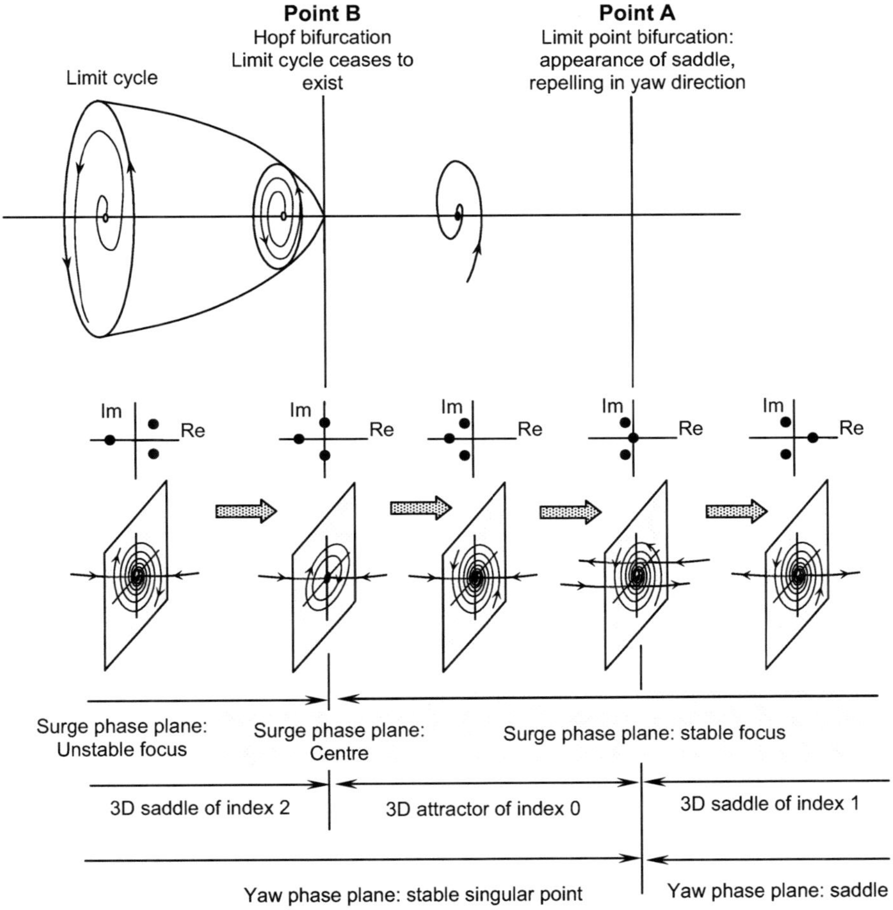

Fig. 6.19 Transition from surge to yaw instability

The region of stability is not large. Shortly after Hopf bifurcation, a negative real eigenvalue, which corresponds to yawing, reaches zero and becomes positive (point A in figures 6.16 and 6.19). Stable focus becomes correspondingly the index 1 saddle. The observed bifurcation also belongs to a type "limit point bifurcation" – it was encountered at point D.

After point A, we again have a saddle (as before at point D), but this saddle repels the ship in the yawing direction. Now we are on the branch that contains equilibria near the wave trough, see fig. 6.16. Further movement brings us to the starting point at zero. The task of our analysis of unsteered vessel equilibria is completed now, since the bottom portion of the curve (locus) is symmetrical.

Summarizing this part of the study, the following comments can be made:

1. Equilibrium at the surf-riding mode is unstable for the vast majority of rudder deflection angles.
2. The equilibrium near the wave crest bears instability in the surging direction, while the equilibrium near the wave trough is unstable in the yawing direction.
3. One of the possible scenarios of transition between surge and yaw instability is capable of creating a short range of stability of equilibrium.

6.5.3 Stability of Equilibria of Steered Ship

We consider stability of the surf-riding equilibrium for steered vessels. Now the autopilot equation is included in the model and rudder deflection angle becomes one of the variables, since it is the result of autopilot actions. Despite the fact direction of ship heading is now controlled by autopilot, still input of commanded or desired heading is necessary. The commanded heading now is the control parameter. The nominal Froude number still plays the roll of the second control parameter.

Following the previous procedure, we start from zero and move along the wave crest branch of the locus. Analogously to the unsteered case, the equilibrium is unstable here; we have an index 1 saddle repelling the system in the surging direction, see fig. 6.20.

The only positive eigenvalue shows the tendency to zero. When it reaches zero, we observe limit point bifurcation. No positive eigenvalues are left over. So, the equilibrium becomes stable: we get an attractor in three-dimensional phase space. This attractor does not exist for a long time. The surging eigenvalue that just left a positive semi-plane joins its pair-mate; they become complex and cross the zero again. As we have seen before, it makes a supercritical Hopf bifurcation. The stable attractor has been transformed into an index 2 saddle. Then it has an unstable focus in one of its phase planes. Analogously to the unsteered vessel's case, the unstable focus is limited by a cycle of periodic surge motions, which are caused by nonlinear properties of the system. Sometime during this range, we got the system to move along a "wave trough" branch.

Then the second supercritical Hopf bifurcation occurs, when the surging pair of complex eigenvalues crosses zero again and we have the attractor back, which makes the equilibrium stable. The stability does not last long. The real eigenvalue corresponding to yawing crosses zero and creates an index 1 saddle that repels the system in the yawing direction.

So far, there are little differences between the steered and unsteered cases with the exception of some details in the bifurcation types. The serious difference comes here. There is a third limit point bifurcation, after which, the system becomes stable and keeps this stability all the way until it reaches zero.

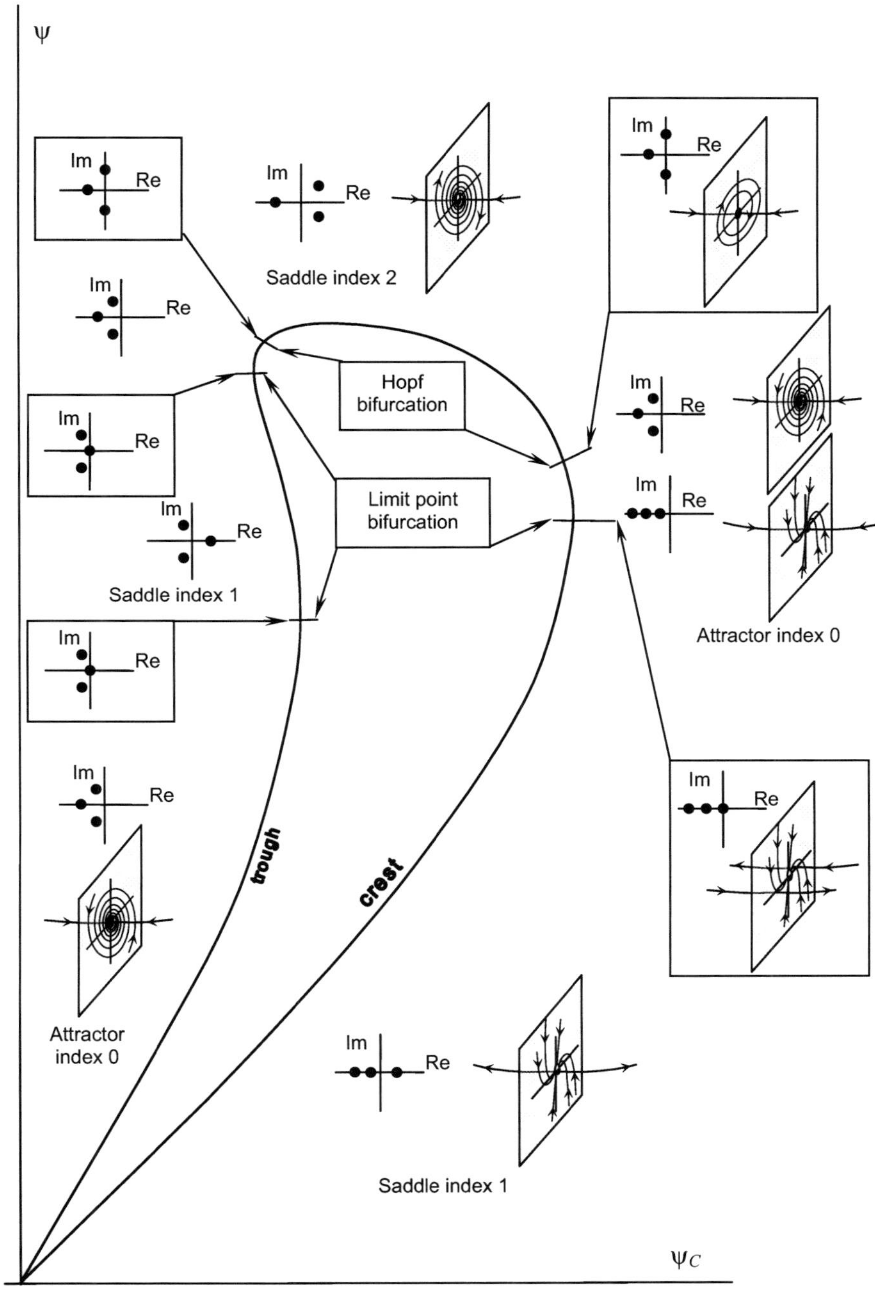

Fig. 6.20 Stability of equilibrium for a steered ship in quartering seas

The main conclusion we can make from our comparison study between steered and unsteered cases is that steering (autopilot or helmsman) is capable of creating a stable surf-riding mode, when a ship is located near a wave trough.

This result is an important. It contains some indications that autopilot settings could play a crucial role in broaching behavior and possibly a means to escape broaching. [Spyrou, 1996] contains more details on how these settings can influence the shape of the equilibrium locus.

6.5.4 Large Ship Motions in Quartering Seas

The study of stability of surf-riding equilibrium gives additional useful information, not only on whether it is possible for a system to stay in a certain equilibrium position. Probably, the most important additional knowledge is the character of the motion in the vicinity of equilibrium. This is actually the first step towards the study of transitions (broaching is one of them), because the equilibria are special (singular) points.

Other special figures are limit cycles or periodical motions. Together with equilibria, they are major factors of phase space topology.

We know that at least one periodic motion mode or limit cycle is possible along with surf-riding (within a certain range of nominal Froude numbers, of course). So we have at least three steady states (limit sets) for a steered ship: stable equilibrium near wave trough, unstable equilibrium near wave crest and periodic motion, see figure 6.21 (assuming that commanded heading angle and nominal Froude number are given).

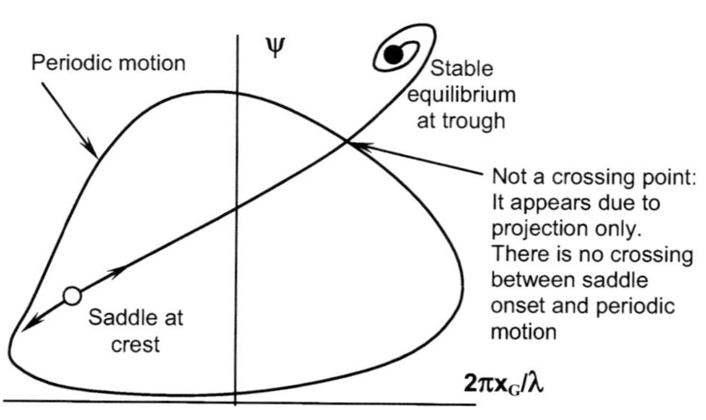

Fig. 6.21 Equilibria and steady state periodic motions in quartering seas

Along these three limit sets, only two have attracting capabilities: stable equilibrium near wave trough and periodic motion. Unstable equilibrium near wave crest repels the system. Basically, an attraction element of the phase space includes singularity (like equilibrium) or special trajectory (like limit cycle or periodic motion) and a certain area of attraction. We shall call it "attraction set" or just "set" during further consideration. It is impossible to visualize it: we are working in eight-dimensional space.

There is one more attractor set that is not shown in fig. 6.21. It actually takes the system so far from the course that the ship can be no longer considered as heading in quartering seas. Spyrou [1996] gives the following physical description of this set.

The yaw wave moment tries to turn a ship away from the direction of wave propagation when the ship is located near wave trough. The autopilot deflects the rudder trying to keep course. Since the yaw wave moment may be large, the rudder may be deflected to a large angle up to its maximum angle of 35^0 or larger if the steering gear permits higher angles. Then the ship reaches wave crest. Here the wave yaw moment changes direction and becomes stabilizing, but the rudder is still deflected at maximum angle and the ship gets involved in quite violent turn. Spyrou [1996] also mentions about the situation when wave yaw moment on wave trough is so large that it makes the ship turn despite the rudder being deflected to its maximum range.

There is one more attractor set: we have seen the short appearance of stable limit cycle as a result of Hopf bifurcation. However, the range of parameters, when these modes are possible, is so narrow that they probably do not hold any practical importance. However, these modes may be important for a general understanding of ship behavior in quartering seas.

So far, let us consider three main attractor sets:

a) Stable surf-riding equilibrium near wave trough,

b) Stable periodical motion around commanded heading,

c) Described above "escape to infinity", when a ship is involved in a turn,

Spyrou [1996] defines broaching as a "transient motion towards (c) from the attractor set (a) and/or (b)".

There are four scenarios identified in the above reference. Here we give a brief review of them, followed by detailed consideration in separate subchapters.

Broaching Scenario #1

Let the ship be involved in periodic motion (attractor set b) then, (as we did for surf-riding) the nominal Froude number is increased so that this mode (periodic motion) ceases to exist. We have seen the same scenario in surf-riding. Limit cycle comes closer to the unstable equilibrium. As soon as it touches, it ceases to exist, see fig. 6.7. The system has a choice: it can be attracted by a nearby stable surf-riding equilibrium (attractor set a) or it may escape to infinity (attractor set c).

As we have seen from our study of stability of surf-riding equilibrium in the case of steered vessels, there are commanded heading angles that do not provide us with stable equilibrium (see fig. 6.20). If such a commanded heading was set up at the moment that periodic motion ceases to exist, then the only attractor left is (c) and broaching is inevitable.

Broaching Scenario #2

If a ship is caught by a wave so the system is in a stable surf-riding equilibrium, then an attempt to escape from surf-riding can be achieved by changing the speed, or changing the commanded heading, or changing both. The system has two alternatives: periodic motion (attractor set b – if available) and broaching (attractor set c). Further study is necessary in order to find the boundaries between these domains of attraction, see subchapter 6.5.5.

Broaching Scenario #3

Let a ship be involved in periodic motion and a gradual increase in the commanded angle occurs. This may result in increasing amplitude of rudder deflection oscillations. The heading of the ship can be changed in quite wide limits $\pm 180^0$; however, the rudder deflection has limits of $\pm 35^0$. These limits and /or insufficient rudder force could lead to a "frozen rudder" situation, which would allow the wave yaw moment to take over and cause broaching. More details are available in [Spyrou, 1996].

Broaching Scenario #4

Let a ship be involved in periodic motion. These oscillations are found to be capable for a jump to a large amplitude mode and even experience period doubling like roll motion. The new amplitude may be large enough to make the turn violent and create a threat of capsizing. This scenario is considered in subchapter 6.5.6.

6.5.5 Global Analysis

The global analysis means that we are looking for boundaries between different areas of attraction and to study changes to these boundaries when control parameters are changed. So, we put a grid in eight-dimensional phase space; each node of this grid is a set of initial conditions. Then, we integrate our system with these initial conditions until the outcome is clear. It means that we are certain which attractor finally takes over for this particular set of initial conditions. Then, the procedure has to be repeated for the next node in the grid and so on. When the calculations are completed, we can fix the boundary between nodes that yield different outcomes. The procedure is meant to be repeated for both control parameters within a given range.

Actually, it is not so difficult to carry out such calculations, but direct interpretation of the results is very complex or even impossible: we cannot visualize the boundary in eight-dimensional space nor its dependence on two control parameters. This is a good example of how the problem of interpretation may render the calculation results useless.

To overcome this difficulty, we have to make all initial values constant with the exception of one. Then our boundary will be just a value and we can get a dependence of this boundary on the control parameter [Spyrou, 1996], see fig. 6.22a (ship is located in wave trough).

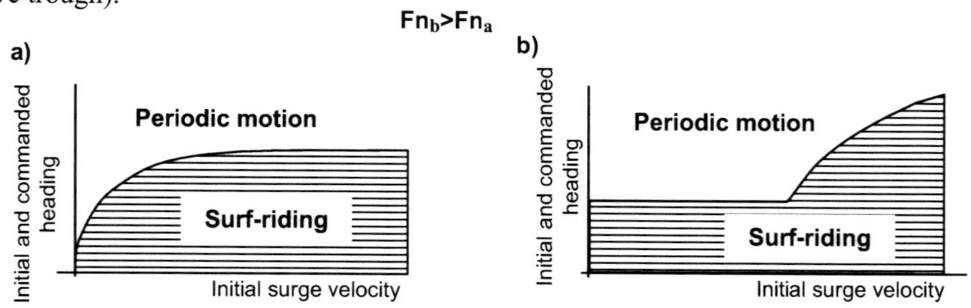

Fig. 6.22 Domains of surf-riding and periodical motions (Froude-Krylov forces only)[Spyrou, 1996]

Despite its simplicity, the curve in fig. 6.22a allows us to draw important conclusions. The area of surf-riding attraction shrinks with increasing heading angle. The surf-riding was inevitable for the following seas, but there is a choice for the system in quartering seas. Increasing nominal Froude number increases the range of commanded heading angles, where surf-riding is the only choice. However, the general principle remains the same –further increasing of commanded heading angle leads to the choice between surf-riding and periodic motion: see fig. 6.22b.

Both cases in figures 6.22 were calculated with Froude-Krylov forces only. Introducing diffraction changes the picture considerably; the broaching area appears, see fig.6.23, (shown in the same scale). Actually, the broaching area was present in figures 6.22 as well, but was not plotted since it happens for headings more than 50 degrees. Such a heading can hardly be called quartering seas.

The significant influence of diffraction forces is explained in [Spyrou, 1996] by increasing swaying and yawing excitation and phase shift mainly due to yawing. Increasing nominal Froude number may lead to interlacing between broaching and surf-riding attraction areas.

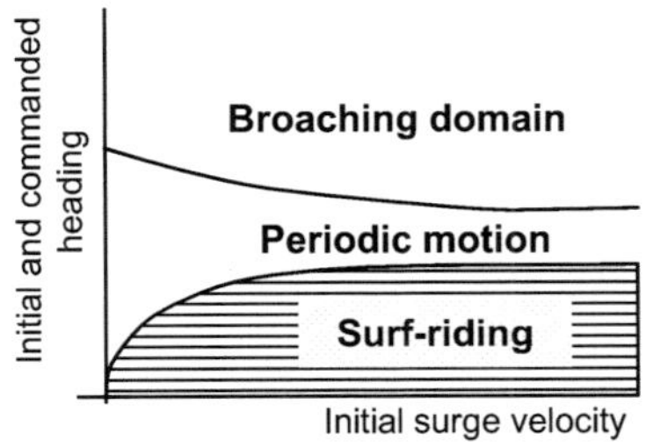

Fig. 6.23 Attraction domains in quartering seas (with diffraction forces) [Spyrou, 1996]

Using mixed initial conditions / control parameters planes allow the results of global analysis to be visible and makes it a useful and even practical tool. It is capable, for example, to show the effect of abrupt changes to speed while surf-riding. Actually, such changes may even lead to broaching [Spyrou, 1996].

6.5.6 Broaching as the Manifestation of Bifurcation of Periodic Motions

The fourth scenario of broaching assumes that a ship is involved in periodic motions that may experience a dramatic increase of amplitude as a result of bifurcation [Spyrou, 1997]. Following the same procedure, let us gradually increase commanded heading, which is our control parameter, and track the changes in amplitude of periodic motions. (Having in mind, of course, that the wave has to be steep enough to make nonlinear effects visible. However, it is not necessary that the nominal Froude number corresponds to wave celerity, in fact it may be lower.) Figure 6.24a shows systematic increase of amplitude of periodic motions while commanded angle is increased.

When commanded angle reaches a certain threshold (0.2 in our example) the motions experience period doubling – the phenomena already known to us from nonlinear roll behavior. The original periodic motion loses stability, but a new type of steady motion appears, it has a period about two times more than the original motion. That is why it is called "period doubling". Another name for this phenomenon is "flip bifurcation", for details see subchapter 4.5.3. Further increases of commanded heading lead to more visual divergence of the first and second semi-periods of motions, (during each semi-period the system has two peaks, as in the full oscillation cycle) see fig. 6.24b. Attempts to increase the commanded angle further lead to a sudden growth of the amplitude; however, it still has a double period, see fig. 6.25a.

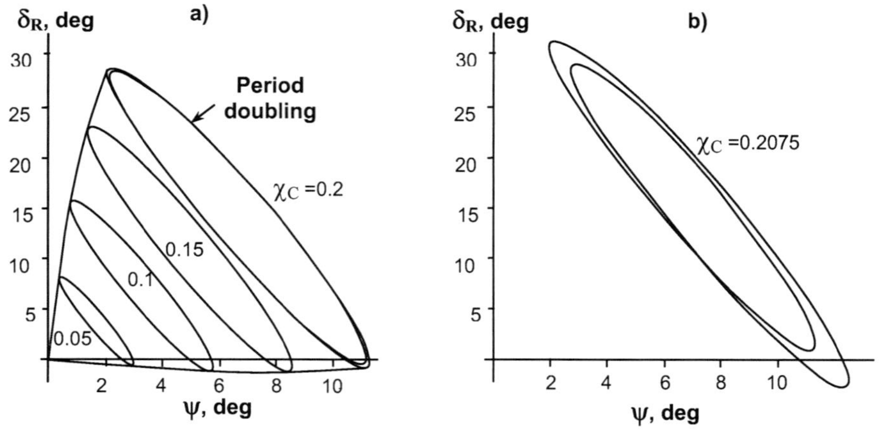

Fig. 6.24 Grow of amplitude of periodic motion with increasing of commanded heading [Spyrou, 1997]

Further increases of commanded heading lead to a gradual decrease of the amplitude (see the case for χ_C=0.3 in fig. 6.25b, and then back to one period mode χ_C =0.45 in fig. 6.25b.

Now let us start to decrease the commanded heading angle from the previously calculated steady state response. (Applying the same technique that was used to study jump in roll motion in beam seas, subchapter 4.5.2.)

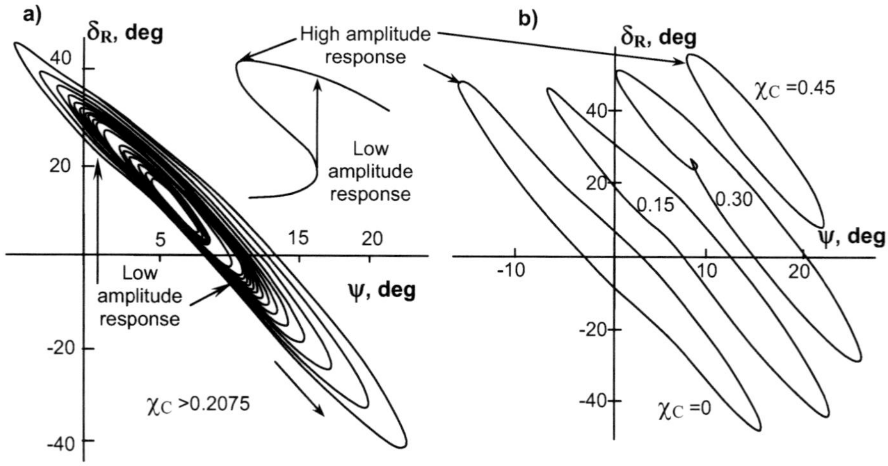

Fig. 6.25 High amplitude response: transition from low amplitudes (a) evolution of steady-state periodic motion with further increase of commanded heading [Spyrou, 1997]

Analogously to rolling in beam seas, there is no "jump back" to a smaller value of amplitude, but a systematic increase, see figure 6.25b. If the amplitude were plotted against commanded heading, we would see a typical jump picture with hysteresis, see fig. 6.26.

The character of change of the periodic motion amplitude clearly indicates the presence of fold bifurcation. So, analogously to nonlinear roll in beam seas, we see a significant increase of the amplitude. It means, when the fold bifurcation occurs, we witness a dramatic increase of heading – a violent turning motion. Finally, fold bifurcation of periodic motion is the fourth mechanism of broaching and has to be avoided.

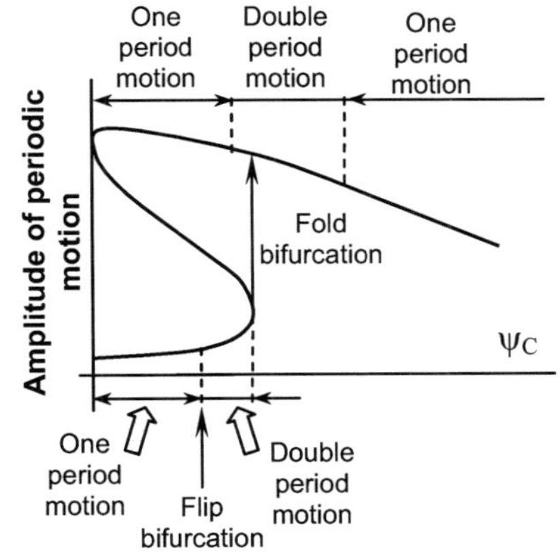

Fig. 6.26 Response curve for periodic motions

One of the ways to avoid it is by changing the autopilot gains. The differential gain is a parameter that determines the shape of the "response curve" – the dependence of amplitudes of periodic motion on commanded headings, see figure 6.27.

As can be clearly seen from this figure, the fold bifurcation disappears at certain values of differential gains. It does not mean that the large amplitudes of the periodic motion are no longer possible, however the transition will be smoother and less violent. Inertial forces caused by turning will be less and that would decrease the chance of capsizing The correct choice of autopilot gains (or more generally of control system gains) is capable of decreasing the danger of broaching. Detailed results of the influence of gains on motion are available from [Spyrou, 1997].

Concluding here the examination of broaching behavior, we went through typical stages of nonlinear dynamics analysis: search for the equilibria, a study of whether these equilibria are stable or not and how all of the above depends on control parameters. In addition to the equilibria we found other types of steady states of the system – periodical motions or limit cycles. Then we considered transitions between these steady states and found which of them could be identified as broaching. Analysis of influence of initial conditions helped us to understand the influence of control parameters on the boundaries between broaching domain and other types of behavior, in other words,

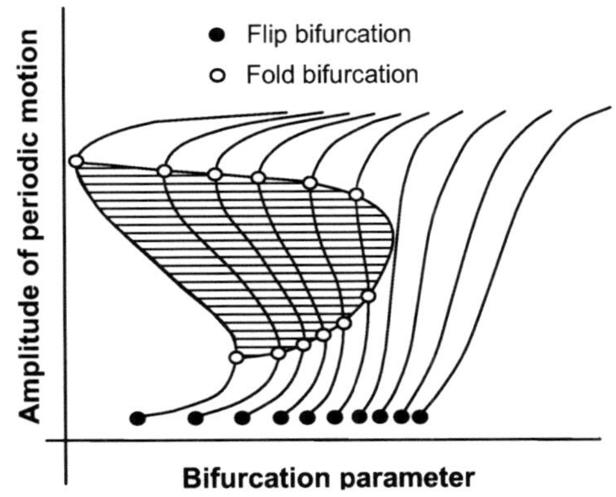

Fig. 6.27 Change of shape of response with differential gain of autopilot [Spyrou, 1997]

how likely broaching can occur for the a given nominal speed and course. Finally, we looked at the influence of autopilot gains on broaching likelihood. The latter holds most of the practical importance: it gives a tool to change autopilot design to make a ship safer at sea.

6.6 Broaching and Capsizing[1]

The above study revealed possible types of behavior of a ship in quartering seas along with the principle mechanics of broaching. Now, we show a different technique based on the same theoretical background of nonlinear dynamics. This technique, however, is especially focused on the capsizing outcome of broaching behavior. Its practical value was proven by a number of model experiments [Umeda, *et al*, 1999], [Umeda and Hamamoto, 1999] and finally it was used as a theoretical background for the design of an anti-broaching device [Umeda and Matsuda, 2000]. This subchapter is mainly based on [Umeda, 1999].

6.6.1 Analysis of Equilibria

Following the previously described procedure, the first step of our study is to learn about equilibrium positions. Since we are working in 8-dimensional space, we can only work with projections. Some of the solution projections are shown on fig. 6.28.

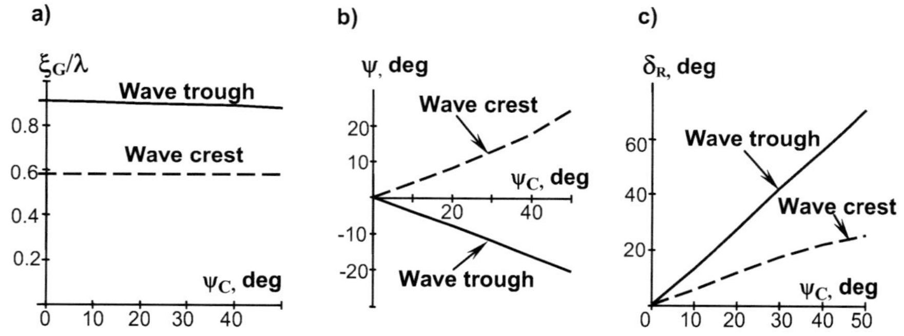

Fig. 6.28 Equilibria in quartering seas [Umeda, 1999]

The next step is to learn if the equilibria we found are stable or not: this procedure was already described in subchapter 6.3, so we go directly to the results from [Umeda, 1999].

Calculation of the eigenvalues of Jacobean matrixes in the vicinity of both equilibria shows one positive real eigenvalue for all of them. So, both these equilibria are unstable. (At least for the ship that was chosen for being a sample – and in specific conditions only.) These eigenvalues are shown at fig. 6.29a and 6.29b. Fig. 6.29c shows behavior of the real positive eigenvalues vs. commanded heading, which is the control parameter.

[1] Written in co-authorship with Prof. Nayoa Umeda of Osaka University

Eigenvalues allow us to see not only that the equilibria are stable or unstable, but also what kind of singularity points will correspond to these equilibria in phase space. The existence of one positive real eigenvalue indicates that we have a singularity point of the "saddle" type of index 1.

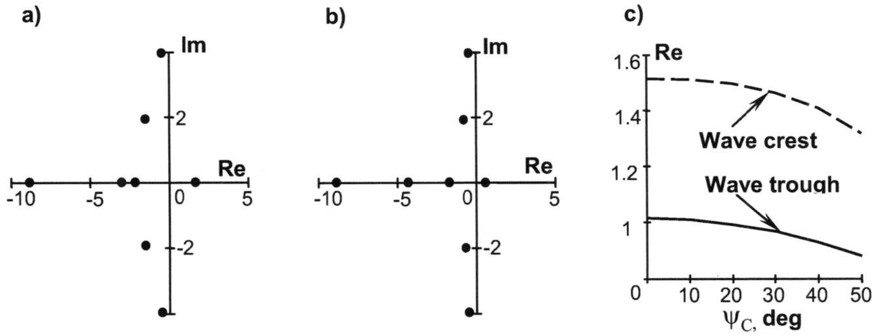

Fig. 6.29 Eigenvalues for equilibria at wave crest (a) and trough (b), real positive part eigenvalues vs. commanded heading (c)

6.6.2 Invariant Manifold

To find out about topology of the phase space we use invariant manifolds. As it was shown in subchapter 5.3.3 invariant manifolds can be calculated if eigenvectors are known.

Let us calculate the eigenvectors and focus on those corresponding to real positive eigenvalues that are "responsible for instability". Looking at fig. 6.29 we see two pairs of complex conjugate eigenvalues and four real eigenvalues – three negative and one positive.

Only one real eigenvalue is positive and "responsible" for the instability; so we have a one-dimensional unstable invariant manifold here. The remaining seven eigenvalues are complex with negative real parts or real negative; so these seven eigenvectors are tangent to the stable invariant manifold, which is a seven-dimensional hyper-surface.

It is difficult to visualize a seven-dimensional hyper-surface, so let us focus on an unstable invariant manifold that is "simply" a line in a eight-dimensional phase space. Then we have to work with projections.

To calculate the unstable invariant manifold, we follow the procedure described in subchapter 5.3.3. So, we put the system into unstable equilibrium and give it a small perturbation in the direction of the eigenvector corresponding to a positive eigenvalue, see Fig. 6.30. This would be our first starting point. To get another one, we perturb the system in the direction that is opposite to the eigenvector.

Having these initial points, it is easy to calculate both branches of the manifold by numerical integration of the dynamical system (6.105).

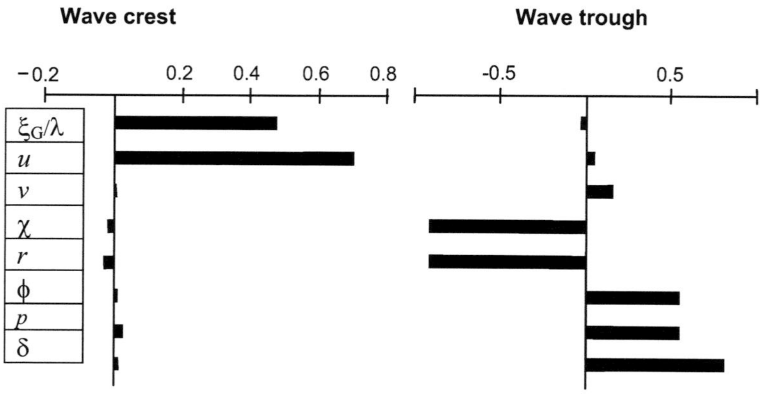

Fig. 6.30 Eigenvectors [Umeda, 1999]

6.6.3 Capsizing

Let us consider projections of an unstable invariant manifold onto phase planes: (ξ_G/λ, u) (ψ, r), (ξ_G/λ, δ_R) and (ϕ, p) – figures 6.31 (a, c, e, g) [Umeda, 1999]; (the scale was changed to make the geometry clear). The manifold is built for an unstable equilibrium at the wave crest with nominal Froude number Fn=0.3248 and desired course $\psi_c=10^0$.

These projections allow a review of the entire broaching picture: ship moves downslope from the wave crest fig. 6.31a, heading angle increases fig. 6.31c, despite the autopilot deflected rudder on opposite side trying to keep course fig. 6.31e, heel angle increases significantly and the ship ends up capsizing fig. 6.31g. If the system were perturbed in the opposite direction, the same picture would be seen at the next incoming wave.

Slight decreasing of number of revolutions (expressed here as nominal Froude number, that would be achieved with this number of revolutions in calm water) leads to dramatic changes in the entire motion of ship behavior. See projections of the unstable invariant manifold: figures 6.31 (b, d, f and h). Now, if the system were perturbed forward, the ship would capsize in the same way as before. However, if perturbed in the opposite side, the ship would experience periodic motion.

As we pointed out before, such drastic changes in phase portrait due to small changes of the control parameter is a result of global bifurcation. As can be seen from all the above figures, a critical point exists somewhere between Froude numbers Fn=0.3248 and Fn=0.3247 (marked as Fn_{cr}, see fig. 6.32). An unstable invariant manifold, being started from an unstable equilibrium at the crest of wave 1, leads exactly to the unstable equilibrium at the crest of wave 2.

So, a trajectory connecting two different saddle points exists in this case. Such a structure is known in nonlinear dynamics as a "heteroclinic connection" (from the Greek word "heteros" – different) and the global bifurcation is known as a "heteroclinic bifurcation". This yields a very important practical outcome: inevitable capsizing due to broaching can be identified with heteroclinic bifurcation. Since the last one is computable by means of nonlinear dynamics, it can be considered as a physical criterion of stability for broaching in regular seas.

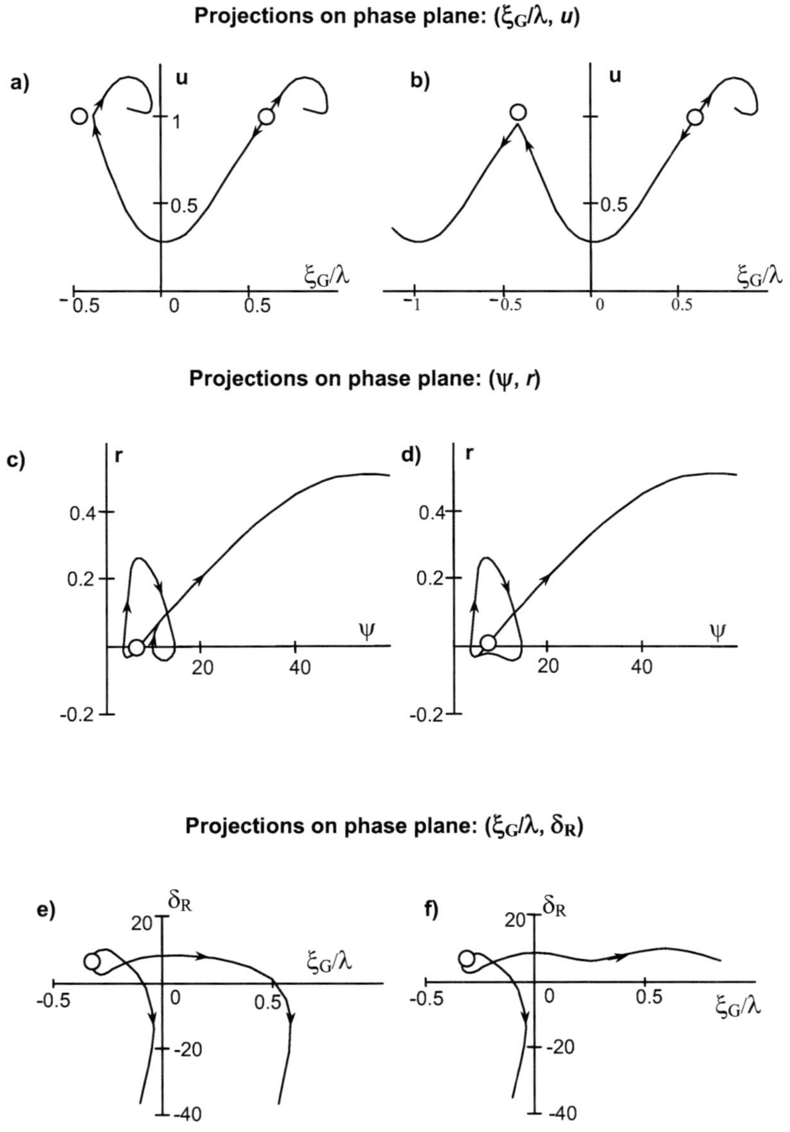

Fig. 6.31 (a through f) Projections of unstable invariant manifold *Fn*=0.3248 (a, c, e) and *Fn*=0.3247 (b, d, f)

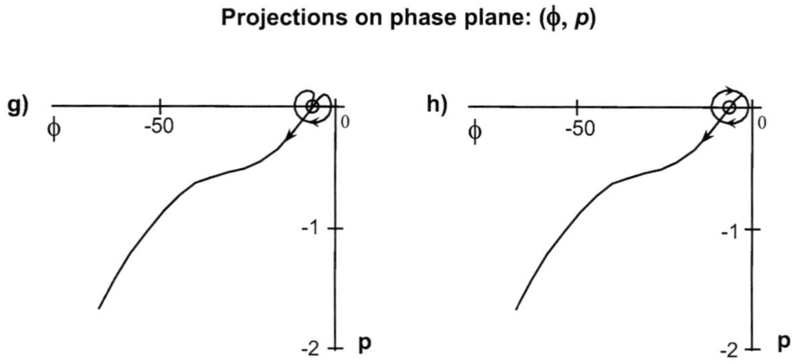

Fig. 6.31 (g and h) Projections of unstable invariant manifold *Fn*=0.3248 (g) and *Fn*=0.3247 (h)

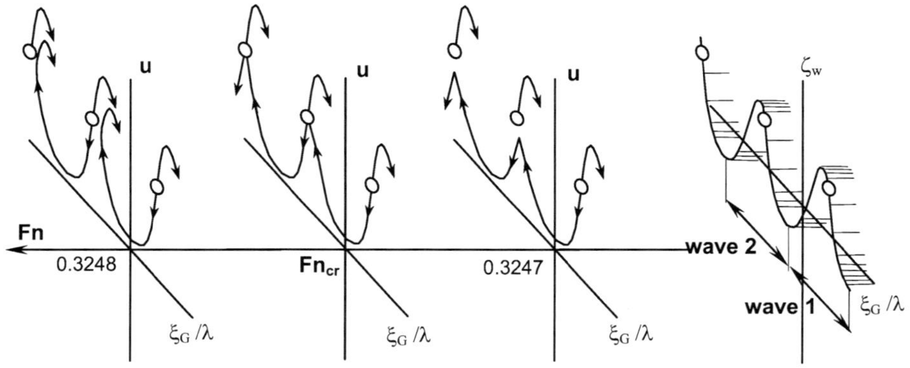

Fig. 6.32 Development of global bifurcation in surging phase plane

Chapter 7

Other Factors Affecting Capsizing

7.1 Aerodynamic Forces and Drift

The basic assumed situation adopted by all stability regulations is the following: a ship is situated in a beam position relative to wind and waves with a sudden squall acting on the ship. All navigation experience leaves no doubt that wind is the major factor of influence on ship safety. The problem of a ship heeling under action of wind is a long-term subject of interest for naval architects. Significant contribution in this area was made in the 1950's when physically based stability criteria were developed see, for example, [Blagoveshchensky, 1951, 1965], [Yamagata, 1959].

An extensive research program on wind–and-drift influence on stability was carried out by the National Laboratory of Seakeeping of Fishing Vessels at Kaliningrad Institute of Technology (Russia). One of the authors was director of this laboratory and supervisor of this program. Only a few results of this program were published in English [Makov and Sevastianov, 1993, 1994a] and [Belenky, 1993a, 1994]. So it would be worthwhile to review these results here; of course there are many more results available on the subject (see, for example [Fediaevsky and Firsov, 1957], [Kinoshita and Okada, 1957] and others), however the scope of the book does not allow us to review them all. The results of the above program are reviewed in subchapters 7.1.1 through 7.1.5, while subchapter 7.1.6 is focused on the background of classical weather criterion.

7.1.1 Steady Drift

Theoretical analysis showed that inertial forces of an aerodynamic nature are small in comparison with aerodynamic forces of a vortex nature when considering the action of a squall on a ship [Makov, *et al* 1987]. Using this conclusion we will consider only non-inertial aerodynamic force. However, the problem of ship motion under a squall or gust of wind remains quite complex both in aerodynamic and hydrodynamic aspects. We begin our discussion from the simplest case: steady state drift of a ship in calm water.

An assumed scheme of force application is given in fig. 7.1. We suppose the principal vector of aerodynamic forces is applied at point A and principal vector of hydrodynamic force is applied at point H. Stable state drift leads to a change in volume displacement by value V_1 in comparison with the state without drift.

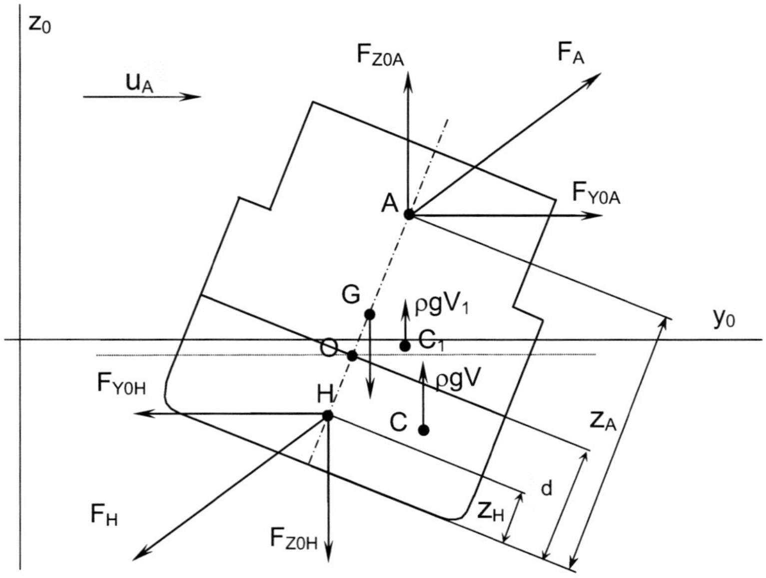

Fig. 7.1 Scheme of application of forces during stable state drift

The gravitational force is applied at point *G,* it is equal numerically to weight displacement, original $\rho g V$ with the addition of $\rho g V_1$. Buoyancy forces are applied at points C and C_1 respectively. Taking into account that $\rho g V = W$, the following system of force equations can be written:

$$\begin{cases} F_{Y0A} - F_{Y0H} = 0 \\ F_{Z0A} - F_{Z0H} + \rho g V_1 = 0 \\ M_{X0A} + M_{X0H} - M_1 = W \cdot GZ(\phi) \end{cases} \tag{7.1}$$

Where M_1 is a moment of the force $\rho g V_1$, following [Makov, *et al* 1987] we neglect terms M_1 and $\rho g V_1$ as they are small values in comparison with other terms of the system (7.1). Moments of aero- and hydrodynamic forces can be expressed as follows:

$$M_{X0A} = F_{Y0A}(z_A - d)\cos\phi - F_{Z0\varsigma}(z_A - d)\sin\phi \tag{7.2}$$

$$M_{X0H} = F_{Y0H}(z_A - d)\cos\phi - F_{Y0H}(z_A - d)\sin\phi \tag{7.3}$$

The whole heeling moment due to the combined action of aero- and hydrodynamic forces is:

$$\begin{aligned} M_{X0} &= M_{X0A} + M_{X0H} = \\ &= F_{Y0A}(z_A - z_H)\cos\phi - F_{Z0A}(z_A - z_H)\sin\phi = \\ &= F_{Y0H}(z_A - z_H)\cos\phi - F_{Z0H}(z_A - z_H)\sin\phi \end{aligned} \tag{7.4}$$

7.1.2 Aerodynamic Forces

Let us consider the aerodynamic components of the equation (7.4) in detail. These forces and moments are the results of the interaction of airflow with the ship sides, superstructures and decks. Contemporary CFD (computational fluid dynamics) technology allows these components to be evaluated numerically, however model tests still play an important role as a source of information, see for example [Brizzolara and Rizzuto 2006].

Here, we review experiments done by Ortiz [1985] in a towing tank and hydrochannel. When modeling was carried out in a hydrochannel or in a towing tank, the airflow was modeled with water flow. The model was placed in an upside down position and flow velocity (or carriage speed) was chosen to avoid wave generation. Similarity by Reynolds number was not kept.

To mitigate scale effect, tests were repeated with different model scales and using different experimental facilities: hydrochannel and towing tank of the Kaliningrad Institute of Technology (KTI) and the large hydrochannel of the Scientific and Industrial Corporation of Commercial Fishery (SICCF, Kaliningrad, Russia), see table 7.1.

Table 7.1 Summary of some model tests to estimate aerodynamic forces

No.	Test data	Wind tunnel [Fediaevsky and Firsov, 1957]	Hydro-channel KTI	Hydro-channel SICCF	Towing tank KTI
1	Year	1957-58	1982	1983	1984
2	Measured width, m	2.25	1.00	3.00	7.00
3	Model scale	1:24	1:65	1:22	1:22
4	Model length , m	1.631	0.602	1.806	1.806
5	Model breadth, m	0.304	0.112	0.336	0.336
6	Velocity, m/s	35	0.45-0.55	0.7-0.9	0.7-0.9
7	Reynolds number	$7.27\ 10^5$	$0.45\ 10^5$ - $0.55\ 10^5$	$2.00\ 10^5$ $-2.80\ 10^5$	$2.00\ 10^5$ - $2.80\ 10^5$

Ortiz [1985] used models of a medium size Russian trawler. All three aerodynamical components were measured (usually such experiments provide only the horizontal and moment components – such as [Fediaevsky and Firsov, 1957]).

The results of these model tests on the horizontal aerodynamic component are shown in fig. 7.2; the vertical component data are shown in fig. 7.3; and the moment component is shown in fig. 7.4. All the results are presented in the non-dimensional form:

$$F_{Y0A} = C_\eta \rho_a \frac{(u_A - v_D)^2}{2} A_W = C_\eta \rho_a \frac{u_A^2}{2} A_W \tag{7.5}$$

$$F_{Z0A} = C_\zeta \rho_a \frac{(u_A - v_D)^2}{2} A_W = C_\zeta \rho_a \frac{u_A^2}{2} A_W \tag{7.6}$$

$$M_{X0A} = C_M \rho_a \frac{(u_A - v_D)^2}{2} A_W (z_A - d) = C_M \rho_a \frac{u_A^2}{2} A_W (z_A - d) \qquad (7.7)$$

Here, u_A is wind speed, v_D is drift speed, A_W is windage area. We have neglected drift speed as a small value in comparison with the wind velocity.

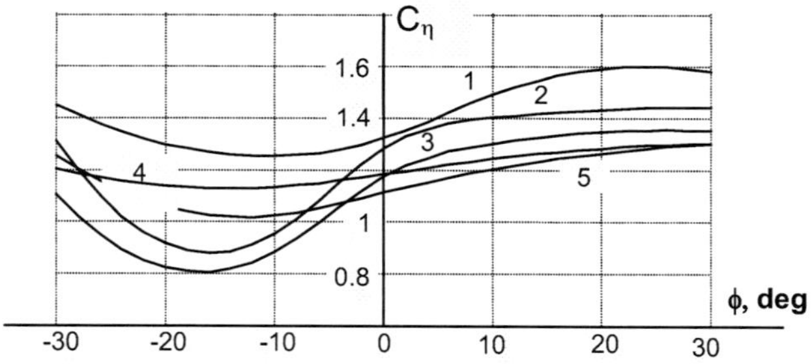

Fig. 7.2 Horizontal aerodynamic component (1-wind tunnel [Fediaevsky and Firsov, 1957], 2-hydrochannel KTI, 3-towing tank KTI, 4-hydrochannel SICCF (1983), 5-hydrochannel SICCF (1984))

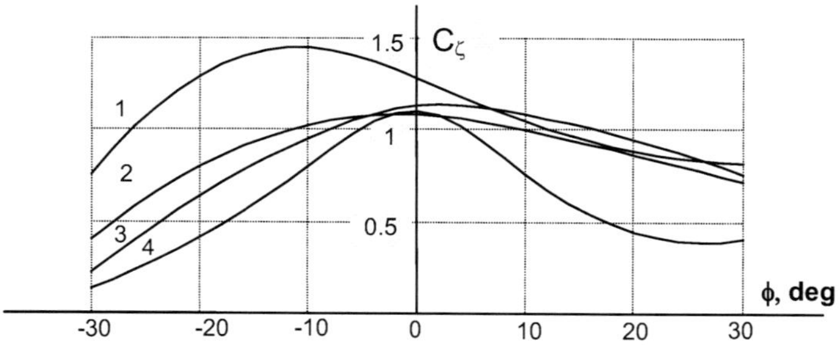

Fig. 7.3 Vertical aerodynamic component (1- hydrochannel SICCF (1983), 2-towing tank KTI, 3-hydrochannel KTI, 4-hydrochannel SICCF (1984))

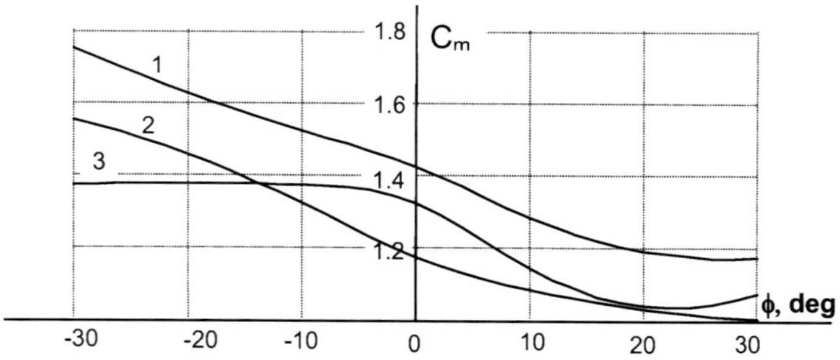

Fig. 7.4 Vertical aerodynamic component (1- hydrochannel SICCF (1984), 2-hydrochannel SICCF (1983)), 3-towing tank KTI)

Test results of Ortiz [1985] show that the vertical component is of the same order as the horizontal component and it should be taken into account. The other result of the aerodynamic research are practical recommendations that were prepared on the background of model tests of a Russian medium fishing trawler, a Cuban fishing vessel and a schematized parallelepiped ship model. The following formulae were proposed for the moment of aerodynamic force:

$$C_M = C_{M0} \cdot C_{M\phi} \quad (7.8)$$

$$C_{M\phi} = \begin{cases} 1.000 - 0.0097 \cdot \phi & for \quad -25^0 \le \phi \le 10^0 \\ 0.945 - 0.0042 \cdot \phi & for \quad \phi > 10^0 \end{cases} \quad (7.9)$$

$$C_{M0} = \{4.0 - 3.25 \cdot [l_S + l_B(2.6 - 1.6 \cdot b_B)]\} \cdot (1 - 0.47 \cdot l_{BW})$$

Here: l_S is total relative length of superstructures, that is the total length of superstructures divided by the length of the ship; l_B is total relative length of bridges, l_{BW} is total relative length of bulwark, b_B is average breadth of bridges divided by the breadth of the ship. The last quantity can be defined by the following formula:

$$b_B = \frac{\sum_{i=1}^{N} l_{Bi} \cdot b_{Bi}}{B \sum_{i=1}^{N} l_{Bi}} \quad (7.10)$$

Where: l_{Bi} is length of bridge i and b_{Bi} is breadth of bridge i.

7.1.3 Hydrodynamic Drift Forces

Garkavy, *et al* [1982] conducted a series of model tests in the towing tank of the National Laboratory of Seakeeping of Fishing Vessels (Russia). The model was towed sidewise in the towing tank; drift force and its moment were measured. The series consists of 11 models; their characteristics are given in Table 7.2, taken from [Garkavy and Kovalenko, 1989].

Table 7.2 Characteristics of models used by Garkavy and Kovalenko [1989]

L	*L/B*	*B/T*	*CM*	*CB*	2(*H-T*)/*B*
1.498	5.095	2.430	0.830	0.552	0.245
1.498	3.500	3.537	0.830	0.552	0.168
1.498	7.000	1.769	0.830	0.552	0.336
1.498	5.095	4.900	0.830	0.552	0.245
1.498	5.095	1.500	0.830	0.552	0.245
1.498	5.095	2.430	0.830	0.415	0.245
1.498	5.095	2.430	0.830	0.654	0.245
1.498	5.095	2.430	0.986	0.552	0.245
1.498	5.095	2.430	0.704	0.553	0.245
1.508	4.710	2.460	0.900	0.520	0.225
1.607	7.372	2.270	0.980	0.637	0.330

All the models were equipped with a bilge keel, with the height according to the following formula:

$$h_{BK} = 0.045\sqrt{(B/2)^2 + d^2} \tag{7.11}$$

Since only the horizontal component was measured, the true coordinate of the point where hydrodynamic force z_H was applied cannot be obtained from these experiments. However, the measured moment is evidently a true moment being created by the whole hydrodynamic force. To avoid error related to neglecting the vertical component, we introduce the concept of a pseudo-centre of hydrodynamic force application. The horizontal component of the force being applied at this point yields the same value as the moment that exists in reality. Therefore, we can predict the elevation of this pseudo-centre from measurements [Garkavy and Kovalenko, 1989]. The results of these measurements are given in tables 7.3 and 7.4. These results are presented in the usual form (elevation of the pseudo-center was presented in a semi-fixed system of coordinates):

$$F_{Y0H} = C_H \frac{\rho v_D^2}{2} A_D ; \quad \eta_{HP} = z_{HP} \cos\phi \tag{7.12}$$

Here A_D is the area of diametrical section of the submerged part of a ship.

Table 7.3 Coefficient of horizontal hydrodynamic component C_H

Model No.	0^0	10^0	20^0	30^0	40^0	50^0
1	0.832	0.948	0.824	0.997	0.971	1.113
2	0.709	0.934	1.117	1.299	1.528	1.945
3	0.927	0.958	1.000	0.868	0.869	0.773
4	0.629	0.959	1.167	1.316	1.631	1.799
5	0.999	0.986	0.893	0.803	0.769	0.785
6	0.887	1.107	1.112	1.136	1.112	1.148
7	0.747	1.071	1.261	1.443	1.715	1.817
8	0.979	1.100	1.227	1.259	1.396	1.468
9	0.835	0.711	1.095	1.067	1.249	1.282
10	0.907	1.050	1.131	1.129	1.180	1.218
11	1.224	1.297	1.327	1.274	1.219	1.145

Formula for coefficient C_H can be obtained from this data by regression analysis. The following formula was proposed in [Belenky, 1994]:

$$C_H(\phi) = C_{H0}(\phi) + \sum_{i=1}^{4} f_i(\phi) \cdot par_i \tag{7.13}$$

Where $C_{H0}(\phi)$ is the coefficient for the basic series of models, $f_i(\phi)$ is the function of influence of geometric parameters and par_i are the geometric parameters. The following geometric characteristics can be used as influencing parameters (index "0" means that these characteristics belong to the basic model of the series):

1. $par_1 = (L/B) - (L/B)_0$
2. $par_2 = (B/d) - (B/d)_0$
3. $par_3 = (d/L) - (d/L)_0$
4. $par_4 = (C_B/C_M) - (C_B/C_M)_0$

1. $(L/B)_0 = 5.095$
2. $(B/d)_0 = 2.43$
3. $(d/L)_0 = 0.0808$
4. $(C_B/C_M)_0 = 0.66$

Where: L is waterplane length; B is breadth; d is draught; C_B is block coefficient; C_M is midship section coefficient. Coefficients for basic model and influence functions are given in tables 7.5 and 7.6 correspondingly.

Table 7.4 Relative elevation of pseudo-centre of hydrodynamic force η_{HP}

Model No.	0^0	10^0	20^0	30^0	40^0	50^0
1	0.0859	-0.1080	0.3660	0.9090	1.5040	1.4390
2	-0.0912	-0.0916	1.1900	2.0530	2.1000	1.7460
3	0.1680	0.0548	-0.0114	0.2350	0.4520	0.7540
4	0.4430	-0.4970	0.9400	2.1260	2.5670	2.7600
5	0.1607	0.1297	0.4200	0.6875	0.9230	1.1950
6	0.2390	-0.0449	0.1976	0.5730	0.9000	1.2350
7	0.1890	-0.0643	0.8580	1.1780	1.0250	0.9840
8	0.1436	-0.0472	0.3590	0.8000	0.8820	0.9040
9	-0.1539	-0.0916	0.6470	1.1320	1.0440	1.1390
10	0.1097	-0.1060	0.2410	0.7770	1.069	1.1470
11	0.0683	0.0147	0.0409	0.2317	0.4740	0.8825

Table 7.5 Coefficients for basic model

ϕ, deg.	0	10	20	30	40	50
$C_{H0}(\phi)$	0.832	0.948	0.874	0.997	0.971	1.113

Table 7.6 Functions of influence

ϕ, deg.	0	10	20	30	40	50
f_1	0.6221	0.3251	0.7319	0.3929	0.2266	0.1603
f_2	-0.09453	-0.001378	0.1216	0.1419	0.2701	0.3026
f_3	0.5686	1.32	1.009	0.8789	0.9936	0.8698
f_4	-0.5608	-0.3085	-0.1047	0.8326	1.736	2.249

Makov [1985] added several tests to the Garkavy series and carried out regression analyses for pseudo centre elevation. Results are also available from [Makov, *et al*, 1987], [Makov and Sevastianov, 1993]. The following formula is proposed:

$$z_{HP}(\phi) = T \cdot \sum_{i=1}^{5} f_i(\phi) \cdot par_i \qquad (7.14)$$

Where $z_{HP}(\phi)$ is the coefficient value for the basic model of the series; $f_i\ (\phi)$ are coefficients of parameters (see Table 7.7); par_i are the parameters:

1 . $par_1 = 1$	1. $par_2 = B / T$
2. $par_3 = \phi_d$	2. $par_4 = C_B / C_M$
3. $par_5 = L / B$	3.

Here, ϕ_d is the angle at which the deck enters into the water while the ship is heeling (in degrees).

Table 7.7 Coefficients for formula (7.14)

ϕ, deg.	0	10	20	30	40	50
f_1	2.548	2.819	3.451	5.355	3.102	1.277
f_2	-0.027	-0.022	-0.103	-0.423	-0.424	-0.487
f_3	0.002	0.009	0.011	0.019	0.007	0.019
f_4	-2.414	-3.635	-6.590	-11.356	-14.111	-12.983
f_5	-0.046	0.037	0.267	0.525	1.198	1.256

The following formula describes the whole heeling moment due to steady state drift:

$$M_{X0} = F_{H\eta}(z_A - z_{HP})\cos\phi \qquad (7.15)$$

This moment depends on heel angle. Being plotted together with the *GZ* curve, it allows finding the angle of heel that corresponds to the new equilibrium position during steady state drift, fig.7.5.

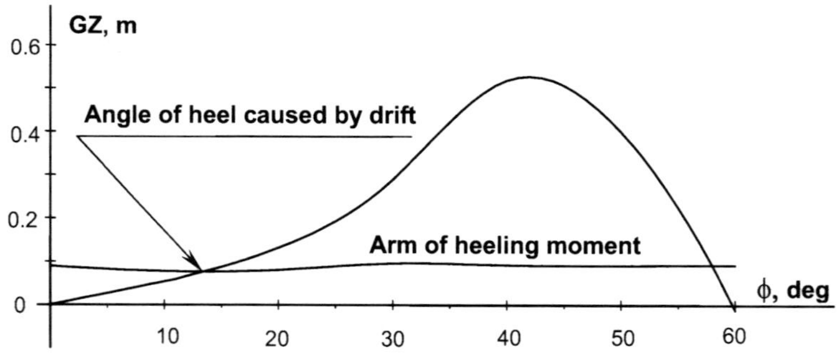

Fig. 7.5 Scheme of definition of angle of heel caused by drift

7.1.4 Sudden Squall of Wind

Sevastianov [1970] considered transition drift behavior under action of a sudden squall of wind. The following equation for horizontal ship motion was used:

$$(m+a_{22})\ddot{y}_{0G}+a_{24}\ddot{\phi}+F_{Y0H}(\dot{y}_{0G})=F_{Y0A}+F_{Y0S} \tag{7.16}$$

Here, F_{Y0S} is the additional horizontal aerodynamic force that appears due to a sudden wind squall. We neglect the term $a_{24}\ddot{\phi}$ and obtain the isolated drift equation:

$$\ddot{y}_{0G}+b\dot{y}_{0G}^{2}=p \tag{7.17}$$

With

$$b=\frac{C_{H}\rho A_{D}}{2(m+a_{22})};\quad p=\frac{F_{Y0A}+F_{Y0S}}{m+a_{22}} \tag{7.18}$$

The expression (7.17) is an ordinary nonlinear differential equation. It can be solved analytically and the relationship between drift velocity and time can be expressed in the following way:

$$t=\frac{1}{\sqrt{bp}}\operatorname{ar\,tanh}\sqrt{\frac{b}{p}}v+\frac{1}{\sqrt{bp}}\operatorname{ar\,tanh}\sqrt{\frac{b}{p}}v_{0} \tag{7.19}$$

Here, $v=\dot{y}_{0G}$ is transient drift velocity and $v_{0}=\dot{y}_{0G}(t_{0})=v_{D}$ is the initial condition at the moment of squall initiation. It is equal to the velocity of stable state drift caused by aerodynamic force F_{A}. Formula (7.19) was obtained by Blagoveshchensky in [1965] with the assumption that squall starts when $v_0=0$ and $F_{Y0A}=0$. The solution by Sevastianov [1970] is free from these assumptions and it can be rewritten as follows:

$$\bar{v}=\frac{(1-\exp(-\varepsilon t))+\bar{v}_{0}(1+\exp(-\varepsilon t))}{(1+\exp(-\varepsilon t))+\bar{v}_{0}(1-\exp(-\varepsilon t))} \tag{7.20}$$

Where: $\bar{v}=v/v_{DS}$; $\bar{v}_{0}=v_{0}/v_{DS}$; $\varepsilon=2bv_{DS}$; v_{DS} is drift velocity of stable state drift caused by aerodynamic forces $F_{A}+F_{S}$ and can be found using equation (7.12) as $v_{SD}=\sqrt{2(F_{A\eta}+F_{S\eta})/C_{H}\rho A_{D}}$. Acceleration of the drift motion can be easily derived:

$$\ddot{y}_{0G}=\frac{dv}{dt}=4\frac{\exp(-\varepsilon t)\cdot(1-\bar{v}_{0}^{2})bv_{DS}^{2}}{[(1+\exp(-\varepsilon t))+\bar{v}_{0}(1-\exp(-\varepsilon t))]^{2}} \tag{7.21}$$

The term in the numerator:

$$\begin{aligned}(1-\bar{v}_{0}^{2})bv_{DS}^{2}&=\frac{C_{H}\rho S_{d}(v_{DS}^{2}-v_{0}^{2})}{2(m+a_{22})}=\\ \frac{F_{Y0A}+F_{Y0S}-F_{Y0H}}{m+a_{22}}&=\frac{F_{Y0S}}{m+a_{22}}=\ddot{y}_{0G}(t_{0})\end{aligned} \tag{7.22}$$

It is can be seen clearly that formula (7.21) expresses drift acceleration at the beginning of the squall. It can also be obtained from equations (7.17) and (7.18) by substitution $v = v_0$. So, we can express acceleration analogously to (7.20):

$$\bar{\ddot{y}}_{0G} = 4\frac{\exp(-\varepsilon t)}{[(1+\exp(-\varepsilon t)) + \bar{v}_0(1-\exp(-\varepsilon t))]^2} \tag{7.23}$$

Where:

$$\bar{\ddot{y}}_{0G} = \ddot{y}_{0G} / \ddot{y}_{0G}(t_0)$$

The following equation models heeling due to wind :

$$\begin{aligned}&(I_{xx} + a_{44})\ddot{\phi} + M_{XD}(\dot{\phi}) + mg \cdot GZ(\phi) + a_{24}\ddot{y}_{0G} + M_{X0H} = \\ &= M_{X0A} + M_{X0S}\end{aligned} \tag{7.24}$$

Here, M_{X0S} is the additional aerodynamic moment caused by a sudden squall. We assume the angle of heel to be small enough to use $\cos\phi = 1$ and collect all independent terms of the equation (7.24) on the right hand side. We consider these terms as the heeling moment due to combined action of aerodynamic and hydrodynamic forces:

$$\begin{aligned}M_{X0} &= M_{X0A} + M_{X0A} - a_{24}\ddot{y}_{0G} - M_{X0H} = \\ &= (F_{Y0A} + F_{Y0S})(z_A - KG) + F_H(KG - z_{HP}) - a_{24}\ddot{y}_{0G}\end{aligned} \tag{7.25}$$

All the moments are written relative to the center of gravity. If we take into account that:

$$F_H = C_H\frac{\rho v^2}{2}A_D = C_H\frac{\rho v_{DS}^2}{2}A_D\bar{v}^2 = (F_{Y0A} + F_{Y0S})\bar{v}^2, \tag{7.26}$$

$$\begin{aligned}\ddot{y}_{0G} &= \bar{\ddot{y}}_{0G}\ddot{y}_{0G}(t_0) = \bar{\ddot{y}}_{0G}\frac{F_{Y0S}}{m + a_{22}} = \bar{\ddot{y}}_{0G}\frac{C_H\rho A_D(v_{DS}^2 - v_0^2)}{2(m + a_{22})} = \\ &= \bar{\ddot{y}}_{0G}\frac{F_{Y0A} + F_{Y0S}}{m + a_{22}}(1 - \bar{v}_0^2)\end{aligned} \tag{7.27}$$

The whole heeling moment con be expressed as:

$$\begin{aligned}M_{X0} &= (F_{Y0A} + F_{Y0S}) \times \\ &\times \left[(z_A - KG) + \bar{v}^2(KG - z_{HP}) - \bar{\ddot{y}}_{0G}\frac{a_{24}}{m + a_{22}}(1 - \bar{v}_0^2)\right]\end{aligned} \tag{7.28}$$

Now it can be clearly seen from formula (7.28) that heeling moment during transient drift is dependent on time. This dependence has an asymptotic exponential character: see formulae (7.20) and (7.23). Calculation of the time when the heeling moment is increasing shows that this figure several times exceeds the natural period. It means that ship capsizing due to a sudden squall does not necessarily occur in the first period of roll motion. The possibility of the capsizing during the second and consequent period of oscillation was proved by special model tests carried out by G. N. Egorov in the towing tank of Kaliningrad Institute of Technology [Sevastianov, 1970].

7.1.5 Method of Energy Balance

Let us consider ship rolling under the action of regular waves and constant wind:

$$\ddot{\phi} + 2\delta\dot{\phi} + \omega_{\phi}^2 f(\phi) = \alpha_E \sin(\omega t) + m_A \tag{7.29}$$

Here: m_A is heeling moment caused by wind.

We rewrite the equation (7.29) as first integrals. For this purpose we can express roll acceleration as:

$$\ddot{\phi} = \frac{d\dot{\phi}}{dt} = \frac{d\dot{\phi}}{d\phi} \cdot \frac{d\phi}{dt} = \frac{d\dot{\phi}}{d\phi}\dot{\phi} \tag{7.30}$$

Then we substitute expression (7.30) in the rolling equation (7.29); multiplying by $d\phi$:

$$\dot{\phi} d\dot{\phi} + 2\delta\dot{\phi} d\phi + \omega_{\phi}^2 f(\phi) d\phi = \alpha_E \sin(\omega t) d\phi + m_A d\phi \tag{7.31}$$

Moving all terms containing $d\phi$ to the left side of the equation, we integrate both parts starting from the initial conditions $\phi_0, \dot{\phi}_0$ until some instance of time t, when heel and angular velocity are equal to values ϕ and $\dot{\phi}$:

$$\int_{\dot{\phi}_0}^{\dot{\phi}} \dot{\phi} d\dot{\phi} = \int_{\phi_0}^{\phi} \alpha_E \sin(\omega t) d\phi + \int_{\phi_0}^{\phi_0} m_A d\phi - \int_{\phi_0}^{\phi} 2\delta\dot{\phi}\, d\phi - \int_{\phi_0}^{\phi} \omega_{\phi}^2 f(\phi)\, d\phi \tag{7.32}$$

We have derived the equation in first integrals, which reflects a balance of changing energies and works of different moments during time range from $t_0(\dot{\phi}_0, \phi_0)$ to $t(\dot{\phi}, \phi)$.

Let us examine the terms of the equation in first integrals focusing on their physical meaning. The first term:

$$\int_{\dot{\phi}_0}^{\dot{\phi}} \dot{\phi}\, d\dot{\phi} = \left[\frac{\dot{\phi}^2}{2} - \frac{\dot{\phi}_0^2}{2}\right] = K(\dot{\phi}_0, \dot{\phi}) \tag{7.33}$$

It is evident from (7.33) that this term expresses a change of kinetic energy. Consider the second term:

$$2\delta \int_{\phi_0}^{\phi} \dot{\phi}\, d\phi = A_D(\dot{\phi}_0, \dot{\phi}) = A_D(t_0, t) \tag{7.34}$$

This is the work of the damping moment, which has been done during the given time range. In other words, this is the energy, which has been dissipated during this time. It is dependent on heel angular velocity, so it is better to write as versus time. The third term:

$$\omega_{\phi}^2 \int_{\phi_0}^{\phi} f(\phi) d\phi = P(\phi_0, \phi) \tag{7.35}$$

This is the work of the restoring moment. It also can be associated with changes of potential energy and expressed through corresponding values of a dynamic stability curve. The fourth term:

$$\int_{\phi_0}^{\phi} \alpha_E \sin(\omega t) d\phi = A_E(\phi_0, \phi, t_0, t) = A_E(t_0, t) \tag{7.36}$$

This is the work of the wave excitation moment. The fifth term:

$$\int_{\phi_0}^{\phi} m_A d\phi = m_A(\phi - \phi_0) = A_A(\phi_0, \phi) \tag{7.37}$$

This is work of the aerodynamic heeling moment.

Using this newly accepted nomenclature, we can rewrite the equation in first integrals (7.32) as an energy work balance equation:

$$K(\dot{\phi}_0) + A_D(t_0, t) + P(\phi_0, \phi) = A_A(\phi_0, \phi) + A_E(t_0, t) \tag{7.38}$$

The weather criterion is based on equalization of potential and kinetic energy. Here, we show what assumptions lead to these kinds of simplifications of the energy balance equation (7.38).

Let us look at the energy balance of the linear system at steady state regime under constant wind. Assume the following form of the solution:

$$\phi = \phi_a \sin(\omega t + \beta) + \phi_W \tag{7.39}$$

Where ϕ_W is the static angle of heel caused by constant wind. Change of kinetic and potential energy does not depend on the particular form of the solution. The derivation of work of constant aerodynamic moment does not pose any problem, so we have to focus on work of damping and excitation moments. Assuming linear or linearized damping, we substitute solution (7.39) into the formula for work of damping moment (7.34) :

$$\begin{aligned} A_D(t_0, t) &= 2\delta \int_{\phi_0}^{\phi} \dot{\phi} d\phi = 2\delta \int_{t_0}^{t} \phi_a^2 \omega^2 \cos^2(\omega t + \beta) dt = \\ &= \delta \phi_a^2 \omega^2 \left\{ (t - t_0) + \frac{1}{2\omega} (\sin(2\omega t + 2\beta) - \sin(2\omega t_0 + 2\beta)) \right\} \end{aligned} \tag{7.40}$$

Let us consider work of wave excitation. As it is well known excitation plays a dual role in periodical motions: it compensates for losses from damping and forces the dynamical system to oscillate with the same period as that of the excitation (synchronization):

$$\begin{aligned} A_E(t_0, t) &= \int_{\phi_0}^{\phi} \alpha_E \sin(\omega t) d\phi = \int_{t_0}^{t} \alpha_E \sin(\omega t) \dot{\phi} dt = \\ &= \phi_a \alpha_E \omega \int_{t_0}^{t} \sin(\omega t + \beta - \beta) \cos(\omega t + \beta) dt = \\ &= \phi_a \alpha_E \omega \left(\sin\beta \int_{t_0}^{t} \cos^2(\omega t + \beta) dt + \cos\beta \int_{t_0}^{t} \cos(\omega t + \beta) \sin(\omega t + \beta) dt \right) \end{aligned} \tag{7.41}$$

Integration is straight forward here:

$$A_E(t_0,t) = -\frac{1}{4}\phi_a\alpha_E\cos\beta\big(\cos(2\omega t+2\beta)-\cos(2\omega t_0+2\beta)\big)-$$
$$-\frac{1}{2}\phi_a\alpha_E\omega\sin\beta\left\{(t-t_0)+\frac{1}{2\omega}\big(\sin(2\omega t+2\beta)-\sin(2\omega t_0+2\beta)\big)\right\}= \quad (7.42)$$
$$= A_{ESync}(t_0,t)+A_{EAct}(t_0,t)$$

Here, A_{Esync} is a portion of the external work that is used for synchronization of excitation and oscillations; A_{EAct} is a portion of excitation work that is used for compensation of damping loses. This part of external work is usually called "active".

Since our system (7.29) is assumed linear or linearized, we can apply well-known formulae for amplitude and phase of linear steady state solution:

$$\phi_a = \frac{\alpha_E}{\sqrt{(\omega^2-\omega_\phi^2)^2+4\delta^2\omega^2}};\quad \tan\beta = \frac{2\delta\omega}{(\omega^2-\omega_\phi^2)} \quad (7.43)$$

This implies:

$$\sin\beta = \frac{2\delta\omega}{\sqrt{(\omega^2-\omega_\phi^2)^2+4\delta^2\omega^2}};\quad \cos\beta = \frac{(\omega^2-\omega_\phi^2)}{\sqrt{(\omega^2-\omega_\phi^2)^2+4\delta^2\omega^2}} \quad (7.44)$$

We substitute formulae (7.43) and (7.44) into equations (7.40) and (7.42):

$$A_D(t_0,t) = \frac{\delta\alpha_E^2\omega^2}{(\omega^2-\omega_\phi^2)^2+4\delta^2\omega^2}\times$$
$$\times\left\{(t-t_0)+\frac{1}{2\omega}\big(\sin(2\omega t+2\beta)-\sin(2\omega t_0+2\beta)\big)\right\} \quad (7.45)$$

$$A_{ESync}(t_0,t) = -\frac{1}{4}\frac{\alpha_E^2(\omega^2-\omega_\phi^2)}{(\omega^2-\omega_\phi^2)^2+4\delta^2\omega^2}\times$$
$$\times\big(\cos(2\omega t+2\beta)-\cos(2\omega t_0+2\beta)\big) \quad (7.46)$$

$$A_{EAct}(t_0,t) = \frac{\delta\alpha_E^2\omega^2}{(\omega^2-\omega_\phi^2)^2+4\delta^2\omega^2}\times$$
$$\times\left\{(t-t_0)+\frac{1}{2\omega}\big(\sin(2\omega t+2\beta)-\sin(2\omega t_0+2\beta)\big)\right\} \quad (7.47)$$

That means that the work of damping equals the active work of excitation, which allows separating the energy balance equation (7.38) in two:

$$A_D(t_0,t) = A_{EAct}(t_0,t) \quad (7.48)$$

$$K(\dot\phi_0,\dot\phi)+P(\phi_0,\phi) = A_A(\phi_0,\phi)+A_{ESync}(\phi_0,\phi) \quad (7.49)$$

Let us expand change of kinetic and potential energy as well as the work of the aerodynamic heeling moment in the same manner we have done for damping and excitation:

$$K(t_0,t)=\left[\frac{\dot{\phi}^2}{2}-\frac{\dot{\phi}_0^2}{2}\right]=\frac{1}{2}\phi_a^2\omega^2\left(\cos^2(\omega t+\beta)-\cos^2(\omega t_0+\beta)\right)= \\ =\frac{1}{4}\phi_a^2\omega^2\left(\cos(2\omega t+2\beta)-\cos(2\omega t_0+2\beta)\right) \tag{7.50}$$

$$P(t_0,t)=\omega_\phi^2\left[\frac{\phi^2}{2}-\frac{\phi_0^2}{2}\right]= \\ =\frac{1}{2}\omega_\phi^2\left((\phi_a\sin(\omega t+\beta)+\phi_W)^2-(\phi_a\sin(\omega t_0+\beta)+\phi_W)^2\right)= \\ =-\frac{1}{2}\omega_\phi^2\left(\frac{1}{2}\phi_a^2\cos(2\omega t+2\beta)-2\phi_a\phi_W\sin(\omega t+\beta)\right)+ \\ +\frac{1}{2}\omega_\phi^2\left(\frac{1}{2}\phi_a^2\cos(2\omega t_0+2\beta)-2\phi_a\phi_W\sin(\omega t_0+\beta)\right) \tag{7.51}$$

$$A_A(t_0,t)=\int_{\phi_0}^{\phi}m_A\,d\phi=m_A(\phi-\phi_0)= \\ =m_A\phi_a\cos(\omega t+\beta)-m_A\phi_a\cos(\omega t_0+\beta) \tag{7.52}$$

Following the previous procedure and substituting (7.43) into (7.50)-(7.52):

$$K(t_0,t)=\frac{1}{4}\frac{\alpha_E^2\omega^2}{(\omega^2-\omega_\phi^2)^2+4\delta^2\omega^2}\times \\ \times\left(\cos(2\omega t+2\beta)-\cos(2\omega t_0+2\beta)\right) \tag{7.53}$$

$$P(t_0,t)=\frac{1}{4}\frac{\alpha_E^2\omega_\phi^2}{(\omega^2-\omega_\phi^2)^2+4\delta^2\omega^2}\times \\ \times\left(\cos(2\omega t_0+2\beta)-\cos(2\omega t+2\beta)\right)- \\ -\frac{\phi_W\alpha_E\omega_\phi^2}{\sqrt{(\omega^2-\omega_\phi^2)^2+4\delta^2\omega^2}}\left(\cos(\omega t+\beta)-\cos(\omega t_0+\beta)\right) \tag{7.54}$$

$$A_A(t_0,t)=\frac{\alpha_E m_A}{\sqrt{(\omega^2-\omega_\phi^2)^2+4\delta^2\omega^2}}\left(\cos(\omega t+\beta)-\cos(\omega t_0+\beta)\right) \tag{7.55}$$

Having in mind that:

$$\phi_W=\frac{m_A}{\omega_\phi^2} \tag{7.56}$$

Expression (7.55) could be rewritten as:

$$A_A(t_0,t)=\frac{\phi_W\alpha_E\omega_\phi^2}{\sqrt{(\omega^2-\omega_\phi^2)^2+4\delta^2\omega^2}}(\cos(\omega t+\beta)-\cos(\omega t_0+\beta)) \tag{7.57}$$

We can see that equation (7.49) is perfectly satisfied. Time histories of all the energies and works are shown in fig. 7.6.

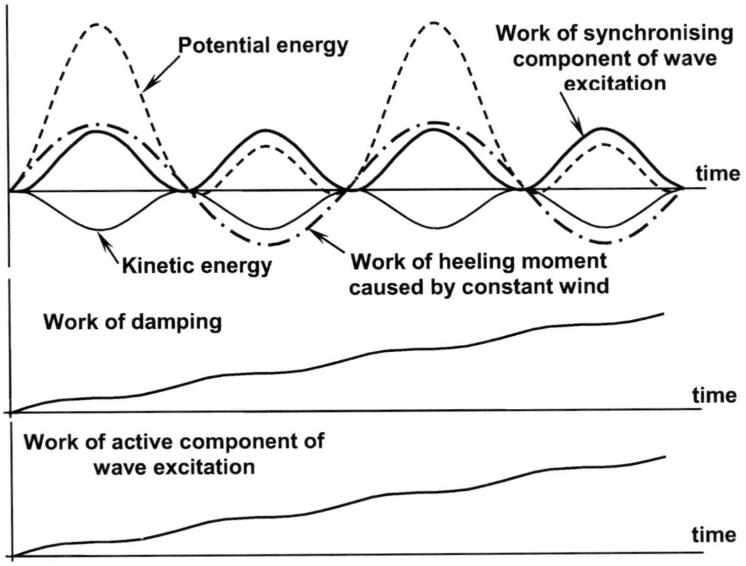

Fig. 7.6 Time histories of work and energies for linear steady-state rolling

Now, let us introduce a nonlinear GZ curve into the equation (7.49). The change of potential energy in this case is expressed as:

$$P(\phi_0,\phi)=\int_{\phi_0}^{\phi}\omega_\phi^2 f(\phi)d\phi=\omega_\phi^2\left(p_f(\phi)-p_f(\phi_0)\right) \tag{7.58}$$

With:

$$\frac{dp_f}{d\phi}=f(\phi) \tag{7.59}$$

We assume that for the nonlinear system there is a separate balance between damping and excitation works and between kinetic and potential energies as it takes place for linear systems in a steady state regime:

$$\left[\frac{\dot{\phi}^2}{2}-\frac{\dot{\phi}_0^2}{2}\right]+\omega_\phi^2\left(p_f(\phi)-p_f(\phi_0)\right)=m_A(\phi-\phi_0) \tag{7.60}$$

Equation (7.60) does not have the term describing the work of the synchronizing component of wave excitation. So it is satisfied only twice during a period, when all the terms become zero (see fig. 7.6), but we are free to choose the initial conditions $\phi_0, \dot{\phi}_0$. Let us start from the negative amplitude:

$$\phi_0 = -\phi_a + \phi_W; \quad \dot{\phi}_0 = 0 \qquad (7.61)$$

Then, the values after the half of a period would be the following:

$$\phi = \phi\left(t_0 + \frac{T}{2}\right) = \phi_a + \phi_W; \quad \dot{\phi} = \dot{\phi}\left(t_0 + \frac{T}{2}\right) = 0 \qquad (7.62)$$

The balance of energy would be described as:

$$\omega_\phi^2\left(p_f(\phi_a + \phi_W) - p_f(-\phi_a + \phi_W)\right) = 2m_A\phi_a \qquad (7.63)$$

Now, let us assume that a wind gust was applied at the moment t_0. This wind gust creates an additional heeling moment:

$$m_A^* = m_A + m_{AG} \qquad (7.64)$$

The work of the wind heeling moment taking into account the gust is:

$$A_A(t_0, t) = \int_{\phi_0}^{\phi} m_A^* \, d\phi = m_A^*(\phi - \phi_0) = (m_A + m_{AG})(\phi + \phi_a) \qquad (7.65)$$

The application of the gust of wind generates a transition process that breaks the conditions for separate energy balance equations (7.48) and (7.49), since they are held only for steady state motion of a linear system. However, if we assume that the transition does not affect the energy balance, we can rewrite the energy balance equation as:

$$\omega_\phi^2\left(p_f(\phi + \phi_W) - p_f(-\phi_a + \phi_W)\right) = (m_A + m_{AG})(\phi + \phi_a) \qquad (7.66)$$

The equation (7.66) allows determining an extreme angle the ship would have after half of a period (when roll velocity would reach zero). Geometrically the equation (7.66) expresses equality of two areas, see fig. 7.7.

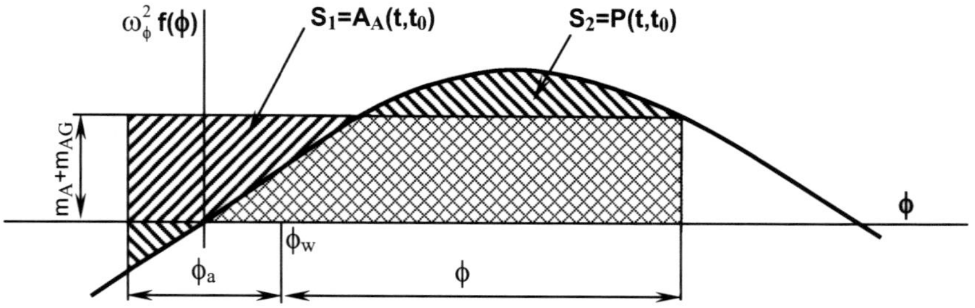

Fig. 7.7 An evaluation of dynamic angle of heel caused by sudden wind gust

We actually derived the weather criterion with a dynamically applied constant wind heeling moment (see also subchapter 3.5.1 of [Kobylinski and Kastner 2003]). Capsizing is associated with the impossibility to find roll angle ϕ that is considered as dynamical heel angle. The reason for this is the lack of the area under the *GZ* curve. Physically it means that potential energy of roll motions is insufficient to balance the work of the heeling moment. The system jumps out of the "potential well" to find the energy balance near another stable equilibrium.

Despite a "physically sound" capsizing definition, we made two serious assumptions to simplify the derivation and facilitate the solution:

- We assumed that a nonlinear system has a balance for wave excitation and damping separate from the work of the wind heeling moment, kinetic and potential energy;
- The transition process does not break the above assumption when the wind gust is applied.

Another possible way to interpret (7.66) is to consider a non-forced and non-damped system. In this case the equation (7.50) describes the energy balance completely. Both of the above assumptions become true and the determination of stability with the weather criterion becomes identical to the classical definition of stability (subchapter 5.1).

7.2 Influence of Freeboard Height and Water on Deck

7.2.1 General

Usually, there is a significant difference regarding capsizing behavior in beam seas of high freeboard and low freeboard vessels. Let us consider the simplest situation: the action of a suddenly applied constant heeling moment in a beam position on a ship. It is known (Sevastianov [1970], Nechaev [1989]) that if a high freeboard ship capsizes, she capsizes in the first semi-period of roll oscillation. A low built vessel withstands several periods of rolling and then capsizes when drift is practically static: see fig. 7.8.

The reason for the difference is that the deck enters the water much earlier in the case of the low freeboard ship. This evident fact changes the whole picture. The deck edge is a sharp body and produces vortexes. This leads to a significant increase of roll damping.

Increasing damping makes the roll amplitude smaller in comparison with the case of a high freeboard ship (where the deck enters into the water too late for increased damping to prevent immediate capsizing). Therefore, the low freeboard ship has time for beam drift to develop. The moment of the hydrodynamic force (subchapter 7.1) combined with the heeling moment increases the chances for a ship to capsize.

For the case of quartering seas, the difference in behavior between high and low freeboard ships is not that significant. The picture of capsizing is more complex.

7.2.2 Experimental Observations. Pseudo-static Heel

When the deck emerges from the water after submerging, some quantity of water is trapped in the deck well. Then, this water flows out through freeing ports in the bulwark. The process of water outflow takes some time, fig. 7.9. This water can be considered as liquid cargo during this time. So, we can consider the stability of a ship with a liquid cargo of varying volume [Rakhmanin, 1966, 1966a, 1971], [Sevastianov, 1970].

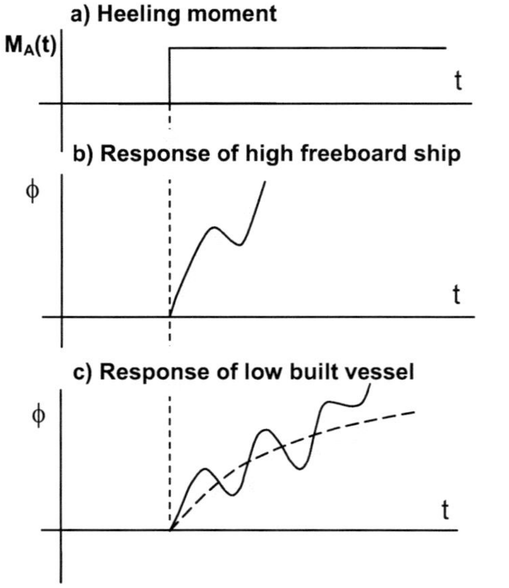

Fig. 7.8 Differences in capsizing scenario of high freeboard and low built ships under action of dynamically applied heeling moment

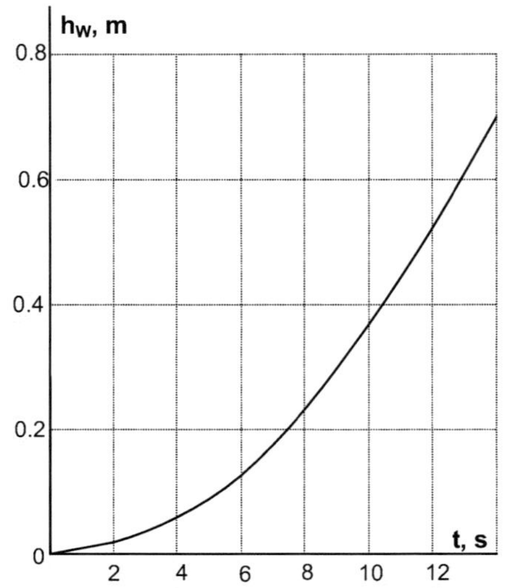

Fig. 7.9 Dependence of outflow time on initial level of deck well flooding

Experiments [Rakhmanin, 1966, 1966a, 1971] showed that rolling of a low freeboard ship is asymmetric (see subchapter 3.3.6 of [Kobylinski and Kastner 2003]). Rolling oscillations occur relative to a certain position that is called a pseudo-static angle of heel. Characteristics of models of small and medium fishing vessels used for these tests were: angle of deck submerging from 6 to 15 degrees, B/T from 1.9 to 3.0 and C_B from 0.4 to 1.0. Models were equipped with superstructures. Qualitative results are presented in fig. 3.14 of [Kobylinski and Kastner 2003].

Three curves show the extreme of roll angles versus excitation frequency along with the value of the pseudo-static angle. The pseudo-static angle of heel could not be determined clearly in the region of principal resonance. It can be explained that the principal rolling resonance is observed in relatively long waves and diffraction is small, so the deck can be flooded from both sides.

Parametric resonance due to heave coupling (subchapter 6.2) was observed in short waves, where diffraction is significant. So, wave height is greater on the "windward" side and less on the "leeward" size. That leads to deck flooding though the "windward" side and respectively to development of pseudo-static angle of heel in the region of parametric resonance.

Presence of the bulwark leads to an increase of pseudo-static heel because of an increased amount of water on deck and a decrease of rolling caused by an increase of roll damping. The value of the pseudo-static angle of heel is very close to the static initial heel caused by liquid cargo on deck.

Garkavy and Ponomarenko [1977] carried out another series of model tests in the small tank of the National Laboratory of Seakeeping of Fishing Vessels at Kaliningrad Institute of Technology (Russia).

A schematized cylindrical model with midship section of a medium trawler had the dimensions $LxBxT$=0.635*x*0.182*x*0.085 *m*. The tank was equipped with a pneumatic wave maker. The model was located in beam waves and was kept in this position by two ropes attached fore and aft. The model was equipped with a bulwark of 0.02 m height. Two KG values were set up KG_1=0.061 m and KG_2=0.056 m. Rolling was recorded with a camera. Results of the model test are presented in fig. 7.10 (for KG_1=0.061 m) and in fig. 7.11 (for KG_2=0.056 m).

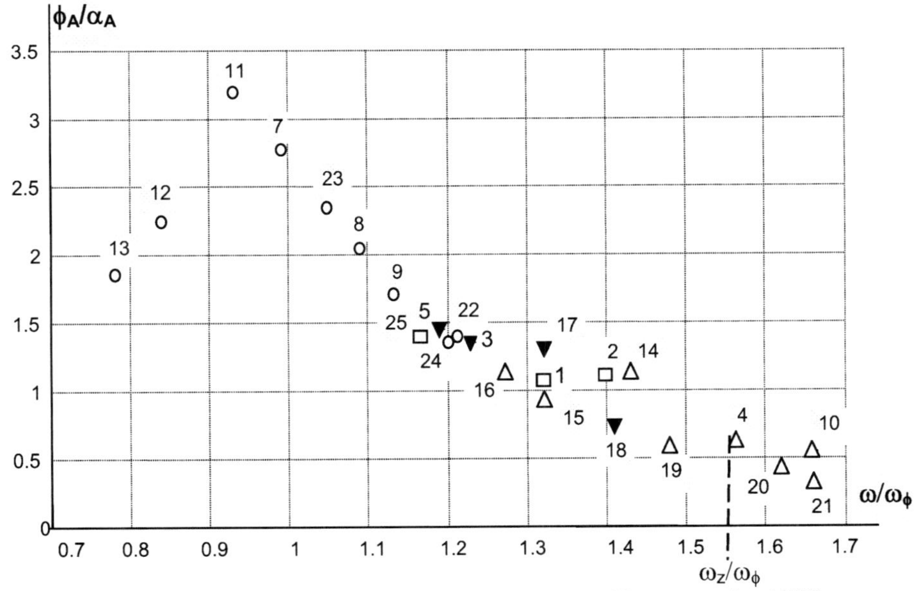

Fig 7.10 Model response for KG=0.061 m [Garkavy and Ponomarenko, 1977]

There were a total of 35 runs, a number near a point in fig. 7.11 and fig. 7.12 marks the number of the run. The legend is the following:

- A circle marks a test where pseudo-static angle could not be determined clearly;
- A black triangle means that the model capsized after several roll oscillations;

- A rectangle means that model behavior was very close to capsizing;
- A white triangle marks the test where pseudo-static angle was observed and determined.

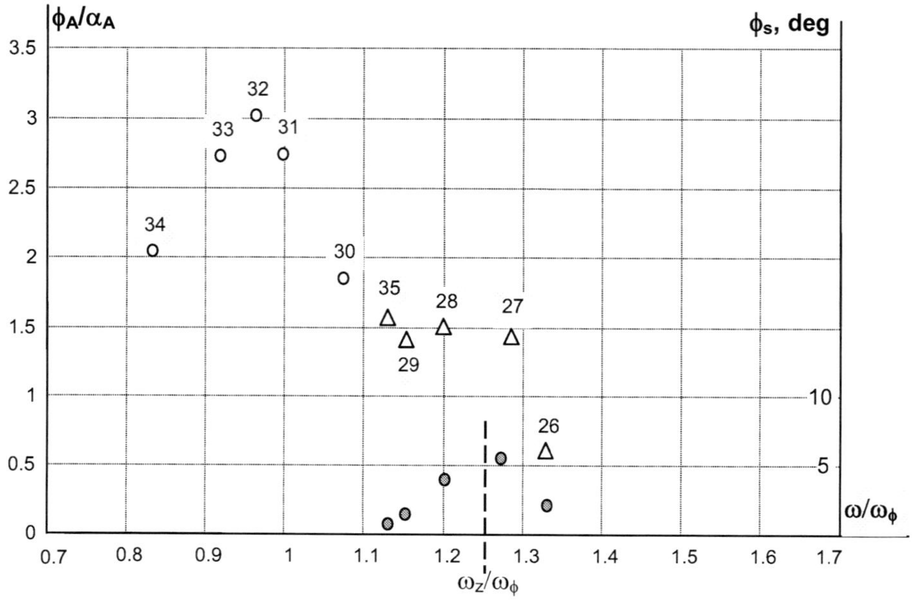

Fig. 7.11 Model response for KG=0.056 m [Garkavy and Ponomarenko, 1977]

The model behavior substantially depended on excitation frequency. If the relative frequency ω/ω_ϕ was less than 0.95, water being trapped on deck at one side immediately flowed out through the other side, since the forces acting on the trapped water were large. When the relative frequency was between 0.95 and 1.15, heaving motion becomes significant and behavior of the trapped water was the same: the model receives water through the "windward" side and outflow is through the "leeward" side.

When the relative frequency reached the range 1.15-1.47, trapped water outflow through the "leeward" side decreased. Collection of the water leads to the appearance of pseudo-static heel. Capsizing of the model was related with the increased amount of trapped water and increase of pseudo-static heel. Heave motion played the main role when the relative frequency was between 1.47-1.7. The amount of trapped water on deck became constant and regime of roll became steady state.

7.2.3 Behavior of Water on Deck[1]

Let us assume first that the bulwark is high and outflow of the trapped water is not significant. Then the water is involved in a complex motion caused by rolling and other

[1] Written in co-authorship with Prof. Stefan Grochowalski of Webb Institute

ship motions. If there is enough water on deck, it influences the motion of a ship, so we have a coupling problem [Dillingham and Falzarano, 1986].

An analogous problem exists for rolling / sloshing coupling for a ship with partially filled tanks (see for example [Francescutto and Contento, 1994], Armenio, *et al* [1996], de Kat [2000]). However, there is a significant difference between sloshing and green water influence on subsequent ship motions. Water on the deck has shallow depth, but large stretching. Fluid in a tank is limited by the size of the tank. Also the main concern associated with low-filling sloshing is focused on hydrodynamic impacts rather than on altering ship behavior (because of low filling, there is not enough fluid to change ship behavior considerably).

There is a more similarity though with behavior of water in a flooded car deck of a Ro-Ro ferry; here, we also have shallow water and large stretch (see for example [Vassalos, 2000], [Naito and Sueyoshi, 2001]). However, consideration of stability of damaged ships is beyond the scope of this book.

Garkavy [1991] proposed the following approximate method for the estimation of heeling moment due to water on deck:

$$M_{XWOD} = W \cdot l_{WOD} \cdot \sin\left[\pi\left(\phi / \phi_d\right)\right] \tag{7.67}$$

With:

$$l_{WOD} = 0.21 \cdot \overline{A}_{DW} \frac{B}{T} \frac{C_B}{C_W} h_W \tag{7.68}$$

Here, $\overline{A}_{DW}$ is relative area of the deck well. For the height of the water on the deck well the following empirical formula is proposed:

$$h_W = \begin{cases} 0 & \phi_A < \phi_B \\ 0.125 h_B \left[1 + 3\dfrac{\phi_A - \phi_B}{\phi_{B1} - \phi_A}\right] & \phi_B \le \phi_A \le \phi_{B1} \\ 0.5 h_B & \phi_A > \phi_{B1} \end{cases} \tag{7.69}$$

Here: ϕ_A is amplitude of the previous semi-period of oscillations, ϕ_B is the angle when bulwark enters water, h_B is bulwark height:

$$\phi_{B1} = \arctan\left[\frac{2(H - T + 1.5 h_B)}{B}\right] \tag{7.70}$$

Amagai, *et al* [1994] proposed classification of the characteristic behavior of trapped water on deck, obtained from experiments with an oscillating tank:

- The shape of the free surface was kept approximately horizontal and behaved as a standing wave;
- The same as above, but with some transient waves;
- The same as above, but when roll reaches the amplitude value, the surface is approximately horizontal;

- A large transient wave appears and the surface becomes nonlinear;
- There are two transient wave profiles in opposite directions.

The above experiments were followed by model tests in a towing tank that were focused on the influence of trapped water on the behavior of the ship roll response [Amagai, *et al*, 1994]. A model of a Japanese fishing vessel with the scale 1:7.6 and dimensions $LxBxHxT$=2.0x0.5x0.195x0.175 *m* was used in these experiments. The deck well of this model had the same form and dimensions as the oscillating tank. The model was tested in regular beam waves and trapped water outflow was not modeled. Results were presented in the following form:

$$\phi_{a1} / \phi_{a2} = f(h_w) \tag{7.71}$$

Where: ϕ_{a1} is the amplitude of ship roll response without water on deck; ϕ_{a2} is the amplitude of ship roll response with water on deck; h_w is the level of water on deck.

A sample of these results is shown in fig. 7.12. One can see from the patterns shown in fig. 7.12 that water on deck can influence rolling in both directions: increasing and decreasing.

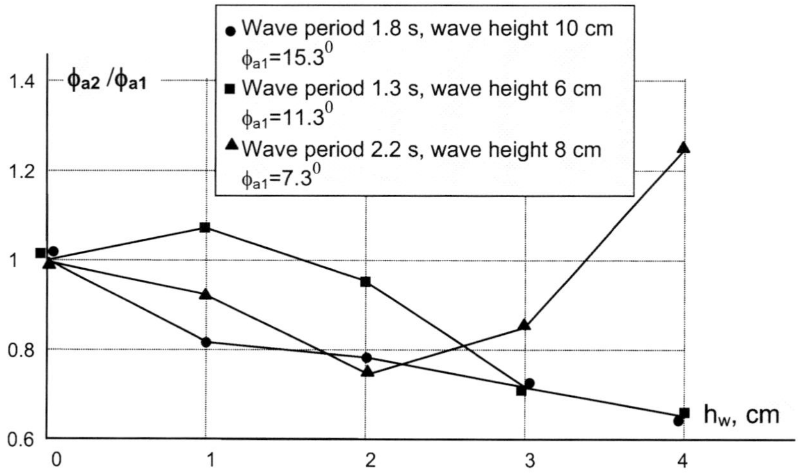

Fig. 7.12 Influence of water on deck level on rolling amplitude response [Amagai, *et al*, 1994]

Huang and Hsiung [1996] considered the problem of greenwater behavior as a shallow-water flow problem. The "Flux Splitting Difference" method was developed for computation forces and moments for the 3-D case. Numerical results were validated with the model test data [Huang, 1995]. An example of the free surface profile is given in fig. 7.13 [Grochowalski, *et al*, 1998], and experimental results were taken from [Adee and Caglayan, 1982].

As can be clearly seen, the water flow makes a bore profile, simulation results are quite close to experimental. One of the main difficulties here was the numerical treatment of hydraulic jump associated with the bore.

Two more problems are related with the influence of greenwater:

- How to estimate the amount of water trapped on deck;
- How and what amount of green water would escape from the deck.

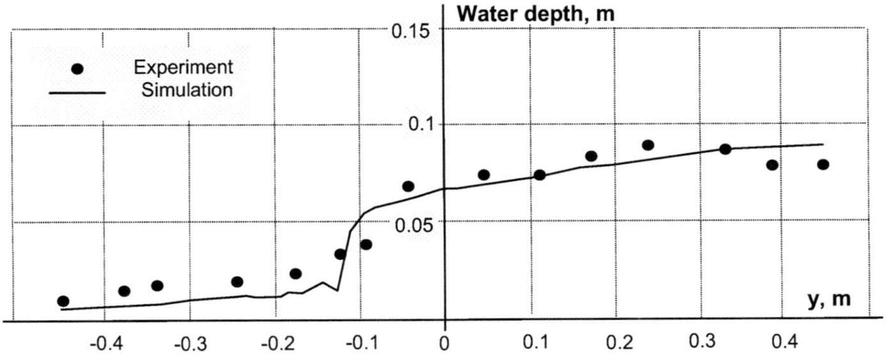

Fig. 7.13 Water on deck: wave profile caused by roll [Grochowalski, *et al*, 1998]

Grochowalski, *et al* [1998] addressed both of these problems. First, it is difficult to get adequate estimates of the amount of trapped water using solely statistical characteristics of relative motion obtained from traditional seakeeping calculations in the frequency domain. This difficulty is caused by the fact that the amount of water trapped on deck depends heavily on the details of the ship's geometry, while any statistical characteristics represent only vertical motion of water surface relative to bulwark edge. In other words "... *different ship geometries or the sea states with the same statistical parameters may lead to different amount of water shipped on deck*" (cited with the above reference).

The alternative is time domain simulation, based on a theoretical model [Grochowalski, 1979, 1981], [Grochowalski, *et al*, 1998], [Huang, *et al*, 1999]. Two conditions have to be satisfied for water to be trapped: wave height has to exceed bulwark or deck edge and relative velocity of liquid particles has to be directed inward onto the deck:

$$\begin{cases} \zeta_W(t) > Z_B(t) \\ v_{Rn} > 0 \end{cases} \tag{7.72}$$

Here, ζ_W is instant wave elevation and Z_B is the elevation of the point *B,* which is the highest point of the bulwark or deck edge in the current section, v_{Rn} is the normal component of the liquid particle relative velocity directed inward onto the deck, see fig. 7.14.

Wave motion can be presented as the sum of two components: incident wave and wave deformation caused by the presence of a ship that includes radiation and diffraction.

$$\zeta_W(t) = \zeta_{W0}(t) + \zeta_{WD}(t) \tag{7.73}$$

Relative horizontal velocity can be presented as:

$$v_R(t) = v_W(t) - v_s(t) + v_D(t) \tag{7.74}$$

$v_W(t)$ is the horizontal component of the liquid particle velocity due to undisturbed wave motion, $v_s(t)$ is the horizontal component of motion of the point *B*, $v_D(t)$ is the additional velocity reflecting influence of wave deformation. The last quantity is modeled using the energy conservation for the hydraulic head [Grochowalski, et al, 1998]:

$$v_D(t) = \sqrt{2g(\zeta_W(t) - Z_B(t))} \tag{7.75}$$

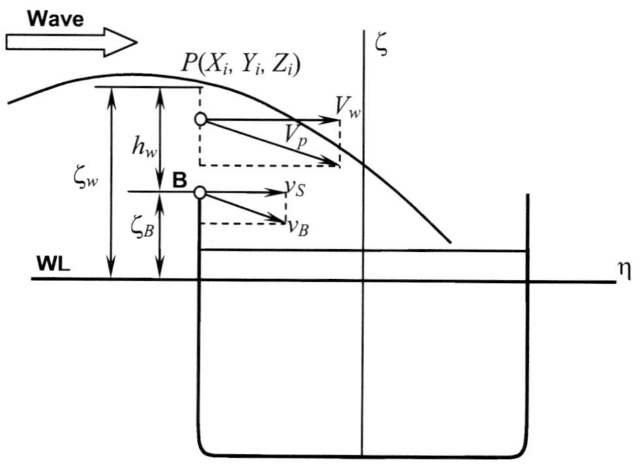

Fig. 7.14 Water shipping on deck [Grochowalski, *et al*, 1998]

To evaluate the amount of water trapped on the deck at the moment, we express elementary mass passing through elementary area $ds \cdot dz$ as:

$$dm = \rho \cdot dV = \rho \cdot ds \cdot dz \cdot v_{Rn}(s,z,t)dt \tag{7.76}$$

The total mass trapped from the moment t_b, when the wave surface reached point *B*, until moment t_e, when the wave receded, can be found by the integration of the expression (7.76) in time and space:

$$m_{GW} = \rho \int_{t_B}^{t_E} \oint_s \int_{Z_B(t)}^{\zeta_W(t)} v_{Rn}(z,s,t)dzdsdt \tag{7.77}$$

Formula (7.77) was validated by model experiments conducted in the towing tank of the Institute of Marine Dynamics (Canada). A cylindrical model was equipped to measure the quantity of trapped water, the deck was open on both the weather and lee side. A comparison of experimental and calculation results are presented in the fig. 7.15.

Another problem is associated with greenwater escape from the deck well. There were a series of model tests carried out in the Center for Marine Vessel Development and Research of the Dalhousie University in Halifax. Water escape from the deck was modeled with an oscillating rectangular tank; two modes of water flow were observed: "waterfall" and "splash". More water escapes with the first mode. The following formulae were developed to calculate the amount of water escaping from the deck:

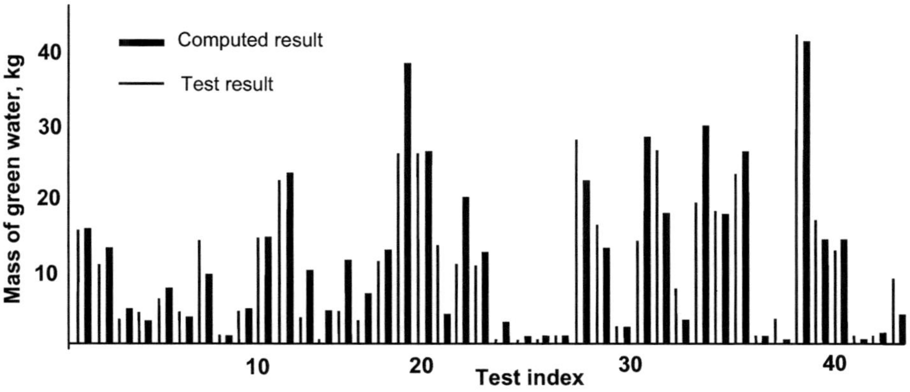

Fig. 7.15 Computed and measured mass of green water [Grochowalski, *et al*, 1998]

$$m_{GWe} = \rho C \int_0^t \oint_s v_e(s,t) \cdot h_e(s,t) ds\, dt \quad (7.78)$$

Here C is a correction coefficient (value $C = 2.5$ was proposed in [Grochowalski, *et al*, 1998]). As in the previous case, s stands for perimeter of the deck well. Height of the water above the bulwark $h_e(s,t)$ can be found as:

$$h_e(s,t) = \begin{cases} h_i(s,t) - h_0(s,t): & h_i > h_0 \\ h_i - h_b: & h_i > h_b \end{cases} \quad (7.79)$$

Here, h_i is height of water surface on the deck, h_0 is the height of water exterior to the bulwark, and h_b is the height of the bulwark. All the co-ordinates are given in a space fixed system.

Velocity of escape for the starboard side is:

$$v_e(s,t) = \begin{cases} \sqrt{|2gb \sin e_1|} + \sqrt{h_e}\left(3.22 + 0.4\dfrac{h_e}{h_{dk}}\right): & e_1 > 0 \\ 0: & e_1 < 0 \end{cases} \quad (7.80)$$

On the portside:

$$v_e(s,t) = \begin{cases} \sqrt{|2gb \sin e_1|} + \sqrt{h_e}\left(3.22 + 0.4\dfrac{h_e}{h_{dk}}\right): & e_1 < 0 \\ 0: & e_1 > 0 \end{cases} \quad (7.81)$$

The velocity of escape through the stern is:

$$v_e(s,t) = \begin{cases} \sqrt{|2gb \sin e_2|} + \sqrt{h_e}\left(3.22 + 0.4\dfrac{h_e}{h_{dk}}\right): & e_2 < 0 \\ 0: & e_2 > 0 \end{cases} \quad (7.82)$$

Here, b is deck width and h_{dk} is bulwark height, e_1 and e_2 are Euler's angles. These formulae were validated with model tests, the comparison is shown in fig. 7.16. The agreement between the experiment and prediction is good [Grochowalski, *et al*, 1998].

More information on trapped water outflow is available from [Calisal, *et al*, 2000, Shin, 2000].

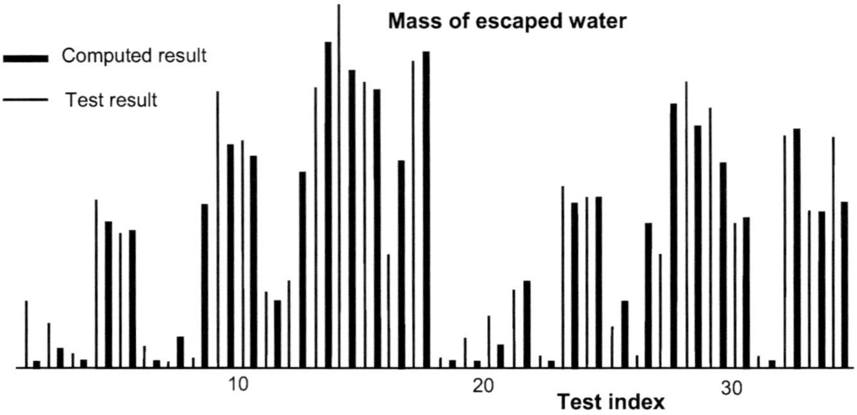

Fig. 7.16 Mass of water escaped form deck [Grochowalski, *et al*, 1998]

7.2.4 Influence of Deck in Water[1]

Earlier researches on ship behavior with water on deck are reviewed in [Caglayan and Storch, 1982].

The comprehensive experimental research of capsizing was undertaken in Canada [Grochowalski, *et al*, 1986, 1994, 1998; Grochowalski, 1989, 1990, 1993, 1993a, 1997, 2000] (see also subchapter 6.5.6 of [Kobylinski and Kastner, 2003]). Here, we briefly revisit the major findings on influence of the deck in the water.

It was found that if the leeward side bulwark becomes submerged and a ship has significant lateral motion towards the immersed side, then a significant hydrodynamic reaction and corresponding additional heeling moment appear.

The hydrodynamic force generated by a submerged deck and bulwark prevents the bulwark and deck edge from coming out of the water and restrains a ship in the heeled position. Her stability reserve is decreased and the ship may be capsized by the next wave if she stays in such a position long enough, see fig. 7.17.

If the weather side bulwark becomes submerged and the ship has lateral motion away from the immersed side, then the hydrodynamic reaction is less than in the previous case.

[1] Written in co-authorship with Prof. Stefan Grochowalski of Webb Institute

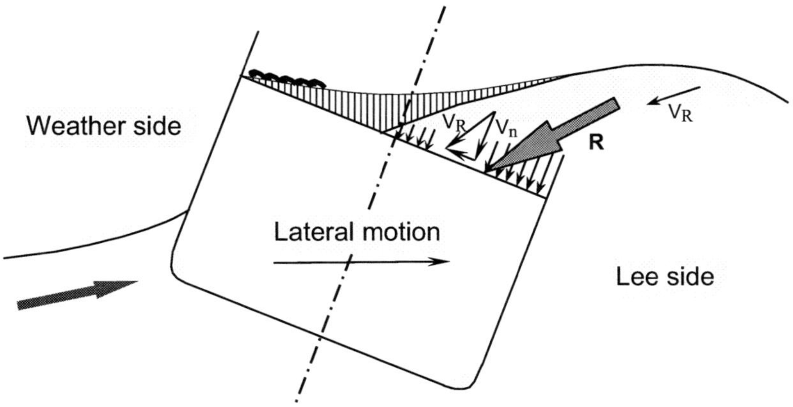

Fig. 7.17 Hydrodynamic reaction on submerged part of the deck [Grochowalski, *et al*, 1998]

This reaction has a dynamic nature: it develops only when there is significant velocity of water flow relative to the submerged deck. This velocity also has to be directed "towards" the deck, otherwise the water would be freed of the deck and the hydrodynamic reaction will not develop. Experimental observations show that swaying and yawing motions are major contributors to the creation of this force, which is, in fact, an addition to conventional hydrostatic, Froude-Krylov and wave diffraction forces.

Grochowalski and Lee [1990a] developed an approximate method of calculation of the hydrodynamic reaction created by a deck in water. It was assumed that the reaction has inertial and gravitational natures, viscosity was neglected; also it was assumed that the bulwark is fully submerged, deformation of the wave profile caused by the presence of a ship was ignored. The above assumptions allow us to present additional dynamic pressure in the following form:

$$p_{dk} = p_{HS} + p_{FK} + p_{RD} \qquad (7.83)$$

The first term in the equation represents the hydrostatic pressure:

$$p_{HS} = -\rho g(\zeta_0 - z) \qquad (7.84)$$

The second term is pressure caused by the wave: it is assumed that diffraction could be ignored in the first expansion, so it is the pressure of an undisturbed wave. Integration of this pressure over the hull surface would produce the Froude-Krylov force:

$$p_{dk} = -\rho \frac{\partial \varphi_0}{\partial t} - \frac{1}{2}\left|\nabla \varphi_0\right|^2 \qquad (7.85)$$

Where φ_0 is the potential of pressures in incident waves.

The third term expresses the dynamic pressure generated by the motion of the immersed part of the deck relative to surrounding water:

$$p_{RD} = k_{DW} f(v_{RD}) \left(\rho \frac{v_{RD}|v_{RD}|}{2} \right) \qquad (7.86)$$

The velocity v_{RD} is the normal component of the total velocity of the fluid in the given point. Here, we consider the immersed part of the deck as a body moving in inviscid fluid, that is why normal relative velocity is considered:

$$v_{RD} = \vec{v}_R \cdot \vec{n} \qquad (7.87)$$

The condition that relative velocity v_{RD} is directed towards the deck is expressed as:

$$f(v_{RD}) = \begin{cases} 1 & v_{RD} < 0 \\ 0 & v_{RD} \geq 0 \end{cases} \qquad (7.88)$$

Forces of a viscous nature are considered to be small here, so tangent components of the pressure are neglected.

Coefficient k_{DW} are to be defined from the model test [Grochowalski, *et al*, 1998].

Additional hydrodynamic forces and moments caused by deck-in-water effects could be found by integration of the pressures p_{RD} over all affected surfaces:

$$\vec{F}_{DW} = k_{DW} \frac{\rho}{2} \iint_A f(v_{RD}) v_{RD} |v_{RD}| dA \qquad (7.89)$$

$$M_{XDW} = k_{DW} \frac{\rho}{2} \iint_A f(v_{RD}) v_{RD} |v_{RD}| y dA \qquad (7.90)$$

Both of the above formulae imply that the domain of integration – area A if instantaneous surface of deck in water. Correction coefficient k_{DW} is determined from model tests to calibrate the model. Grochowalski [1997] carried out such an experiment in the towing tank of the Institute for Marine Dynamics (IMD).

A sample of the calculation of force and moment caused by deck-in-water effect is presented in fig. 7.18. Time histories of ship motions (roll and yaw motions, velocities of heave and sway) that are shown there are taken from the above model test. The dashed line plotted along with roll motion time history is the result of simulation: it represents anticipated rolling without deck-in-water effect. Numbers in circles mark specific moments; fig. 7.19 shows simulated position of a ship and wave crest along with immersed portion of deck at these moments. Finally, the calculated moment caused by deck-in-water effect is shown in fig 7.18 at the same time scale as time histories.

Fig. 7.18 clearly shows that deck-in-water influence on rolling (difference in dashed and solid line at roll time history plot) coincides in time with the development of significant hydrodynamic moment on deck calculated with formula (7.90). This can be considered as an indirect proof of the adequacy of model deck-in-water effect described in [Grochowalski, *et al*, 1998].

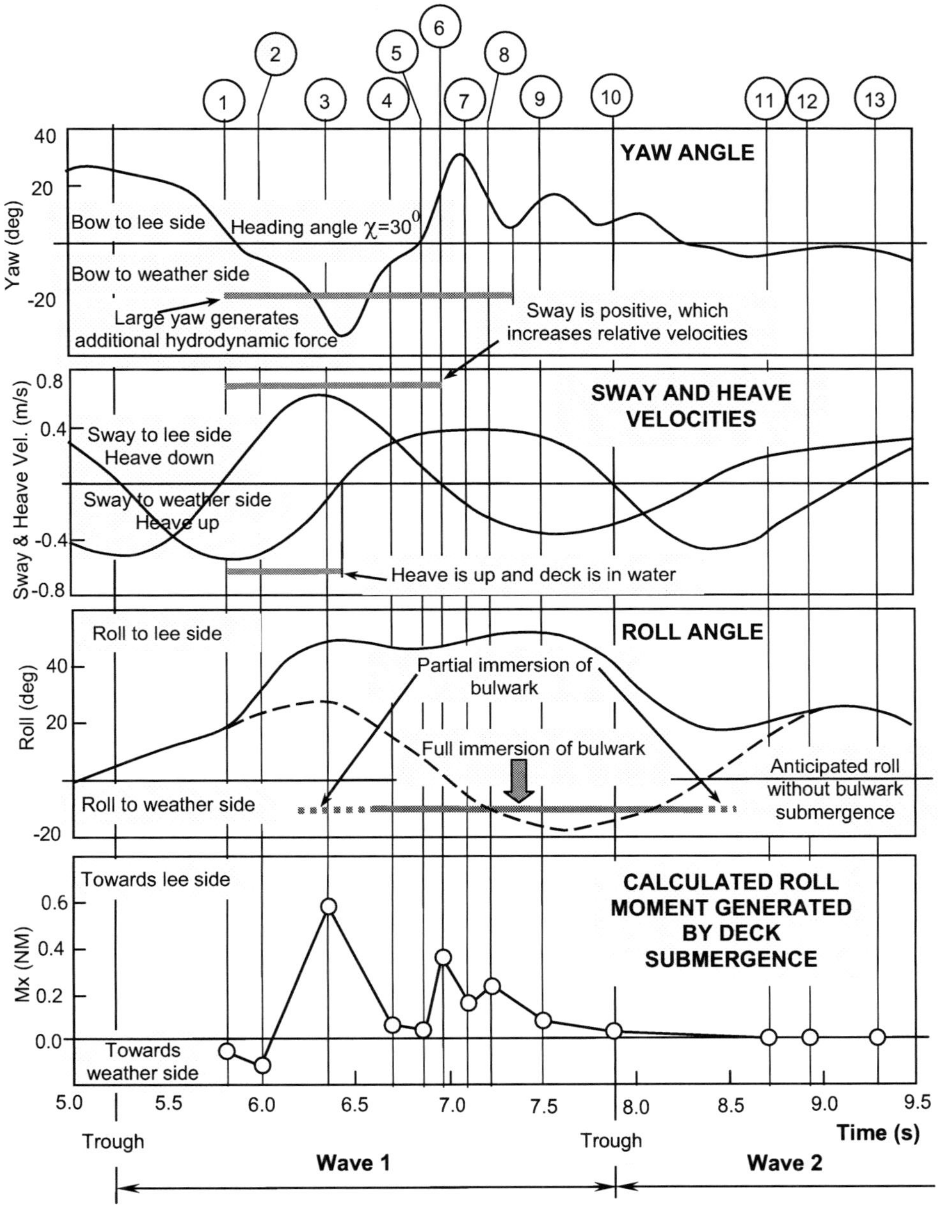

Fig. 7.18 Influence of deck submergence on roll motions. Fragment of time record of free running model test #25 and the calculated roll moment generated by submerged part of the deck [Grochowalski, 1993, 1993a]

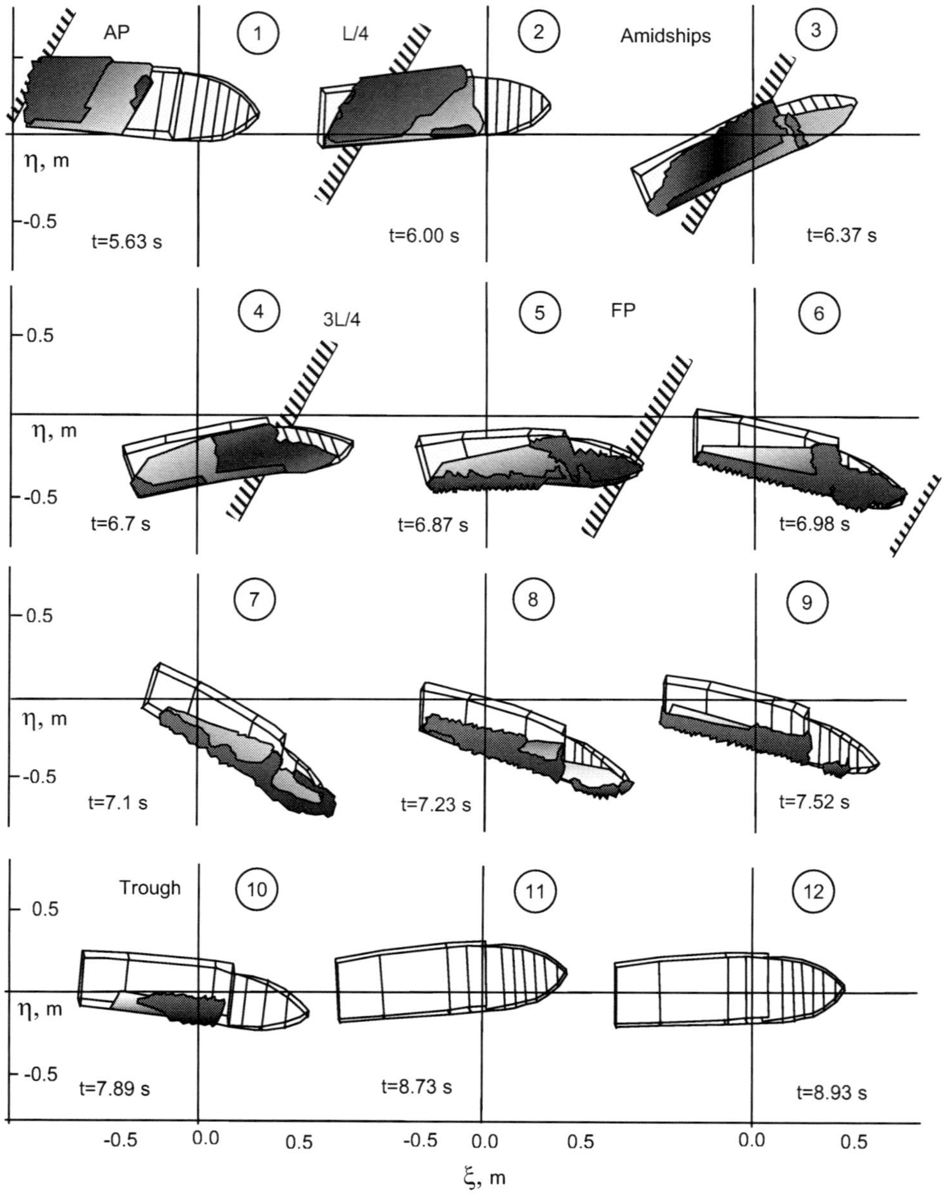

Fig. 7.19 Deck submergence in a free-running model test in breaking quartering waves (projection in horizontal plane, run no. 25). Time points correspond to those given in fig. 7.18 [Grochowalski, 1993, 1993a]

7.2.5 Model of Ship Motions

Let us examine the following case: ship in beam seas, fitted with a bulwark and this bulwark is periodically submerged. We assume that the ship is equipped with freeing ports.

Following Lee and Adee [1994] we consider a system with three nonlinear differential equations reflecting swaying, heaving and rolling:

$$\begin{cases} (m+a_{22})\ddot{y}_G + b_{22}\dot{y}_G + F_{Yd} + a_{24}\ddot{\phi} + b_{24}\dot{\phi} + F_{YR} = \\ \qquad = F_{YE} + F_{YW} + F_{YF} \\ (m+a_{33})\ddot{z}_G + b_{33}\dot{z}_G + F_{Zd} + F_{ZR} + F_{YH} = F_{ZE} + F_{YW} + F_{ZF} \\ (I_{xx} + a_{44})\ddot{\phi} + b_{22}\dot{\phi} + M_{Xd} + a_{24}\ddot{y}_G + b_{24}\dot{y}_G + M_{XH} + M_{XR} \\ \qquad = M_{XE} + M_{XW} + M_{XF} \end{cases} \tag{7.91}$$

We consider the terms included in the system (7.68).

m and I_x are mass and transverse moment of ship inertia without water on deck.

a_{ij} and δ_{ij} with i=2,3,4 and j=2,3,4 are added masses and coefficients of wave damping of a ship. Elis [1980, 1980a] showed that influence of submerging of the deck edge can be quite significant: the difference may be up to ten times, see fig. 7.20.

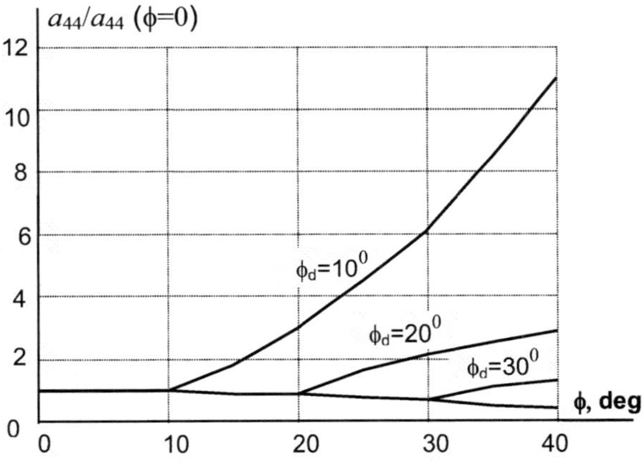

Fig. 7.20 Influence of deck edge submergence on value on added mass 44 for some ship section, Elis [1980, 1980a]

F_{Yd}, F_{Zd}, M_{Xd} are additional damping forces and moments caused by deck edge or bulwark submergence. The following expression for M_{Xd} is available from Sevastianov [1977]:

$$M_{Xd} = 2 \cdot mg \cdot GM \cdot \delta_\phi \cdot k_d \cdot \dot{\phi} \cdot (1 + a \cdot |\dot{\phi}|) - b_{44} \cdot \dot{\phi} \tag{7.92}$$

k_d is a coefficient, taking into account bulwark/deck edge submergence; the appearance is shown in fig. 7.21:

$$k_d = 1 + b \cdot \exp\left[-3 \cdot (\phi/\phi_d)^{-6}\right] \tag{7.93}$$

Here, b is a coefficient determined by the model test. If the bulwark is present b=8, and if the bulwark is absent b=2 [Sevastianov, 1977]. Coefficient a takes into account the viscous component of roll damping. Rakhmanin [1995] proposes to consider additional damping as:

$$M_{Xd} = 2 \cdot \delta_d \cdot (I_{xx} + a_{44}) \cdot \dot{\phi}; \quad \delta_d = f(p) \tag{7.94}$$

p is a measure of deck edge / bulwark submergence (ϕ_a is roll amplitude, ϕ_S is angle of static heel, ϕ_b – angle when bulwark enters water, ϕ_d –angle when deck enters water) :

$$p = \begin{cases} (\phi_a + \phi_S)/\phi_b \\ (\phi_a + \phi_S)/\phi_d \end{cases} \tag{7.95}$$

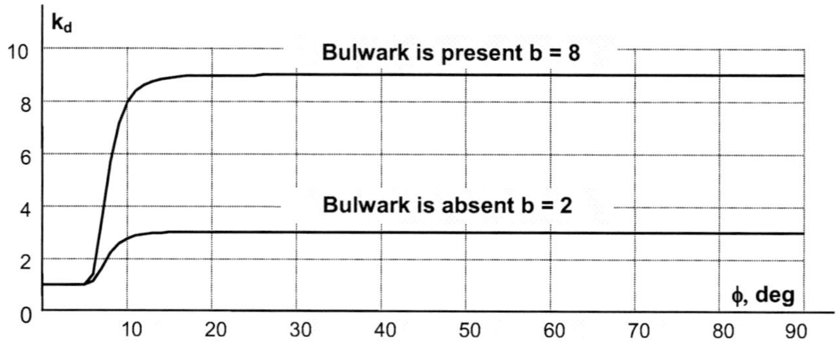

Fig. 7.21 Influence of deck edge / bulwark submergence on vortex component of roll damping

F_{YF}, F_{ZF}, M_{XF} are moment and components of the force caused to water outflow. We have considered outflow of trapped water on deck including the outflow problem discussed in subchapter 7.2.3. We were interested in estimation of the amount of water on deck. However, a significant amount of water out-flowing from the deck is capable of creating a jet force.

F_{YR}, F_{ZR}, M_{XR} are moment and components of the force of drift resistance. This drift is caused by the reaction to the jet force created by water outflow from the deck. Rakhmanin [1966a] assumes that:

$$M_{XF} = M_{XR} \tag{7.96}$$

Analogously, we can assume that:

$$F_{YF} = F_{YR}\,; \quad F_{ZF} = F_{ZR} \tag{7.97}$$

F_{ZH}, M_{XH} are restoring terms for heaving and rolling. Rakhmanin [1966a] proposes to take the roll restoring term as the hydrostatic moment of stability (KG_1 is vertical position of centre of gravity taking into account water on deck):

$$M_{XH} = mg \cdot GZ(\phi, KG_1) \tag{7.98}$$

F_{YE}, F_{ZE}, M_{XE} are wave excitation forces and moment. Garkavy [1979] derived the theoretical solution for the Froude-Krylov components using series by powers of two small parameters. Lugovsky [1966] originally proposed this method of bi-parametric expansion for a wall-sided ship. Lee and Adee [1994] have used the method of hydrodynamic singularities for the calculation of these terms within a linear approach.

F_{YW}, F_{ZW}, M_{XW} are forces and moment caused by either influence water on deck (see subchapter 7.2.3) or deck-in-water effect (see subchapter 7.2.4).

7.2.6 Behavior of Ship with Water on Deck

The dynamical system (7.91) is quite complex. Numerical integration of the system does not pose a problem, but interpretation of the results may be difficult and simplification of the model is a necessary step. Garkavy [1979, 1985] analyzed the rolling equation with periodical submersion of the deck using the equivalent linearization method (see subchapter 4.2.1). The effect of water on deck was not taken into account and it was shown that periodical deck submersion leads to decreasing rolling amplitudes and to simultaneous extension of the instability zone, see fig. 7.22.

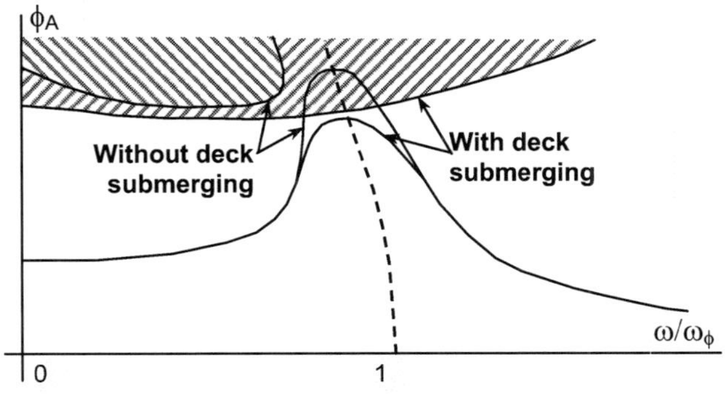

Fig. 7.22 Change of response curve and instability zone caused by periodical deck edge submersion, [Garkavy, 1985]

Falzarano and Troesch [1990], Falzarano, *et al* [1992] proposed to transform the water on deck into a fixed weight. Such a method leads to consideration of ship rolling with negative initial stability. Equilibrium at the origin of the coordinate system becomes unstable and stable equilibrium is shifted. As a result, we have two separatorices: one starts from the unstable equilibrium at the origin and the other starts from the angle of vanishing stability, see fig. 7.23.

The first separatrix usually is called a homoclinic connection (from Greek word "homios" – the same). This term emphasizes that the separatrix starts and ends on the same saddle point, associated with unstable equilibrium. The second separatrix is usually called a heteroclinic connection (from Greek word "heteros" – different). The term emphasizes that the separatrix starts from one saddle point and ends on different one. Both separatorices for the damped case are shown in fig. 7.24.

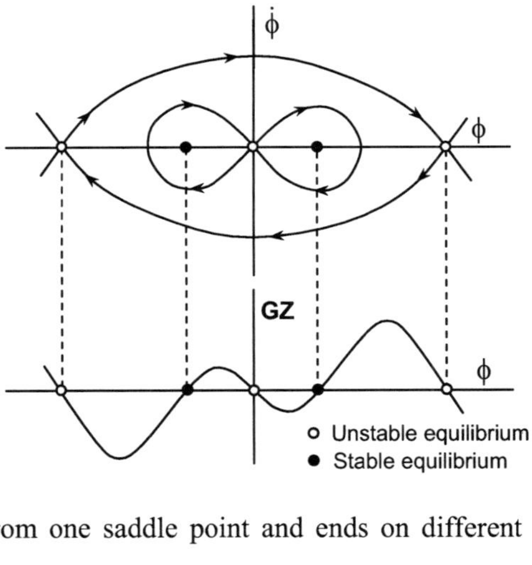

Fig. 7.23 Separatorices for negative initial stability

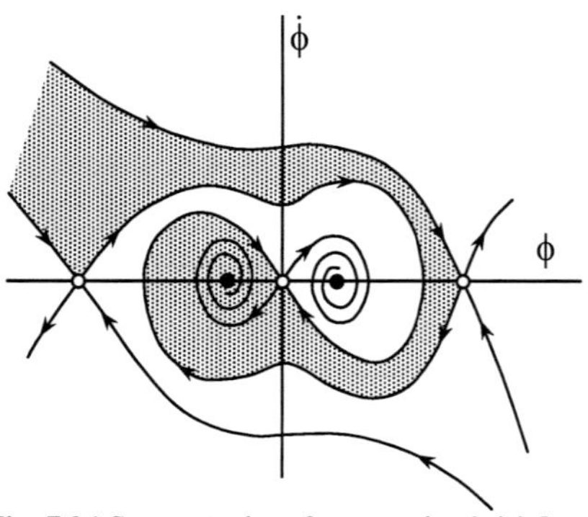

Fig. 7.24 Separatorices for negative initial stability with damping [Falzarano, *et al*, 1992]

Following Falzarano, et al [1992], we consider ship behavior in the region of the homoclinic separatrix:

$$\ddot{\phi}+2\delta\dot{\phi}-\omega_{\phi}^{2}\phi+a_{3}\phi^{3}= \\ =\alpha_{E}\sin(\omega t+\varphi_{E}) \qquad (7.99)$$

Since the system (7.99) is forced, we need to calculate invariant manifolds to determine the area of attraction for different stable equilibria. The method of calculation of the invariant manifold for a heteroclinic connection was described in subchapter 5.3.3. There is no principal difference for calculation of the homoclinic invariant manifold. The formula for the Melnikov function for the homoclinic connection is follows [Falzarano, *et al*, 1992] :

$$M(\tau_0)=-\frac{\alpha_{E0\tau}\pi\omega_{\tau}\sqrt{2}}{\sqrt{a_3}}\operatorname{sech}\left(\frac{\omega_{\tau}\pi}{2}\right)\sin(\omega_{\tau}\tau_0)+\frac{4\mu_0}{3a_3} \qquad (7.100)$$

With:

$$\mu=\frac{\delta}{\omega_{\phi}};\quad \alpha_{E\tau}=\frac{\alpha_E}{\omega_{\phi}^2};\quad \omega_{\tau}=\frac{\omega}{\omega_{\phi}} \qquad (7.101)$$

$$\alpha_{E0\tau}=\varepsilon\alpha_{E\tau};\quad \mu_0=\varepsilon\mu \qquad (7.102)$$

ε -is a small parameter $\varepsilon \ll 1$.

Zeros of the Melnikov function correspond to the intersection between stable and unstable manifolds. As we have seen from subchapter 5.3, if even one such crossing occurs, there will be an infinite number of them. The basin of attraction to stable equilibrium experiences erosion and the system shows chaotic response. The critical value of excitation amplitude that makes the Melnikov function to cross the zero is as follows:

$$\alpha_{Cr}=\frac{4\mu_0}{3\sqrt{2a_3}\pi\omega_{\tau}\cosh\left(\frac{\omega_{\tau}\pi}{2}\right)} \qquad (7.103)$$

Normally, a ship with negative initial stability has a static hill angle, so she rolls near a non-zero stable equilibrium. Erosion of area of attraction also means that a ship is capable of sudden changes of stable equilibrium, such a regime of rolling is called erratic rocking, two samples of which are shown in fig. 7.25. The behavior of heteroclinic invariant manifolds do not differ from the case of positive initial stability, see subchapter 5.3.

More information on the subject is available from [Ananiev 1989], [Kan 1992a], [Kan and Taguchi 1992b, 1993], [Lee and Adee 1994]. Garkavy [1991] observed deterministic chaos and sub-harmonic response in model tests.

There were a number of numerical simulations of ship motions with water on deck as well: see, for example [Huang, and Hsiung 1997]; some of these simulations used advanced time domain codes: LAMP was applied for study of small fishing vessel behavior [Belenky, *et al* 2002, 2003].

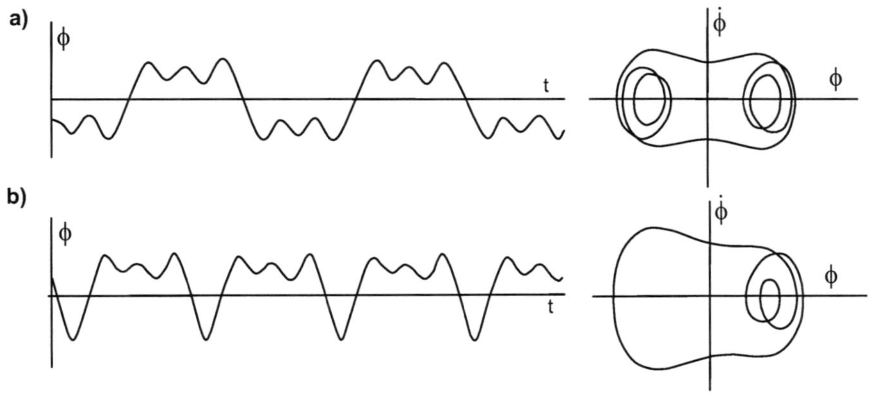

Fig. 7.25 Samples of erratic rocking [Kan and Taguchi, 1992b, 1993]

7.3 Stability in Breaking Waves

7.3.1 General

All the above discussions were based on a harmonic presentation of waves. This assumption is derived from the theory of small waves. So far, we did not raise the question: what is the influence of the shape of the wave on ship stability? There are papers focused on consideration of ship response in nonlinear waves (see, for example, [Cardo, *et al*, 1985], [Trincas, 1986]). Nonlinearity of the wave is expressed as a non-sinusoidal shape of the wave profile and, first of all, in the absence of symmetry.

Here action of the breaking wave is considered. Blagoveshchensky was one of the first who suggested the breaking wave was a reason for capsizing [Nechaev, 1989]. He investigated accidents with two drifters in the Caspian Sea during the winter of 1955. Both shipwrecks took place not very far from the harbor of Kianly, near the city of Krasnovodsk. Two sister ships, "Morlovets" and "Kuuli-Mayak" were small wooden ships, length 18 m, breadth 4.4 m, depth 2.56 m and full load displacement 64 metric tons with the draught 1.57 m. Both ships were built in 1950.

"Morlovets" was lost in January 1955. After some time, she was found lying on the sea bottom at a depth of 8-9 m, only 1.5 nautical miles from the harbor of Kianly. The drifter, "Kuuli-Mayak", was spotted by a storm while returning from a fishing ground on February 23rd 1955. Northwest winds reached 9 on Beaufort scale. Waves were about 6 m high with a length of 60 m. It is known that high waves break when they reach shallow water. One of these waves covered the deck of "Kuuli-Mayak". The wave was so strong that it knocked out the wheelhouse door. Two crewmen were washed out; one crewman was washed into the machinery space through the open hatch. The ship capsized and floated keel up about two days, after which she was found on the shoreline. The *GZ* curve of "Kuuli-Mayak" was normal and met requirements for this type of vessel. Investigators believed that a breaking wave was the reason for the capsizing in both cases [Nechaev, 1989].

M/V Helland-Hansen was capsized by a breaking wave in September 1976 near the Norwegian coastline [Dahle and Kjaerland, 1980]. The ship was returning to the harbor with empty holds. North wind with force 8 on Beaufort scale had the same direction as the waves. Significant wave height was about 3.5 m. There was a sea current in the opposite direction to the waves and wind. Norwegian fisherman knew this place for the occurrence of large breaking waves. The ship had a speed of about 6 knots, when the captain noticed a breaking wave 5-m high approaching from the portside. Then, the wave hit the ship. After the impact, the ship was heeled up to 60 degrees within 5-6 seconds. Succeeding waves increased the heel up to 80 degrees and the ship was flooded through hatches, ventilation holes and wheelhouse windows. The ship sunk after 20 minutes with 2 crewmen killed, and the remaining 10 crewmen rescued by a helicopter. Again, the stability of the Helland-Hansen met all the requirements.

7.3.2 Geometry and Classification of Breaking Waves

A breaking wave profile is shown in fig. 7.26. The usual geometric characteristics are insufficient to describe this wave. The following geometric characteristics of an asymmetric wave of finite height are used to describe the breaking wave [Kjeldsen and Myrhaug, 1978]:

Crest front steepness: $\varepsilon_f = \eta_{W1} / \lambda_1$ (7.104)

Crest rear steepness: $\varepsilon_R = \eta_{W1} / \lambda_2$ (7.105)

Vertical asymmetry factor: $\delta_V = \lambda_2 / \lambda_1 = \varepsilon_f / \varepsilon_R$ (7.106)

Horizontal asymmetry factor: $\delta_V = \eta_{W1} / h$ (7.107)

Theoretical criteria for breaking waves are summarized in table 7.8 [Kjeldsen and Myrhaug, 1978].

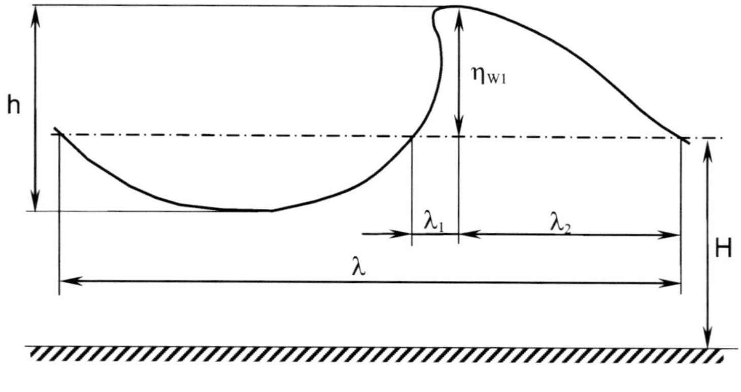

Fig. 7.26 Geometry of asymmetric wave of finite height

Table 7.8 Theoretical breaking wave criteria for gravity water waves [Kjeldsen and Myrhaug, 1978]

Condition	State		Symmetry		Action of wind		Water		Criteria	Reference
	steady	un-steady	Yes	No	Yes	No	Deep	Shallow		
Geo-metrical	X		X			X	X		The angle at the top becomes 120 degree	Stokes 1880
	X	(X)	X	(X)		X	X		The steepness h/λ reaches limiting value 0.142	Michell 1983
		X		X		X	X	X	The front face of the wave becomes vertical	-
	X	X	X	X		X		X	h/H =0.78	McCown 1981
Cinematic		X		X		X	X	X	The horizontal particle velocity at the surface exceeds the phase velocity	-
		X		X	X		X	X	$h_{max} = \frac{c^2}{2g}\left(1-\frac{q^2}{c}\right)$ q-surface drift	Banner & Phillips 1974
Dynamic	X		X			X	X		The download acceleration in the wave crest exceeds 0.5g	Lonuett-Higgins 1969
		X		X		X	X	X	The vertical upward acceleration in the wave crest exceeds g or $\frac{D_w}{D_t} \leq -\frac{1}{\rho}\cdot\frac{\partial p}{\partial z}+g$ no vertical momentum flux	Smith 1976

There is a difference between breaking waves generated in shallow and deep waters. Breaking waves are similar in their height and length in shallow water conditions contrary to deep water, [Nechaev, 1989]. For shallow water, the wave profile changes significantly when the wave approaches the shore. Steepness of the wave increases, the wave becomes asymmetric and front crest steepness increases faster than rear crest steepness, see fig. 7.27. The wave becomes unstable and breaks itself with foam and noise.

Fig. 7.28 shows the classification of deep water breaking waves [Kjeldsen and Myrhaug, 1978], where a detailed study of spilling and plunging breakers can be found. The reference also contains recommendations for the generation of these types of wave in a towing tank.

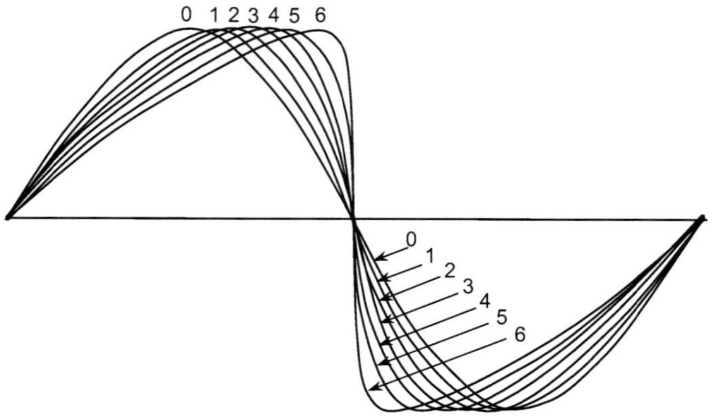

Fig. 7.27 Scheme of wave transformation in shallow water [Balitskaya, 1965]

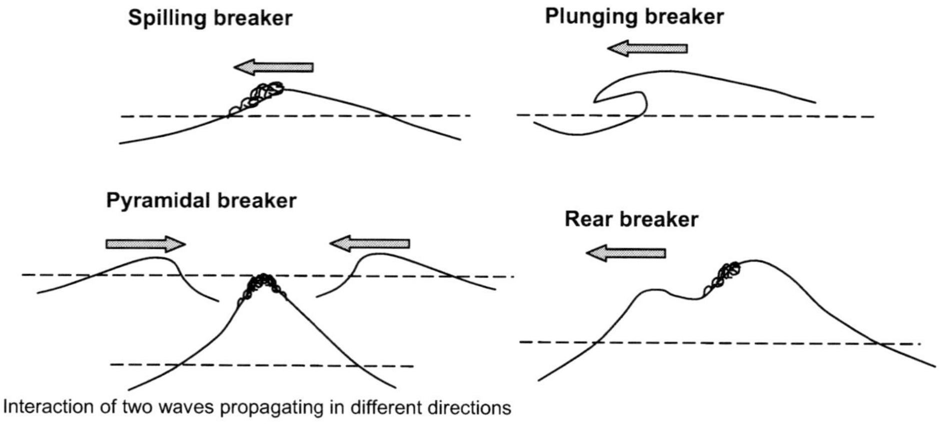

Fig. 7.28 Classification of breaking waves in deep water waves [Kjeldsen and Myrhaug, 1978]

7.3.3 Impact of Breaking Wave: Experiment and Theory

A model test by Balitskaya [1965] was one of the first experiments on stability in breaking waves; it was carried out in the towing tank of Saint-Petersburg Marine Technology University by the initiative of S. N. Blagoveshchensky in the early 1960's (see also subchapter 6.5.5 in [Kobylinski and Kastner 2003]).

The towing tank was equipped with a floating bottom; see fig. 7.29. The position of this floating bottom relative to the level of calm water could be easily changed. The towing tank was also equipped with a buoy transducer for wave profile measurement in deep water and level transducers for wave profile measurements in shallow water.

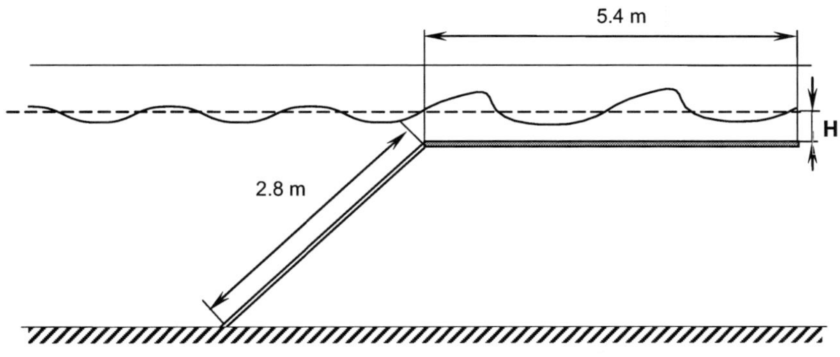

Fig. 7.29 Scheme of floating bottom for wave transformation [Balitskaya, 1965]

A schematized pontoon model was used for this experiment. The model was free to drift and roll, first under the action of symmetric regular waves in deep water and then under the action of breaking waves in shallow water. The model was equipped with pressure gauges for measuring impact pressures caused by breaking waves. Roll angles were measured by filming. The model's characteristics are given in table 7.9 and its appearance is shown in fig. 7.30.

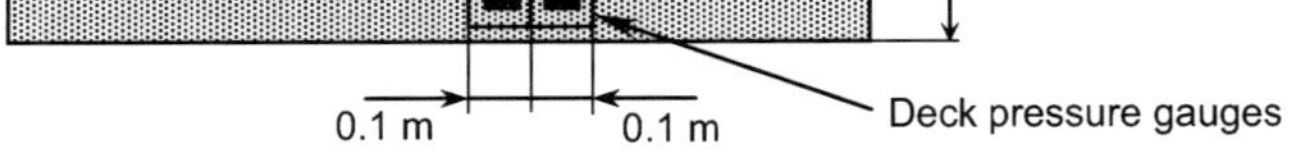

Fig. 7.30 Scheme of pontoon model [Balitskaya, 1965]

Table 7.9 Characteristics of the models for experiment in Saint-Petersburg University of Marine Technology

Model No.	Length, m	Breadth, m	Depth, m	Description	Year
49563	1.38	0.288	0.137	Tug, V=1680 m^3, P=1700 bhp, scale 1:40	1960
71621	2.00	0.350	0.195	Pontoon	1962
70590	2.00	0.250	0.137	Pontoon	1962
71622	2.00	0.175	0.096	Pontoon	1962

The characteristics of the wave in deep water were: period 1.2 s, length 2.2 m, height 0.2 m. Examples of time histories of model rolling and capsizing under the action of breaking waves are presented in fig. 7.31.

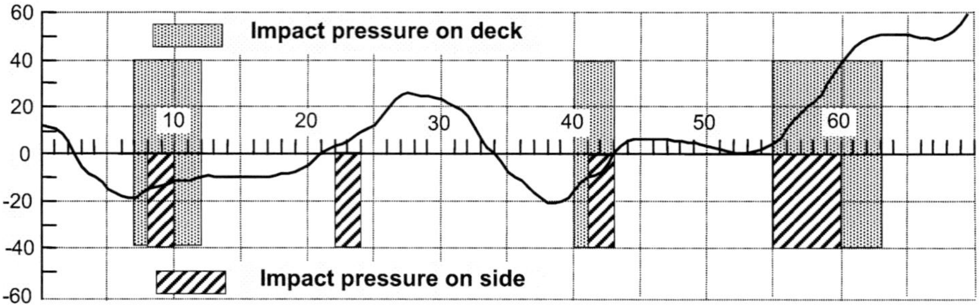

Fig. 7.31 Time history of rolling and capsizing under action of breaking waves. Model No. 70590, [Balitskaya, 1965]

Following Kholodilin [1963], Balitskaya [1965], Kholodilin and Tovstikh [1969], we assume that all action of breaking waves on a ship can be expressed as impact:

$$(I_{xx} + a_{44})\ddot{\phi} + M_{XD}(\dot{\phi}) + mg \cdot GZ(\phi) = M_{XE}(t) + M_{XI}(t) \tag{7.108}$$

Where $M_{XI}(t)$ is the heeling moment due to a breaking wave impact. We integrate it during the time of a single impact from moment t_1 until t_2. Because time of impact is small, the integrals of finite quantities $M_{XD}(\dot{\phi}), W \cdot GZ(\phi), M_{XE}(t)$, can be neglected and the equation (7.108) can be significantly simplified:

$$\int_{t_1}^{t_2} \frac{d}{dt}\left[(I_{xx} + a_{44})\dot{\phi}\right]dt = \int_{t_1}^{t_2} M_{XI}(t)dt \tag{7.109}$$

Assuming that $\dot{\phi}(t_1) = 0$ we obtain:

$$\frac{(I_{xx} + a_{44})\dot{\phi}^2(t_2)}{2} = \frac{1}{2(I_{xx} + a_{44})}\left[\int_{t_1}^{t_2} M_{XI}(t)dt\right]^2 \tag{7.110}$$

The right hand term can be calculated using model test data, the left-hand-side term is kinetic energy. Capsizing occurs when kinetic energy exceeds potential energy:

$$\frac{1}{2(I_{xx} + a_{44})}\left[\int_{t_1}^{t_2} M_{XI}(t)dt\right]^2 \geq W \int_{\phi_1}^{\phi_2} GZ(\phi)d\phi \tag{7.111}$$

A comparison of theoretical and model test results is summarized in fig. 7.32, showing satisfactory agreement. Further research of ship behavior in breaking waves in shallow water was carried out in the St. Petersburg University of Marine Technology and was reflected in [Kholodilin and Mirokhin, 1972]. Experiment with the model of Japanese fishing vessel was carried out by Ishida and Takaishi [1990].

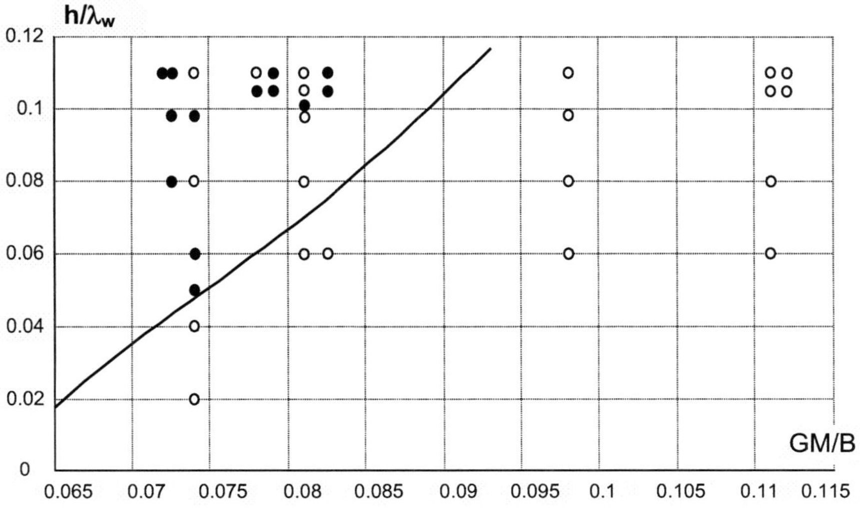

Fig. 7.32 Results of model test (Model 71622) and theoretically predicted capsizing boundary (white circles denotes surviving and black one denotes capsizing) [Kholodilin, 1963]

7.3.4 Probabilistic Approach to Capsizing in Breaking Waves

Ship behavior in deep water breaking waves was studied in the towing tank of the National Laboratory for Seakeeping of Fishing Vessels (Kaliningrad Institute of Technology, Russia) [Sevastianov, 1984], [Cerka and Batuev, 1985], [IMO, 1984].

Deep-water breaking waves were created with a pneumatic wave maker. It generated a sequence of waves with a continuously increasing period from 1 to 3 seconds. The sequence is generated during one control cycle with duration of 30 seconds. The waves with a large period have larger phase velocity. Short-period waves are slower. Long-period waves pass short period waves and interfere with them. The resulting wave becomes too steep and breaks (fig. 7.33). The control program was written to provide wave breaking at the location where the model was tested [Cerka and Batuev, 1985].

A particular purpose of these model tests was to study the influence of superstructures on ship stability in breaking waves. A small stern Russian trawler "Baltika" was chosen for these tests. The model scale was 1:6. The ship characteristics are given in table 7.10.

The model was schematized and had no keel and deckhouses, variants of superstructure types are shown in fig. 7.34, while the stability characteristics are given in table 7.11. The loading condition for all the models was in a critical state. This means that all the stability criteria of the Russian Register of Shipping were satisfied without lack and one criterion was satisfied without surplus and lack. *GZ* curves of these models are shown in fig. 7.35.

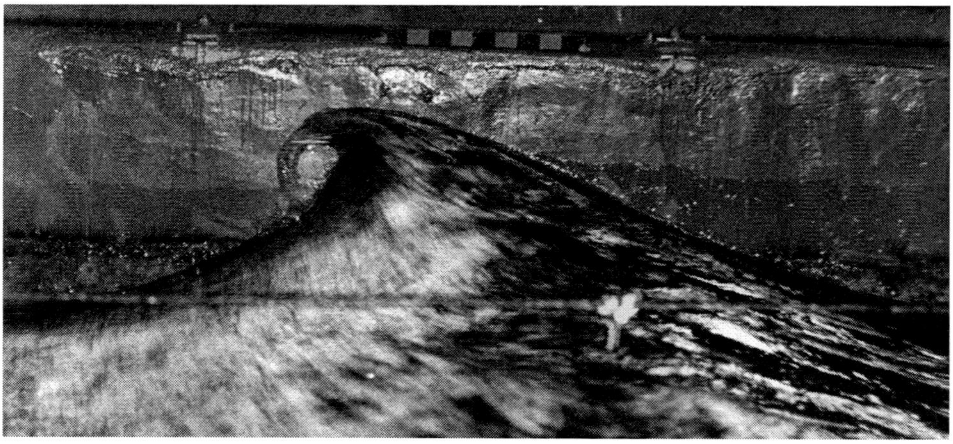

Fig. 7.33 Breaking wave in the towing tank of National Laboratory of Seakeeping at Kaliningrad Institute of Technology, Photo by A .A. Adriashkin [Nechaev, 1989]

Table 7.10 Characteristics of small trawler "Baltika"

Length between perpendiculars, m	22
Waterplane breadth, m	6.8
Draft, m	2.35
Block coefficient	0.458
Total depth, m	3.3
Windage area, m^2	87.95
Windage area centre elevation, m	1.89

The model was placed in the towing tank in a beam position to waves. Two ropes were attached to the bow and stern allowing maintaining the desired position in the towing tank. The model was released in a few seconds before the impact of the breaking wave. So, it had freedom to drift immediately before the breaking wave impact and then could capsize or not. There were 70 runs for every model. The probabilistic nature of this test is that the initial phase of the rolling before impact was random. Results of the model tests are summarized in fig. 7.36.

Table 7.11 Stability characteristics of the tested models

Number of variant	GM/B, m	GZ_{max}/B, m	ϕ_{GZmax}, deg	ϕ_V, deg	Weather criterion	$\zeta_{CR}=KG_{cr}/D$	Critical criterion
1	0.110	0.050	30	61	4.285	0.773	ϕ_{GZmax}
2	0.110	0.050	30	61	3.091	0.773	ϕ_{GZmax}
3	0.070	0.056	67	90	2.779	0.985	GM
4	0.079	0.037	30	84	2.830	0.836	GZ_{max}
5	0.079	0.037	30	84	2.258	0.836	GZ_{max}
6	0.119	0.082	30	72	7.402	0.754	ϕ_{GZmax}

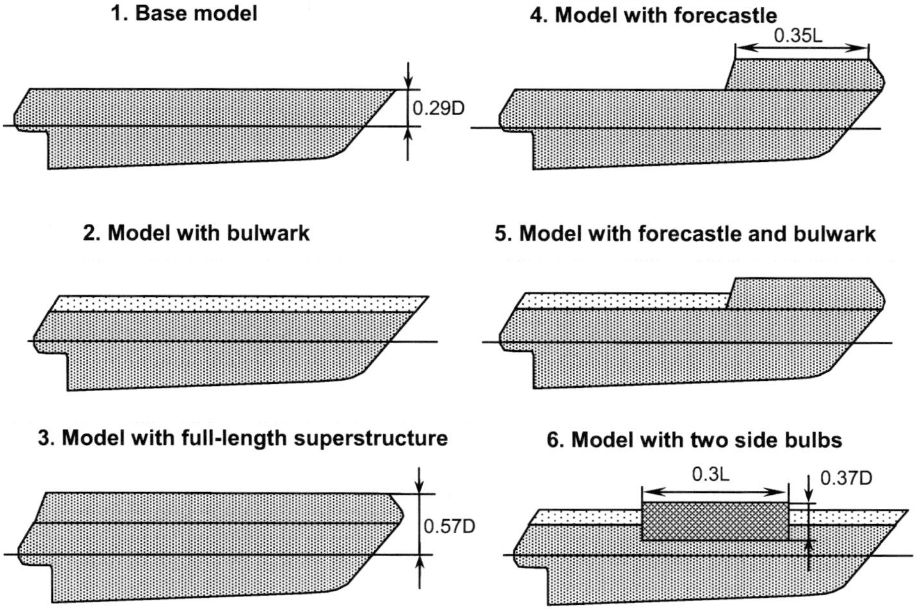

Fig. 7.34 Variants of superstructure type

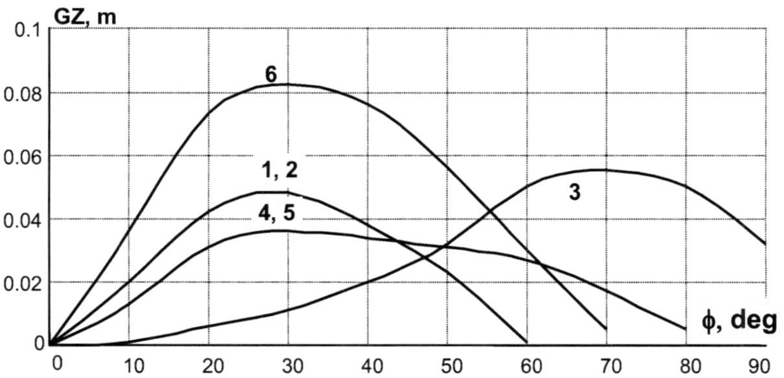

Fig. 7.35 *GZ* curves of the models. Number of curve corresponds to the number of model.

Observations revealed there were three different types of model behavior. The model without structures and bulwark had a significant amount of water on deck, but this water moved with the same speed as that of the breaking wave moves in absence of the model. The model did not track the wave and did not have a significant heel angle.

The bulwark changed this picture dramatically. It kept some quantity of the water on deck and the model practically did not roll and could be capsized by the next wave.

The model with a superstructure had significant heel due to breaking wave action (sometimes up to 45 degrees), but capsizing was very rare. The model with a forecastle was turned 20-49 degrees toward the direction of wave proliferation. Side bulbs decreased stability in breaking waves. One can see that the influence of superstructure may be quite significant for the stability in breaking waves.

Further development of both models of breaking wave action on a vessel is described in [Nechaev, *et al* 2006].

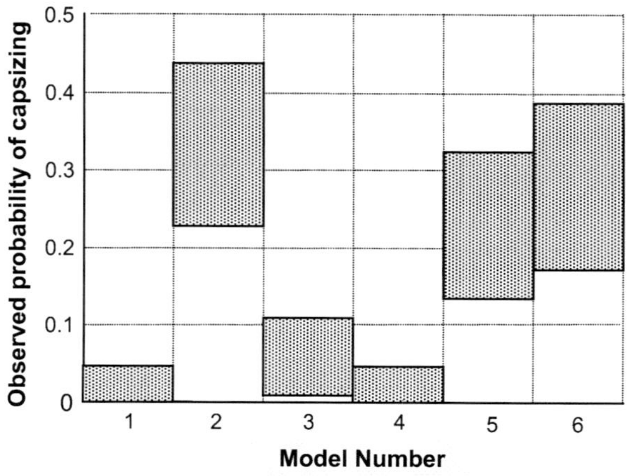

Fig. 7.36 Probability of capsizing with confidence interval

Chapter 8

Nonlinear Roll Motions in Irregular Seas

8.1 Fundamentals of Stochastic Processes

The reader is expected to be familiar with the principal concepts of probability theory including stochastic processes. Nevertheless, we will review and emphasize certain aspects of theory that is critical for our further considerations of probabilistic models of wind, waves, and nonlinear ship motions and capsizing.

8.1.1 General

A stochastic process is a random variable changing in time. So, instead of having one random variable with an infinite number of possible values, we have an infinite number of random variables, each of which corresponds to a moment of time. If we choose one realization of each variable, we get one realization of the whole process. Since there are an infinite number of realizations for each variable, we have an infinite number of realizations for the whole process. The set of all these realizations is usually called the "ensemble of realizations".

If the time is fixed, we get a "usual" random variable with infinite number of realizations, which is called "time section". Stochastic process may be considered as having "double infinity": in time and in number of realizations (in other terms in "probability space"), see fig. 8.1.

This random variable may represent the wind velocity, wave elevation or roll angle of a ship. Despite the different physical nature of these figures, they represent a system with a certain inertia; instant changes are impossible since that would mean an infinite acceleration and infinite inertia force. Therefore, there are always transitions between extremes.

These transitions represent relationships between the current value and the value in another moment. It is evident that this relationship will be weaker with increasing length of time. That means that random variables in time sections of stochastic processes are dependent in a probabilistic sense:

$$f(x_1 \mid x_2) = \frac{f(x_1, x_2)}{f(x_2)} \qquad (8.1)$$

Here, x_1 is a random variable at the section $t = t_1$ and x_2 is a random variable at the section $t = t_2$, symbol $f(...)$ stands for the probability density function that is frequently abbreviated as PDF. Having in mind that, if we have a joint probability density $f(x_1, x_2)$ for two random variables x_1 and x_2, we can always derive the PDF for each of them:

$$f(x_1) = \int_{-\infty}^{\infty} f(x_1, x_2) dx_2 \qquad (8.2)$$

We can continue and introduce the joint PDF for three and even more time sections. Theoretically, a complete probabilistic description of a stochastic process is provided by the joint distribution of all time sections, but we will not need a joint PDF of more than two sections.

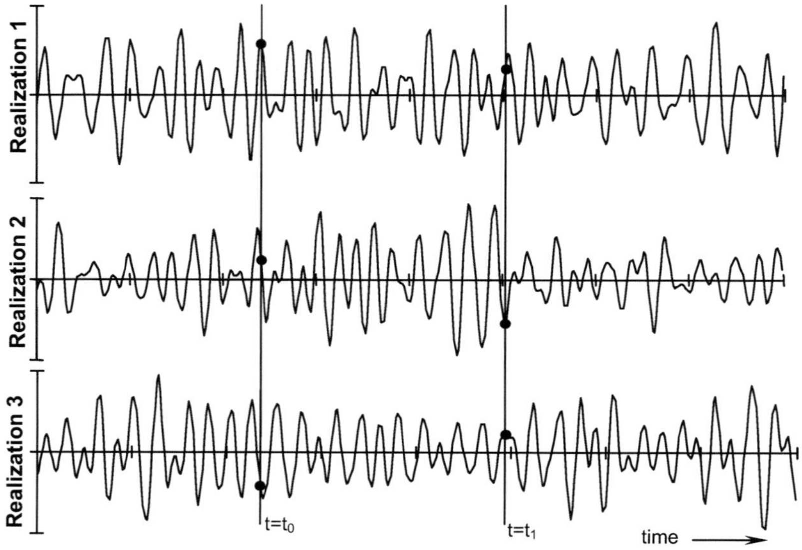

Fig. 8.1 Stochastic process: realizations and time sections

8.1.2 Moments of Stochastic Process. Autocorrelation

Analogously to the case of a random variable, a stochastic process has moments: mean value, variance and so on. (These moments are also called "moments of distribution"; the difference between them and spectral moment will be demonstrated later.)

A stochastic process is a set of random variables each of which correspond to a certain moment of time. If we calculate the mean value for each variable, we receive a set of mean values: each one would correspond to the moment of time. In other words, the mean value of a stochastic process is a function of time. Analogously, the variance and higher moments of distribution are also functions of time.

The same consideration can be applied to PDF. It can be considered as a function of two variables: the current ordinate of the process and time. So, we will be using the following expression :

$$f(x(t)) = f(x,t) \tag{8.3}$$

The mean value and variance of a stochastic process can be expressed correspondingly:

$$m(t) = \int_{-\infty}^{\infty} x(t) f(x,t)dx \tag{8.4}$$

$$V(t) = \int_{-\infty}^{\infty} (x(t) - m(t))^2 f(x,t)dx \tag{8.5}$$

Since every time section is a random variable, we can rewrite the joint distribution of any two of these variables as a function of four arguments:

$$f(x(t_1),x(t_2)) = f(x_1,x_2,t_1,t_2) \tag{8.6}$$

The simplest measures of dependence between two random variables are correlation moment and correlation coefficient. However, if they equal zero, the variables may be still dependent (dependence may be revealed with the correlation moments of higher orders). Vice versa: if the variables are independent, the correlation moment and coefficient will surely equal zero. If two random variables are time sections of a stochastic process, their correlation moment is a function of two arguments:

$$C(t_1,t_2) = \int_{-\infty}^{\infty}\int_{-\infty}^{\infty} (x_1 - m_1(t))(x_2 - m_2(t)) f(x_1,x_2,t_1,t_2)dx_1dx_2 \tag{8.7}$$

This function has a special name: it is called the "Autocorrelation function". It expresses the fact that there are transitions between extremes. It is the simplest indicator of "memory" of a stochastic process; it shows how strong the current state's influence is on the future state. If we fix time t_1, autocorrelation becomes a function of one argument t_2; a sample of such an autocorrelation function is shown in fig. 8.2.

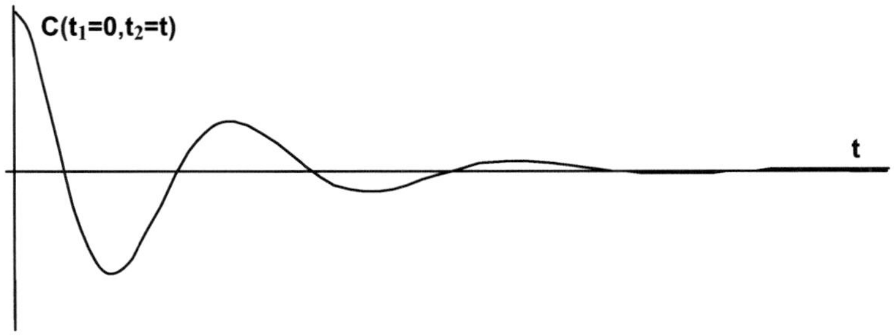

Fig. 8.2 Typical appearance of autocorrelation function

8.1.3 Stationary and Non-stationary Processes

Stochastic processes, for which probabilistic characteristics do not depend on time, are called "stationary". Processes of this type are especially important for practical applications in Naval Architecture: descriptions of irregular waves and gusty wind use an assumption that the processes are stationary.

Stationary processes possess the following qualities, which make their application simpler:

- Their probability distribution does not depend on time. Any section may be used to evaluate PDF:

$$f(x(t_1)) = f(x(t_2)) = f(x) \tag{8.8}$$

- Moments of the distribution do not depend on time either:

$$m = \int_{-\infty}^{\infty} x(t_1) f(x,t_1)dx = \int_{-\infty}^{\infty} x(t_2) f(x,t_2)dx \int_{-\infty}^{\infty} xf(x)dx \tag{8.9}$$

$$\begin{aligned} V &= \int_{-\infty}^{\infty} (x(t_1) - m)^2 f(x,t_1)dx = \int_{-\infty}^{\infty} (x(t_2) - m)^2 f(x,t_2)dx = \\ &= \int_{-\infty}^{\infty} (x - m)^2 f(x)dx \end{aligned} \tag{8.10}$$

- Autocorrelation function does not depend on choice of the moments t_1 and t_2. It depends only on the difference between these two moments of time $\tau = t_2 - t_1$:

$$C(t_1, t_2) = C(\tau) = \int_{-\infty}^{\infty}\int_{-\infty}^{\infty} (x_1 - m)(x_2 - m_2) f(x_1, x_2) dx_1 dx_2 \tag{8.11}$$

Further, we will be dealing with stationary processes only. At the same time, non-stationary processes are also used in practical applications, when a character of changing external conditions can be identified. For example, Dimentberg [1980] considered the problem of stochastic oscillation of an aircraft's front wheel during the take-off. Stochastic oscillations are caused by irregularities of runway surface; the process is non-stationary, because the speed of the aircraft is increasing.

Another possible application of a non-stationary stochastic process can be found for a ship turning in irregular seas. Changing of course causes ship motion processes to be non-stationary.

8.1.4 Ergodicity

Some stationary processes possess an ergodic quality or ergodicity. It makes work simpler than with the general stationary process. Ergodicity means that we can evaluate any probabilistic characteristics using only one realization only if it is long enough. However there is no formal definition what is "long enough".

How are these characteristics defined for one realization? In this case, we have one value at each moment of time, so we just average these figures; the mean value and variance can be defined as:

$$m = \lim_{T\to\infty}\frac{1}{T}\int_0^T x(t)dt$$
$$V = \lim_{T\to\infty}\frac{1}{T}\int_0^T (x(t)-m)^2 dt \qquad (8.12)$$

Formulae (8.12) and (8.13) give only a formal definition for the mean value and variance, which is why we carry out integration to infinite limits. In practice, our data are always a finite number of values obtained within a range from zero to *T*. We can still use these formulae, but the results obtained are only estimates of the true values. ::

$$\langle m\rangle = \int_0^T x(t)dt$$
$$\langle V\rangle = \int_0^T (x(t)-m)^2 dt \qquad (8.13)$$

The autocorrelation function is formally defined as:

$$C(\tau) = \lim_{T\to\infty}\left[\frac{1}{T-\tau}\int_0^{T-\tau}(x(t)-m)(x(t+\tau)-m)\,dt\right] \qquad (8.14)$$

An estimate of the autocorrelation function

$$\langle C(\tau)\rangle = \frac{1}{T-\tau}\int_0^{T-\tau}(x(t)-m)(x(t+\tau)-m)\,dt \qquad (8.15)$$

Evaluation of the estimate of the autocorrelation function using one realization only encounters certain difficulties as amount of available statistical data decreases with increase of the argument τ. As a result, the initial part of the estimate of autocorrelation function has better accuracy than the rest of the estimate. However, this may be enough for practical purposes, because this initial part contains the most important information about "memory" or "inertia" of the process. Therefore, a numerical method is needed to find out where the calculations have to stop or where the estimate of autocorrelation function has to be cut-off. The alternative is to mitigate the error by statistical weighting [Belenky 2004].

More details will follow in subchapter 8.6.1, where ergodicity of nonlinear ship roll is addressed.

8.1.5 Spectrum and Autocorrelation Function

Spectrum represents distribution of energy over frequencies. To obtain the spectrum, we presented a stochastic process in a form of a series with trigonometric functions, known as a Fourier series:

$$x(t)=\sum_{i=1}^{n} c_i \cos(\omega_i t+\varepsilon_i)=\sum_{i=1}^{n} a_i \sin\omega_i t + b_i \cos\omega_i t \qquad (8.16)$$

Coefficients a_i and b_i are related with amplitude c_i and phase ε_i with the following evident formulae:

$$a_i=-c_i \sin\varepsilon_i\,;\quad b_i=c_i\cos\varepsilon_i\,;\quad c_i=\sqrt{a_i^2+b_i^2}\,;\quad \varepsilon=-\arctan\frac{a_i}{b_i} \qquad (8.17)$$

As it is well known, coefficients a_i and b_i can be calculated with the Fourier transform formulae:

$$a_i=\frac{2}{T}\int_0^T x(t)\sin(\omega_i t)dt;\quad b_i=\frac{2}{T}\int_0^T x(t)\cos(\omega_i t)dt \qquad (8.18)$$

Provided that:

$$\omega_i=\frac{\pi i}{T} \qquad (8.19)$$

Our presentation of stochastic processes in a form Fourier series (8.16) also is called the "inverse Fourier transform" in contrast with the direct Fourier transform (8.18), when we calculate coefficients with a given realization of the process.

As can be seen from (8.18), only one realization was used to obtain coefficients of the Fourier series. It means that the process $x(t)$ is assumed to be ergodic. Therefore, the inverse Fourier transform also yields an ergodic process.

However, it does not mean that the spectrum could be used only for ergodic processes. There is a general relationship between the autocorrelation function and spectrum (more precisely – spectral density) defined by direct and inverse cosine Fourier transforms:

$$S(\omega)=\frac{1}{2\pi}\int_0^\infty C(\tau)\exp(-i\omega\tau)d\tau \qquad (8.20)$$

$$C(\tau)=\int_0^\infty S(\omega)\exp(i\omega\tau)d\omega \qquad (8.21)$$

Here: $i=\sqrt{-1}$.

Formula (8.21) reveals a relationship between variance and spectrum. Variance has a clear and simple relationship with the autocorrelation function:

$$V=C(0) \qquad (8.22)$$

Therefore, the variance can be found as an area under the spectrum directly from (8.21); it is also called the spectral moment of zero order:

$$V=C(0)=\int_0^\infty S(\omega)d\omega \qquad (8.23)$$

Its physical meaning (for mechanical stochastic processes) is averaged total energy (see also subchapter 9.6 [Kobylinski and Kastner 2003]). Some probabilistic qualities of spectral representation will be revisited in the subchapter 8.2.3 while some numerical aspects are considered in the subchapter 8.3.6.

8.1.6 Envelope of Stochastic Process

If a stochastic process is stationary and has a Gaussian distribution, it can be presented (following [Sveshnikov, 1968] in this subchapter) as:

$$x(t) = A(t)\cos(\Phi(t)) \tag{8.24}$$

Amplitude $A(t)$ and phase $\Phi(t)$ are stochastic processes. Why is the presentation (8.24) simpler than (8.16), if we express one stochastic process via two?

First, quite often we are interested in the peak value of the process and the presentation (8.24) expresses process of extreme values explicitly. Secondly, trigonometric form (8.24) makes it easy to use as an excitation for the method of multiple scales and other asymptotic methods. We have considered this method for regular roll in subchapter 4.2.4 and will apply it for irregular nonlinear roll in subchapter 8.3.5.

The method is called an "envelope", because the amplitude process "envelopes" the original stochastic process, see fig. 8.3.

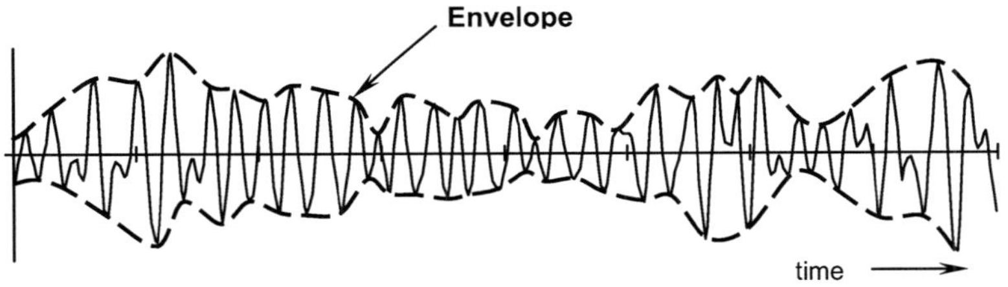

Fig. 8.3 Envelope of stochastic process

Further we assume that $x(t)$ has a zero average and normal distribution. The last assumption imposes certain limits on the application of the envelope method for nonlinear irregular roll (since it may not be Gaussian, as we will see in subchapter 8.4). However, this limitation does not really affect our presentation of irregular waves.

Let us first consider distribution of both amplitude and phase processes. To facilitate it, we introduce another process:

$$y(t) = A(t)\sin(\Phi(t)) \tag{8.25}$$

We can consider the processes $x(t)$ and $y(t)$ as projections of a stochastic vector on perpendicular axes. We consider $y(t)$ as a stationary Gaussian stochastic process, which is independent of $x(t)$ at the same instance of time. (They are still dependent if we consider different moments of time.) We also assume that both processes have an identical auto-correlation function.

Let us consider $x(t)$ and $y(t)$ as co-ordinates and find a probability that a point will be within a small ring stretched from radius a to $a+da$:

$$P(A \in [a,a+da]) = \iint\limits_{a \le \sqrt{x^2+y^2} \le a+da} f(x,y)dx,dy \qquad (8.26)$$

Since we assumed that the processes are Gaussian and independent:

$$f(x,y) = \frac{1}{2\pi V}\exp\left(-\frac{x^2+y^2}{2V}\right) \qquad (8.27)$$

Here, $V = V_x = V_y$ is a variance: since we assumed that both processes have the same auto-correlation functions, indeed their variances are the same. Let us rewrite (8.26) in polar co-ordinates (a, φ) which would correspond to amplitude $A(t)$ and phase $\Phi(t)$ in envelope presentation (8.24):

$$P(A \in [a,a+da]) = f(a)da = \int\limits_a^{a+da}\int\limits_0^{2\pi}\frac{a}{2\pi V}\exp\left(-\frac{a^2}{2V}\right)dad\varphi \qquad (8.28)$$

The integrand in the formula (8.28) has a meaning of joint probability function:

$$f(a,\varphi) = \frac{a}{2\pi V}\exp\left(-\frac{a^2}{2V}\right) \qquad (8.29)$$

There is no variable φ in the right hand side of (8.29), which means that amplitude $A(t)$ does not depend on phase $\Phi(t)$. It also means that phase $\Phi(t)$ has a uniform distribution from 0 to 2π:

$$f(\Phi) = \begin{cases} \frac{1}{2\pi} & \Phi \in [0;2\pi] \\ 0 & \Phi \notin [0;2\pi] \end{cases} \qquad (8.30)$$

Then, the probability density of amplitudes can be expressed as follows:

$$f(a) = \int\limits_0^{2\pi}\frac{a}{2\pi V}\exp\left(-\frac{a^2}{2V}\right)d\varphi = \frac{a}{V}\exp\left(-\frac{a^2}{2V}\right) \qquad (8.31)$$

Formula (8.31) is known as a Rayleigh distribution. So any stationary Gaussian process can be presented in the envelope form with the uniformly distributed phase and amplitude having a Raleigh distribution.

The above presentation yields even more exact analytical expressions for probabilistic figures of amplitude and phase. The derivations are rather bulky, so we provide final formulae. Since amplitude and phase actually are stochastic processes themselves, auto-correlation functions of amplitudes and phase contain important information:

$$C_A(t) = V\left(2\mathbf{E}(1-p^2) - p^2\mathbf{K}(1-p^2) - \frac{\pi}{2}\right) \qquad (8.32)$$

The formula for the auto-correlation of $\cos\Phi(t)$ is more convenient and concise than that one for the phase itself:

$$C_{\cos\Phi}(t)=\frac{C(t)}{2\left(1-p^2\right)}\left(2\mathbf{E}\left(1-p^2\right)-p^2\mathbf{K}\left(1-p^2\right)\right) \tag{8.33}$$

Here V and $C(t)$ are the variance and autocorrelation function of the original process $x(t)$, **E** and **K** are elliptic integrals of the first and the second kind:

$$\mathbf{E}(x)=\int_0^{\pi/2}\frac{dz}{\sqrt{1-x\cdot\sin^2 z}};\quad \mathbf{K}(x)=\int_0^{\pi/2}\sqrt{1-x\cdot\sin^2 z}\,dz \tag{8.34}$$

Parameter p is expressed through the autocorrelation coefficient and the joint correlation coefficient between $x(t)$ and $y(t)$:

$$p^2=1-c^2(t)-r^2(t) \tag{8.35}$$

Where: $c(t)$ is the auto-correlation coefficient $r(t)$ is joint correlation coefficient:

$$c(t)=\frac{C_x(t)}{V};\quad r(t)=\frac{R_{xy}(t)}{V} \tag{8.36}$$

As we mentioned above, the joint correlation function of processes $x(t)$ and $y(t)$ equals zero if we consider ordinates of both processes at the same time (since they are independent). However, it is non-zero if we take the processes at different time moments. It can be derived using a spectral presentation of both processes. Here we included only the final result:

$$R_{xy}(t)=2\int_0^{\infty}S_x(\omega)\sin\omega t\,d\omega \tag{8.37}$$

The joint distributions of derivatives of amplitude and phase are other important analytical results available for envelope presentation:

$$f(A,\dot{A},\Phi,\dot{\Phi})=\frac{A^2}{4\pi^2(m_1^2-m_2^2)V^2}\exp\left(-\frac{\dot{A}^2+A^2\left(\dot{\Phi}^2-2m_1\dot{\Phi}+m_2^2\right)}{2V\left(m_2^2-m_1^2\right)}\right) \tag{8.38}$$

Here, m_1 and m_2 are non-dimensional spectral moments, expressed via derivatives of autocorrelation $c(t)$ and joint correlation $r(t)$ functions:

$$m_2^2=-\ddot{c}(t=0)=-\frac{1}{V}\frac{d^2}{dt^2}\left(\int_0^{\infty}e^{-i\omega t}S(\omega)d\omega\right)=\frac{1}{V}\int_0^{\infty}\omega^2S(\omega)d\omega \tag{8.39}$$

$$m_1=\dot{r}(t=0)=\frac{1}{V}\frac{d}{dt}\left(\int_0^{\infty}S(\omega)\sin\omega t d\omega\right)=\frac{1}{V}\int_0^{\infty}\omega S(\omega)d\omega \tag{8.40}$$

Any combination of one- or two-dimensional distributions of the amplitude, phase and their derivatives can be deduced from (8.38). Further, we will need particularly, the distribution of derivative of phase and amplitude separately:

$$f(\dot{\Phi})=\frac{(m_1^2-m_2^2)}{2\sqrt{\left((\dot{\Phi}-m_1)^2+(m_1^2-m_2^2)\right)^3}} \tag{8.41}$$

$$f(\dot{A}) = \frac{1}{\sqrt{2\pi V\left(m_2^2 - m_1^2\right)}} \exp\left(-\frac{\dot{A}^2}{2V\left(m_2^2 - m_1^2\right)}\right) \tag{8.42}$$

If the spectrum has a clearly pronounced maximum and is narrow, (definition of a narrow spectrum, see subchapter 9.7 of [Kobylinski and Kastner 2003]) the envelope presentation (8.25) can be further simplified. Consider the difference between squares of ω_1 and ω_2 in the details. Having in mind that ω_1 does not depend on frequency and:

$$\frac{1}{V}\int_0^{\infty} S(\omega)d\omega = 1$$

The above difference can be presented as:

$$\begin{aligned} m_2^2 - m_1^2 &= \frac{1}{V}\int_0^{\infty}\omega^2 S(\omega)d\omega - m_1^2 = \frac{1}{V}\int_0^{\infty}\omega^2 S(\omega)d\omega - 2m_1^2 + m_1^2 = \\ &= \frac{1}{V}\int_0^{\infty}\omega^2 S(\omega)d\omega - \frac{2}{V} m_1 \int_0^{\infty}\omega S(\omega)d\omega + \frac{1}{V}\int_0^{\infty} m_1^2 S(\omega)d\omega = \\ &= \frac{1}{V}\int_0^{\infty}\left(\omega^2 - 2\omega m_1 + \omega_1^2\right)S(\omega)d\omega = \frac{1}{V}\int_0^{\infty}(\omega - m_1)^2 S(\omega)d\omega \end{aligned} \tag{8.43}$$

Again, in terms of moments, m_1 is the first initial spectral moment, m_2^2 is the second initial moment, and value $\Delta^2 = m_2^2 - m_1^2$ is the second central moment. It corresponds to the variance for a random value, if the spectrum is considered analogous to a distribution density.

Now, if the spectrum is narrow, the second central spectral moment has to be small, analogously as a random value with a small variance has a narrow probability density function. This means:

$$\Delta^2 << m_1^2 \tag{8.44}$$

Looking at the distribution of envelope derivative (8.42) we see that it is, in fact, a Gaussian law:

$$f(\dot{A}) = \frac{1}{\sqrt{2\pi V\Delta^2}} \exp\left(-\frac{\dot{A}^2}{2V\Delta^2}\right) \tag{8.45}$$

With a zero average and variance that is equal to $V\Delta^2$. This distribution becomes more pointed with decreasing Δ and hence a narrowing of the spectrum. This means, large deviations of $\dot{A}$ from zero have a small probability. In other words, amplitude A is a slowly changing value.

Consider the distribution of the derivative of the phase, rewriting (8.42):

$$f(\dot{\Phi}) = \frac{\Delta}{2\sqrt{\left((\dot{\Phi} - m_1)^2 + \Delta^2\right)^3}} \tag{8.46}$$

It is evident from (8.46) that the phase derivative distribution has a peak when $\dot{\Phi}$ is close to ω_1 and this peak becomes sharper with decreasing Δ. Then is makes sense to present a phase as:

$$\Phi(t) = m_1 t + \varphi(t) = \omega_a t + \varphi(t) \tag{8.47}$$

In fact, we consider now not the phase itself, but its deviation from one of pure sinusoidal motion. It is clear that:

$$f(\dot{\varphi}) = \frac{\Delta}{2\sqrt{\left(\dot{\varphi}^2 + \Delta^2\right)^3}} \tag{8.48}$$

Finally, in the case of a narrow spectrum or narrow banded process, we can present it as:

$$x(t) = A(t)\cos(\omega_a t + \varphi(t)) \tag{8.49}$$

Where amplitude $A(t)$ is a slowly changing value and ω_a is the spectrum averaged frequency.

8.2 Probabilistic Models of Wind and Waves

8.2.1 Gusty Wind

We define wind gust as a short time variation of wind velocity (see also subchapter 9.2 of the [Kobylinski and Kastner 2003]). Therefore, it is convenient to present instantaneous horizontal wind velocity as a sum of averaged wind velocity and gust fluctuation:

$$u(t) = u_A + u_G(t) \tag{8.50}$$

Mean wind velocity is to be obtained over several hours. We could assume that the wind velocity is a stationary process and the weather does not experience significant changes.

Since there is spectrum available ([Davenport, 1964], also subchapter 9.2 of [Kobylinski and Kastner 2003]), we can present the fluctuating component of the wind velocity in the form of a Fourier series:

$$u_G(t) = \sum_{i=1}^{n} c_i \cos(\omega_i t + \varepsilon_i) \tag{8.51}$$

Here, amplitudes c_i are defined from the spectrum exactly in the same way it was described in subchapter 8.1.5. Phases ε_i are random numbers, which have uniform distribution from 0 to 2π. A sample of a time history is shown in fig 8.4.

The application of formula (8.51) implies that wind velocities have a Gaussian distribution. Actually, this hypothesis works well, see [Lugovsky, 1976].

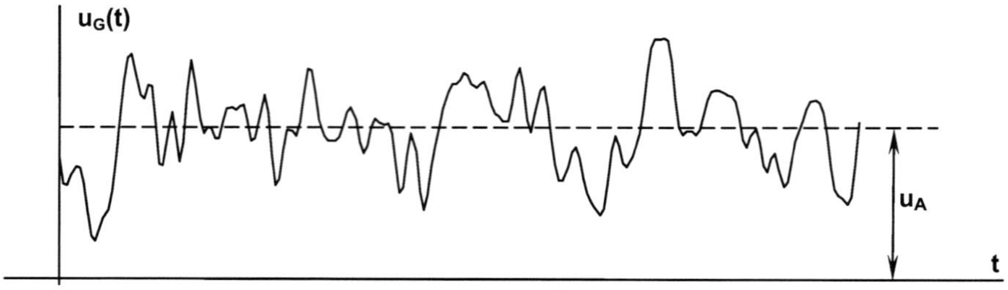

Fig. 8.4 Sample of time history of gusty wind

The next question to be addressed is probabilistic qualities of gust velocity extremes. It is convenient to introduce a non-dimensional measure of the maximum wind velocity in a gust that is called a "gust coefficient":

$$C = \frac{u_{G\max}}{u_a} \tag{8.52}$$

Here, $u_{G\max}$ is maximal wind gust velocity. The spectrum of wind gust velocities is not narrow-banded as the spectrum for waves (see subchapter 9.7 of [Kobylinski and Kastner 2003], where the concept of spectral breadth parameter ε is described). Lugovsky [1976] reported the value of the parameter $\varepsilon = 0.89$ based on observation of wind on the Caspian Sea. A little wider spectrum was evaluated above the Antarctic with the value of about 0.9.

Since the spectrum of gust velocities is wide, the gust coefficient has Rice distribution:

$$f(z) = \frac{1}{\sqrt{2\pi}}\left[\varepsilon \exp\left(-\frac{z^2}{2\varepsilon^2}\right) - z\sqrt{1-\varepsilon^2}\exp\left(-\frac{z^2}{2}\right)\int_{-\infty}^{z\sqrt{1-\varepsilon^2}/\varepsilon} \exp\left(-\frac{x^2}{2}\right)dx\right] \tag{8.53}$$

With: $z = (C-1)/\sqrt{V_C}$ and spectral breadth $\varepsilon = \sqrt{\left(m_0 m_4 - m_2^2\right)/m_0 m_4}$ (m_n is spectral moment of order n), V_C is the variance of the gust coefficient.

Lugovsky [1976] offers two values for the variance of the gust coefficient:

- Caspian Sea: $V_C = 0.0167$
- Antarctica: $V_C = 0.0206$

The mean value of the gust coefficient can be derived from equation (8.53) [Lugovsky, 1976]:

$$m_C = 1 + 0.55\sqrt{V_C} \tag{8.54}$$

Yamagata [1959] recommended a mean value of the gust coefficient of 1.23.

8.2.2 Squalls[1]

We have used the above spectral presentation of the stochastic process of horizontal components of wind velocity. Such a presentation takes into account wind gusts that are described as the fluctuating component of wind velocity. However, this model does not take into account squalls. Zinkovsky-Gorbatenko [1965] proposed that wind gusts and squalls are to be distinguished by duration. Wind gusts are considered as stationary stochastic fluctuations. Squalls are defined as single, rare wind speed increases with increasing wind velocity growing to some maximum value and with a long velocity decrease of random duration, see fig. 8.5.

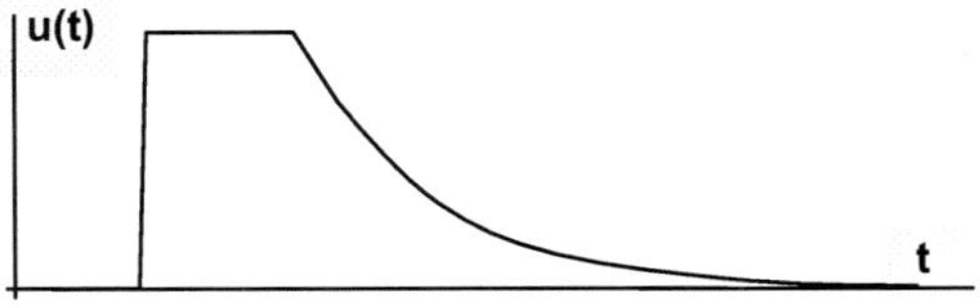

Fig. 8.5 Scheme of squall

Such rare (or even single) non-stationary events could be the main reason of ship capsizing or dangerous heeling. However, these squalls cannot be reflected in the wind speed spectrum, even if they occur during wind speed recording. This can be explained by the fact that the time duration of the squall is too small in comparison with the whole recording period and so the statistical characteristics of these additional non-stationary singular events will equal zero.

This can easily be demonstrated by the following consideration. We take two stochastic wind velocity processes: the first is the stationary presented by formula (8.50):

$$u(t) = u_A + u_G(t)$$

The second process is the same as the previous one in all moments of time excluding the time range from t_1 to t_2, where the squall wind velocity $u_S(t)$ is added, see fig. 8.6.

$$u_1(t) = u_A + u_G(t) + u_S(t) \tag{8.55}$$

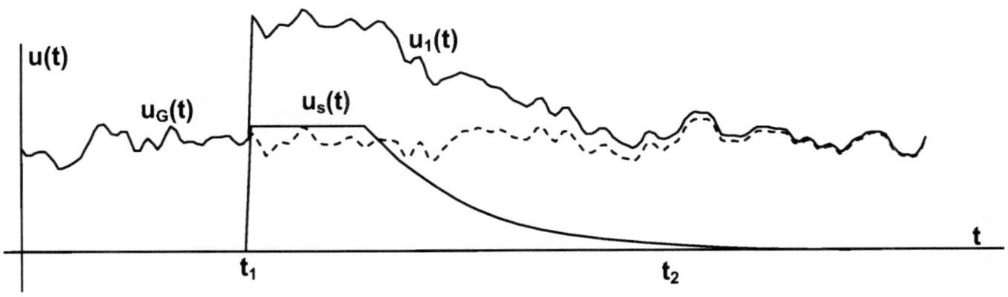

Fig. 8.6 Sample of time history of gusty wind with squall

The limit equals zero because the integral $\int_{t_1}^{t_2} u_S(t)\,dt$ is a finite quantity and $T \to \infty$. A similar result can be obtained for the variance:

$$m\{(u(t) - m[u(t)])^2\} = m[u_G^2(t)] = V \tag{8.56}$$

---

[1] Unpublished paper of N. B. Sevastianov was used in this subchapter

$$m\left[(u_1(t)-m[u_1(t)])^2\right]=\lim_{T\to\infty}\frac{1}{T}\left[\int_0^T(u_A+u_G(t)+u_S(t)-u_A)^2\,dt\right]=$$
$$=V+\lim_{T\to\infty}\frac{1}{T}\left[2\int_{t_1}^{t_2}u_S(t)\cdot u_G(t)\,dt+\int_{t_1}^{t_2}u_S^2(t)\,dt\right]=V \qquad (8.57)$$

It means that a single squall with the finite value of the integral in time range from t_1 to t_2 cannot be detected by spectral analysis. Taking into account that a single squall may be quite dangerous for the ship, we should supplement ship stability assessment under the action of a stationary stochastic process (irregular waves plus gusty wind) with a probability assessment of capsizing due to action of single non-stationary squalls.

8.2.3 Spectral Model of Irregular Waves

As we mentioned in the subchapter 8.1.5, Fourier series or inverse Fourier transform can be used to model a process with a given spectrum and the result is stationary and ergodic. Let us repeat formula (8.16) here:

$$\zeta_W(t)=\sum_{i=1}^{n}c_i\cos(\omega_i t+\varepsilon_i) \qquad (8.58)$$

The wave elevation is a sum of many random components (since phase ε is a random number, components are random numbers too). Usually most of the contribution comes from the frequencies around the peak of the spectrum and usually there are enough of them, so we can say that we consider the sum of a large number of approximately equal random components. The Central Limit Theorem establishes that a sum of equally contributing components tends to a normal distribution with an increase of the number of components; the distribution of indicial components does not affect the result. Therefore, the distribution of a resulting wave elevation is Gaussian. Since sea waves are known to be Gaussian, this is considered an adequate model as far as distribution is concerned.

Application of the inverse Fourier transform, however, is not limited by normally distributed processes.

8.2.4 Method of Envelope

We have seen in subchapter 8.1.6 that a stochastic process can be presented with a single trigonometric function, if both amplitude and phase are considered as random quantities. If the assumption of a narrow spectrum is acceptable, the presentation is significantly simplified. Its amplitude is a slowly changing figure and it is possible to define a dominant frequency. However, phase shift stays random: see formula (8.49):

$$\zeta_W(t)=A(t)\cos(\omega_a t+\varphi(t)) \qquad (8.59)$$

The spectrum of seaways can be considered narrow and presented as an envelope [Francescutto, 1991]. This assumption seems to be adequate also because a dynamical system has a limited band of response and thus works as a filter.

The most common way to introduce spectrum into the envelope presentation is by application of a linear filter, described by the following differential equation:

$$\ddot{\zeta}_W + \gamma\dot{\zeta}_W + \omega_a^2\zeta_W = \omega_a\sqrt{\gamma}W(t) \tag{8.60}$$

Here, γ is a "damping" parameter, defining the sharpness of the spectrum. $W(t)$ is a stochastic process that does not have memory at all. It is called a "white noise" or Wiener process. Its current ordinate does not depend on any other, so its autocorrelation function is a delta-function. It is usually modelled with a uniform spectrum in the given frequency range $[\omega_1;\omega_2]$, which is also shown in fig. 8.7:

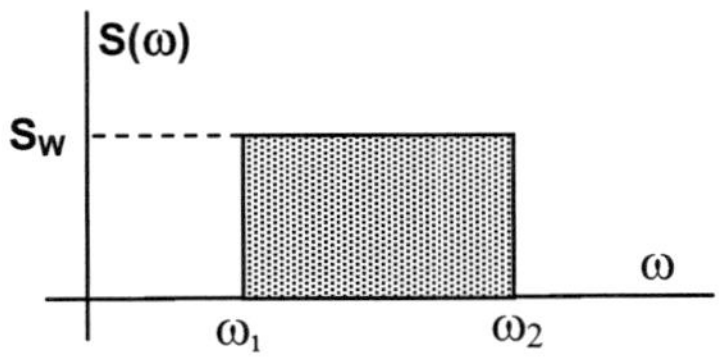

Fig. 8.7 Spectrum of a "white noise" or Wiener process

$$S(\omega) = \begin{cases} 0 & \omega \notin [\omega_1;\omega_2] \\ S_W & \omega \in [\omega_1;\omega_2] \end{cases} \tag{8.61}$$

The spectrum of the response (which is wave elevation) is evident:

$$S_\zeta(\omega) = \frac{\gamma\omega_a^2 S_W}{(\omega_a^2 - \omega^2)^2 + \omega^2\gamma^2} \tag{8.62}$$

The parameters of the above spectrum (8.62): S_W, γ and ω_a have to be fitted with some of the known spectra. ITTC and JONSWAP spectra fitting is available from [Francescutto and Nabergoj, 1990], see table 8.1.

Table 8.1 Envelope Presentation Parameters

Spectrum	White noise spectral level	Frequency	Damping factor
ITTC	0.35	1.01	0.50
JONSWAP (sharpness magnification factor 7)	1.3	1.16	0.14

Stratonovich [1963] proposed an alternative form that is more convenient to present excitation while applying perturbation methods:

$$\zeta_W(t) = A_{\zeta C}(t)\cos\omega_a t + A_{\zeta S}(t)\sin\omega_a t \tag{8.63}$$

Two amplitudes $A_{\zeta C}(t)$ and $A_{\zeta S}(t)$ are slowly varying independent Gaussian stochastic processes that are defined by the following differential equations:

$$\begin{cases} \dot{A}_{\zeta C} + \frac{\gamma}{2}A_{\zeta C} = \sqrt{\frac{\gamma}{2}}W_C(t) \\ \dot{A}_{\zeta S} + \frac{\gamma}{2}A_{\zeta S} = \sqrt{\frac{\gamma}{2}}W_S(t) \end{cases} \tag{8.64}$$

More details are available from Stratonovich [1963].

8.2.5 Autoregression Model[1]

The Fourier series and method of envelope are not the only ways irregular waves may be presented. A different method, the auto-regression model, should be mentioned. This model is convenient for simulation; its parameters can be easily identified from measurement, which makes it important for application in intelligent on-board safety systems. Such systems use simulation to deliver safety judgment, see [Nechaev and Degtyarev, 2000], [Nechaev, *et al* 2001].

The main idea of the autoregression model is presenting a stochastic process as a series of consequent ordinates, each of which depends on the previous one. This dependence is meant to be dying with the increasing of the distance in time. The auto-regression model therefore directly simulates the most important quality of a stochastic process, which is "memory".

We already discussed the Wiener process, where the current value does not depend on the time history. If the current value depends on the previous one only, the process is called a Markov process:

$$\zeta(t_i) = H \cdot \zeta(t_{i-1}) + W(t_i) \tag{8.65}$$

Here, $\zeta(t)$ is the value of the stochastic process; H is a coefficient, which takes into account the time history and $W(t)$ is realization of the white noise at time moment t. We will return to the Markov process later in subchapter 8.6.4.

The auto-regression model takes into account dependence between a time section further back in time:

$$\zeta(t_i) = \sum_{j=1}^{\infty} H_j \zeta(t_{i-j}) + W(t_i) \tag{8.66}$$

Coefficients H_j are parameters of autoregression, they show the dependence between the value at the present moment of time and time history. White noise or Weiner process, $W(t)$ is assumed to be a zero mean value.

The events that have taken place long ago cannot influence values now, so we can reduce the number of terms of series (8.65) up to a certain quantity N:

$$\zeta(t_i) = \sum_{j=1}^{N} H_j \zeta(t_{i-j}) + W(t_i) \tag{8.67}$$

Formula (8.67) has a very clear physical meaning: the term $\sum_{j=1}^{N} H_j \zeta(t_{i-j})$ expresses the inertial character of the sea surface while white noise introduces some stochastic contribution and simulates reaction for wind and other disturbances.

Coefficients H_j can be determined using the autocorrelation function [Gurgenidze and Trapeznikov, 1988]. For a sample of the application of this model, see [Belenky, *et al*, 1995, 1998].

[1] Written in co-authorship with Prof. Alexander B. Degtyarev and Dr. Alexander V. Boukhanovsky of Institute for High Performance Computing and Information Systems (Russia)

One of the most important advantages of the autoregression model is its ability to represent 3-D waves – stochastic surface, keeping both time and spatial autocorrelation functions. Moreover, this stochastic field may be both non-stationary and non-homogeneous. More information is available from [Boukhanovsky, *et al*, 1998a, 2001].

8.2.6 Non-Canonical Presentation[1]

A non-canonical presentation of a stochastic process was proposed by Cherentsky [1968]. The idea of the method is to substitute a stochastic process with a finite nonlinear combination of deterministic functions of random arguments and time. The number of terms of the non-canonical model has to be chosen to equalize certain statistical characteristics of the model and the process. This is the main difference between non-canonical and canonical models of stochastic processes. Canonical models usually use an infinite number of deterministic functions: presentation with Fourier series is one of the typical examples of canonical models.

Stationary 3-D sea waves can be presented in the following form [Belenky, *et al*, 1997]:

$$\zeta(t) = \Lambda_1 \cos\omega t + \Lambda_2 \sin\omega t \qquad (8.68)$$

Random variables Λ_1 and Λ_2 have Gaussian distributions and their mean values are equal to zero. Their variances are determined from the spectrum:

$$V[\Lambda_{1,2}] = \int_{-\infty}^{\infty}\int_{-\infty}^{\infty} S_\zeta(u,v)\,du\,dv \qquad (8.69)$$

Here, $S_\zeta(u,v)$ is a 2D spectrum of waves and u and v are wave numbers in both directions. Random frequency, ω depends on two wave numbers u and v, which are defined from their joint distribution $f(u,v)$:

$$\omega = \sqrt[4]{g^2(u^2+v^2)}, \quad f(u,v) = \frac{S_\zeta(u,v)}{V[\Lambda_{1,2}]} \qquad (8.70)$$

Formula (8.68) completely describes a stochastic process (up to the moments of the second order –moments higher than the variance cannot be reproduced) along with its derivative and primitive. This is especially important for simulation of dynamical systems that are described by differential equations.

Even moments of distribution of non-canonical presentation (8.68)-can be expressed as:

$$M[\zeta_W^k(t)] = (k-1)!!\sigma_\zeta^k; \quad \sigma_\zeta = \sqrt{V_\zeta} \qquad (8.71)$$

This shows that the process presented by (8.68) has a Gaussian distribution. Unfortunately, it does not hold for joint distributions of several time sections. This imposes certain limits on using this presentation. More details are available from [Degtyarev, 1994; Belenky, *et al*, 1998].

However, this is the only model available that allows simulating time section without calculating time history. This is a very important advantage for the study of non-ergodic systems, where we cannot limit ourselves with only one realization but have to consider the whole ensemble.

---

[1] Written in co-authorship with Prof. Alexander B. Degtyarev and Dr. Alexander V. Boukhanovsky of Institute for High Performance Computing and Information Systems (Russia)

8.3 Irregular Roll in Beam Seas

This subchapter is focused on how a ship, being a dynamical system, transforms irregular waves into roll motion. Most of our attention is paid to nonlinearity and how does it affect probabilistic characteristics of roll.

8.3.1 Linear System. Weiner–Khinchin Theorem

We start from the simplest linear differential equation of roll motion. Irregular waves are presented in the from of a Fourier series:

$$\ddot{\phi} + 2\delta_\phi \dot{\phi} + \omega_\phi^2 \phi = \sum_{i=1}^{N} \alpha_{Ei} \sin(\omega_i t + \varepsilon_i) \qquad (8.72)$$

Here, α_{Ei} is the amplitude of the i-th component of a Fourier presentation of wave excitation. If we consider Froude-Krylov forces only (see subchapter 3.5), it can be expressed as:

$$\alpha_{Ei} = \omega_\phi^2 \alpha_{Ami} = \omega_\phi^2 \kappa_\phi(\omega_i) \cdot \alpha_{Awi} = \omega_\phi^2 \kappa_\phi(\omega_i) \cdot k_i \zeta_{Ai} = \omega_\phi^2 \kappa_\phi(\omega_i) \cdot \frac{\omega_i^2}{g} \zeta_{Awi} \qquad (8.73)$$

Here, α_{Ami} is effective angle of wave slope (reduced to take into account finite size of a ship), α_{Awi} is angle of wave slope for i-th component, $\kappa(\omega_i)$ is a coefficient for reduction of roll excitation caused by a finite sized ship in comparison with the wave estimated for the given frequency ω_i [Blagoveshchensky, 1962], it equals 1 for the long waves; k_i is the wave number for the same frequency and ζ_{Awi} is amplitude of the i-th component calculated from the spectrum.

Phase shift ε_i is a random variable distributed uniformly from 0 to 2π.

For the solution of the differential equation (8.72), the fact that excitation is a stochastic process does not play any role. Since the equation is linear, its solution consists of a general solution of the homogeneous equation that represents the transition process and a partial solution of the heterogeneous equation, which represents the steady state solution after the transition would die out:

$$\phi(t) = \phi_{A0} e^{-\delta t} \sin(\omega_{A0} t + \varepsilon_A) + \sum_{i=1}^{N} \phi_{Ai} \sin(\omega_i t + \beta_i + \varepsilon_i) \qquad (8.74)$$

Amplitude ϕ_{A0} and phase ε_A are arbitrary constants and depend on initial conditions. Frequency of the free damped oscillations ω_{A0} is related with eigenvalues. All these figures are deterministic and do not differ from the case for regular beam waves (see Chapter 4).

Since equation (8.72) is linear, its partial solution can be searched in the same form that excitation is presented. That is why we use Fourier series as a partial solution of the heterogeneous equation.

The excitation itself is a sum of sine functions with individual amplitude, frequency and phase. We demonstrate briefly, how a linear system "treats" such excitation. Assuming the solution in a form Fourier series we find both derivatives:

$$\phi(t) = \sum_{i=1}^{N} \phi_{Ai} \sin(\omega_i t + \beta_i + \varphi_{0i}) \tag{8.75}$$

$$\dot{\phi}(t) = \sum_{i=1}^{N} \phi_{Ai} \omega_i \cos(\omega_i t + \beta_i + \varphi_{0i}) \tag{8.76}$$

$$\ddot{\phi}(t) = -\sum_{i=1}^{N} \phi_{Ai} \omega_i^2 \sin(\omega_i t + \beta_i + \varphi_{0i}) \tag{8.77}$$

Let us substitute (8.75)-(8.77) into the equation (8.72):

$$\begin{gathered} -\sum_{i=1}^{N} \phi_{Ai} \omega_i^2 \sin(\omega_i t + \beta_i + \varphi_{0i}) + 2\delta_\phi \sum_{i=1}^{N} \phi_{Ai} \omega_i \cos(\omega_i t + \beta_i + \varphi_{0i}) + \\ + \omega_\phi^2 \sum_{i=1}^{N} \phi_{Ai} \sin(\omega_i t + \beta_i + \varphi_{0i}) = \sum_{i=1}^{N} \alpha_{Ei} \sin(\omega_i t + \varphi_{0i}) \end{gathered} \tag{8.78}$$

We regroup (8.78):

$$\begin{gathered} \sum_{i=1}^{N} \left(-\phi_{Ai} \omega_i^2 \sin(\omega_i t + \beta_i + \varphi_{0i}) + 2\delta_\phi \phi_{Ai} \omega_i \cos(\omega_i t + \beta_i + \varphi_{0i}) \right. \\ \left. + \omega_\phi^2 \phi_{Ai} \sin(\omega_i t + \beta_i + \varphi_{0i})\right) = \sum_{i=1}^{N} \alpha_{Ei} \sin(\omega_i t + \varphi_{0i}) \end{gathered} \tag{8.79}$$

If we equalize trigonometric functions with the same frequency, the equation (8.79) becomes a system of N equations, each of which contains the single frequency:

$$\begin{cases} -\phi_{A1} \omega_1^2 \sin(\omega_1 t + \beta_1 + \varphi_{01}) + 2\delta_\phi \phi_{A1} \omega_1 \cos(\omega_1 t + \beta_1 + \varphi_{01}) + \\ \quad + \omega_\phi^2 \phi_{A1} \sin(\omega_1 t + \beta_1 + \varphi_{01}) = \alpha_{E1} \sin(\omega_1 t + \varphi_{01}) \\ -\phi_{Ai} \omega_i^2 \sin(\omega_i t + \beta_i + \varphi_{0i}) + 2\delta_\phi \phi_{Ai} \omega_i \cos(\omega_i t + \beta_i + \varphi_{0i}) + \\ \quad + \omega_\phi^2 \phi_{Ai} \sin(\omega_i t + \beta_i + \varphi_{0i}) = \alpha_{Ei} \sin(\omega_i t + \varphi_{0i}) \\ -\phi_{AN} \omega_N^2 \sin(\omega_N t + \beta_N + \varphi_{0N}) + 2\delta_\phi \phi_{Ai} \omega_N \cos(\omega_N t + \beta_N + \varphi_{0N}) + \\ \quad + \omega_\phi^2 \phi_{AN} \sin(\omega_N t + \beta_N + \varphi_{0N}) = \alpha_{EN} \sin(\omega_N t + \varphi_{0N}) \end{cases} \tag{8.80}$$

We see now that the linear equation "deals" with each component separately, so we can rewrite the regular solution for the case of irregular waves:

$$\phi_{Ai} = \frac{\alpha_{Ei}}{\sqrt{(\omega_\phi^2 - \omega_i^2)^2 + 4\omega_i^2 \delta_\phi^2}}; \quad \beta_i = \arctan\left(\frac{4\omega_i \delta_\phi}{\omega_\phi^2 - \omega_i^2}\right) \tag{8.81}$$

It is very clear from the solution (8.51) that the response of a linear system is proportional to excitation: this fact is well known from classical seakeeping analysis. This allows presenting the spectrum of the response using RAO (Response Amplitude Operator):

$$S_{\phi}(\omega) = |\Phi(\omega)|^2 S_{\zeta}(\omega) \tag{8.82}$$

With:

$$|\Phi(\omega)| = \frac{\omega^2 \omega_{\phi}^2 \kappa_{\phi}(\omega)}{g\sqrt{(\omega_{\phi}^2 - \omega^2)^2 + 4\omega^2 \delta_{\phi}^2}} \tag{8.83}$$

Formula (8.82) is also known as the Wiener-Khinchin theorem and it works for linear systems only! This method was first applied to ship roll by St. Denis and Pierson [1953].

8.3.2 Correlation of Irregular Roll

The Wiener-Khinchin theorem provides the relationship between spectra of excitation and response; both these processes are dependent. Let us look at their correlation.

Irregular waves are known to be an ergodic stochastic process. As it is clearly seen from the Wiener-Khinchin theorem (8.82), the linear system response is a result of action of a linear operator. It is known that linear transformation does not break ergodicity; therefore, both linear irregular roll and its excitation are also ergodic stochastic processes. This allows working in a time domain with one realization and the use of a Fourier series to present it.

The correlation moment is defined as:

$$C(\phi, \alpha_E) = \int_{-\infty}^{\infty}\int_{-\infty}^{\infty} (\alpha_E - m_{\alpha})(\phi - m_{\phi}) f(\alpha_E, \phi) d\alpha_E d\phi \tag{8.84}$$

Taking into account that the mean value of wave elevation and, correspondingly, wave excitation is always zero and both waves and roll are ergodic processes, the correlation moment can be expressed as follows:

$$C(\phi, \alpha_E) = \frac{1}{T}\int_0^T (\phi - m_{\phi}) \alpha_E dt \tag{8.85}$$

Using a Fourier presentation for both excitation and roll processes:

$$\begin{aligned} C(\phi, \alpha_E) &= \frac{1}{T}\int_0^T \left(\sum_{i=1}^{N} \phi_{Ai} \sin(\omega_i t + \beta_i + \varphi_{0i}) \times \sum_{j=1}^{N} \alpha_{Ei} \sin(\omega_j t + \varphi_{0j}) \right) dt = \\ &= \int_0^T \left(\sum_{i=1}^{N}\sum_{j=1}^{N} \phi_{Ai} \alpha_{Ei} \sin(\omega_i t + \beta_i + \varphi_{0i}) \sin(\omega_j t + \varphi_{0j}) \right) dt \end{aligned} \tag{8.86}$$

Having in mind that:

$$\begin{aligned} &\sin(\omega_i t + \beta_i + \varphi_{0i}) \sin(\omega_j t + \varphi_{0j}) = \\ &= \frac{1}{2}\left(\cos(\omega_i t + \beta_i + \varphi_{0i} - \omega_j t - \varphi_{0j}) - \cos(\omega_i t + \beta_i + \varphi_{0i} + \omega_j t + \varphi_{0j})\right) \end{aligned}$$

Formula (8.56) can be presented as:

$$C(\phi,\zeta)=\frac{1}{2T}\int_0^T\left(\sum_{i=1}^{N}\sum_{j=1}^{N}\phi_{Ai}\alpha_{Ei}\cos\left(\omega_i t+\beta_i+\varphi_{0i}-\omega_j t-\varphi_{0j}\right)\right)dt-$$
$$-\frac{1}{2T}\int_0^T\left(\sum_{i=1}^{N}\sum_{j=1}^{N}\phi_{Ai}\alpha_{Ei}\cos\left(\omega_i t+\beta_i+\varphi_{0i}+\omega_j t+\varphi_{0j}\right)\right)dt=$$
$$=\frac{1}{2T}\sum_{i=1}^{N}\sum_{j=1}^{N}\int_0^T\phi_{Ai}\alpha_{Ei}\cos\left((\omega_i-\omega_j)t+\beta_i+\varphi_{0i}-\varphi_{0j}\right)dt-$$
$$-\frac{1}{2T}\sum_{i=1}^{N}\sum_{j=1}^{N}\int_0^T\phi_{Ai}\alpha_{Ei}\cos\left((\omega_i+\omega_j)t+\beta_i+\varphi_{0i}+\varphi_{0j}\right)dt \tag{8.87}$$

To perform the integration, we have to consider two cases: for equal and non-equal indexes i and j. If the indexes are not equal, both integrals in (8.87) are mean values of the cosine function. Since both realizations are long:

$$\frac{1}{T}\int_0^T\phi_{Ai}\alpha_{Ei}\cos\left((\omega_i-\omega_j)t+\beta_i+\varphi_{0i}-\varphi_{0j}\right)dt\approx$$
$$\approx\lim_{T\to\infty}\frac{1}{T}\int_0^T\phi_{Ai}\alpha_{Ei}\cos\left((\omega_i-\omega_j)t+\beta_i+\varphi_{0i}-\varphi_{0j}\right)dt=0$$
$$\frac{1}{T}\int_0^T\phi_{Ai}\alpha_{Ei}\cos\left((\omega_i+\omega_j)t+\beta_i+\varphi_{0i}+\varphi_{0j}\right)dt\approx \tag{8.88}$$
$$\approx\lim_{T\to\infty}\frac{1}{T}\int_0^T\phi_{Ai}\alpha_{Ei}\cos\left((\omega_i+\omega_j)t+\beta_i+\varphi_{0i}+\varphi_{0j}\right)dt=0$$

If the indexes are equal, one of the integrals has a non-zero value:

$$\frac{1}{T}\int_0^T\phi_{Ai}\alpha_{Ei}\cos\left((\omega_i-\omega_j)t+\beta_i+\varphi_{0i}-\varphi_{0j}\right)dt=$$
$$=\frac{1}{T}\int_0^T\phi_{Ai}\alpha_{Ei}\cos(\beta_i)dt=\phi_{Ai}\alpha_{Ei}\cos(\beta_i)$$
$$\frac{1}{T}\int_0^T\phi_{Ai}\alpha_{Ei}\cos\left((\omega_i+\omega_j)t+\beta_i+\varphi_{0i}+\varphi_{0j}\right)dt\approx \tag{8.89}$$
$$\approx\lim_{T\to\infty}\frac{1}{T}\int_0^T\phi_{Ai}\alpha_{Ei}\cos\left(2\omega_i t+\beta_i+2\varphi_{0i}\right)dt=0$$

Therefore, after multiplication and integration of both Fourier series, only terms with the same indexes have a non-zero value. This quality of trigonometric functions is known as orthogonality. Finally:

$$C(\phi,\alpha_E)=\frac{1}{2}\sum_{i=1}^{N}\phi_{Ai}\alpha_{Ei}\cos(\beta_i) \tag{8.90}$$

The correlation moment depends on both amplitude and phase angle between excitation and roll. Actually, formula (8.90) can be applied to any two stochastic processes that can be presented by a Fourier series. Let us check the correlation between all the processes we are dealing with including roll velocity (8.76) and acceleration (8.77).

Since the phase angle between the sine and cosine functions equals $\pi/2$:

$$C(\phi,\dot{\phi}) = \frac{1}{2}\sum_{i=1}^{N}\phi_{Ai}^2\omega_i\cos\left(\frac{\pi}{2}\right) = 0 \tag{8.91}$$

This means that roll velocity is not correlated with roll. Moreover, roll velocity does not depend on roll angle. This is also true for nonlinear response [Gerasimov, 1979, Belenky, 1993a]. Independence of roll angle and velocity means that if a ship has a certain value of roll angle, it does not have an affect on the probability to have any value for roll velocity at this moment.

The same is true for correlation between roll velocity and roll acceleration:

$$C(\ddot{\phi},\dot{\phi}) = -\frac{1}{2}\sum_{i=1}^{N}\phi_{Ai}^2\omega_i^3\cos\left(\frac{\pi}{2}\right) = 0 \tag{8.92}$$

At the same time, there is a strong negative correlation between roll angle and acceleration, the correlation moment equals the variance of roll velocity taken with the opposite sign:

$$C(\ddot{\phi},\phi) = -\frac{1}{2}\sum_{i=1}^{N}\phi_{Ai}^2\omega_i^2\cos(0) = -\frac{1}{2}\sum_{i=1}^{N}\phi_{Ai}^2\omega_i^2 = -V_{\dot{\phi}} \tag{8.93}$$

Both roll velocity and roll acceleration are correlated with excitation:

$$C(\dot{\phi},\alpha_E) = \frac{1}{2}\sum_{i=1}^{N}\phi_{Ai}\omega_i\alpha_{Ei}\cos\left(\beta_i + \frac{\pi}{2}\right) = -\frac{1}{2}\sum_{i=1}^{N}\phi_{Ai}\omega_i\alpha_{Ei}\sin(\beta_i) \tag{8.94}$$

$$C(\ddot{\phi},\alpha_E) = -\frac{1}{2}\sum_{i=1}^{N}\phi_{Ai}\omega_i^2\alpha_{Ei}\cos(\beta_i) \tag{8.95}$$

Despite pairs of processes, roll angle – roll velocity and roll velocity – roll acceleration are not correlated, all these processes are dependent on and correlated with excitation. Gerasimov [1979] considered excitation as a combination of two parts. One part is correlated with roll angle and the other one with roll velocity; both parts are not correlated with each other:

$$\begin{aligned}\alpha_E(t) &= \sum_{i=1}^{N}\alpha_{Ei}\sin(\omega_i t + \varphi_{0i}) = \sum_{i=1}^{N}\alpha_{Ei}\sin(\omega_i t + \varphi_{0i} + \beta_i - \beta_i) = \\ &= \sum_{i=1}^{N}\alpha_{Ei}\sin(\omega_i t + \varphi_{0i} + \beta_i)\cos\beta_i - \sum_{i=1}^{N}\alpha_{Ei}\cos(\omega_i t + \varphi_{0i} + \beta_i)\sin\beta_i \\ &= \alpha_{E\phi}(t) + \alpha_{EV}(t)\end{aligned} \tag{8.96}$$

It is easy to see (taking into account orthogonality of trigonometric functions):

$$\begin{aligned}C(\phi,\alpha_{E\phi}) &= \frac{1}{T}\int_0^T\left(\sum_{i=1}^{N}\phi_{Ai}\alpha_{Ei}\sin^2(\omega_i t + \beta_i + \varphi_{0i})\cos(\beta_i)\right)dt = \\ &= \frac{1}{2T}\sum_{i=1}^{N}\int_0^T\phi_{Ai}\alpha_{Ei}\cos(\beta_i)(1-\cos 2(\omega_i t + \beta_i + \varphi_{0i}))\,dt = \\ &= \frac{1}{2}\sum_{i=1}^{N}\phi_{Ai}\alpha_{Ei}\cos(\beta_i)\end{aligned} \tag{8.97}$$

Analogously:

$$C(\dot{\phi},\alpha_{E\dot{\phi}})=\frac{1}{T}\int_0^T\left(\sum_{i=1}^{N}\phi_{Ai}\alpha_{Ei}\cos^2(\omega_i t+\beta_i+\varphi_{0i})\sin(\beta_i)\right)dt=$$

$$=\frac{1}{2T}\sum_{i=1}^{N}\int_0^T\phi_{Ai}\alpha_{Ei}\sin(\beta_i)(1+\cos 2(\omega_i t+\beta_i+\varphi_{0i}))dt= \qquad (8.98)$$

$$=\frac{1}{2}\sum_{i=1}^{N}\phi_{Ai}\omega_i\alpha_{Ei}\sin(\beta_i)$$

It is also easy to see that both parts are not correlated:

$$C(\alpha_{E\phi},\alpha_{E\dot{\phi}})=\frac{1}{T}\int_0^T\sum_{i=1}^{N}\alpha_{Ei}^2\cos(\omega_i t+\beta_i+\varphi_{0i})\sin(\beta_i)\times$$

$$\times\sin(\omega_i t+\beta_i+\varphi_{0i})\cos(\beta_i)\,dt= \qquad (8.99)$$

$$=\frac{1}{T}\sum_{i=1}^{N}\alpha_{Ei}^2\sin(2\beta_i)\int_0^T\sin 2(\omega_i t+\beta_i+\varphi_{0i})\,dt=0$$

Division of the excitation into two parts has physical meaning, we already discussed it in subchapter 7.1.5. Excitation of any periodic process plays two roles: synchronizing (forcing the dynamical to follow with the excitation frequency) and active (compensation energy dissipation by the damping term). The first part of excitation that is correlated with roll angle is responsible for synchronizing and the other one is active. This consideration can be extended for nonlinear roll as well, if we assume ergodicity [Gerasimov, 1979].

8.3.3 Statistical Linearization

As we mentioned in Chapter 4, there is no general analytical solution for a nonlinear differential equation describing roll motion of a ship. Introducing stochastic excitation makes the problem more complex, since we no longer can take advantage of periodic qualities of excitation. We cannot use the previously described linearization technique since it is related with periodic excitation, and we need to know the amplitude of response in order to complete the procedure, see subchapter 4.2.

However, there is another linearization technique available for stochastic excitation. The idea of linearization remains the same: we substitute the original nonlinear system with the linear equivalent. The equivalence could be built using different approaches: the simplest one is to require that linearization error be minimized [Vassilopoulos, 1971]. More information on statistical linearization in general is available from a comprehensive book by Roberts and Spanos [2003].

So let us consider the nonlinear roll equation:

$$\ddot{\phi}+2\delta\dot{\phi}+\omega_\phi^2 f(\phi)=f_E(t) \qquad (8.100)$$

Here, $f_E(t)$ is stochastic excitation. It is usually assumed as a stationary ergodic stochastic process with a Gaussian distribution. Following our previous examples, we assume the restoring moment to be approximated with cubic parabola:

$$f(\phi) = \phi - a_3\phi^3 \tag{8.101}$$

So, we will be looking to substitute the nonlinear system:

$$\ddot{\phi} + 2\delta\dot{\phi} + c_1\phi + c_3\phi^3 = f_E(t) \tag{8.102}$$

Where $c_1 = \omega_\phi^2;\quad c_3 = \omega_\phi^2 a_3$ with the linear system:

$$\ddot{\phi} + 2\delta\dot{\phi} + c_L\phi = f_E(t) \tag{8.103}$$

We try to find such a value for c_L that would make the mean square of an error of such a substitute minimal:

$$V\left[c_1\phi + c_3\phi^3 - c_L\phi\right] \rightarrow \min \tag{8.104}$$

This can be achieved with:

$$\frac{\partial\left(V\left[c_1\phi + c_3\phi^3 - c_L\phi\right]\right)}{\partial c_L} = 0 \tag{8.105}$$

Consider the variance operator first, assuming average of the roll angle equals zero, then the variance is just a mean value of the square of the value:

$$\begin{aligned} V\left[c_1\phi + c_3\phi^3 - c_L\phi\right] &= M\left[\left(c_1\phi + c_3\phi^3 - c_L\phi\right)^2\right] = \\ &= c_1^2 V_\phi + c_3^2 M_\phi^4 + c_L^2 V_\phi - 2c_1c_L V_\phi - 2c_1c_3 M_\phi^4 - 2c_Lc_3 M_\phi^4 \end{aligned} \tag{8.106}$$

Here, M_ϕ^4 is the fourth moment of the distribution:

$$M_\phi^4 = \int_{-\infty}^{\infty} (\phi - m_\phi)^4 f(\phi)d\phi = \int_{-\infty}^{\infty} \phi^4 f(\phi)d\phi \tag{8.107}$$

Following [Vassilopoulos, 1971], we assume that roll has a Gaussian distribution, such as the excitation (in subchapter 8.6.2 we consider nonlinear roll distribution issues). Gaussian distribution provides the following relationship between variance and the fourth moment:

$$M_\phi^4 = 3 \cdot V_\phi^2 \tag{8.108}$$

Substitution of (8.78) and (8.76) into (8.75) leads to the following equation:

$$2c_L V_\phi - 2c_1 V_\phi - 6c_3 V_\phi^2 = 0 \tag{8.109}$$

With the final expression for the unknown coefficient of the linear system:

$$c_L = c_1 + 3c_3 V_\phi \tag{8.110}$$

8.3.4 Energy-Statistical Linearization

The statistical linearization technique described in the previous subchapter sometimes also is referred to as equivalent statistical linearization. Originally, this technique was developed for applications in electronics, where dynamical systems are characterized with relatively small inertia.

Gerasimov [1979] proposed an alternative method of statistical linearization based on equivalence of statistical characteristics of work/energy balance. To emphasize the energy conservation based approach, this method was called "energy-statistical linearization". We consider it briefly in this subchapter.

We already considered the energy balance of roll equation as a background of the weather criterion in subchapter 7.1.5. Here, we repeat these derivations using a different technique. Also following [Gerasimov, 1979] we consider nonlinear damping as well as the nonlinear restoring moment:

$$\ddot{\phi} + r(\dot{\phi}) + f(\phi) = \alpha_E(t) \tag{8.111}$$

Here, $r(\dot{\phi})$ is the nonlinear roll damping and $f(\phi)$ is the nonlinear restoring moment. We multiply this equation first by roll angle, then by roll velocity and finally apply averaging over time (it means that we assume roll is an ergodic process):

$$m_t \left\{ \begin{array}{l|l} \ddot{\phi} + r(\dot{\phi}) + f(\phi) = \alpha_E(t) & \times \phi \\ \ddot{\phi} + r(\dot{\phi}) + f(\phi) = \alpha_E(t) & \times \dot{\phi} \end{array} \right\} \tag{8.112}$$

If the mean value of roll equals zero (we can always subtract it from both sides of the roll equation if it is non-zero), we obtain a system of equations with correlation moments:

$$\begin{cases} C(\ddot{\phi},\phi) + C(r(\dot{\phi}),\phi) + C(f(\phi),\phi) = C(\alpha_E,\phi) \\ C(\ddot{\phi},\dot{\phi}) + C(r(\dot{\phi}),\dot{\phi}) + C(f(\phi),\dot{\phi}) = C(\alpha_E,\dot{\phi}) \end{cases} \tag{8.113}$$

Let us consider each term of (8.79) in detail [Gerasimov, 1979].

As we have seen from the equation (8.63), the correlation moment between roll angle and acceleration is equal to the variance of roll velocity taken with the opposite sign. This value also can be interpreted as the average of change of kinetic energy (compare with equation 7.33):

$$C(\ddot{\phi},\phi) = -V_{\dot{\phi}} = -m[\dot{\phi}^2] = -\frac{1}{T}\int_0^T \dot{\phi}^2 dt = -2m[K(\dot{\phi})] \tag{8.114}$$

The correlation moment between nonlinear damping and roll angle equals zero. Nonlinear damping can be considered as a deterministic function of a random argument – roll velocity. Naturally, such a deterministic function is also a stochastic process, but it cannot be dependent, if its argument is independent:

$$C(r(\dot{\phi}),\phi) = 0 \tag{8.115}$$

The correlation moment between the restoring moment and roll angle is proportional to the mean value of potential energy.

The correlation moment between roll acceleration and roll velocity equals zero.

The correlation moment between nonlinear damping and roll velocity expresses the average value of work of the damping moment.

The correlation moment between the restoring moment and roll velocity is zero (see analogous consideration above). Finally:

$$\begin{cases} C(f(\phi),\phi)-V_{\dot\phi}=C(\alpha_E,\phi) \\ C(r(\dot\phi),\dot\phi)=C(\alpha_E,\dot\phi) \end{cases} \tag{8.116}$$

As we have seen in subchapter 8.3.2, the excitation can be presented as a sum of two parts, one of each is correlated with roll angle and the other – with roll velocity:

$$\begin{cases} C(f(\phi),\phi)-V_{\dot\phi}=C(\alpha_{E\phi},\phi) \\ C(r(\dot\phi),\dot\phi)=C(\alpha_{EV},\dot\phi) \end{cases} \tag{8.117}$$

Now, we have two equations for the energy/work balance as averaged values. The first equation describes the balance of kinetic and potential energy with synchronizing action of the excitation. The second equation shows that a certain portion of excitation power is used for the work of damping forces. This system is very similar to the energy balance equation for a linear system. The difference is that instead of change of energy or work, we have corresponding correlation moments, which can be treated as averaged values of these figures.

The system (8.117) expresses the law of energy conservation and its structure is the same for linear and nonlinear systems. Gerasimov [1979] used it for energy statistical linearization. He introduced a concept of energy equivalent cycle for this purpose. It is free oscillations of a non-damped, non-forced linear system with the frequency equal to the average frequency of the original nonlinear system:

$$\ddot\phi+\omega_a^2\phi=0 \tag{8.118}$$

Here, ω_a is the average frequency of irregular nonlinear roll described by differential equation (8.111). The definition of averaged frequency of a stochastic process is based on the upcrossing theory, which is discussed in Chapter 9. For a Gaussian stochastic process it is expressed as:

$$\omega_a=\sqrt{\frac{V_{\dot\phi}}{V_\phi}} \tag{8.119}$$

Initial conditions that are necessary to define a cycle of the system (8.118) are defined as:

$$\phi_0=\frac{2\sqrt{V_{\dot\phi}}}{\omega_a};\quad \dot\phi_0=0 \tag{8.120}$$

The resulting cycle is described as:

$$\phi=\phi_{AE}\cos\omega_a t \tag{8.121}$$

The amplitude ϕ_{AE} is defined through the initial conditions (8.120). Gerasimov [1979] called it "energy equivalent amplitude":

$$\phi_{AE} = \phi_0 = \frac{2\sqrt{V_{\dot{\phi}}}}{\omega_a} \tag{8.122}$$

If we calculate formally the average value of kinetic energy of this cycle or variance of its roll velocity:

$$V_{\dot{\phi}E} = \phi_{AE}^2 \omega_a^2 \int_0^{\infty} \sin^2(\omega_a t)\,dt = \frac{1}{2}\phi_{AE}^2 \omega_a^2 = V_{\dot{\phi}} \tag{8.123}$$

Therefore, the energy equivalent cycle formally has the same variance of roll velocity and the frequency is equal to the averaged frequency of nonlinear irregular roll.

The idea of energy-statistical linearization is to find a linear system with the same energy equivalent cycle as the nonlinear one. Consider the linear system:

$$\ddot{\phi} + 2\delta_L \dot{\phi} + \omega_L^2 \phi = \alpha_E(t) \tag{8.124}$$

All statistical characteristics of energies for a linear system are known in analytical form (using formulae (8.98) and (8.99) as well):

$$\begin{aligned}
&C(\phi, f(\phi)) = C(\phi, \omega_L^2 \phi) = \omega_L^2 V_\phi \\
&C(\phi, \alpha_{E\phi}) = \frac{1}{2}\sum_{i=1}^{N} \alpha_{Ei}\phi_{ai}\cos\beta_i \\
&C(\phi, \alpha_{E\dot{\phi}}) = \frac{1}{2}\sum_{i=1}^{N} \alpha_{Ei}\phi_{ai}\omega_i\sin\beta_i \\
&C(\dot{\phi}, r(\dot{\phi})) = C(\dot{\phi}, 2\delta_L\dot{\phi}) = 2\delta_L V_{\dot{\phi}}
\end{aligned} \tag{8.125}$$

Substitution of these formulae into the energy balance equation (8.117) yields:

$$\begin{cases}
\omega_L^2 V_\phi - V_{\dot{\phi}} = \dfrac{1}{2}\displaystyle\sum_{i=1}^{N} \alpha_{Ei}\phi_{ai}\cos\beta_i \\
2\delta_L V_{\dot{\phi}} = \dfrac{1}{2}\displaystyle\sum_{i=1}^{N} \alpha_{Ei}\phi_{ai}\omega_i\sin\beta_i
\end{cases} \tag{8.126}$$

The variance of roll is related with the variance of roll velocities through averaged frequency (8.89). The pair, $V_{\dot{\phi}}$ and ω_a, completely defines the statistics of energy balance of a linear system.

To complete energy-statistical linearization, we have to find a relationship between nonlinear functions $r(\dot{\phi})$ and $f(\phi)$ and characteristics of the energy equivalent cycle: energy equivalent amplitude and average frequency (or variance of roll velocities and average frequency). This was done with the assumption of a Gaussian distribution of roll angles and velocities. (We will consider the distribution of nonlinear roll as a separate problem in subchapter 8.6.2.) Since the derivations are lengthy, we provide the results only. Details can be found in [Gerasimov, 1973, 1974, 1979]:

$$2\delta_L(V_{\dot\phi}) = \frac{1}{\sqrt{2\pi V_{\dot\phi}^3}} \int_{-\infty}^{\infty} \dot\phi r(\dot\phi) \exp\left(-\frac{\dot\phi}{2V_{\dot\phi}}\right) d\dot\phi \qquad (8.127)$$

$$\omega_L = \omega_\phi(\phi_{AE}); \quad \phi_{AE} = \frac{2\sqrt{V_{\dot\phi}}}{\omega_a} \qquad (8.128)$$

Function $\omega_\phi(\phi_A)$ is the backbone line that we have considered in subchapter 4.1. The only difference is that it has to be calculated with energy equivalent amplitude as an argument.

Formulae (8.127) and (8.128) use characteristics of the energy equivalent cycle ω_a and ϕ_{AE} that can be calculated only when variances of linearized roll angle and velocities are known. These variances can be calculated only when coefficients of the linearized equation are known, so an iteration technique is required.

When calculations are completed, we obtain a linear system, for which the energy balance has the same average as the one of the nonlinear system.

8.3.5 Method of Multiple Scales

As we mentioned above, perturbation methods can be successfully applied for the case of irregular roll if excitation is presented with the envelope. Here, we demonstrate this technique, following Rajan and Davis [1988]. They considered the Duffing equation, which we used as the simplest model of nonlinear roll:

$$\ddot\phi + 2\delta\dot\phi + \omega_\phi^2\phi - a_3\phi^3 = \alpha_{EC}(t)\cos\omega_a t + \alpha_{ES}(t)\sin\omega_a t \qquad (8.129)$$

Here, amplitudes $\alpha_{EC}(t)$ and $\alpha_{ES}(t)$ are independent Gaussian stochastic processes which vary slowly in comparison with both roll excitation and roll response, see subchapter 8.2.4. These amplitudes are defined by two linear differential equations (8.64), rewritten to filter roll excitation instead of wave elevation:

$$\begin{cases} \dot\alpha_{EC} + \frac{\gamma}{2}\alpha_{EC} = \sqrt{\frac{\gamma}{2}}W_C(t) \\ \dot\alpha_{ES} + \frac{\gamma}{2}\alpha_{ES} = \sqrt{\frac{\gamma}{2}}W_S(t) \end{cases} \qquad (8.130)$$

Here, $W_C(t)$ and $W_S(t)$ are two independent Gaussian Wiener processes, their spectral level adjusted to accommodate differences between wave elevation and roll excitation.

We consider the frequency domain near the main resonance; detuning parameter is introduced slightly differently in comparison with our previous consideration in subchapter 4.2.4:

$$\omega_a = \omega_\phi + \varepsilon\cdot\sigma \qquad (8.131)$$

Following established procedure:

$$\delta = \varepsilon\nu; \quad \alpha_{EC} = \alpha_{EC0}\varepsilon; \quad \alpha_{ES} = \alpha_{ES0}\varepsilon; \quad a_3 = a_{30}\varepsilon \qquad (8.132)$$

We look for the solution as a function of time in different scales. Here, we limit ourselves with the first expansion only:

$$\phi = \phi_0(T_0, T_1) + \varepsilon\phi_1(T_0, T_1) \qquad (8.133)$$

T_0, and T_1 are different time scales. They correspond to different orders of expansion:

$$T_0 = t; \quad T_1 = \varepsilon t \qquad (8.134)$$

Here, these scales have a direct physical meaning: "fast" time T_0 has the scale of changes of roll excitation and response processes, "slow" time T_1 corresponds to the pace of slow varying of the amplitude process. Considering both time scales as independent variables, we express their derivatives (following the procedure from subchapter 4.2.4 almost exactly) as:

$$\frac{d}{dt} = D_0 + \varepsilon D_1 + ...; \quad \frac{d^2}{dt^2} = D_0^2 + 2\varepsilon D_1 D_0 + ... \qquad (8.135)$$

With $D_i = \dfrac{\partial}{\partial T_i}$ is a partial derivative operator.

Using expression (8.131), taking into account that amplitudes are slowly varying function of time and finally substituting formulae (8.134), the stochastic excitation can be presented as:

$$\begin{aligned} \alpha_E(t) &= \alpha_{EC}(\varepsilon t)\cos(\omega_a t) + \alpha_{ES}(\varepsilon t)\sin(\omega_a t) = \\ &= \alpha_{EC}(\varepsilon t)\cos(\omega_\phi t + \sigma\varepsilon t) + \alpha_{ES}(\varepsilon t)\sin(\omega_a t + \sigma\varepsilon t) = \\ &= \alpha_{EC}(T_1)\cos(\omega_\phi T_0 + \sigma T_1) + \alpha_{ES}(T_1)\sin(\omega_a T_0 + \sigma T_1) \end{aligned} \qquad (8.136)$$

We substitute formulae (8.132), (8.133), (8.135) and (8.136) into original equation (8.129) and transform it to the system according to powers of book keeping parameter ε:

$$\varepsilon^0: \quad D_0^2\phi_0 + \omega_\phi^2\phi_0 = 0 \qquad (8.137)$$

$$\begin{aligned} \varepsilon^1: \quad & D_0^2\phi_1 + \omega_\phi^2 2\phi_1 = \alpha_{EC0}(T_1)\cos(\omega_\phi T_0 + \sigma T_1) + \alpha_{ES0}(T_1)\sin(\omega_a T_0 + \sigma T_1) \\ & - 2D_0 D_1\phi_0 - 2\nu D_0\phi + a_{30}\phi_0^3 \end{aligned} \qquad (8.138)$$

Comparing equations (8.137) and (8.138) with analogous equations (4.68) and (4.69), it is necessary to note the slight difference between them. It is caused by differences in the way the detuning parameter was introduced.

Considering the solution of the first expansion, the following form is expected:

$$\phi_0 = A(T_1)\exp(i\omega_\phi T_0) + CC \qquad (8.139)$$

Substitution of (8.139) into the equation of the second expansion (8.138) and presenting all trigonometric functions in exponential form allows extracting the secular term:

$$\begin{aligned}&D_0^2\phi_1+\omega_\phi^2\phi_1=\\&=\left[0.5(\alpha_{EC0}(T_1)-i\alpha_{ES0}(T_1))\exp(i\sigma T_1)-\right.\\&\left.-2i\omega_\phi D_1A-2\nu i\omega_\phi A+3a_{30}A^2\overline{A}\right]\exp(i\omega_\phi T_0)+\\&+a_{30}A^3\exp(3i\omega_\phi T_0)+CC\end{aligned}\tag{8.140}$$

The secular term appears as the coefficient at $\exp(i\omega_\phi T_0)$. Following the procedure described in subchapter 4.2.4, we set it to zero:

$$0.5(\alpha_{EC0}(T_1)-i\alpha_{ES0}(T_1))\exp(i\sigma T_1)-2i\omega_\phi D_1A-2\nu i\omega_\phi A+3a_{30}A^2\overline{A}=0\tag{8.141}$$

Expression (8.141) is an ordinary differential equation relative to slowly varying amplitude of roll motion. We are looking for the solution in the following form:

$$A(T_1)=0.5(x-iz)\exp(i\sigma T_1)\tag{8.142}$$

Substitution of the expected solution into (8.141) allows separation of the real and imaginary parts. We obtain the following system of differential equations:

$$\begin{cases}-\omega_\phi D_1x-0.5\alpha_{ES0}(T_1)-\omega_\phi\nu x-\omega_\phi\sigma z-\dfrac{3}{8}a_{30}z(x^2+z^2)=0\\-\omega_\phi D_1z+0.5\alpha_{EC0}(T_1)-\omega_\phi\nu z+\omega_\phi\sigma x+\dfrac{3}{8}a_{30}x(x^2+z^2)=0\end{cases}\tag{8.143}$$

System (8.143) contains slowly varying excitation amplitudes α_{ES0} and α_{EC0}, which can be defined from the system (8.130). Let us add these definitions to (8.143) and consider a system of four equations (all the terms in these definitions are of the same order, so we can simply multiply by the bookkeeping parameter ε):

$$\begin{cases}D_1x=-\dfrac{1}{2\omega_\phi}\alpha_{ES0}-\nu x-\sigma_e z=0\\D_1z=\dfrac{1}{2\omega_\phi}\alpha_{EC0}-\nu z+\sigma_e x=0\\D_1\alpha_{EC0}+\dfrac{\gamma}{2}\alpha_{EC0}=\sqrt{\dfrac{\gamma}{2}}W_{C0}(T_1)\\D_1\alpha_{ES0}+\dfrac{\gamma}{2}\alpha_{ES0}=\sqrt{\dfrac{\gamma}{2}}W_{S0}(T_1)\end{cases}\tag{8.144}$$

Here: $W_{S0}=\varepsilon W_S$; $W_{C0}=\varepsilon W_C$, and

$$\sigma_e=\sigma+\frac{3}{8\omega_\phi}a_{30}(x^2+z^2)\tag{8.145}$$

Since we are working with stochastic figures, our goal here is to calculate the variance. To get it we need to rewrite equations (8.144), so they will be in the form of squared terms of the searched quantities and then we shall average them.

We multiply the first equation of (8.144) by x and the second one by z; then we consider their sum:

$$xD_1x + zD_1z = \frac{1}{2\omega_\phi}(z\alpha_{EC0} - x\alpha_{ES0}) - \nu(x^2 + z^2) \tag{8.146}$$

Let us introduce new variables:

$$Y = \frac{1}{2}(x^2 + z^2); \quad P = z\alpha_{EC0} - x\alpha_{ES0} \tag{8.147}$$

With the new variables, equation (8.146) has the following appearance:

$$D_1Y = \frac{1}{2\omega_\phi}P - \frac{1}{2}\nu Y \tag{8.148}$$

Equation (8.148) contains values of the second order. Now, we multiply the first equation of (8.144) by $-\alpha_{ES0}$, the second one by α_{EC0}, the third one by $-x$ and the fourth one by z. Taking into account (8.147), their sum can be expressed as:

$$D_1P = \frac{1}{\omega_\phi}F - \left(\frac{\gamma}{2} + \nu\right)P + \sigma_e Q + \sqrt{\frac{\gamma}{2}}R_1 \tag{8.149}$$

Here, new variables are:

$$F = \frac{1}{2}(\alpha_{EC0}^2 + \alpha_{ES0}^2); \quad Q = x\alpha_{EC0} + z\alpha_{ES0}; \quad R_1 = zW_{C0} - xS_{S0} \tag{8.150}$$

The third equation can be derived by the following multiplication and summation: the first equation of (8.144) has to be multiplied by α_{EC0}, the second one by $-\alpha_{ES0}$, the third one by z and the fourth one by x:

$$D_1Q = \sigma_e P - \left(\frac{\gamma}{2} + \nu\right)Q + \sqrt{\frac{\gamma}{2}}R_2 \tag{8.151}$$

Here:

$$R_2 = zW_{S0} + xW_{C0} \tag{8.152}$$

Now, instead of the system of four equations (8.144), we have a system of three equations, with all of them in the form of the second order of our original variables:

$$\begin{cases} D_1 Y = \dfrac{1}{2\omega_\phi} P - \nu Y \\ D_1 P = \dfrac{1}{\omega_\phi} F - \left(\dfrac{\gamma}{2} + \nu\right) P + \sigma_e Q + \sqrt{\dfrac{\gamma}{2}} R_1 \\ D_1 Q = \sigma_e P - \left(\dfrac{\gamma}{2} + \nu\right) Q + \sqrt{\dfrac{\gamma}{2}} R_2 \end{cases} \tag{8.153}$$

The next step is to average the system (8.153) over a period corresponding to the modal frequency of excitation ω_a. Quadratic variables of the system (8.153) yield mean square of slowly varied response.

The input process of white noise has a zero average. It is not correlated with components of response amplitude x and y, so averaging R_1 and R_2 produces zero:

$$m[R_1] = 0; \quad m[R_2] = 0 \tag{8.154}$$

All other terms produce a non-zero average:

$$m[P] = P_a; \quad m[Q] = Q_a \tag{8.155}$$

Y and F are squared values of slowly varied amplitudes of the response and excitation respectively. Being averaged over a period, they produce estimates for the corresponding variances:

$$m[Y] = V_\phi; \quad m[F] = V_E \tag{8.156}$$

The value σ_e contains a quadratic response, see equation (8.145), using notation (8.147):

$$\sigma_e = \sigma + \frac{3}{4\omega_\phi} a_{30} Y \tag{8.157}$$

Therefore, this term also has to be averaged:

$$m[\sigma_e] = \sigma + \frac{3}{4\omega_\phi} a_{30} V_\phi \tag{8.158}$$

Averaged system (8.153) is expressed as:

$$\begin{cases} D_1 V_\phi = \dfrac{1}{2\omega_\phi} P_a - 2\nu V_\phi \\ D_1 P_a = \dfrac{1}{\omega_\phi} V_E - \left(\dfrac{\gamma}{2} + \nu\right) P_a + m[\sigma_e] Q_a \\ D_1 Q_a = m[\sigma_e] P_a - \left(\dfrac{\gamma}{2} + \nu\right) Q_a \end{cases} \tag{8.159}$$

System (8.159) is the stochastic analogue of the system (4.75). The difference is that variables are the mean square of the response and filter equations are included. The system (8.159) actually can be used to track the transition process, but here we are interested in the stationary (steady state in deterministic terms) solution. So, all the derivatives are set to zero and the system of differential equations (8.159) is converted to the following system of algebraic equations:

$$\begin{cases} \dfrac{1}{2\omega_\phi} P_a - 2\nu V_\phi = 0 \\ \dfrac{1}{\omega_\phi} V_E - \left(\dfrac{\gamma}{2} + \nu\right) P_a + m[\sigma_e] Q_a = 0 \\ m[\sigma_e] P_a - \left(\dfrac{\gamma}{2} + \nu\right) Q_a = 0 \end{cases} \tag{8.160}$$

The system (8.158) can be resolved relative to V_ϕ:

$$V_\phi = \frac{V_E(\gamma + 2\nu)}{2\omega_\phi^2 \nu \left(4(m[\sigma_e])^2 - (\gamma + 2\nu)^2\right)} \tag{8.161}$$

Taking into account (8.131) and assuming $\varepsilon = 1$, the value $m[\sigma_e]$ can be presented as:

$$m[\sigma_e] = (\omega_a - \omega_\phi) + \frac{3}{4\omega_\phi} a_3 V_\phi \tag{8.162}$$

It is assumed that the averaged frequency of excitation is close to resonance: the following approximation is introduced [Rajan and Davies, 1988], [Francescutto, 1992, 1998]:

$$\omega_a^2 - \omega_\phi^2 \approx 2\omega_\phi(\omega_a - \omega_\phi) \tag{8.163}$$

The above approximation leads to the following:

$$m[\sigma_e] \approx \frac{1}{2\omega_\phi}\left(\omega_a^2 - \omega_\phi^2 + \frac{3}{2} a_3 V_\phi\right) = \frac{1}{\omega_\phi}\left(\omega_a^2 - \omega_x^2\right) \tag{8.164}$$

Here (compare with (4.79)):

$$\omega_x^2(V_\phi) = \omega_\phi^2 - \frac{3}{2} a_3 V_\phi \tag{8.165}$$

Finally, (8.161) can be presented as:

$$V_\phi = \frac{V_E(\gamma + 2\delta)}{2\delta\left(\left(\omega_a^2 - \omega_x^2(V_\phi)\right)^2 - \omega_\phi^2(\gamma + 2\delta)^2\right)} \tag{8.166}$$

It is very clear that in the deterministic case $(\gamma = 0)$ this formula (8.166) coincides with (4.80) having in mind that for the sinusoidal oscillation, the variance formally is equal to half of the square amplitude: $V = A^2/2$.

Formula (8.166) describes the reaction of the system (8.129) on a narrow band excitation in the vicinity of the resonance area. However, we have seen that a nonlinear system is capable for sub- and super-harmonic response to sinusoidal excitation (subchapter 4.6). A similar response can be found for the case of narrow banded excitation as well. High-order resonance solutions for the system (8.129) are available from [Davies and Rajan, 1988].

For sub-harmonic response $\omega_a \approx 3\omega_\phi$, the detuning parameter is set to satisfy:

$$\omega_a = 3\omega_\phi + \varepsilon\sigma \tag{8.167}$$

The remaining coefficients are presented exactly as in the case of main resonance (8.132), with the exception of excitation amplitude:

$$\delta = \varepsilon\nu; \quad a_3 = a_{30}\varepsilon \tag{8.168}$$

The solution is presented in the form (8.133), time is scaled in accordance with (8.134), and partial derivatives are expressed in the form (8.135). The equations, rewritten in accordance with the degree of bookkeeping parameter, are slightly different from (8.137) and (8.138):

$$\varepsilon^0: \ D_0^2\phi_0 + \omega_\phi^2\phi_0 = \alpha_{EC}(T_1)\cos(\omega_\phi T_0) + \alpha_{ES}(T_1)\sin(\omega_a T_0) \tag{8.169}$$

$$\varepsilon^1: \ D_0^2\phi_1 + \omega_\phi^2 2\phi_1 = -2D_0D_1\phi_0 - 2\nu D_0\phi + a_{30}\phi_0^3 \tag{8.170}$$

A solution of the first expansion is searched for in the following form:

$$\phi_0 = A(T_1)\exp(i\omega_\phi T_0) + H(T_1)\exp(i\omega_a T_0) + CC \tag{8.171}$$

The response consists of two components: the first one describes oscillations with the natural frequency and the second one with the excitation frequency (compare with formula (4.225) from subchapter 4.6.3). Amplitudes of both responses are slow varying stochastic processes. Since the oscillations are far from the main resonance region, the second component is considered linear.

In accordance with conventional multiple scales method, the elements of the solution (8.171) can be found from the condition of elimination of the secular term, which appears with the frequency ω_ϕ in the equation (8.170):

$$2i\omega_\phi D_1 A + 2i\omega_\phi \nu A - 6a_{30}H\bar{H}A - 3a_{30}\bar{A}^2 A - 3a_{30}H\bar{A}^2\exp\left(i\frac{\sigma}{3}T_1\right) = 0 \tag{8.172}$$

The amplitude of sub-harmonic response is expected to be in the following form:

$$A(T_1) = \frac{1}{2}(x - iz)\exp\left(i\frac{\sigma}{3}T_1\right) \tag{8.173}$$

Amplitudes of the second component of the solution (8.171) are defined in the same manner as excitation amplitude was defined in the above case of main resonance. Since it is linear, it is presented with its sine and cosine components by filtering of white noise through the following differential equations:

$$\begin{cases} D_1 H_C + \frac{\gamma}{2}H_C = \sqrt{\frac{\gamma}{2}}W_C(T_1) \\ D_1 H_S + \frac{\gamma}{2}H_S = \sqrt{\frac{\gamma}{2}}W_S(T_1) \end{cases} \tag{8.174}$$

Further derivations do not make so much a difference as with the main resonance case. Equivalent detuning parameter σ_e is defined as:

$$\sigma_e = \sigma + \frac{3}{8\omega_\phi} a_{30}\left(x^2 + z^2 + 8\left(H_c^2 + H_s^2\right)\right) \tag{8.175}$$

The next step is to switch to a squared response equation (8.172) along with (8.174) and then averaging it to get variance estimates. Averaging of the equivalent detuning parameter includes using the following approximation:

$$\left(\frac{\omega_a}{3}\right)^3 - \omega_\phi^2 \approx 2\omega_\phi\left(\frac{\omega_a}{3} - \omega_\phi\right) \tag{8.176}$$

and:

$$m[\sigma_e] \approx \frac{1}{\omega_\phi}\left(\left(\frac{\omega_a}{3}\right)^3 - \omega_x^2(V_\phi)\right) \tag{8.177}$$

Here:

$$\omega_x^2(V_\phi) = \omega_\phi^2 - \frac{3}{2}a_3\left(V_\phi + V_{\phi L}\right) \tag{8.178}$$

Here, $V_{\phi L}$ is the variance estimate of the linear roll response with frequency ω_a, defined by averaging its amplitude:

$$V_{\phi L} = m\left[\frac{1}{2}\left(H_c^2 + H_s^2\right)\right] \tag{8.179}$$

The final sub-harmonic solution for the equation (8.129) [Davies and Rajan, 1988] is:

$$V_\phi = \frac{9a_3^2 V_\phi^2 V_{\phi L}(\gamma/3 + 2\delta)}{8\delta\left(\left((\omega_a/3)^3 - \omega_x^2(V_\phi)\right)^2 - \omega_\phi^2(\gamma/3 + 2\delta)^2\right)} \tag{8.180}$$

Formula (8.180) does not give an explicit expression for the roll variance; it is, actually, a quadratic equation relative to it. In the limiting case of deterministic excitation when $\gamma = 0$, we get the deterministic solution we derived in subchapter 4.6.3, again taking into account the relationship between amplitude and variance $V = A^2/2$.

Completing our consideration of the envelope method, we put here the most general result for nonlinear roll developed by Francescutto [1991, 1991a, 1992, 1998], Francescutto and Nabergoj [1990].

The following equation was considered in the above references:

$$\ddot{\phi} + (2\delta + \delta_1\phi)\dot{\phi} + \delta_3\dot{\phi} + \omega_\phi^2\phi + a_3\phi^3 = \alpha_m(t) \tag{8.181}$$

This equation has an advanced model of roll damping that includes the influence of roll angle. A cubic presentation of the *GZ* curve allows consideration of both small and large freeboard ships (only the initial part, however, is included in the latter case).

The following implicit expression for the roll variance estimate was derived for the case of main resonance excitation:

$$V_E = \frac{V_\phi}{1+\gamma/2\delta_{eq}} \cdot \left(\left(\omega_a^2 - \omega_\phi^2 - \frac{3}{2} a_3 V_\phi \right)^2 + \left(2\omega_\phi \delta_{eq}\right)^2 \left(1+\gamma/2\delta_{eq}\right)^2 \right) \quad (8.182)$$

Here, δ_{eq} is the equivalent damping coefficient:

$$\delta_{eq}(V_\phi) = \left(\delta + \frac{1}{4} V_\phi \left(\delta_1 + 3\omega_\phi^2 \delta_3 \right) \right) \quad (8.183)$$

Considering application of the envelope presentation in conjunction with the multiple scale method, we were able to derive formulae for variance estimates. These estimates include certain nonlinear phenomena: sub-harmonic response, for example. Moreover, the form of equations (8.166), (8.180) and (8.182) allow multiple responses for the same averaged wave frequency, which, as we have seen for the deterministic nonlinear roll, may be an indication of possible bifurcations.

8.3.6 Monte-Carlo Method

Monte-Carlo method was named after the city (and principality) of Monte-Carlo. Famous for its casinos, early probability laws were discovered there, while observing outcomes of gambling, in particular, dice. Dice were the earliest random generator. Method Monte-Carlo is actually a kind of stochastic simulation that always uses a random generator of a certain kind; this explains the name of the method.

Monte-Carlo method application for irregular nonlinear roll usually consists of two parts: generation of irregular waves and a numerical solution of the roll equation. Since a numerical solution of ordinary differential equations (see subchapter 4.2.5) does not imply any limitation of type of terms in the roll equation, the Monte-Carlo method is the most general way to treat nonlinear irregular roll. Results, however, depend strongly on adequacy of presentation of stochastic excitation.

Conventional application of Monte-Carlo method uses a Fourier presentation for simulation of irregular seas. As we have seen while considering different ways of presentation of stochastic processes, any of these models requires a random number as an input. For Fourier presentation, phase of component plays a role of input random number; see formula (8.58) repeated here again:

$$\zeta_W(t) = \sum_{i=1}^{N} c_i \cos(\omega_i t + \varepsilon_i)$$

Phase ε_i is considered to be a random variable with uniform distribution from 0 to 2π. It has to be calculated with a random generator, the piece of software, which nowadays is included into all math packages and standard libraries of subroutines supplied with compilers for all major programming languages. These types of generators, however, do not provide a really random number. Given exactly the same input data, any computer software always reproduces the same outcome. So, it is more correct to call them generators of pseudo-random numbers.

The algorithms for generation of pseudo random numbers usually involve the remainder of division of some number that can be user-supplied, or taken from the system timer. More information on the generation of pseudo-random numbers is available from a classical monograph [Knuth, 1997]. The majority of available pseudo-random generators returns a variable with a uniform distribution from 0 to 1. In order to get the required range, the result has to be multiplied by 2π. A sample outcome of one such generator is presented in fig. 8.8 in the form of a histogram.

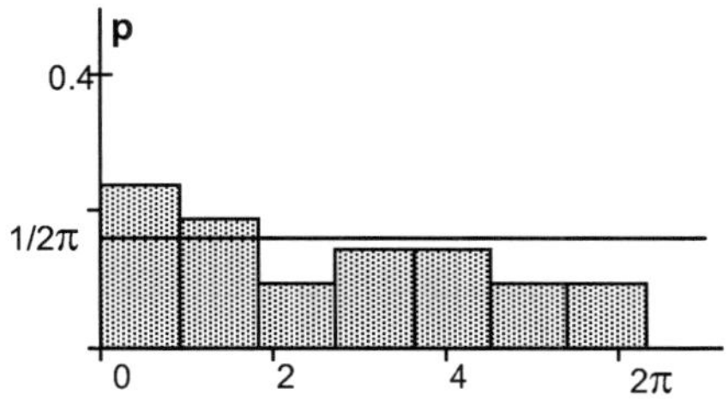

Fig. 8.8 Distribution of pseudo-random values –output of random generator

Next, a set of frequencies has to be chosen. So, we have to determine frequency band and number of components. There are several factors affecting the choice. First, a minimal time step, which still provides meaningful statistical information, is related with the width of frequency band:

$$\Delta t = \frac{1}{2(\omega_N - \omega_1)} \tag{8.184}$$

This value is called Nyquist interval and this relationship is a part of the sampling [Bendat and Piersol, 1986] and Kotelnikov theorems [Korn and Korn, 1968]. If the time step of simulation is less than defined by the formula (8.129), the result of the simulation may not be statistically representative.

Another important factor is frequency spacing. It is known from simulation practice that if equal frequency spacing is used, time history reconstructed with (8.58) nearly repeats itself. This effect is the most evident on an autocorrelation function plot, see fig. 8.9 (a). The peaks of autocorrelation function show strong dependence between different instances of time that normally should not be related that closely; this constitutes "self-repeating effect".

The conventional technique to mitigate this "self-repeating effect" is to use unequal spacing for frequency set. However, it does not completely eliminate the error, just spreading it over the time, see fig. 8.9 (b). Both autocorrelation functions in fig. 8.9 were calculated with cosine Fourier transform (8.21) with the frequency set considered: with equal and unequal spacing for 8.9 (a) and 8.9 (b), respectively.

It is shown in [Belenky 2004] that time domain integration (8.15) delivers almost identical results, but the effect is combined with statistical error, unavoidable for time domain simulations. It is not clear, if this error is acceptable for practical calculations and what effect it may have on the results of the simulations

It is shown in [Belenky 2005] that the reason for "self-repeating effect" is accumulation of numerical errors as both the inverse Fourier transform (8.58) and cosine Fourier transform (8.21) are essentially numerical integration procedures.

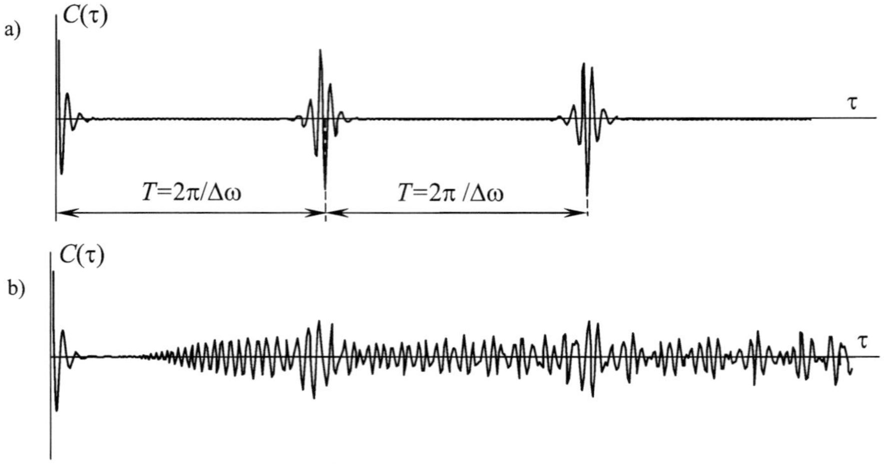

Fig. 8.9 Autocorrelation function calculated with cosine Fourier transform for the stochastic process presented with Fourier series (a) equal frequency spacing (b) unequal frequency spacing

A time period between two peaks corresponds to frequency step; it provides the time while we can consider (8.58) as a representation of stochastic process. This time duration has to be taken into account for correct choice of frequency set.

Physical considerations are also important for this choice. It is probably worthy to limit the highest frequency with the wave equal to half of ship breadth. Waves of higher frequency hardly create any heeling moment. An analogous approach has to be used when the lowest frequency is being chosen. In other words, frequency band has to cover the range of dynamic ship response.

Component amplitude c_i is calculated in accordance with the following formula:

$$c_i = \sqrt{2 \int_{\omega_i - 0.5\Delta\omega}^{\omega_i + 0.5\Delta\omega} S_\zeta(\omega) d\omega} \qquad (8.185)$$

Where $S_\zeta(\omega)$ is spectral density. With the integral calculated by substitution of segments of the function with rectangles, this formula converts into $c_i = \sqrt{2S_\zeta \Delta\omega}$. This procedure is illustrated in fig. 8.10.

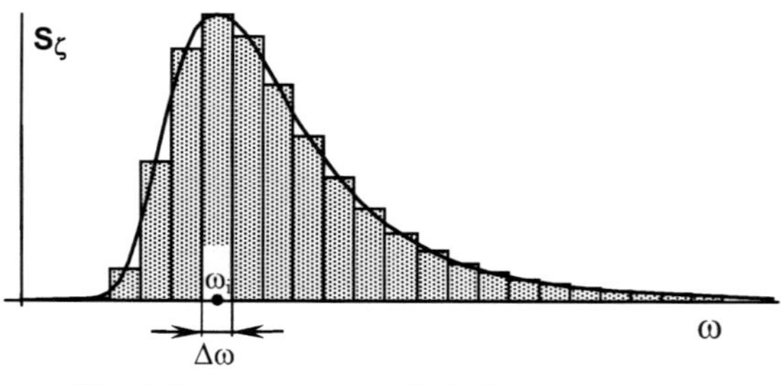

Fig. 8.10 Procedure of calculation of component amplitude

Calculation of amplitudes completes the definition of input for stochastic process of irregular waves. Wave elevations then have to be calculated using formula (8.58) for every step of simulation. However, one set of random phases provides us with only one realisation of irregular waves. Simulation with another realization of waves requires generation of another set of random phases (with each set containing N values). We would need as many of these sets for as many realizations of waves are required.

8.3.7 Non-Canonical Presentation and Monte-Carlo Method[1]

We considered non-canonical presentation of a stochastic process in subchapter 8.2.5. As we mentioned, this presentation is known for it ability to simulate time section of the process. Therefore, the simulation procedure has certain specific features that we will discuss here.

First, we have to determine input random numbers. Formula (8.68) contains three independent random variables:

$$\zeta(t) = \Lambda_1 \cos \omega t + \Lambda_2 \sin \omega t$$

These variables are "amplitudes" Λ_1 and Λ_2 with a Gaussian distribution and ω with the distribution density equal to the normalized spectrum.

The average value of both Λ_1 and Λ_2 has to be zero. Their variance is equal to the variance of waves. How do we create a normally distributed variable with the given mean and variance? We use commercially available software packages, which contain functions to generate normally distributed random variables. However, if such a generator is not available, it is very easy to create one. The central limit theorem states that for an infinite sum of random variables, each of which has an equal contribution, the distribution tends to be Gaussian. The theorem works regardless of the distribution of each individual component.

Actually, it is sufficient to sum up six independent random variables, which are supplied by the pseudo-random generator and have a uniform distribution. Frequently, this number is increased to 12, just to be on the safe side and because such a sum results in a variance equal to one. As it can be seen, this is the variance of a random variable with uniform distribution from 0 to 1 (having in mind that the mean value equals 0.5):

$$V(x) = \int_{-\infty}^{\infty} (x - m_x)^2 f(x) dx = \int_0^1 \left(x - \frac{1}{2} \right)^2 dx = \frac{1}{12} \tag{8.186}$$

Since variance of a sum of independent components equals the sum of the variances, the resulting variance of a random variable with a Gaussian distribution equals 1:

$$V(y) = \sum_{i=1}^{12} V(x_i) = 12 \cdot \frac{1}{12} = 1; \quad m(y) = \sum_{i=1}^{12} m(x_i) = 12 \cdot \frac{1}{2} = 6 \tag{8.187}$$

This is convenient for further simulations. Its average value has to be subtracted from the result, in order to get the required distribution.

A simple algorithm is also available for generation of random "frequency" ω. As we mentioned above, its distribution is identical to a normalized spectrum (spectrum divided by the variance):

$$f(\omega) = \frac{S_\zeta(\omega)}{V_\zeta} \tag{8.188}$$

---

[1] Written in co-authorship with Prof. Alexander B. Degtyarev and Dr. Alexander V. Boukhanovsky of Institute for High Performance Computing and Information Systems (Russia)

We introduce two scales: for frequencies $\omega_i\big|_{i=1,K}$ and probabilities $p_j\big|_{j=1,M}: p_1 = 0$ and $p_M = 1$. Say we need N values for ω. We calculate the number of expected occurrences of the value ω_i:

$$n_i = N \cdot \int_{\omega_i - 0.5\Delta\omega}^{\omega_i + 0.5\Delta\omega} f(\omega) d\omega \quad : \quad i = 1 \ldots N \tag{8.189}$$

Now, we run the generator of pseudo-random numbers. It produces the values ξ_i, with uniform distribution in the range from 0 to 1. First, we transform the range:

$$x_j = \omega_1 + (\omega_K - \omega_1) \cdot \xi_j : \quad \xi \in [0; 1] \tag{8.190}$$

Then, we build a simple filter: it passes through the current value of x_j only if their quantity is in a range $[\omega_i - 0.5\Delta\omega\,;\, \omega_i + 0.5\Delta\omega]$ (provided $x_j \in [\omega_i - 0.5\Delta\omega\,;\, \omega_i + 0.5\Delta\omega]$ does not exceed n_i). As it is exceeded, no more x-values in the range are passed through. The procedure stops when the total number of the values passed through reaches N.

Now we have completed our simulation of the excitation. The last step is rather simple: for each realization of wave elevation, we have a nonlinear roll equation:

$$\ddot{\phi} + 2\delta\dot{\phi} + f(\phi) = \omega_\phi^2 \frac{\omega^2}{g} (\Lambda_1 \cos \omega t + \Lambda_2 \sin \omega t) \tag{8.191}$$

This equation is solved with any approximate analytical method described in subchapter 4.2.

There is a problem associated with the fold bifurcation. There are two stable solutions at the same frequency. Which one should we take? There is no single answer yet. However, it is intuitively clear that if we need a upper estimate of roll variance, we would rather take high amplitude. We will return to this problem during our discussion of ergodic qualities of nonlinear roll.

8.3.8 Parametric Resonance in Irregular Beam Seas

We conclude our consideration of roll in irregular beam seas with the method of calculating parametric roll developed by Boroday and Morenshildt [1986], (available also from [Boroday, *et al*, 1989]).

We have considered parametric resonance in subchapter 6.2. Changes of the *GZ* curve in following seas were meant to be a possible physical reason for parametric excitation. We also mentioned that coupling with pitch and heave is another source of the parametric resonance. The last one may exist in beam seas as well if above-the-waterline geometry makes significant changes to the *GZ* curve.

To generate parametric oscillation, natural frequency of heaving should be about twice of the natural frequency of roll.

We use the approximate formulae for natural periods of heaving and rolling [Boroday, *et al*, 1989]:

$$T_z = 2.4\sqrt{d}; \quad T_\phi = \frac{0.8B}{\sqrt{GM}} \tag{8.192}$$

It is not difficult to express the condition of parametric resonance:

$$B = 6\sqrt{d \cdot GM} \tag{8.193}$$

The other condition to be satisfied is the substantial change of the *GZ* curve with the draft to provide a significant value of amplitude of parametric excitation, which can be expressed through the derivative of *KM* on draft *d*.

We introduce the concept of "equivalent" wave: it has parametric resonance frequency $2\omega_\phi$ and its amplitude is defined by the energy that is contained in the frequency band of parametric excitation, see fig. 8.11.

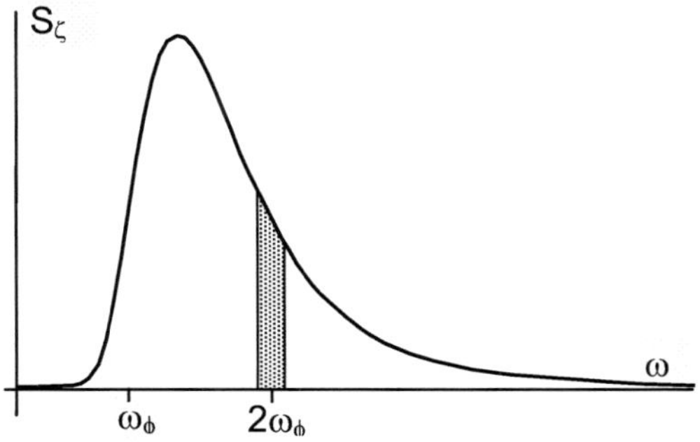

Fig. 8.11 An equivalent wave concept for beam seas

The amplitude of equivalent wave r_e as well as the amplitude of parametric roll a_p can be obtained by numerical solution of the following system of nonlinear algebraic equations:

$$\begin{cases} r_e = \sqrt{2V_{we}} \\ V_{we} = \int_{2\omega_\phi\sqrt{1-a_p/2}}^{2\omega_\phi\sqrt{1+a_p/2}} s_w(\omega)d\omega \\ V_{we} = \omega_\phi a_p s_w(\omega_{pm}) \\ a_p = \dfrac{1}{GM_0} \cdot \left|\dfrac{\partial KM}{\partial d}\right| \cdot \Phi_{wz}(\omega_{pm}) \cdot r_e \end{cases} \tag{8.194}$$

Here: V_{we} variance of equivalent wave; $s_w(\omega)$ wave spectrum density; Φ_{wz} heaving RAO; ω_{pm} is mean frequency of equivalent waveband, (see fig. 8.11).

The next step is the calculation of the RAO from wave height to roll parametric response:

$$\Phi_{wp\phi}(\omega) = \frac{\phi_p(\omega)}{a_p} \cdot \Phi_{wp}(\omega) \tag{8.195}$$

That includes the RAO from wave height to parametric excitation:

$$\Phi_{wp}(\omega) = \begin{cases} \dfrac{1}{GM_0} \cdot \left|\dfrac{\partial KM}{\partial d}\right| \cdot \Phi_{wz}(\omega_{pm}) & \text{if} \quad \omega \in \left[2\omega_\phi\sqrt{1-\dfrac{a_p}{2}}, 2\omega_\phi\sqrt{1+\dfrac{a_p}{2}}\right] \\ 0 & \text{if} \quad \omega \notin \left[2\omega_\phi\sqrt{1-\dfrac{a_p}{2}}, 2\omega_\phi\sqrt{1+\dfrac{a_p}{2}}\right] \end{cases} \tag{8.196}$$

As we have already mentioned in subchapter 6.2.2, parametric oscillation in a damped system can be generated if the excitation exceeds the threshold. The following formula is available from [Boroday, *et al*, 1989]:

$$\delta_\phi < \frac{\pi}{8}\omega_\phi^2 \cdot \left|\Phi_{wp}(\omega_{pm})\right|^2 \cdot s_w(\omega_{pm}) \tag{8.197}$$

If the inequality (8.197) is satisfied, the spectrum of parametric roll response can be calculated:

$$s_{p\phi}(\omega) = \left|\Phi_{wp\phi}(\omega)\right|^2 \cdot s_w(\omega) \tag{8.198}$$

This spectrum further has to be added to the spectrum of non-parametric roll response forming a picture shown in fig. 8.12.

The result of such an addition rather has to be understood as a pseudo-spectrum of roll motion because it shows the distribution of energy by the excitation frequencies. At the same time the real spectrum should represent energy distribution by frequencies of oscillations.

The pseudo-spectrum can be only used for calculation of variance but not for comparison with spectrums obtained from simulations or model tests. Boroday and Morenshildt [1986] recommend showing parametric roll response on half of the frequency in order to get the real spectrum eligible for comparison, see fig. 8.13.

For further development on parametric roll in beam seas, see [Ikeda, *et al*, 2005], [Munif, *et al*, 2006].

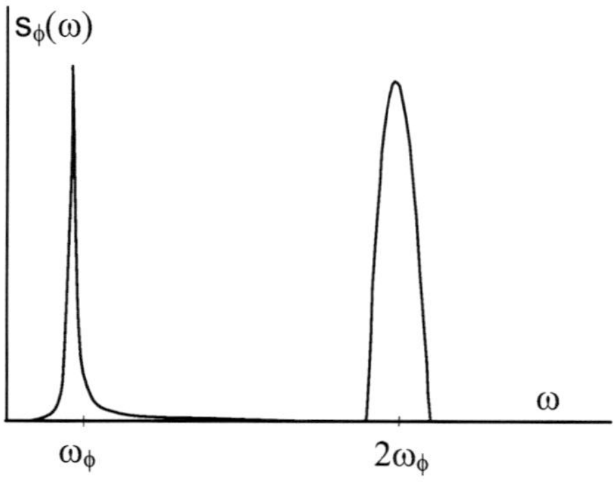

Fig. 8.12 Pseudo-spectrum of roll response with parametric component

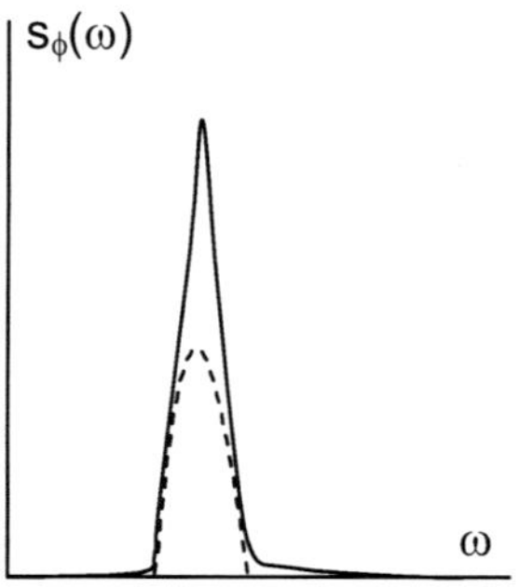

Fig. 8.13 True spectrum of ship roll response with contribution of parametric component (The last is shown by dashed line.)

8.4 Roll in Irregular Longitudinal Seas

8.4.1 Probabilistic Model of Irregular Longitudinal Seas

The coordinate system we are going to use was introduced in subchapter 6.4.2 and shown in fig. 6.13. It is the conventional combination of immovable and semi-movable systems: a semi-movable system experiences translating motion with average heading speed v_S.

Wave forces acting on a ship in following and quartering seas are dependent on the position of the ship in relation to the wave crest, (see subchapter 6.4). Therefore, we need to include wave position into the model of irregular seas. The rest is the same: irregular waves are presented as a Fourier series, having amplitudes defined from the spectrum and initial phases being random numbers with uniform distribution from 0 to 2π:

$$\zeta_W(t,\xi,\eta) = \sum_{i=1}^{n} r_i \cos\left(k_{W\,i}(\xi\cos\chi + \eta\sin\chi) - \omega_{ei}t + \varphi_i\right) \tag{8.199}$$

Here, ω_{ei} and k_i are encounter frequency and wave number corresponding to true frequency ω_i of i-th component respectively. More details are available in subchapter 10.14 of [Kobylinski and Kastner 2003].

8.4.2 Surging in Irregular Seas

As is well known, surging motion could be important for stability in quartering and following seas because the ship spends more time in the vicinity of the wave crest where stability is decreased. Here, we use the surging model (6.81) with small modifications expressing irregularity of waves:

$$(m + a_{11})\ddot{\xi}_G + R(v_s + \dot{\xi}_G) - T(v_s + \dot{\xi}_G) = F_w(t,\xi_G) \tag{8.200}$$

Here: $R(v_s + \dot{\xi}_G)$ is resistance of a ship in irregular seas. $T(v_s + \dot{\xi}_G)$ is propeller thrust. $F_w(t,\xi_G)$ is wave excitation force.

Following [Umeda, *et al*, 1995], we retain only the Froude–Krylov component of the wave excitation force. Its derivation does not meet any difficulties: direct integration of wave pressures along the hull surface [Belenky, 2000b]:

$$F_w(t,\xi_G) = \sum_{i=1}^{N} F_{axi} \sin\left[k_{W\,i}\xi_G(t)\cos\chi - \omega_{ei}t + \varphi_i + \gamma_i\right] \tag{8.201}$$

$$F_{axi} = -\rho g r_i k_{W\,i} \cos\chi \times$$
$$\times\sqrt{\left[\int_L I(x)\sin(k_{W\,i}x\cos\chi)dx\right]^2 + \left[\int_L I(x)\cos(k_{W\,i}x\cos\chi)dx\right]^2} \tag{8.202}$$

$$\gamma_i = \arctan\frac{\int_L I(x)\sin(k_{W\,i}x\cos\chi)dx}{\int_L I(x)\cos(k_{W\,i}x\cos\chi)dx} \tag{8.203}$$

$$I(x)=\begin{cases}\dfrac{2}{k_i\sin\chi}\displaystyle\int_0^{d(x)}e^{-k_{W_i}z}\sin\left[0.5k_{W_i}xb_H(x,z)\right]dx & \text{if: } \chi\neq 0\\ e^{-k_{W_i}d(x)}S(x) & \text{if: } \chi=0\end{cases}\qquad(8.204)$$

Here, $b_H(x,z)$ is hull breadth at abscissa x and depth z, $S(x)$ – is area of submersed section at abscissa x.

8.4.3 Changing Stability in Longitudinal Irregular Seas

The next problem to be addressed is how to take into account *GZ* curve changes in irregular seas. There are several approaches available. Boroday [1967] found an analytical solution for probabilistic characteristics of the restoring moment and its work in irregular seas. Then, he used the energy balance method, see [Boroday, 1968], [Boroday and Netsvetaev, 1969].

Another approach is based on the effective wave concept originally proposed by Grim [1961], fig. 8.14. The idea is that an irregular wave should be substituted by a regular one with the length equal to the length of a ship and its crest or trough situated at the centre of gravity. The effective wave concept was used by Umeda, *et al* [1990, 1993] for calculating probability of ship capsizing due to pure loss of stability in quartering seas. A comparison of the application of effective wave concept with direct stability calculations has validated such an approach.

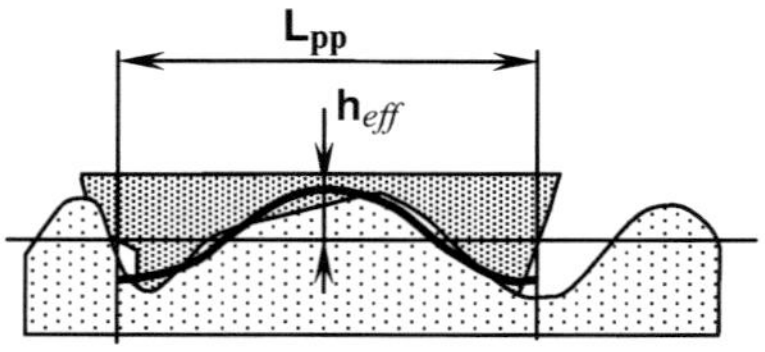

Fig. 8.14 A concept of effective wave [Grim, 1961]

This effective wave is unmovable, its crest is always amidships, but its random amplitude is set to have the same effect on stability as that of the original irregular profile:

$$\zeta_{eff}(t)=h_{eff}(t)\cos\left[\frac{2\pi}{L}x\right]\qquad(8.205)$$

The random amplitude of the effective wave $h_{eff}(t)$ is presented as a Fourier series in the same way we used for waves:

$$h_{eff}(t,\xi_G)=\sum_{i=1}^{n}h_i\cos(k_{W_i}\xi_G(t)\cos\chi-\omega_{ei}t+\varphi_i)\qquad(8.206)$$

Amplitudes h_i are to be found with a transfer function that was derived to minimize the difference between the effective wave and the original profile of an irregular wave along the ship's waterplane:

$$h_i=\frac{k_{W_i}L\cos\chi\sin(0.5k_{W_i}\cos\chi)}{\pi^2-(0.5k_{W_i}L\cos\chi)^2}\cdot r_i\qquad(8.207)$$

As it can be clearly seen from formula (8.206), surging is included into phase of effective wave amplitude. This is the way to take into account surging influence on stability in following and quartering seas. Now we can express changing of the *GZ* curve in quartering seas in a very simple way:

$$GZ_w(\phi, t, \xi_G) = GZ(\phi) + \Delta GZ_w(\phi, h_{eff}(t, \xi_G), \lambda = L, v_s) \quad (8.208)$$

Here, $\Delta GZ_w(\phi, h_{eff}(t, \xi_G), \lambda = L, v_s)$ is changing of the *GZ* curve due to heading in quartering or following seas. It does not depend on course directly: course influence is already included in effective wave height – formulae (8.206) and (8.207). It is assumed also that positive values of $h_{eff}(t)$ correspond to wave crest and negative ones to wave trough.

Alternatively, the *GZ* curve can be evaluated at every time instance using instantaneous waterline. This approach is implemented in LAMP, FREDYN and other time simulation codes.

Once the *GZ* curve is known, we can write the roll equation as:

$$\ddot{\phi} + 2\delta\dot{\phi} + \frac{\omega_\phi^2}{GM} GZ(\phi, t, \xi_G) = \frac{\omega_\phi^2}{GM} \cdot \alpha_E(t) \quad (8.209)$$

This equation can be solved by any method considered for roll in irregular beam seas. We will be using this model for estimation of capsizing probability in following seas in Chapter 9.

8.4.4 Parametric Resonance in Irregular Longitudinal Seas[1]

We have seen from our previous consideration (subchapter 6.2) that significant time should pass for parametric oscillation to be induced. In addition, it is quite frequency sensitive. Therefore, it seems that parametric excitation is unlikely to meet in irregular seas, where instantaneous wave characteristics are changing all the time.

Nevertheless, parametric resonance in irregular seas is not that rare. Phomenko [1967] observed this phenomenon during full scale trials Nechaev [1972] observed parametric excitation during semi-full scale model tests in natural waves in the Kursches gulf (Southeast Baltic Sea). Catastrophic parametric resonance in head seas caused damage and loss of cargo on a post-Panamax containership during her voyage in the Pacific [France, *et al*, 2003].

The post-Panamax C11 class containership was on route from Kaoshing (Taiwan) to Seattle and encountered a storm (wind 11 on Beaufort scale, sea state 9) for about 12 hours duration. The ship experienced parametric roll being in head seas. Roll amplitudes (according to the crew) reached 35-40 degrees, pitch amplitude was described as extremely high, based upon frequent overspeed trips of the main engine and significant shaft vibrations. Yaw amplitude reached 20 degrees making steering almost impossible. Because of extreme roll motion, one third of the 1300 containers on deck were lost.

Investigation of the accident included extensive weather modeling. It was shown that wind speed reached 29.5 m/s and significant wave height was around 13.4 m with a peak period of 15.4 seconds. The natural roll period of the ship was estimated to be 25.7 s, which is close to twice the peak period.

[1] Written in co-authorship with Prof. Alexander B. Degtyarev and Dr. Alexander V. Boukhanovsky of Institute for High Performance Computing and Information Systems (Russia)

Results of weather modeling were used in a series of model tests carried out in a seakeeping basin of MARIN. Parametric roll was successfully reproduced during these tests on both regular and irregular (long and short crested) waves. Experimental results confirmed observations of the crew: 33 degrees roll amplitude was recorded during runs on short crested irregular seas (significant wave height 12.63 m, peak period 14.65 s in full scale) with the calm water speed 11 knots. Model tests also had shown significant dependence of the likelihood parametric resonance on heading speed: it has to fall below a certain threshold for parametric roll to occur.

Model testing was supplemented with numerical simulations using codes FREDYN and LAMP. Both programs are nonlinear, time domain simulation software developed by MARIN [Hooft, 1987] and SAIC [Lin and Salvesen, 1998] correspondingly. FREDYN simulations reproduced parametric roll in the model test conditions for both regular and irregular seas with some underestimation of parametric roll amplitude.

LAMP simulations have shown good agreement with experiments when roll damping was tuned with results of a decoy test. The simulations have shown the importance of roll damping for correct prediction of parametric roll amplitude. We already mentioned in subchapter 6.2.3 that the nonlinear term of the roll equation is the factor limiting the amplitude of parametric response (parametric resonance in a linear system generates unlimited response). Therefore, adequate modeling of damping is the critical for parametric resonance simulation (another nonlinear factor, the *GZ* curve is known exactly). The LAMP simulation also demonstrated that the above-waterline geometry is a major factor to generate roll parametric response [France, *et al*, 2003].

The reason why the parametric resonance can be encountered in irregular seas is the group structure of irregular waves, see fig. 8.15 (see also [Blocky, 1980], [Tikka and Paulling 1990]). As we have discussed it earlier in subchapter 8.2.4, the envelope of irregular seas can be presented as a slowly changing function, which is another exhibition of this group structure. It is also known that frequency in the wave group does not change much (it is also the result of a relatively narrow wave spectrum). So, if a ship encounters a group of waves, which are long enough, parametric roll oscillations can build up to significant amplitudes.

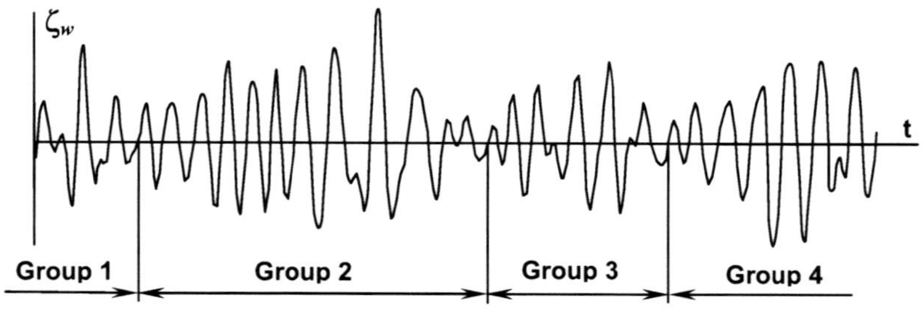

Fig. 8.15 Group structure of irregular seas

Group structure of waves also shows why roll damping has significant influence on roll variance and maximum amplitude in irregular seas. An increase of roll damping (for example, due to increase of speed) leads to an increase of excitation threshold (see subchapter 6.2), and then groups with smaller waves are no longer capable to produce parametric excitation, as they are below the threshold.

Adequate numerical simulation of parametric roll, therefore, requires a good model of roll damping. Some of the numerical codes allow tuning to reproduce results of a roll decay test [France, *et al,* 2003], [Belenky, *et al,* 2006]. ABS [2004] requires such a tuning to be performed when numerical simulations are used for development of ship-specific on-board guidance.

We will briefly discuss some of the statistical issues related to group structure of waves and their correlation with the group structure of roll response in subchapter 8.6.3. Here, we consider some outcomes of this group approach [Boukhanovsky and Degtyarev, 1996] [Degtyarev and Boukhanovsky, 2000]:

- Influence wave height in-group does not have a significant influence on amplitude of parametric response. However, periods of wave in-group are very important.
- Parametric response occurs in long groups with a large number of waves.
- Sequence of short groups with similar periods is more dangerous for a ship than one long wave group.
- If a ship encounters a group of wave without static bias, the parametric response is independent of initial conditions. However, it is not correct for real (asymmetrical) waves.

Considering parametric response for a different type of wave environment (wind waves, swell and complex seas) the following outcomes were reached:

- Swell is the most dangerous wave regime for parametric resonance. Usually such a regime produces groups with 13-14 waves, although wave height and steepness are small.
- Wind waves are less dangerous than swell since they have less pronounced group structure. Nevertheless, period coincidence makes stormy seas dangerous (see above description of post-Panamax containership), though group size is small: less than 10. For moderate sea periods, they are more dispersed and length of a group is up to 12. However, since period (or repetition of consequence group periods) has more effect, the stormy sea is more dangerous than a moderate one.
- Complex sea is characterized by very weak group structure. The number of waves in a group is less than 8. Periods are dispersed significantly, so significant parametric excitation is unlikely.

8.5 Influence of Gusty Wind

8.5.1 Distribution of Aerodynamic Pressures

The next step towards a more realistic model of ship motions in irregular seas is taking into account the stochastic character of a gusting wind [Sevastianov, 1994]. We present a horizontal component of wind speed as we have described it in subchapter 8.2.1, formula (8.50):

$$u(t) = u_A + u_G(t)$$

Here, u_A is a mean value of wind horizontal speed and $u_G(t)$ is the fluctuating part of the process. We assume wind velocity has a Gaussian distribution and look for wind pressures or wind forces.

The relationship between wind velocities and pressures is nonlinear:

$$|p| = \frac{\rho_a u^2}{2} = \frac{\rho_a u_A^2 u_f^2}{2} \tag{8.210}$$

Here, we introduce non-dimensional wind velocity u_f as:

$$u_f = \frac{u}{u_A} \tag{8.211}$$

It can be expressed via aerodynamic pressure as:

$$|u_f| = \sqrt{\frac{2}{\rho_a u_A^2}}\sqrt{|p|} = \sqrt{K}\sqrt{|p|} \tag{8.212}$$

Since we are interested in positive wind speed, we shall carry out our further analysis without a sign of absolute value.

Using the well known formula for the distribution of monotonic deterministic function of a random argument, we can obtain a formula for probability density of aerodynamic pressures caused by horizontal component of wind speed:

$$g(p) = f[\Psi(p)]\frac{d\Psi(p)}{dp} \tag{8.213}$$

Here, $\Psi(p)$ is the expression for wind velocity as a function of aerodynamic pressure, which is formula (8.212) in our case. Evidently:

$$g(p) = \sqrt{\frac{K}{8\pi p V_{uf}}}\exp\left[-\frac{(\sqrt{K}\sqrt{p}-1)^2}{2V_{uf}}\right] \tag{8.214}$$

Here, V_{uf} is a variance of non-dimensional wind velocity u_f .

It can be seen clearly that an exponential term is present in (8.214). The distribution of aerodynamic pressure is different from a Gaussian distribution. The appearance of this distribution is shown in fig. 8.16.

As it can be seen from equation (8.214) and from fig. 8.16 the probability density of the pressure has a singularity at p=0. However, the following integral $\int_{-\infty}^{\infty} g(p)dp$ converges and is equal to 1. Also:

$$\int_0^\infty g(p)dp = 1 - \beta \tag{8.215}$$

Where β is a small value. This can be proved by the following substitution:

$$z = \frac{\sqrt{kp} - 1}{\sqrt{V_u}} \quad \text{and} \quad dz = \frac{1}{2}\sqrt{\frac{K}{V_u}}\frac{dp}{\sqrt{p}}$$

We shall get:

$$\int_0^\infty g(p)dp = \int_{-1/V_U}^\infty \exp(-z^2/2)dz = \frac{1}{2}\left[\Phi(\infty) - \Phi\left(-1/\sqrt{V_u}\right)\right] \tag{8.216}$$

Where Φ is the Laplace function: $\Phi(x) = \int \exp(0.5 \cdot z^2)dx$

The value of β here is very small: in our example $\beta < 10^{-5}$.

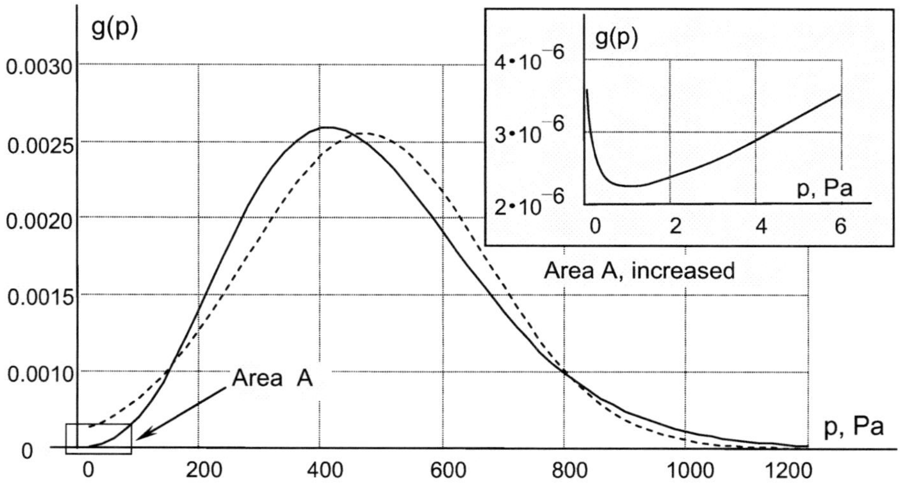

Fig. 8.16 Probability density of pressure (dashed line is Gaussian distribution), u_{Am} = 27 m/s , V_u = 0.046, ρ_A= 1.25 kg/m^3 [Sevastianov, 1994]

Let us estimate how large an approximation error we will make if the true distribution of gusty wind pressure (8.214) is substituted by a Gaussian distribution. This can be done in a more convenient way by introducing a dimensionless wind pressure that can expressed as:

$$C_P = \frac{p(t)}{\rho_a \cdot \frac{u_{Am}^2}{2}}$$

Consequently:

$$C_P = u^2 \tag{8.217}$$

"True" and Gaussian distributions of non-dimensional wind pressures can be expressed as follows, correspondingly:

$$g(C_P) = \frac{1}{\sqrt{8\pi V_{fu} C_P}} \exp\left[-\frac{\left(\sqrt{C_P} - 1\right)^2}{2V_{fu}} \right]$$

$$f(C_P) = \frac{1}{\sqrt{8\pi V_{fu}}} \exp\left[-\frac{\left(C_P - 1\right)^2}{8V_{fu}} \right]$$

As it can be clearly seen, if the Gaussian law is applied, then the mean of the dimensionless wind pressure is equal to unity and its variance is:

$$V_{Cp} = 4V_{fu} \tag{8.218}$$

Relative error can be expressed as:

$$err(C_P) = \frac{f(C_P) - g(C_P)}{g(C_P)} \tag{8.219}$$

Weighted mean square error of the given range [C_{P1}; C_{P2}] (using the distribution density as a weight function) looks like:

$$\overline{err} = \sqrt{\frac{\int_{C_{P1}}^{C_{P2}} g(C_P) \cdot \left[err(C_P)\right]^2 dC_p}{\int_{C_{P1}}^{C_{P2}} g(C_P) dC_p}} \tag{8.220}$$

A numerical example of the weighted mean error calculation is given below in table 8.2. The error cannot be estimated for V_{fu}=0.046 when the range is $1 \pm 3\sqrt{V_{Cp}}$ because the beginning of the range has a negative value.

Table 8.2 Numerical example of error estimation

Variance Vu	0.0169			0.046		
Range, σ_p	$1-\sigma_p$ $1+\sigma_p$	$1-2\sigma_p$ $1+2\sigma_p$	$1-3\sigma_p$ $1+3\sigma_p$	$1-\sigma_p$ $1+\sigma_p$	$1-2\sigma_p$ $1+2\sigma_p$	$1-3\sigma_p$ $1+3\sigma_p$
Range, C_p	0.74-1.26	0.48-1.52	0.22-1.78	0.57-1.43	0.14-1.86	-
Error, %	1.7	7.0	7.4	8.2	6.0	-

These examples show very clearly that the Gaussian approximation for the "true" law of pressure distribution is more accurate for small variances of relative wind speed. Further, we shall assume a Gaussian distribution of aerodynamic pressure in the first expansion.

8.5.2 Fourier Presentation for Aerodynamic Forces

Aerodynamic forces and moments caused by gusty wind also can be presented with a Fourier series using the same frequencies that were set up for wave presentation. However, the set of initial phase angles should be different since we assume that wind and waves are not synchronous:

$$F_{YA}(t) = F_{YAm} + \sum_{i=1}^{N} a_{Yi} \sin(\omega_i t + \varphi_{Ai}) \tag{8.221}$$

$$M_{XA}(t) = M_{XAm} + \sum_{i=1}^{N} a_{Mi} \sin(\omega_i t + \varphi_{Ai}) \tag{8.222}$$

These presentations presume a Gaussian distribution for aerodynamic forces and moments caused by gusty wind. We have discussed how applicable such an assumption is in relation to wind pressures (see subchapter 8.5.1) and found the error acceptable. Because the integration operation is linear, it does not change the character of a Gaussian distribution and we can assume a Gaussian distribution of aerodynamic forces and moments as well without any additional stipulations.

This means that using the formula we neglect the second power of the fluctuating part of velocities u:

$$F_{YA}(t) = C_y \frac{\rho_a}{2} A_W (u_A^2 + 2u_A u_G) \tag{8.223}$$

$$M_A(t) = C_{mx} \frac{\rho_a}{2} A_W (u_A^2 + 2u_A u_G)(z_A - KG) \tag{8.224}$$

Expressions (8.223) and (8.224) yield formulae for Fourier amplitudes and mean values in (8.221) and (8.222):

$$a_{Yi} = C_y \rho_a A_W u_A u_{0i} \tag{8.225}$$

$$a_{Mi} = C_{mx} \rho_a A_W u_A (z_A - KG) u_{0i} \tag{8.226}$$

$$F_{YAm} = C_y \frac{\rho_a}{2} A_W u_A^2 \tag{8.227}$$

$$M_{XAm} = C_{mx} \frac{\rho_a}{2} A_W u_A^2 (z_A - KG) \tag{8.228}$$

Where u_{0i} is to be defined from an appropriate wind spectrum with an accepted set of frequencies.

8.5.3 Swaying and Drift in Beam Irregular Seas

We have considered heeling wind action on a ship in subchapter 7.1.3. We have seen that wind causes drift. This leads to the generation of an additional heeling moment by a pair of wind horizontal forces and hydrodynamic reaction on drift. To include this factor into the roll model in irregular seas, let us consider an equation of horizontal motion.

Following [Belenky, 1995] we will use a horizontal motion equation of the following form (assumption of its independence from roll was addressed in [Belenky, 1990]) :

$$(m+a_{22})\ddot{y}+F_{Y0H}(\dot{y})=F_{YE}(t)+F_{YA}(t) \quad (8.229)$$

Here, $F_{Y0H}(\dot{y})$ is hydrodynamic drift reaction, $F_{YE}(t)$ is wave excitation and $F_{YA}(t)$ is aerodynamic force considered in subchapter 8.5.2.

We present velocity of horizontal motion as a sum of a constant and time varying components analogously to wind velocity:

$$\dot{y}(t)=v_D+\dot{y}_t(t) \quad (8.230)$$

Using formula (7.12) for the hydrodynamic reaction:

$$F_{Y0H}=C_H\frac{\rho(\dot{y}(t))^2}{2}A_D \quad (8.231)$$

The equation (8.229) is nonlinear. We linearized it for further use with the piece-wise linear approach. It is necessary only in order to get a solution for capsizing probability in closed form. If numerical simulation is used, linearization is not needed. Linearization can be done with any appropriate method. Here, we just drop the second power of fluctuating horizontal velocity, assuming it is small in comparison with a constant drift:

$$(m+a_{22})\ddot{y}_t+C_H\rho A_d v_D\cdot\dot{y}_t+0.5C_H\rho A_D v_D^2=F_{YE}(t)+F_{YA}(t) \quad (8.232)$$

Equalizing time independent terms, the constant horizontal velocity component v_D can be found:

$$v_D=\sqrt{\frac{2F_{YAm}}{C_H\rho A_D}} \quad (8.233)$$

Now (8.232) is a linear differential equation and its solution can be found:

$$\dot{y}_t(t)=C\exp(-\delta_y t)+\sum_{i=1}^{N}b_i\sin(\omega_i t+\gamma_i+\vartheta_i) \quad (8.234)$$

Here: b_i - harmonic amplitude of horizontal velocities:

$$b_i=\frac{f_i}{\sqrt{\omega_i^2+\delta_y^2}} \quad (8.235)$$

δ_y is relative horizontal damping coefficient:

$$\delta_y=\frac{C_H\rho A_D v_D}{(m+a_{22})} \quad (8.236)$$

f_i is amplitude of total horizontal excitation:

$$f_i=\frac{\sqrt{(a_{EYi}^2+a_{Yi}^2+4a_{EYi}a_{Yi}\cos(\varphi_{Ei}-\varphi_{Ai})}}{m+a_{22}} \quad (8.237)$$

γ_i is phase angle of total horizontal excitation:

$$\gamma_i = \arctan \frac{a_{EYi} \sin \varphi_{Ei} + a_{Yi} \sin \varphi_{Ai}}{a_{EYi} \cos \varphi_{Ei} + a_{Yi} \cos \varphi_{Ai}} \quad (8.238)$$

Here, a_{EYi} is the amplitude of wave excitation for swaying. For the first expansion it can be taken in the following form [Belenky, 1990, 1995]:

$$a_{EYi} = Mgk_Y(\omega_i)\alpha_{Ai} \quad (8.239)$$

With $k_Y(\omega)$ being the reduction coefficient taking into account finite ship dimension in comparison with the wave. It tends to 1 if a ship can be considered small in comparison with the wave. α_{Ai} is amplitude of the wave slope for component i.

ϑ_i is initial phase angle of swaying motion velocities:

$$\vartheta_i = \arctan\left(-\frac{\omega_i}{\delta_y}\right) \quad (8.240)$$

Summarizing, we can write a final general solution of the horizontal motion equation:

$$\dot{y}(t) = C\exp(-\delta_y t) + v_D + \sum_{i=1}^{N} b_i \sin(\omega_i t + \gamma_i + \vartheta_i) \quad (8.241)$$

Here, C is the arbitrary constant of horizontal motion depending on initial horizontal velocity $\dot{y}(t_0)$:

$$C = \dot{y}(t_0) - v_D - \sum_{i=1}^{N} b_i \sin(\gamma_i + \vartheta_i) \quad (8.242)$$

An example of time history for horizontal velocity is given in fig 8.17.

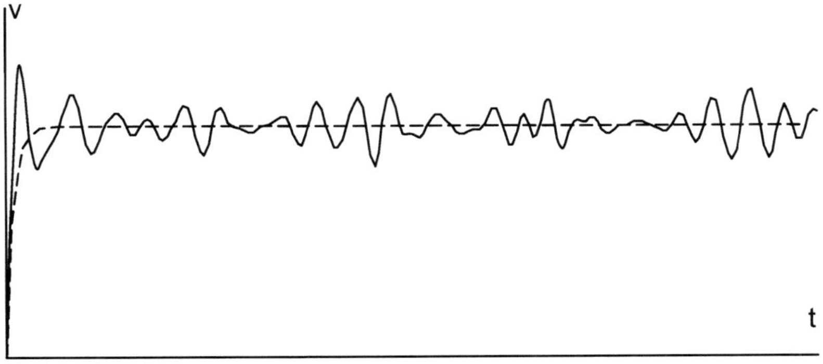

Fig. 8.17 Transition process of horizontal velocity, dashed line is drift velocity without swaying influence

8.5.4 Roll Under Action of Beam Irregular Seas and Gusty Wind

We continue consideration of gusty wind effect on roll in irregular seas. Let us supplement the equation of horizontal motion with the roll equation from [Belenky, 1995]. We consider the following system of equations; they describe ship motion under combined action of gusty wind and irregular seas:

$$\begin{cases}(m+a_{22})\ddot{y}+F_{Y0H}(\dot{y})=F_{YE}(t)+F_{YA}(t)\\ (I_{xx}+a_{44})\ddot{\phi}+N_{\phi}\dot{\phi}+mg\cdot GZ(\phi)+a_{24}\ddot{y}+M_{X0H}(\dot{y})=M_{XE}(t)+M_{XA}(t)\end{cases}\qquad(8.243)$$

Here N_{ϕ} is linearized roll damping coefficient and $F_{YE}(t)$, $M_{XE}(t)$, $F_{YA}(t)$ and $M_{XA}(t)$ are wave and wind excitation force and moment correspondingly. They are presented as a Fourier series using random phase set φ_{Ei} and φ_{Ai}. Since we do not consider roll influence on drift and sway, it is enough to consider only the roll equation: coupling with horizontal motion may be included into the excitation:

$$(I_{xx}+a_{44})\ddot{\phi}+N_{\phi}\dot{\phi}+mg\cdot GZ(\phi)=M_{XE}(t)+M_{XA}(t)-a_{24}\ddot{y}-M_{X0H}(\dot{y})\qquad(8.244)$$

We present this hydrodynamic moment M_{X0H} similar to horizontal motion velocity, see formula (8.230):

$$M_{X0H}=M_{X0Hm}+M_{X0Hf}(t)\qquad(8.245)$$

We already considered this moment for regular seas in subchapter 7.1.3:

$$M_{X0Hm}=0.5\cdot\rho C_H A_D v_D^2\cdot(KG-z_{HP}(\phi))\cdot\cos\phi\qquad(8.246)$$

$$M_{X0Hf}(t)=0.5\cdot\rho C_H A_D v_D\cdot(KG-z_{HP}(\phi))\cdot\dot{y}_t\cdot\cos\phi\qquad(8.247)$$

Here, z_{HP} is pseudo centre of elevation (see subchapter 7.1.3). Actually, hydrodynamic moment is dependent on heel angle and it is a nonlinear term, so it is placed on the left side of the roll equation:

$$(I_{xx}+a_{44})\ddot{\phi}+N_{\phi}\dot{\phi}+mg\cdot GZ(\phi)+M_{X0H}(\phi,t)=M_{XE}(t)+M_{XA}(t)-a_{24}\ddot{y}\qquad(8.248)$$

It can be linearized; however, to facilitate the application of the piecewise linear model of capsizing, we use substitution $\phi=\phi_D$, where ϕ_D is static heel angle caused by constant component of the wind.

Descriptions of further development of models for roll motions under action of wind can be found in [Paroka, *et al,* 2006, Paroka and Umeda 2006].

8.6 Probabilistic Qualities of Nonlinear Irregular Roll

8.6.1 Ergodicity of Nonlinear Irregular Roll[1]

We have considered ergodicity as a general quality of a stochastic process in subchapter 8.1.4. We also discussed in subchapter 8.2.3 that a stochastic process of irregular sea waves is assumed ergodic.

It is known from probability theory that the output of a linear system is always ergodic if its input possesses ergodicity. Evidently, linear system response to the external excitation is precisely described by the RAO. This operator changes amplitude and phase of each component of the Fourier presentation leaving the frequency set untouched, which is an essence of the Weiner-Khinchin theorem (see subchapter 8.3.1). It means that we get a Fourier presented process on output as an exact response. This kind of response cannot deviate from ergodicity. A strict proof can be found in [Kramer and Leadbetter, 1967].

However, there is nothing proved concerning the response of general nonlinear systems. Consequently, the response of a nonlinear system may be non-ergodic. We need to know how significant is the error for application purposes. Precisely, what kind of error do we make if we neglect this absence of ergodicity for the nonlinear roll response.

Usually the presence or absence of an ergodic quality is fixed relative to a certain probabilistic characteristic. We will work with the variance, since it is the most common way to describe intensity of ship motions.

How do we check the ergodicity? Let us obtain realization and calculate its variance after the first 50 seconds (or other interval). Then, repeat the calculation for the first 100 s, 150 s, 200 s and so on. We receive a sequence of variance estimates that theoretically should converge towards the true value of the variance. This sequence is shown in fig. 8.18. The case shown there is purely illustrative, we just took the process presented by a Fourier series and with a known variance.

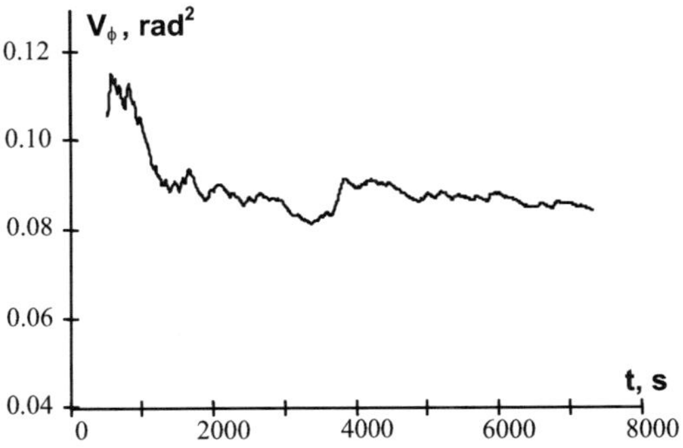

Fig. 8.18 Sequence of variance estimates plotted against time for one realization

As it can be clearly seen from fig. 8.18, this sequence makes an oscillating curve converge to the true value of the variance. The amplitude of the oscillations is decreasing with time, so the convergence of the estimate to true value can be visualized.

Now let us consider the linear model of roll:

$$\ddot{\phi} + 2\delta\dot{\phi} + \omega_{\phi}^{2}\phi = \alpha_{E}(t) \tag{8.249}$$

Excitation presented with the Fourier series:

---

[1] Written in co-authorship with Prof. Alexander B. Degtyarev and Dr. Alexander V. Boukhanovsky of Institute for High Performance Computing and Information Systems (Russia)

$$\alpha_E(t) = \sum_{i=1}^{n} \alpha_{Ei} \cos(\omega_i t + \varepsilon_i)$$

Despite the linear system, (8.249) does have an analytical solution the Monte-Carlo method is used (see subchapter 8.3.6) in order to be consistent with further nonlinear study. We generated several sets of initial phases, used each of them to create a set of realizations of the excitation. Then we applied the Runge-Kutta method (see subchapter 4.2.5) to get the roll response. The variance of each realization was estimated in the above manner for the interval increasing with a step of 50 seconds. The resulting sequences are plotted against time in fig. 8.19a [Belenky, *et al*, 1995, 1998].

Another way to visualize absence or presence of ergodicity is to calculate the confidence interval for variance estimates for each realization and plot them along with the estimates, see insert in fig. 8.19a. Method of calculation of confidence interval for realization of ergodic is briefly described in Appendix 2 [Belenky 2004].

To see the effect of nonlinearity on ergodicity, a nonlinear restoring term is introduced.

$$\ddot{\phi} + 2\delta\dot{\phi} + \omega_\phi^2 f(\phi) = \alpha_m(t) \tag{8.250}$$

Exactly the same procedure is applied to the system (8.250), and the result is shown in fig. 8.19b. The GZ curve is shown in the second insert of fig. 8.19b.

Comparing cumulative variances in fig. 8.19 (a) and (b), it can be see that the tendency to convergence is visually better in the linear case. The cumulative variances of a nonlinear system in fig. 8.19b show a tendency to go almost parallel to each other, so convergence is visually worse for the nonlinear system.

Comparing the plots of confidence intervals; it can be seen that in the case of a linear system in fig 8.19a, the boundaries of confidence are lined up in a more or less orderly fashion. At the same time, in the case of a nonlinear system, these boundaries are shifted against each other.

It is visually clear from fig. 8.19 (a) and (b) that the difference between variance estimates is larger for a nonlinear system. A measure was proposed in [Belenky 2004] to quantify this difference. This measure is essentially a shift between confidence intervals averaged over available realizations and presented in terms of probability

$$K_{Un} = P_\beta^2 - \frac{1}{(n-1)\cdot n}\left(\sum_{i=1}^{n}\sum_{j=1}^{n} P_{ij}P_{ji} - \sum_{i=1}^{n} P_{ii}^2\right)$$
$$P_{ij} = P(V_i \in [a_i, b_i] \cap [a_j, b_j]) \tag{8.251}$$

Here P_β is the confidence probability, n available number of realizations, P_{ij} is a probability that the true value of variance for the realization i belongs to an intersection between confidence intervals of realization i and j, see fig. 8.20. If there is no shift between the confidence interval, $P_{ij}= P_\beta$ and the measure equals zero.

This measure is meant to be used relatively, comparing its value for nonlinear roll with the similar value calculated for the process of excitation of a similar linear system, as they are known to be ergodic.

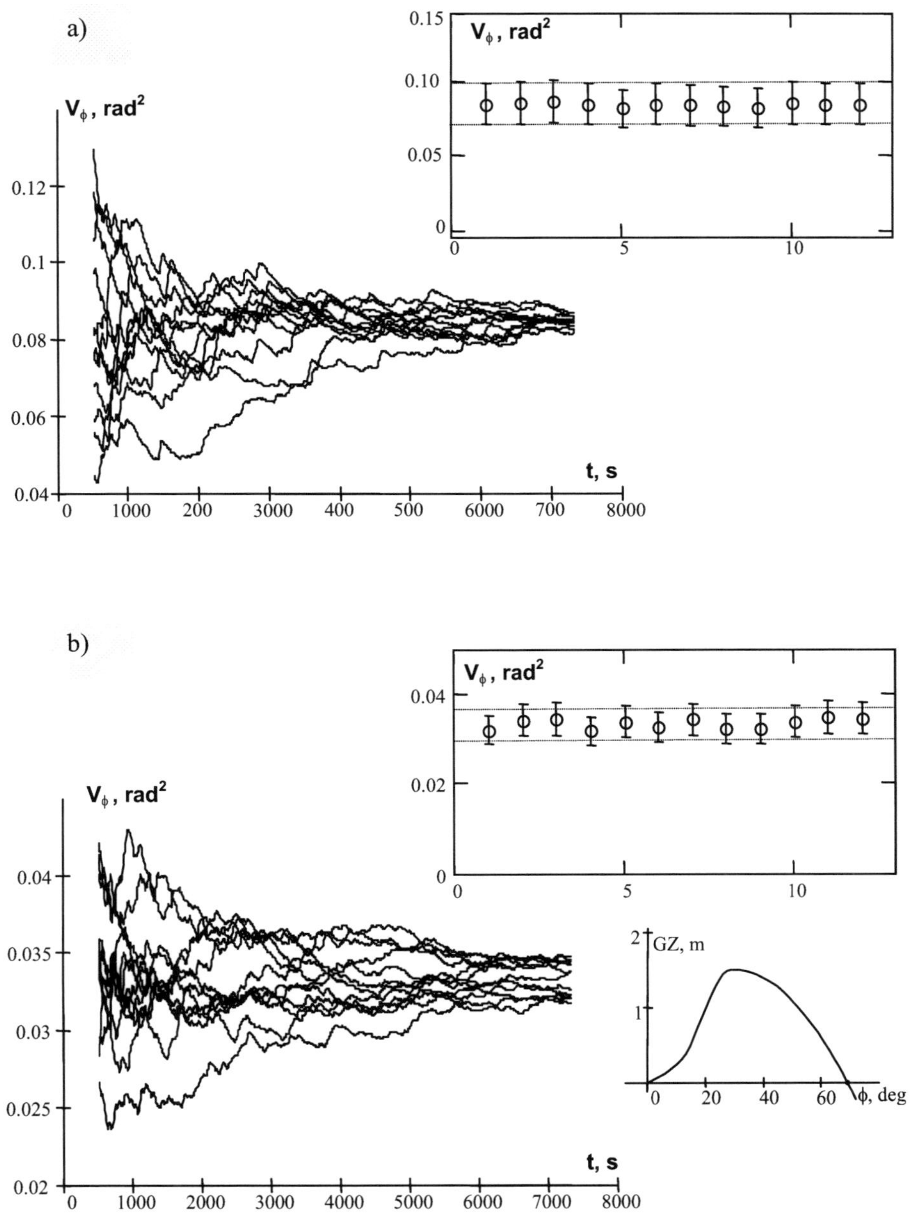

Fig. 8.19 Cumulative variances and confidence intervals of variance estimates for linear system (a) and nonlinear system (b)

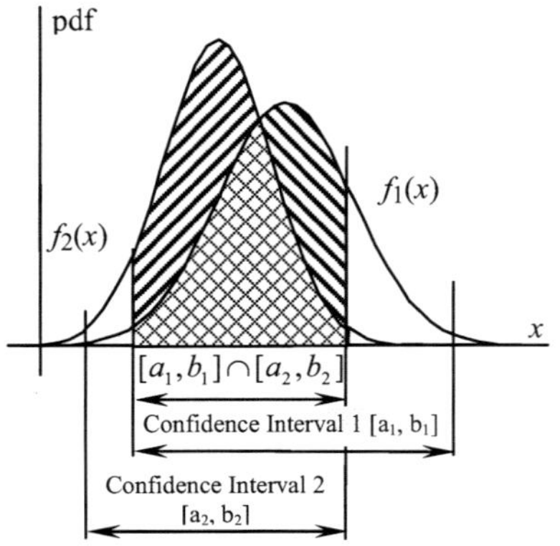

Fig. 8.20 On measure of ergodicity

A significant increase in the numerical value of this measure over the linear system indicates possible practical non-ergodicity and the necessity to consider several realizations. In the considered example, the measure for the linear system was 0.08 while for the nonlinear system, it reached 0.25. A detailed description of this measure is placed in Appendix 2.

This analysis, however, does not necessarily establish absence of ergodicity in a strict theoretical sense. The time duration is limited. Maybe if we increase time duration, the difference between the values of this measure would disappear? It is possible; however, it does not matter for practical calculations, as the time is always limited while the wave conditions still could be considered stationary. If we cannot use an ergodic assumption during this time, then the process is not ergodic for our practical purposes. (This is why the term "practical non-ergodicity" is being used; another term used for the same concept is "cyclic non-stationarity".)

The physical reasons of this phenomenon were addressed in [Belenky, *et al*, 1995, 1997, 1998]. It was suggested that fold bifurcation may be one of these reasons. It was found that the measure of non-ergodicity increases as the system reaches conditions where expected response domain covers frequency band with fold bifurcation possible. Model tests carried out in the towing tank of the National Research Institute of Fisheries Engineering [Belenky, *et al*, 2001] have shown, however, that fold bifurcation may not be the sole reason for practical non-ergodicity.

8.6.2 Distribution of Nonlinear Irregular Roll

It is known that the wave surface elevation can be presented as a Gaussian stochastic process as well as the process of angles of wave slope. There is also a well known proof that a linear system having Gaussian input produces a Gaussian output as well.

Here we address the question: if the difference between the real distribution of output of a nonlinear system with a Gaussian distribution strong enough, that makes it impossible to assume a Gaussian distribution of the output for practical calculation purposes? We focus on the Monte-Carlo method. An alternative approach is based on the application of Markov processes, its background is considered in subchapter 8.6.4.

Nechaev [1989] obtained distribution of irregular nonlinear roll using the Monte-Carlo method. The roll response process had a non-zero mean value. The resulting histograms are given in fig. 8.21. It is quite clear that the distributions in fig. 8.21 are not Gaussian. It is easy to check it using standard methods of checking distribution hypothesis as χ^2 or Kholmogorov criterion. It is necessary to remember, however, that if a method is intended for random variables rather then process, it may require independent data. Time history of realization of a stochastic process is not necessarily present in such a data, as dependence may be quite strong between two consecutive points. Therefore some points need to be skipped to allow the autocorrelation function to die out.

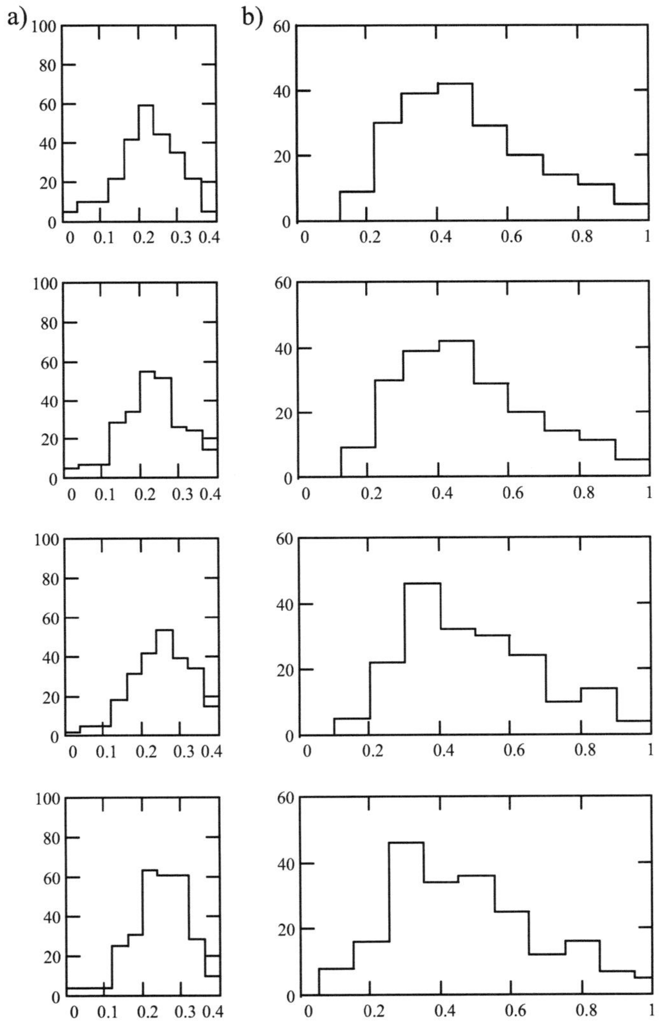

Fig. 8.21 Histograms of nonlinear roll angles for the ship with initial bias 13 degrees (a) and 8.8 degrees (b) [Nechaev, 1989]

Belenky [1993, 1994], Belenky, *et al,* [1998] contains results of simulation using the system (8.243). Taking into account absence of ergodicity, distributions were calculated using a whole ensemble of realizations. The appearance of the *GZ* curve is given in fig. 8.22. Resulting distributions are given in fig. 8.23.

As it can be clearly seen from fig. 8.23, the discrepancies between the Gaussian and actual distribution are visually negligible. The application of a Gaussian type of distribution seems reasonable.

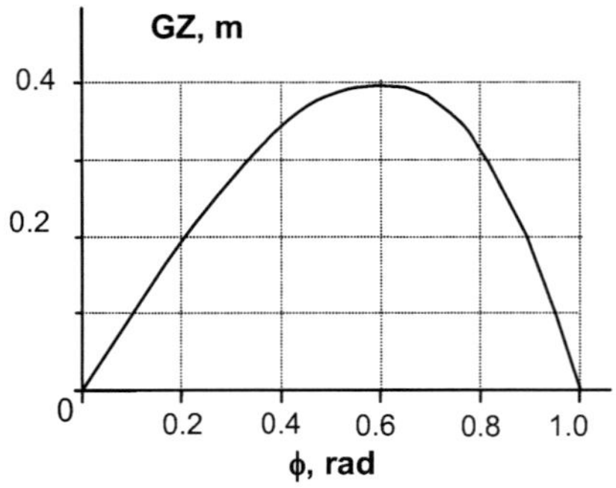

Fig. 8.22 "Normal" *GZ* curve

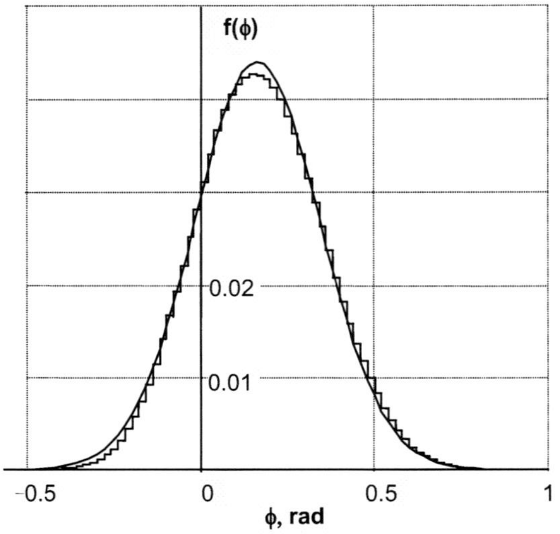

Fig. 8.23 Histogram of severe roll. Solid line is Gaussian distribution.

Let us see now whether the form of the *GZ* curve can influence the distribution, appearance of *GZ* curve is given in fig. 8.24 and the distributions obtained are shown in fig. 8.25.

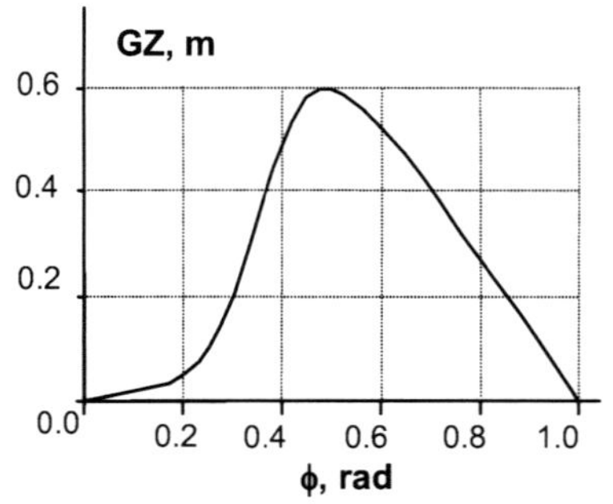

Fig. 8.24 Exaggerated S-shaped *GZ* curve

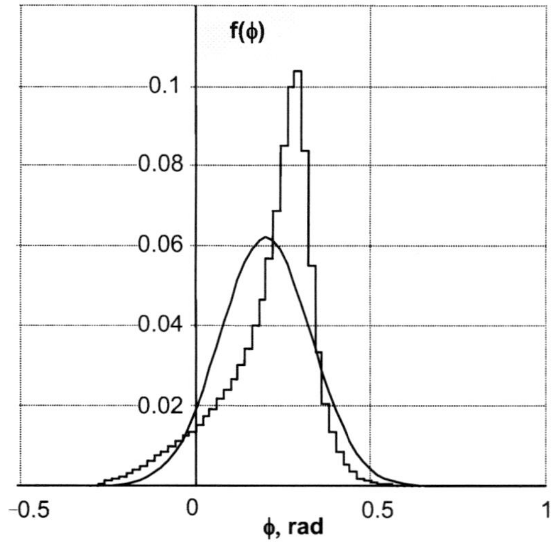

Fig. 8.25 Histogram of severe roll velocities for exaggerated S-shape *GZ* curve. Solid line is Gaussian distribution.

It can be seen clearly from fig. 8.25 that the distribution for roll with an S-shaped *GZ* curve is not Gaussian.

The difference between these cases can be explained in the following way. Nonlinearity in the first case becomes large when some of roll angles reach the peak of the *GZ* curve. Such roll angles are rare, so nonlinearity does not have a large influence on the distribution. Therefore, the character of the distribution mainly is defined by the most statistically frequent. Roll angles that are located near the origin, where the Nonlinearity is not the significant.

Contrary to the first case, the second *GZ* curve has a significant nonlinearity in the vicinity of the origin, where most of the roll angles in the set are concentrated. Statistical influence of Nonlinearity is quite significant there, which can be clearly seen in the resulting distribution.

Probability density of the processes of roll angles and angular velocities can be approximated with Tchebysheff-Hermite polynomials [Sveshnikov, 1968] for moderate Nonlinearity at the origin. Then, the simulation may be limited with calculation of the third and fourth central moments. This technique is called Gram-Charlier or Egenworth asymptotic expansions. This approach was used by Haddara and Zhang [1994], Belenky [1994].

The Tchebysheff-Hermite polynomial can be defined as:

$$H_l(x) = (-1)^l \cdot \exp(x^2) \cdot \frac{d^l}{dx^l}\left[\exp(-x^2)\right] \qquad (8.252)$$

We search the series in the following form:

$$f(x) = \exp(-x^2) \cdot \sum_{i=0}^{\infty} b_i \cdot H_i(x) \qquad (8.253)$$

This series converges if $f(x)$ has a finite number of discontinuities and:

$$\int_{-\infty}^{\infty} f(x) \cdot \exp\left(\frac{(x-\bar{x})^2}{4\sigma^2}\right) dx < \infty \qquad (8.254)$$

The series (8.254) converges if probability density $f(x)$ decreases quickly enough when argument x is increasing.

To obtain coefficients b_i, it is sufficient to multiply both sides of the equation (8.253) by $H_l(x)$ and then integrate it by x for infinite limits. This operation yields:

$$f(x) = \frac{1}{\sigma^2} \cdot \varphi\left(\frac{(x-\bar{x})}{\sigma}\right) + \frac{1}{\sigma} \cdot \sum_{l=3}^{\infty} a_l \cdot \frac{1}{\sigma^l} \varphi^{(l)}\left(\frac{(x-\bar{x})}{\sigma}\right) \qquad (8.255)$$

Where:

$$\varphi(x) = \frac{1}{\sqrt{2\pi}} \exp\left(-\frac{x^2}{2}\right) \qquad (8.256)$$

$\varphi^{(l)}(x)$ is the derivative of $\varphi(x)$ of the l-th order and a_l are the coefficients, which can be defined as:

$$a_l = \sum_{j=0}^{\mathrm{int}(l/2)} \frac{(-1)^{j+l} \cdot \sigma^{2j} \cdot M_{l-2j}}{(l-2j)! \cdot j! \cdot 2j!}; \quad (l = 3, 4, ...) \tag{8.257}$$

Where: M_l is the l-th central moment of the distribution. Proceeding up to the fourth order derivative, we obtain:

$$f(z) = \frac{1}{\sigma \cdot \sqrt{2\pi}} \exp\left(-\frac{z^2}{2}\right) \cdot \left(1 - \frac{Sk}{3!}(3z - z^3) + \frac{Ex}{4!}(z^4 - 6z^2 + 3)\right) \tag{8.258}$$

Here: $z = (x - \bar{x})/\sigma$; σ is standard deviation; M_3 is the third central moment of the distribution; M_4 is the fourth central moment of the distribution; Ex is excess kurtosis of the distribution, Sk is skewness of the distribution:

$$Sk = \frac{M_3}{\sigma^3}; \quad Ex = \frac{M_4}{\sigma^4} - 3 \tag{8.259}$$

Formula (8.258) can be applied for the moderate case of S-shaped *GZ* curves. Increasing nonlinearity at the origin may cause (8.258) to oscillate.

The assumption of the Gaussian distribution should be applied carefully for nonlinear roll: special attention has to be paid to the behavior of the *GZ* curve in the vicinity of the equilibrium.

8.6.3 Group Structure of Irregular Roll[1]

We previously mentioned the group structure of the roll process, while considering parametric resonance in irregular seas (subchapter 8.4.4). Here, we look at this subject closer, also in conjunction with probabilistic qualities of parametric roll response.

Group structure of waves makes analogous the group structure of ship roll. Since the ship is a dynamical system, it generally acts as a filter. The group structure of roll motions experiences the following transformation:

- Large motion inertia, i.e. the relation between any and the largest peak in the group more than 0.8, the number of cycles in group 8 and more;
- More frequent encounters of intensive groups with excitation increasing;
- Growth of the number of cycles in the group with excitation growth.

Parametric response is narrower banded than in roll resonant mode, therefore parametric oscillations have stiffer group structure than waves.

For consideration of the correlation between waves group structure and ship roll, we introduce a new statistic. It represents the maximum amplitude in the group and it depends on the number of waves in-group (n), maximum wave height (h^0), period (T^0) and initial heel with which ship encounters wave group (ϕ_0):

[1] Written in co-authorship with Prof. Alexander B. Degtyarev and Dr. Alexander V. Boukhanovsky of Institute for High Performance Computing and Information Systems (Russia)

$$\phi_{max} = \phi_{max}(n, h^0, T^0, \phi_0) \tag{8.260}$$

Degtyarev and Boukhanovsky [1995, 2000] carried out numerical simulation using an auto-regression model of waves (we considered the auto-regression model in subchapter 8.2.5) and subdivided ship oscillations into three classes using statistics (8.260).

1. Ordinary linear oscillations. Their characteristics are:

   - The height of the maximum wave in the group is higher than anywhere else and it increases with the above-mentioned statistics increasing.
   - In general, wave period is greater than the period of parametric excitation and it increases with statistics increasing.
   - Distribution of wave numbers in the group is sufficiently compact (from 3 to 6-7).
   - The statistics are independent on initial heel angle at the entrance.

2. "Primary parametric oscillations". The values of the chosen statistics are in boundaries from "quasi-linear oscillations" up to $\phi_{1/3}$ approximately. These oscillations are characterized by the following features:

   - The height of the maximum wave in the group is somewhat less than anywhere else and it is always equal to $h°$ for waves in general. It means that parametric oscillations could be held by waves of any height. Only the period value is important.
   - Wave period is nearly equal to parametric exciting period. Variance of T^0 is not large.
   - Number of the waves in the group increases from 3 to 8-12.
   - The statistics are independent on initial heel angle at the entrance.

3. "Secondary parametric oscillations". We can find such kinds of oscillations when one wave group generates parametric excitation and then another similar wave group approaches the ship. Here, the values of ϕ_{max} are greater than $\phi_{1/3}$. The main characteristics of this class are:

   - The same characteristic of maximum wave height and period in the group as for primary parametric excitation.
   - Distribution of the wave numbers in the group is compact again. It is not necessary to have a large length of the group for large roll to continue.
   - Statistics depend on heel. This heel is equal to static heel that appears during parametric resonance. This is a criterion that the ship enters a new group with developed parametric oscillations.

All these qualities are summarized in table 8.3.

Table 8.3 Three types of oscillations in parametric excitation case

Quasi-linear oscillations	Primary parametric oscillations	Secondary parametric oscillations
$\Theta_{max} \le \theta_{ql}$	$\theta_{ql} < \Theta_{max} \le \theta_{1/3}$	$\Theta_{max} > \theta_{1/3}$
$\overline{h^\circ} > \overline{h^\circ}_{gen}$	$\overline{h^\circ} \cong \overline{h^\circ}_{gen}$	$\overline{h^\circ} \cong \overline{h^\circ}_{gen}$
$\overline{\tau}^\circ > \tau_{ex}$	$\overline{\tau}^\circ \cong \tau_{ex}$	$\overline{\tau}^\circ \cong \tau_{ex}$
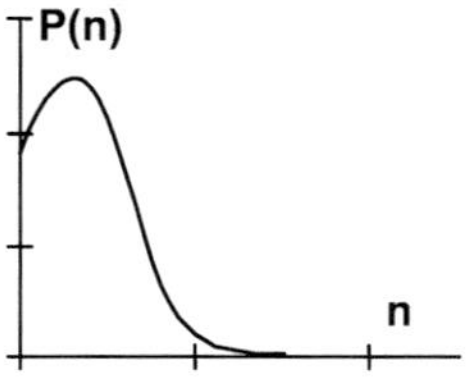 $3 \le n \le 7$	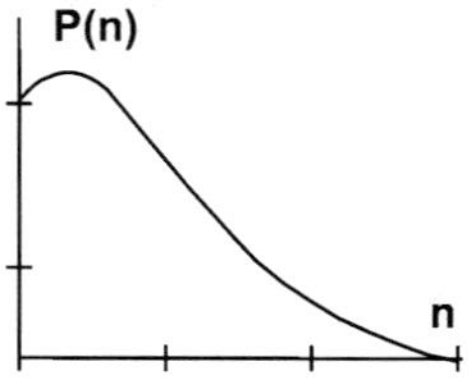 $3 \le n \le 12$	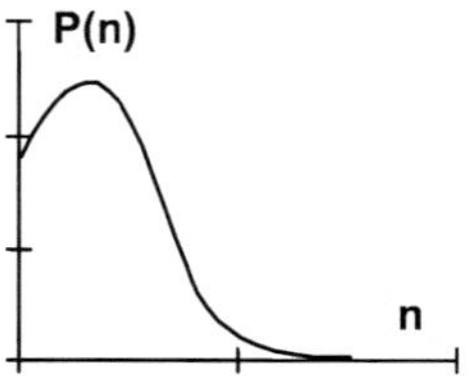 $3 \le n \le 8$
$\overline{\theta}_0 \approx 0$	$\overline{\theta}_0 \approx 0$	$\overline{\theta}_0 \approx \overline{\theta}_{eq}$

8.6.4 Application of Markov Processes

It is possible to obtain at times an analytical solution for the distribution of the stochastic response of a dynamical system, if we consider it as Markov process. What kind of stochastic process is called 'Markov'? The current value of the Markov process depends only on the value in a previous moment of time and does not depend on the previous history of the process.

To formulate this quality more strictly, let us consider the conditional probability density $f(x_k \mid x_1,...,x_{k-1})$ where x_k is the current value of the process and there are known values in previous moments x_1 ,..., x_{k-1} , [Sveshnikov, 1968]. For the Markov process:

$$f(x_k \mid x_1,...,x_{k-1}) = f(x_k \mid x_{k-1}) \qquad (8.261)$$

The Markov process can be fully characterized by the conditional distribution at two subsequent moments of time.

It is more convenient to consider such a conditional distribution as a function of four arguments $f(t,x;\tau,y)$, two subsequent moments of time t and τ and corresponding values of the process $x(t)$ and $y(\tau)$. We consider some arbitrary moment of time t_1, that lies between t and τ ($t< t_1< \tau$). The above mentioned conditional distribution $f(t,x;\tau,y)$ can be expressed then through two other conditional distributions of value z that has taken place in the moment t_1:

$$f(t,x;\tau,y) = \int_{-\infty}^{\infty} f(t,x;t_1,z) f(t_1,z;\tau,y) dz \tag{8.262}$$

The equation (8.262) is known as the generalized Markov equation or Markov-Smoluchowski-Chapman equation.

The principal advantage that could be reached by using the Markov process is that the function of the conditional distribution $f(t,x;\tau,y)$ can be considered as a solution of one of two differential equations in partial derivations:

$$\frac{\partial f}{\partial t} + a(t,x)\frac{\partial f}{\partial x} + \frac{1}{2} b(t,x)\frac{\partial^2 f}{\partial x^2} = 0 \tag{8.263}$$

$$\frac{\partial f}{\partial \tau} + \frac{\partial}{\partial y}\left[a(\tau,y)\cdot f\right] - \frac{1}{2}\frac{\partial^2}{\partial y^2}\left[b(\tau,y)\cdot f\right] = 0 \tag{8.264}$$

The difference between these two equations is that (8.263) concerns the previous moment of time and (8.264) concerns the current moment of time. The equation (8.263) is called the first Kholmogorov equation. The equation (8.264) is called the Fokker-Plank-Kholmogorov (or the second Kholmogorov equation). Functions a and b can be expressed as the following limits:

$$a(t,x) = \lim_{\tau \to t} \frac{1}{\tau - t} \langle Y - X \mid X = x \rangle \tag{8.265}$$

$$b(t,x) = \lim_{\tau \to t} \frac{1}{\tau - t} \langle (Y - X)^2 \mid X = x \rangle \tag{8.266}$$

Here, $X = x(t)$ and $Y = y(\tau)$, symbol $\langle \; \rangle$ means averaging procedure.

The physical meaning of these functions is the following: the function a characterizes a velocity of changing ordinate of the process and the function b characterizes velocities of changing of conditional variance of the ordinate.

The concept of the Markov process can be generalized for the aggregate of several stochastic processes or, in other words, for a multi-dimensional stochastic process; formulae (8.263)-(8.266) have the following appearance in this case:

Markov -Smoluchowski-Chapman equation:

$$\begin{aligned} & f(t,x_1,\ldots,x_n;\tau,y_1,\ldots y_n) = \\ & = \int_{-\infty}^{\infty}\ldots\int_{-\infty}^{\infty} f(t,x_1,\ldots,x_n;t_1,z_1,\ldots z_n) f(t_1,z_1,\ldots,z_n;\tau,y_1,\ldots y_n)\, dz_1 \ldots dz_n \end{aligned} \tag{8.267}$$

The first Kholmogorov equation:

$$\frac{\partial f}{\partial t} + \sum_{i=1}^{n} a_i \frac{\partial f}{\partial x_i} + \frac{1}{2}\sum_{i=1}^{n}\sum_{j=1}^{n} b_{i,j} \frac{\partial^2 f}{\partial x_i \partial x_j} = 0 \tag{8.268}$$

Fokker-Plank-Kholmogorov equation:

$$\frac{\partial f}{\partial t}+\sum_{i=1}^{n}\frac{\partial}{\partial x_i}[a_i f]-\frac{1}{2}\sum_{i=1}^{n}\sum_{j=1}^{n}\frac{\partial^2}{\partial x_i \partial x_j}[b_{i,j} f]=0 \tag{8.269}$$

Functions a_i and b_{ij} have the same sense as in the previous case:

$$a_i(t,x_1,...,x_n)=\lim_{\tau\to t}\frac{1}{\tau-t}\langle Y_i-X_i \mid x_1,...,x_n\rangle \tag{8.270}$$

$$b_{i,j}(t,x_1,...,x_n)=\lim_{\tau\to t}\frac{1}{\tau-t}\langle (Y_i-X_i)(Y_j-X_j) \vert x_1,...,x_n\rangle \tag{8.271}$$

To define the Fokker-Plank-Kholmogorov equation, it is necessary to find functions a_i and b_{ij}. It can be done if the components of the multi-dimensional Markov process satisfy a system of differential equations (which must be in the form of the roll equation in our case):

$$\frac{dx_i}{dt}=\Psi_i(t,x_1,...,x_n)+\sum_{k=1}^{n}g_{ik}(t,x_1,...,x_n)\xi_k(t);\quad i=1,...,n \tag{8.272}$$

Here $\Psi_i(t,x_1,...,x_n)$ and $g_{ik}(tx_1,...,x_n)$ are arbitrary functions of time, x_i are components of the Markov process, and $\xi_k(t)$ are mutually independent Wiener processes (or white noise processes, see subchapter 8.2.4).

If the relationship between the components of the Markov process is expressed by the system (8.272), then the following formulae can be applied:

$$a_i(t,x_1,...,x_n)=\Psi_i(t,x_1,...,x_n) \tag{8.273}$$

$$b_{i,j}(t,x_1,...,x_n)=\sum_{k=1}^{n}g_{ik}(t,x_1,...,x_n)\cdot g_{jk}(t,x_1,...,x_n) \tag{8.274}$$

To solve the problem of distribution of large roll, we introduce a two-dimensional Markov process: $x_1=\phi$; $x_2=\dot{\phi}$. We consider the following roll equation in a general form:

$$\ddot{\phi}+r(\dot{\phi})+s(\phi)=\alpha_E(t) \tag{8.275}$$

Here, $r(\dot{\phi})$ is the nonlinear damping term; $s(\phi)$ is restoring term and $\alpha_E(t)$ is excitation.

To be able to apply the Markov process analysis we should make some special assumption concerning excitation. We assume $\alpha_E(t)$ is a stationary stochastic process with the qualities of white noise. Substitution of Markov process components $x_1=\phi$ and $x_2=\dot{\phi}$ transforms the equation (8.275) into a system of differential equations of the first order:

$$\begin{cases} \dfrac{dx_1}{dt} = x_2 \\ \dfrac{dx_2}{dt} = -r(x_2) - s(x_1) + \alpha_E(t) \end{cases} \tag{8.276}$$

To find functions a_i and b_{ij} in our case, we can use formulae (8.273) and (8.275):

$$a_1(t, x_1, x_2) = x_2\,; \quad a_2(t, x_1, x_2) = -(r(x_2) + s(x_1)) \tag{8.277}$$

$$\begin{aligned} &b_{11}(t.x_1, x_2) = 0; \quad b_{12}(t.x_1, x_2) = 0 \\ &b_{21}(t.x_1, x_2) = 0; \quad b_{22}(t.x_1, x_2) = 1 \end{aligned} \tag{8.278}$$

Consequently, the Fokker-Plank-Kholmogorov equation can be rewritten as follows:

$$\frac{\partial f}{\partial t} + \frac{\partial}{\partial x_1}[x_2 f] - \frac{\partial}{\partial x_2}[(r(x_2) + s(x_1))f] - \frac{1}{2}\frac{\partial^2}{\partial x_2^2}[f] = 0 \tag{8.279}$$

Expression (8.279) is the differential equation in partial derivatives of the parabolic type. To solve it both initial conditions and boundary conditions need to be specified.

Roberts [1980, 1982] presented a review of the application of the Markov process for nonlinear roll in irregular seas. The general problem of application of Markov processes analysis to nonlinear dynamic system excited by white noise is considered in [Caughey, 1963]. The stationary solution of Fokker-Planck-Kholmogorov equation for the case of nonlinear roll was reviewed in [Haddara, 1974].

Caughey [1964] showed that an analytical solution of (8.279) can be derived for the case when the only nonlinear term is damping. The Galerkin method was applied to solve the Fokker-Plank-Kholmogorov equation for the roll equation with nonlinear damping and restoring [Haddara and Zhang, 1994].

The combination of averaging method with the Markov process analysis was used by Roberts [1980, 1980a, 1982, 1982a], Haddara and Nassar [1986]. Application of the averaging technique allows consideration of slowly varying amplitude and phase as the components of a two-dimensional Markov process.

The principle advantage of the averaging method is that the Fokker-Plank-Kholmogorov equation can be reduced to one dimension. This equation further can be solved for any type of nonlinear terms and it is not necessary to assume input excitation as white noise: the spectrum of the excitation can be arbitrary [Roberts, 1980, 1982]. Haddara and Nassar [1986] combined the averaging technique with the Galerkin method.

Francescutto [1998] applied the Markov process to derive the probability density function for the response envelope. There are a number of applications of the Markov process in [Shlesinger and Swean, 1998].

Chapter 9

Probability of Capsizing

9.1 Application of Upcrossing Theory

9.1.1 General

We considered several models of capsizing in Chapter 5. In many cases, capsizing is associated with the crossing of a certain boundary. For example, the classical definition of capsizing considers crossing of the separatrix by the phase trajectory. Another example is the nonlinear dynamics approach (subchapter 5.3), where stability is considered as insufficient when the Melnikov function crosses zero level.

When we are talking about capsizing in irregular seas, the crossing becomes a random event. The probability of this event then can be perceived as a probability of capsizing within assumptions of the particular capsizing model.

To "randomise" a capsizing model we need to:

- Introduce a model stochastic excitation (described in subchapter 8.2).
- Calculate probabilistic characteristics of roll, roll velocities and other ship motion processes included in the model of capsizing.
- Calculate probabilistic characteristics of the "carrier process" - the process, by which crossing of the boundary is associated with the capsizing.
- Calculate the probability of crossing.

We already considered the importance of time for practical use of probability of capsizing in subchapter 1.3. So, whenever possible, we have to look for the probability of capsizing during a given time interval.

The theory of upcrossing fits the best for the above purpose. It looks for probabilistic characteristics of different random variables related with the crossing: as number of crossings, time above the boundary, etc.

The following assumptions are involved in our further consideration:

The "Carrier" process is considered stationary. This assumption is adequate: we consider the duration of time short enough for environmental conditions to be constant. Change of these conditions is already taken into account with the introduction of stochastic vectors of assumed situations and loading conditions, see (subchapter 1.2 and Chapter 2) with further combining of those changes into long term probabilistic criterion described in subchapter 1.5.

The "Carrier" process is considered differentiable. Almost all processes derived from motion of a mechanical object are differentiable.

Additional assumptions concerning distribution will be considered later in this chapter.

9.1.2 Averaged Number of Crossings

The number of crossings of the given level by a stochastic process is a random variable. Here, we are looking for the mean value of this variable. Following [Sveshnikov, 1968], consider a stationary, differentiable process *x*(*t*), crossing the level *a*, sometimes between moment t and $t+dt$:

$$\begin{cases} x(t) < a \\ x(t+dt) > a \end{cases} \tag{9.1}$$

Taking into account differentiability of the process *x*(*t*), the system of inequalities can be rewritten:

$$\begin{cases} x(t) < a \\ x(t) + \dot{x}dt > a \end{cases} \Leftrightarrow \begin{cases} x(t) < a \\ x(t) > a - \dot{x}dt \end{cases} \tag{9.2}$$

Consider a probability so that the conditions of (9.2) are satisfied:

$$P(a - \dot{x}dt < x(t) < a) = \int_{0}^{\infty} \int_{a-\dot{x}dt}^{a} f(x,\dot{x})dxd\dot{x} \tag{9.3}$$

Here, $f(x,\dot{x})$ is the joint probability density of the process *x*(*t*) and its derivative $\dot{x}(t)$.

The internal integral in (9.3) has limits that differ only on an infinitely small value $\dot{x}dt$, therefore:

$$\int_{a-\dot{x}dt}^{a} f(x,\dot{x})dx = \dot{x}dt \cdot f(a,\dot{x}) \tag{9.4}$$

Substitution of (9.4) into (9.3) yields:

$$P(a - \dot{x}dt < x(t) < a) = dt \int_{0}^{\infty} \dot{x} f(a,\dot{x})d\dot{x} \tag{9.5}$$

Formula (9.5) shows that the probability of upcrossing of the level *a* at specific time *t* is indefinitely small. Let us consider time density of upcrossing as:

$$p_U(t) = \frac{P(a - \dot{x}dt < x(t) < a)}{dt} = \int_{0}^{\infty} \dot{x} f(a,\dot{x})d\dot{x} \tag{9.6}$$

Time density of upcrossings can be treated as probability of upcrossing per unit of time.

Now let us take the time period from 0 to T and divide it in N small intervals, say Δt_i, centred at moments t_i. We introduce an auxiliary random variable n_i defined on each of these intervals as:

$$n_i = \begin{cases} 1 & a \in [x(t_i - 0.5\Delta t_i), x(t_i + 0.5\Delta t_i)[\text{ and } \dot{x}(t_i) > 0 \\ 0 & a \notin [x(t_i - 0.5\Delta t_i), x(t_i + 0.5\Delta t_i)[\text{ or } \dot{x}(t_i) < 0 \end{cases} \tag{9.7}$$

In other words, the variable n_i equals 1 if there was an upcrossing and 0 if there was not (upcrossing is defined as crossing the level with a positive derivative). We chose Δt_i small enough, so only one upcrossing could happen at this time, see fig. 9.1.

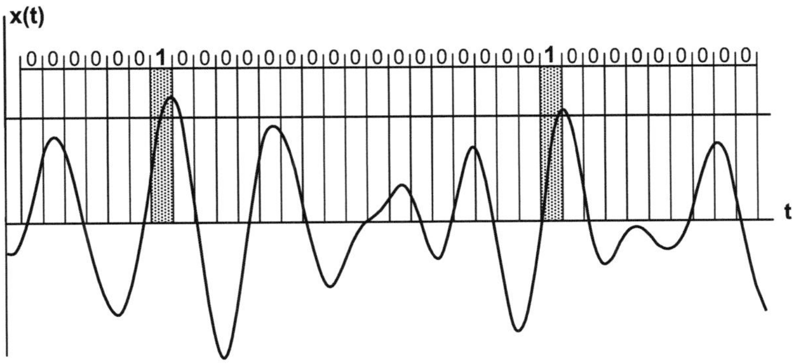

Fig. 9.1 Auxiliary variable for upcrossings

The number of upcrossings during time T equals the sum of auxiliary variables n_i:

$$n_a = \sum_{i=1}^{N} n_i \tag{9.8}$$

Now, let us make Δt_i infinitely small and simultaneously average both parts of (9.8). We have obtained the averaged number of upcrossings on the left-hand side. Average value of each auxiliary variable n_i is equal to the probability of upcrossing during small interval Δt_i, which also can be expressed by the time density of upcrossing p_U:

$$m(n_a) = \lim_{\Delta t_i \to 0}\left(m\left[\sum_{i=1}^{N} n_i\right]\right) = \lim_{\Delta t_i \to 0}\left(\left[\sum_{i=1}^{N} P_a(t_i)\right]\right) = \lim_{\Delta t_i \to 0}\left(\left[\sum_{i=1}^{N} p_U(t_i)\Delta t_i\right]\right) \tag{9.9}$$

Once the interval Δt_i is infinitely small, the probability of upcrossing at the moment t is equal to $p_U dt$ and the sum is converted to an integral:

$$m(n_a) = \lim_{\Delta t_i \to 0}\left(\left[\sum_{i=1}^{N} p_U(t_i)\Delta t_i\right]\right) = \int_0^T p_U(t)dt \tag{9.10}$$

Substitution of (9.6) into (9.10) yields the general formula for averaged number of crossings:

$$m(n_a) = \int_0^T \int_0^\infty \dot{x} f(a, \dot{x}) d\dot{x} dt \tag{9.11}$$

So far, we have not yet used the assumption that x is a stationary process. It means that formula (9.11) works even for non-stationary processes. For the stationary process, it can be simplified, since the distribution does not depend on time:

$$m(n_a) = T \int_0^\infty \dot{x} f(a, \dot{x}) d\dot{x} \tag{9.12}$$

For the stationary process, the averaged number of upcrossings does not depend on where we have chosen the period from 0 to T on the time axis. It depends on the duration of this period, e.g. on T. So let us consider the averaged number of crossings per unit of time:

$$m_T(n_a) = \int_0^\infty \dot{x} f(a, \dot{x}) d\dot{x} \tag{9.13}$$

If the process has a Gaussian distribution, the integral in the formula (9.13) can be evaluated:

$$m_T(n_a) = \frac{1}{2\pi} \sqrt{\frac{V_{\dot{x}}}{V_x}} \exp\left(-\frac{(a - m_x)^2}{2V_x} \right) \tag{9.14}$$

Similar considerations can be given to downcrossings; the only difference is that the derivative is assumed negative.

9.1.3 Crossings as Poisson Flow

Here, we look for probability of upcrossing during a given period of time. Consider a crossings as a flow of random events. Flow means that these events may happen at random moments of time within a given time period. We do not know *apriori* how many of these events are going to happen; neither do we know when they are going to happen. Graphically, the flow event could be presented as points that may appear in random places on the axis, see fig. 9.2.

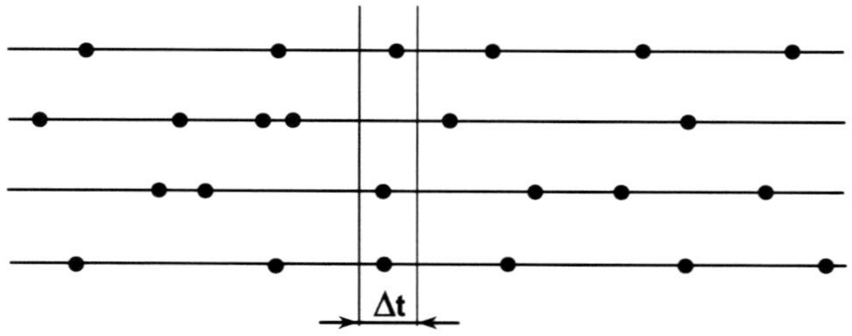

Fig. 9.2 Flow of random events

We assume that the crossings are rare events satisfying conditions of Poisson flow. This way we can use probabilistic theory of rare events, which is widely applied in the theory of mass service. The following statements are formally included in this concept:

- Only one crossing can happen at the given time. This is trivial for ship motions and mentioned here only to meet the formal definition of flow of rare events: this is the so-called ordinary condition.
- The probability that a crossing happens at the particular moment of time is infinitely small. We meet this condition without any additional assumption (see subchapter 9.1.2).
- Crossings are independent of each other. This is the most questionable assumption. We can meet it only if the level is high enough and the autocorrelation function dies out before the next crossing happens. Otherwise, the previous crossing will affect the next one, which is a violation of this condition.

Assuming these three conditions are satisfied, consider the probability that there will be n_T events within the time period from 0 to T. We will make a special note every time we are going to use any of the above conditions.

As in the previous subchapter, we divide our time period T into N small intervals with duration Δt each, centered at the moment t_i.

Let us start from the sample: we divided our period into 10 intervals $N = 10$ and we are looking for the probability that a crossing will happen precisely during the second, fifth and ninth intervals. Using our assumption that crossings are not dependent on each other (third condition) we can write:

$$P(10,3) = q_1 p_2 q_3 q_4 p_5 q_6 q_7 q_8 p_9 q_{10} \tag{9.15}$$

Here, the probability that upcrossing happens in the i-th interval is defined by formula (9.5):

$$p_i = P\left(a - \dot{x}dt < x(t) < a\right) = p_U \cdot \Delta t_i = \frac{p_U T}{N} \tag{9.16}$$

The probability that upcrossing does not happen in the i-th interval is defined as an opposite to (9.16):

$$q_i = 1 - p_i = 1 - \frac{p_U T}{N} \tag{9.17}$$

The probability that the event will happen during the second interval does not differ from the probability that the event will happen during the fifth interval: the probability that the event will happen during i-th event does not depend on i. The same can be stated about the probability of the opposite event:

$$P(10,3) = q^7 p^3 \tag{9.18}$$

Generalizing our sample, we can write the probability that there would be n_T events in n_T specific moments out of N moments total:

$$P^*(N,n_T)=q^{N-n_T}p^N \tag{9.19}$$

Here, the asterisk mark indicates the events happen at specific moments. Formula (9.19) expresses the assumption that crossings do not depend on each other (third condition) and there should be enough time between them for the autocorrelation function to die out and for the dynamical system to "forget" the previous crossing.

Now let us look for the probability that n_T upcrossings happen at any n_T moments out of N moments total. To find it, we have to find all possible variants and sum their probabilities.

How many ways can we choose n_T moments out of N without repetition (we formally used our ordinary condition)? The answer is known from combinatorics: it is given by the formula for the number of combinations without repetitions [Bronshtein and Semendyayev, 1997]:

$$C(N,n_T)=\frac{N!}{n_T!(N-n_T)!} \tag{9.20}$$

Finally, the probability that happen exactly n_T upcrossings out of N possible (or during time T divided by N intervals):

$$P(N,n_T)=C(N,n_T)p^{n_T}q^{N-n_T} \tag{9.21}$$

The formula (9.21) expresses the well-known binomial distribution. The reason why we choose to repeat all the derivation is to see where and how these three assumptions are included in the final results.

Now let us make Δt infinitely small, which means $N=\infty$ and look for the probability that it would be exactly n_T upcrossings during time T:

$$P_T(n_T)=\lim_{N\to\infty}(P(N,n_T))=\lim_{N\to\infty}\left(C(N,n_T)p^{n_T}q^{N-n_T}\right) \tag{9.22}$$

We substitute formula (9.20) into (9.22) and expand some of factorials:

$$P_T(n_T)=\lim_{N\to\infty}\left(\frac{1\cdot 2\cdot ...(N-n_T)\cdot(N-n_T+1)..(N-1)\cdot N}{\left(1\cdot 2\cdot ...(N-n_T)\right)\cdot n_T!}p^{n_T}q^{N-n_T}\right) \tag{9.23}$$

After dividing both parts of the fraction by $\left(1\cdot 2\cdot ...(N-n_T)\right)$, we substitute expressions for probability p and q from the formulae (9.16) and (9.17) correspondingly:

$$\begin{aligned}P_T(n_T)&=\lim_{N\to\infty}\left(\frac{(N-n_T+1)..(N-1)\cdot N}{n_T!}\left(\frac{p_U T}{N}\right)^{n_T}\left(1-\frac{p_U T}{N}\right)^{N-n_T}\right)=\\&=\lim_{N\to\infty}\left(\frac{(N-n_T+1)..(N-1)\cdot N}{N^{n_T}}\cdot\left(1-\frac{p_U T}{N}\right)^{-n_T}\cdot\left(\frac{(p_U T)^{n_T}}{n_T!}\right)\cdot\left(1-\frac{p_U T}{N}\right)^{N}\right)\end{aligned} \tag{9.24}$$

Now consider the limits of each component:

$$\lim_{N\to\infty}\left(\frac{(N-n_T+1)..(N-1)\cdot N}{N^{n_T}}\right)=1 \tag{9.25}$$

$$\lim_{N\to\infty}\left(\left(1-\frac{p_U T}{N}\right)^{-n_T}\right)=1 \tag{9.26}$$

$$\lim_{N\to\infty}\left(\frac{(p_U T)^{n_T}}{n_T!}\right)=\frac{(p_U T)^{n_T}}{n_T!} \tag{9.27}$$

$$\lim_{N\to\infty}\left(\left(1-\frac{p_U T}{N}\right)^{N}\right)=\exp(-p_U T) \tag{9.28}$$

Taking into account (9.25) through (9.28) we finally receive:

$$P_T(n_T)=\frac{(p_U T)^{n_T}}{n_T!}\cdot\exp(-p_U T) \tag{9.29}$$

The formula (9.29) is known as the Poisson distribution for random variable n_T. The expression $p_U T$ is simultaneously the mean value and variance for this random variable.

Considering capsizing related problems, we are interested usually in the probability of at least one upcrossing, because this event is the opposite of event "no upcrossings". To find the probability of at least one upcrossing, it is enough to find the probability that there will be no upcrossings during given time *T*. This can be easily done using formula (9.29):

$$P_T(n_T=0)=\exp(-p_U T) \tag{9.30}$$

The probability of at least one upcrossing can be, therefore, expressed as:

$$P_T(n_T\neq 0)=1-\exp(-p_U T) \tag{9.31}$$

Taking into account formulae (9.6) and (9.12), we can rewrite (9.21) and (9.31) using average number of upcrossings:

$$P_T(n_T=0)=\exp(-m_T(n_a)T) \tag{9.32}$$

$$P_T(n_T\neq 0)=1-\exp(-m_T(n_a)T) \tag{9.33}$$

The results (9.32) and (9.33) are very important for our further study, because they provide the direct relationship between the event and time period, within which it can happen. However, these formulae are derived with the assumption of independence of the upcrossing. Therefore, its application requires consideration of the applicability of the above assumption.

9.1.4 Time before Crossing

Since the upcrossing can happen at any moment within a given time period, the time from the beginning before it actually happens is a random variable. Let us find its distribution.

We assume here that upcrossing flow is Poisson flow and upcrossings are not dependent on each other. Consider random time T_0, such, that no upcrossing would happen during this time. Its probability could be found with formula (9.32):

$$P_T(n_T = 0) = \exp(-m_T(n_a)T)$$

According to the definition of the cumulative probability distribution function:

$$F(t) = P(T < t) \tag{9.34}$$

We also can express cumulative distribution through probability of the opposite event:

$$1 - F(t) = P(T \geq t) \tag{9.35}$$

Formula (9.34) expresses the probability that there are no crossings during time t that is less than T, so formula (9.33) can be applied:

$$1 - F(t) = P(T \geq t) = \exp(-m_T(n_a) \cdot t) \tag{9.36}$$

Finally, the cumulative distribution can be expressed as:

$$F(t) = 1 - \exp(-m_T(n_a) \cdot t) \tag{9.37}$$

Distribution density is defined through the derivative of cumulative distribution:

$$f(t) = m_T(n_a)\exp(-m_T(n_a) \cdot t) \tag{9.38}$$

Formulae (9.37) and (9.38) constitute exponential distribution. This is another important feature of Poisson flow: the time before the first event has an exponential distribution.

The same distribution actually applies for time between events. We can always place the origin right after the first event. At the same time the presence of the event at the origin of the time axes cannot affect the time passing until the next event, since Poisson flow presumes independence of the events.

9.2 Probability of Capsizing in Beam Seas

9.2.1 Mathematical and Physical Modeling

In the absence of an exact analytical solution of a system of differential equations describing rolling and capsizing, the natural way to solve the problem is by physical experiment and mathematical modeling: digital and analogue. The main obstacle to this approach is that capsizing is a very rare random event for a ship with normal stability. To have enough statistics to judge about the risk function value, it is necessary to have a very large quantity of repetitive tests in physical experiments or calculations in digital modeling.

Analogue modeling allows easy time scaling; and it was the only method a few decades ago, when fast computers were expensive and were not available on every engineer's desk. Pham Ngock Hoeh [1981] made one of the first attempts. He used the following system describing a ship in beam irregular seas and a gusty wind:

$$\begin{cases} (m + a_{22})\ddot{y}_{1G} + K \cdot \dot{y}_{1g}^2 = F_A(t) \\ (I_{xx} + a_{44})\ddot{\phi} + N_\phi \cdot k(\phi) \,|\, \dot{\phi} \,|\, \dot{\phi} + mg \cdot GZ(\phi) = M_{XE}(t) + M_{XA}(t) \end{cases} \tag{9.39}$$

Here: N_ϕ is a dimensional coefficient of roll damping, $k(\phi)$ is a coefficient, which takes into account the bulwark entering water during heeling, see fig. 9.3. External aerodynamic force was modeled as follows:

$$F_A(t) = 1.3v^2(t)\frac{A_w}{16} \tag{9.40}$$

Velocity of gusty wind was assumed in the form of stochastic pulses shown in fig. 9.4

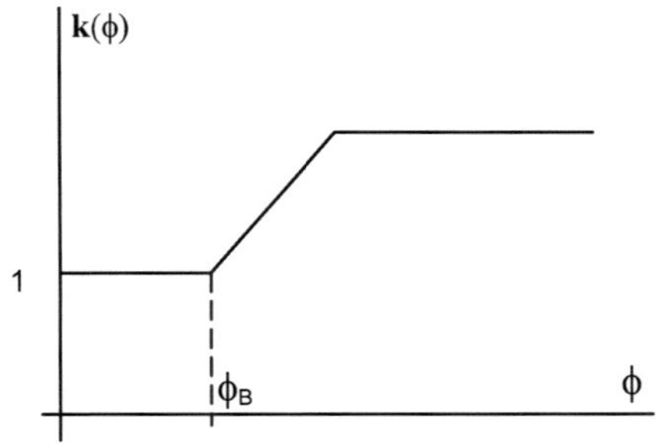

Fig. 9.3 Model of influence of bulwark entering into water on roll damping

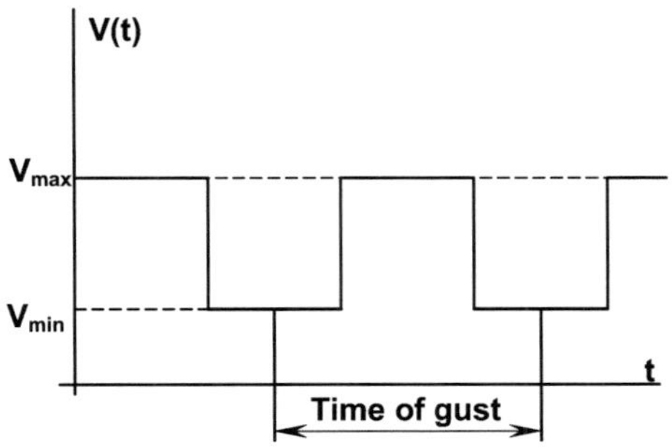

Fig. 9.4 Model of gusty wind

The study was done for a mid-size Russian trawler, SRT-840. The distribution density of time before capsizing and an estimate of capsizing risk function were produced with analogue modeling. Corresponding numerical data is given in table 9.1.

Table 9.1 Numerical data for analogue modeling (Region 1 – North Sea)

Sea state, points	8-9	Waterplane coefficient *CW*	0.83
Mean wind velocity, m/s	45	Mid-ship section coefficient *CM*	0.82
Length *L*, m	34.80	Windage area A_W, m²	129
Breadth *B*, m	7.30	Height of centre of windage area x_A, m	2.35
Draught *T*, m	2.92	Angle of bulwark flooding, ϕ_B, deg	18
Mass displacement *m*, tonne	447	KG_1, m	2.68
Block coefficient *CB*	0.57	KG_2, m	2.98

The probability to meet such a sea state is 0.04% in the chosen region [Register of USSR, 1974]. Also, it was assumed that the chosen ship had twenty years of service, so she encountered about 40 hours of storms since she entered service.

For this modeling, a Russian analogue computer MN18 was used. Such a computer has maximal time scaling M_t=50 and a limit of integration time T=1000. This time T was assumed as trial time. The number of the "tested" vessel was set up as N=30. The general scheme of analogue model is given in fig. 9.5.

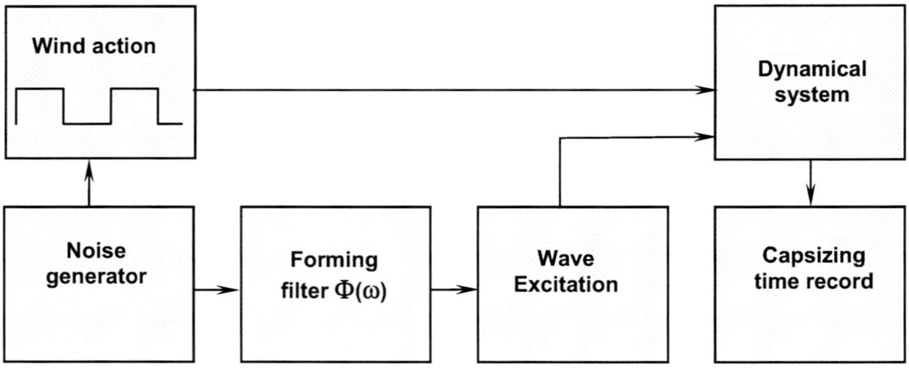

Fig. 9.5 General scheme of analogue model

The result of analogue modeling is the time before capsizing t_i. The statistical estimate of the risk function could be evaluated as:

$$\langle \lambda \rangle = \frac{d(T)}{S(T)} \tag{9.41}$$

Where $d(T)$ is cases of capsizing and $S(T)$ is total time before capsizing:

$$S(T) = \sum_{i=1}^{d(T)} t_i + [N - d(T)] \cdot T \tag{9.42}$$

Numerical values of these estimates were the following:

$$\begin{aligned} \langle \lambda \rangle &= 0.109 \quad 1/\text{hour} \quad \text{for} \quad KG = 2.98\,\text{m} \\ \langle \lambda \rangle &= 0.0106 \quad 1/\text{hour} \quad \text{for} \quad KG = 2.68\,\text{m} \end{aligned} \tag{9.43}$$

The confidence interval for the last case was $0.004 < \langle \lambda \rangle < 0.02$. Empirical and theoretical distributions are shown in fig. 9.6.

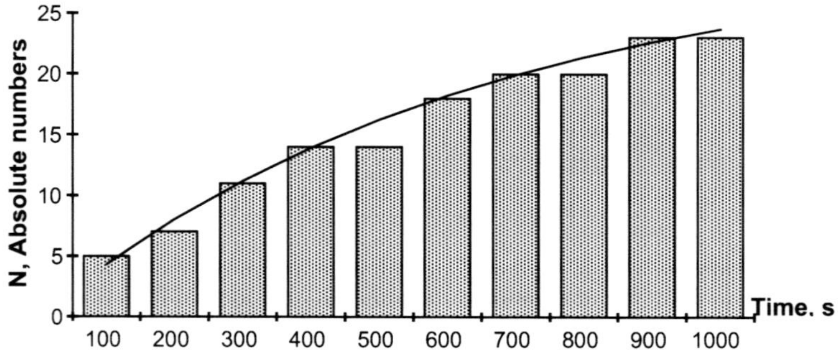

Fig. 9.6 Cumulative distribution of time before capsizing (in absolute numbers) for *KG* =2.98 m

It is interesting to notice that a change in KG of only 0.3 m leads to dramatic increasing in the probability of non-capsizing during the vessel's life time (40 hours of storms):

$$P_1(40) = 1 - e^{-0.109\cdot 40} = 0.013$$

$$P_2(40) = 1 - e^{-0.0106\cdot 40} = 0.654$$

Let us compare 0.654/0.013 = 50.3, the ratio is more than 50!

The above analogue modeling assumed an exponential character of the distribution of time ranges before capsizing suggested on the background of Poisson flow for capsizing.

Ananiev and Savchuck [1982] validated this exponential character of the distribution with model tests carried out in the towing tank of Kaliningrad Institute of Technology.

Such an experiment could be carried out only in quite special conditions: stability of the model should be very low, otherwise the time required for the trials becomes very large and the experiments become impossible. A schematic model with U-shape sections was used for the experiment. Characteristics of the model are given in table 9.2. Waves were created by a pneumatic wavemaker, which can generate irregular waves with a given spectrum. The form of the spectrum is given in fig 9.7.

Table 9.2 Characteristics of schematic model

Length L, m	2.2
Breadth B, m	0.40
Draught T, m	0.20
Depth H, m	0.22
Block coefficient CB	0.638
Metacentric height GM, m	0.00256
Transverse moment of inertia I_X, kg m^2	0.17
Angle of vanishing stability, ϕ_V, degree	28.5
Angle of max. GZ curve, ϕ_{max}, degree	11
Maximal value of GZ curve, l_{max} m	0.0037
Natural frequency ω_ϕ, s^{-1}	0.4

Experiments were carried out in the following way: the schematic model was allowed to drift freely under the action of the waves. The beam position of the model was maintained by ropes, so the model had no opportunity to yaw. The moment in time, when the model capsizes was recorded, as well as roll and drift motion. Results of experiments are given: in table 9.3, in fig. 9.8 and 9.9.

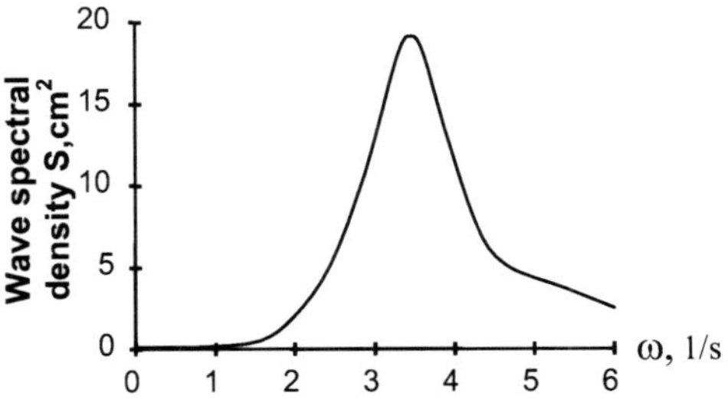

Fig. 9.7 Spectral density of waves

Table 9.3 Numerical results of capsizing test [Ananiev and Savchuck 1982]

Total number of trials N	94
Number of trials without capsizing N_{CAPS},	1
Averaged time before capsizing T_a, s	66.6
Estimate of the risk function $\hat{\lambda}$, s^{-1}	0.01502
Time of running without capsizing, s	326
Length of measuring part of towing tank, m	29
Variance of waves V_ζ, mm^2	3400
Averaged velocity of drift, cm/s	10.1
Velocity of drift, max, cm/s	2.0
Velocity of drift, min, cm/s	5.0
Probability of agreement by Pierson criterion	0.25
Probability of agreement by Kholmogorov criterion	0.999

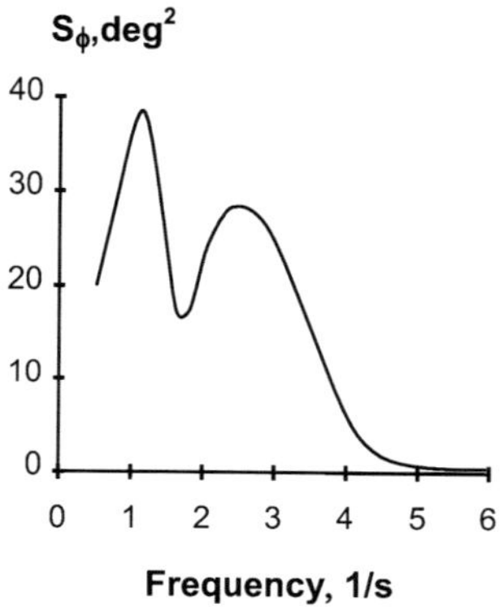

Fig. 9.8 Roll spectral density calculated from model test measurements

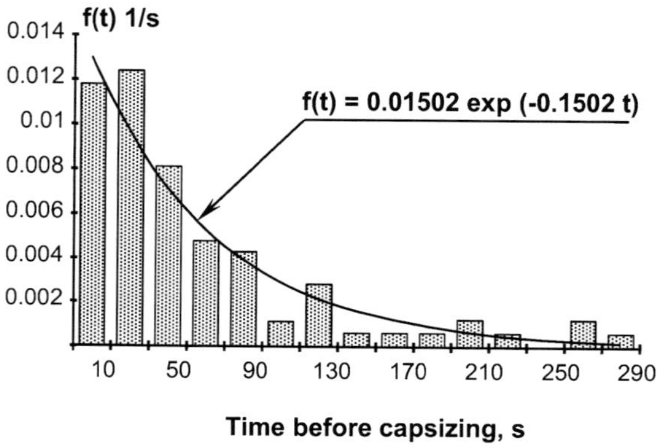

Fig 9.9 Probability density distribution for time before capsizing: bars -experimental values, solid line – fitted exponential law

A good agreement of theoretical and empirical distributions does not disapprove the hypothesis of exponential distribution for the time before capsizing. Ananiev and Savchuck [1982] indicated that the same results were obtained for trials with other spectra as well as during a semi-full scale test in natural seas in the Kursches Gulf (South Baltic).

Techniques using numerical simulation with a digital computer do not pose a serious problem nowadays. There are a number of codes available, we already mentioned in subchapter 8.2.4: LAMP [Lin and Salvesen, 1998] and FREDYN [Hooft, 1987]; both of these codes are suitable for these kinds of simulations. McTaggart and de Kat [2000] used FREDYN and Gumbel statistics [Gumbel, 1958] to estimate capsizing risk for frigates.

Numerical simulation as a practical tool works fine for the cases when stability is dangerously low or environmental conditions are extremely severe. However, exposure time rises exponentially when we are dealing with a normal ship in hard but not in extreme conditions. Direct calculation probability of capsizing may be very computationally expensive in such conditions, since a normal vessel may have a very small numerical value of risk function [Belenky, 1985].

As a result, we cannot limit ourselves to numerical tools only; there is still room for analysis of probability of capsizing that is supposed to complement these numerical tools [Belenky, 2000].

9.2.2 Classical Definition of Stability

We already considered the classical definition of capsizing in subchapter 5.1. We have seen that if the classical definition is used, the event of capsizing is associated with crossing of the separatrix by the phase trajectory of forced motion. Now we have to find the probability of such an event.

The simplest answer is the following: if we assume that the joint distribution density of roll angles and roll velocity $f(\phi,\dot{\phi})$ is known, we can write the expression for capsizing probability as:

$$P(X) = 1 - \iint_{Sd} f(\phi,\dot{\phi})\, d\phi\, d\dot{\phi} \qquad (9.44)$$

The value - $f(\phi,\dot{\phi})d\phi d\dot{\phi}$ - is an infinitely small probability that the image point has the coordinates $(\phi,\dot{\phi})$ on phase plane at any moment in time. Integration of these small probabilities on the safe domain *Sd* (see fig. 9.10) yields a probability that the image point will stay within the safe domain, or in other words, probability of non-capsizing.

Unfortunately, the simplest answer cannot be recognized as fully satisfactory. We can calculate probability of capsizing, but the value obtained will not be related with time, because we address the probability of crossing of the separatrix in any moment of time.

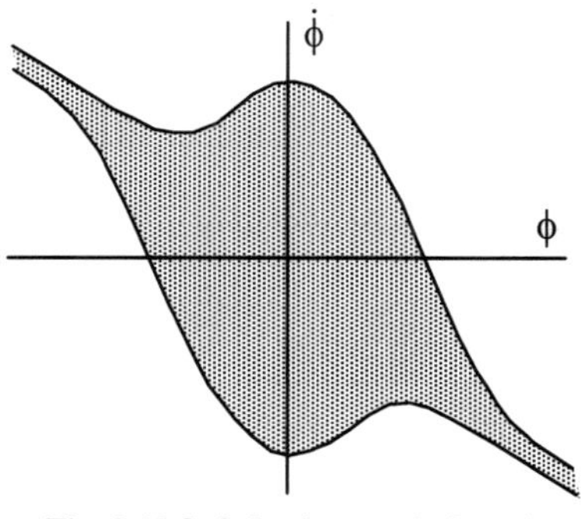

Fig. 9.10 Safe basin or safe domain

A possible solution was proposed by Umeda, et al [1992] for combined action of irregular waves and gusty wind with a non-zero mean value.

The idea was to use a distribution of amplitudes and relate capsizing with exceeding of the points, where the separatrix crosses axis ϕ. Because the roll angle can reach the amplitude value only twice during the oscillation period, we can relate the obtained probability with the duration of one cycle of roll oscillation. If we assume that roll is a Gaussian stochastic process (we considered the matter in subchapter 8.6), the distribution density of amplitudes or, maxima and minima of roll angle (taking into account possible asymmetry of roll caused by the constant component of wind) is:

$$f_{\pm}(\phi)=\frac{\varepsilon}{\sqrt{2\pi M_0}}\exp\left\{-\frac{\phi^2}{2M_0\varepsilon^2}\right\}\pm$$
$$\pm\frac{\sqrt{1-\varepsilon^2}}{m_0}\exp\left\{-\frac{\phi^2}{2M_0}\right\}\left[\frac{1}{2}\pm\Phi\left\{\frac{\phi}{\varepsilon}\sqrt{\frac{(1-\varepsilon)^2}{M_0}}\right\}\right] \tag{9.45}$$

The positive and the negative sign should be applied to the maxima and minima of the roll angle respectively. Here, M_n is the spectral moment of n-th order:

$$M_n=\int_0^{\infty}\omega^n S_{\phi}(\omega)d\omega \tag{9.46}$$

ε is the width of the roll motion spectrum S_{ϕ}:

$$\varepsilon=1-\frac{M_2^2}{M_0M_4} \tag{9.47}$$

$\Phi(x)$ is the Laplace function:

$$\Phi(x)=\frac{1}{\sqrt{2\pi}}\int_0^x\exp\left(-\frac{z^2}{2}\right)dz \tag{9.48}$$

The probability of capsizing during one period of roll oscillation can be calculated as follows:

$$P(X)=\int_{-\infty}^{\phi_a}f_-(\phi)d\phi+\int_{\phi_b}^{\infty}f_+(\phi)d\phi \tag{9.49}$$

Values ϕ_a and ϕ_b are points where the separatrix crosses axis ϕ. When the mean value of the wind equals zero and roll is symmetrical, these values become equal to the angle of vanishing stability.

We considered upcrossing theory in subchapter 9.1, it seems to be quite useful tool for the problem. We have a stochastic process, which is the phase trajectory and the boundary, which is the separatrix. The average number of the upcrossings of this boundary would serve the purpose. It is directly related with time and can be actually identified with the risk function for this particular assumed situation. The only thing to do is to generalize the one-dimensional case considered in 9.1 for the phase plane, so there will be two dimensions.

Sevastianov [1977] considered this problem for an even more complex case: drift was included. So, we are working in a three-dimensional phase space.

Upcrossing is still defined as a simultaneous satisfaction of two conditions. The first, phase trajectory should approach the separatrix. The second, image point should have enough velocity to cross the separatrix i.e. projection of image point velocity, that is normal to separatrix, should be positive.

So, we are working with the following coordinates: roll angle, roll velocity and drift velocity:

$$x_1 = \phi; \quad x_2 = \dot{\phi}; \quad x_3 = \dot{y}_G \tag{9.50}$$

The separatrix is a surface that could be described by an equation:

$$x_{1S} = f_S(x_{2,}, x_3) \tag{9.51}$$

The derivatives of the above coordinates are expressed as follows:

$$\dot{x}_1 = \dot{x}_1(t) = \frac{dx_1(t)}{dt} = \dot{\phi} \tag{9.52}$$

$$\dot{x}_2 = \dot{x}_2(t) = \frac{dx_2(t)}{dt} = \ddot{\phi} \tag{9.53}$$

$$\dot{x}_3 = \dot{x}_3(t) = \frac{dx_3(t)}{dt} = \ddot{y}_{G1} \tag{9.54}$$

The absolute value of the vector of image point velocity is:

$$|\vec{v}| = v = \sqrt{\dot{x}_1^2 + \dot{x}_2^2 + \dot{x}_3^2} \tag{9.55}$$

The image point velocity is a vector in phase space. The angles of this vector with the coordinate axis are the following:

$$\cos\alpha_v = \cos(v, \hat{x}_1) = \frac{\dot{x}_1}{v} \tag{9.56}$$

$$\cos\beta_v = \cos(v, \hat{x}_2) = \frac{\dot{x}_2}{v} \tag{9.57}$$

$$\cos\gamma_v = \cos(v, \hat{x}_3) = \frac{\dot{x}_3}{v} \tag{9.58}$$

We obtain angles between the internal normal projection and coordinate axis:

$$\cos\alpha_n = \cos(n, \hat{x}_1) = -\frac{1}{\sqrt{1 + \left(\frac{\partial f_S}{dx_2}\right)^2 + \left(\frac{\partial f_S}{dx_3}\right)^2}} \tag{9.59}$$

$$\cos\beta_n = \cos(n, \hat{x}_2) = \frac{\frac{\partial f_S}{dx_2}}{\sqrt{1 + \left(\frac{\partial f_S}{dx_2}\right)^2 + \left(\frac{\partial f_S}{dx_3}\right)^2}} \tag{9.60}$$

$$\cos\gamma_n = \cos(n, \hat{x}_3) = \frac{\frac{\partial f_S}{dx_3}}{\sqrt{1 + \left(\frac{\partial f_S}{dx_2}\right)^2 + \left(\frac{\partial f_S}{dx_3}\right)^2}} \tag{9.61}$$

The angle between the vector of image point velocity and normal projection can be expressed as:

$$\cos(\hat{v,n}) = \cos\alpha_n \cos\alpha_v + \cos\beta_n \cos\beta_v + \cos\gamma_n \cos\gamma_v \tag{9.62}$$

The projection of the vector $\vec{v}$ on the internal normal direction of the separatrix surface is:

$$v_n = v\cdot\cos(\hat{v,n}) = \frac{-\dot{x}_1 + \dot{x}_2\dfrac{\partial f_S}{dx_2} + \dot{x}_3\dfrac{\partial f_S}{dx_2}}{\sqrt{1+\left(\dfrac{\partial f_S}{dx_2}\right)^2+\left(\dfrac{\partial f_S}{dx_3}\right)^2}} \tag{9.63}$$

The numerator of equation (9.63) defines the sign of normal projection of the image point velocity. So, we can write one of the crossing conditions $(v_n \geq 0)$ as:

$$\dot{x}_1 \leq \dot{x}_2\frac{\partial f_S}{dx_2} + \dot{x}_3\frac{\partial f_S}{dx_2} \tag{9.64}$$

We add the condition that the phase trajectory should approach the separatrix surface:

$$x_1 = f_s(x_2, x_3) \tag{9.65}$$

We have received the system of equations that determines conditions of upcrossing. This system can be easily generalized for the n-dimensional case:

$$\begin{cases} \dot{x}_1 = f_S(x_{2,}x_3,...x_n) \\ \dot{x}_1 \leq \sum\limits_{i=2}^{2} \dot{x}_i \dfrac{\partial f_S}{\partial x_i} \end{cases} \tag{9.66}$$

Consider the probability of these conditions to be satisfied at a given point on the separatrix surface and at the given moment of time. Let us introduce the combined distribution of all the components of phase trajectory and their first derivatives:

$$\varphi = \varphi_t(x_{1,}x_2,...x_n, \dot{x}_{1,}\dot{x}_2,...\dot{x}_n) \tag{9.67}$$

The above probability can be expressed as follows:

$$dp_t(x_1 = f_S(x_{2,}x_3,..x_n), x_2,..x_n, \dot{x}_{1,}\dot{x}_2,..\dot{x}_n) =$$
$$\int_{-\infty}^{\infty} d\dot{x}_n ... \int_{-\infty}^{\infty} d\dot{x}_i ... \int_{-\infty}^{\dot{x}_{1\max}} \varphi_t(x_1 = f_S(x_{2,}x_3,..x_n), x_2,..x_n, \dot{x}_{1,}\dot{x}_2,..\dot{x}_n) dx_1 dx_2 ... dx_n d\dot{x}_1 \tag{9.68}$$

The integration has to be done for all derivatives where $v_n \geq 0$. The upper limit of the internal integral is expressed as:

$$\dot{x}_{1\max} = \sum_{i=2}^{n} \dot{x}_i \frac{\partial f_S}{\partial x_i} \tag{9.69}$$

Here, n is the number of dimensions. For roll and drift it equals 3. Then we integrate the probability (9.68) over the separatrix surface S taking into account $dx_1 = \dot{x}_1 dt$:

$$dp_t(S) = dt \int_{-\infty}^{\infty} dx_n ... \int_{-\infty}^{\infty} dx_2 \int_{-\infty}^{\infty} d\dot{x}_n ... \int_{-\infty}^{\dot{x}_{1\max}} F d\dot{x}_1 \qquad (9.70)$$

With:

$$F = \dot{x}_1 \varphi_t (x_1 = f_S(x_2, x_3, ... x_n), x_2, ... x_n, \dot{x}_1, \dot{x}_2, ... \dot{x}_n) \qquad (9.71)$$

The probability of upcrossing per unit of time can be expressed as follows:

$$\lambda(t) = \frac{dp_t(S)}{dt} \qquad (9.72)$$

Finally, we can write a formula for probability of the upcrossings per unit of time, which is the same as the averaged number of upcrossings:

$$\lambda(t) = \int_{-\infty}^{\infty} dx_n ... \int_{-\infty}^{\infty} dx_2 \int_{-\infty}^{\infty} d\dot{x}_n ... \int_{-\infty}^{\dot{x}_{1\max}} \dot{x}_1 \varphi_t (x_1 = f_S(x_2, x_3, ... x_n), x_2, ... x_n, \dot{x}_1, \dot{x}_2, ... \dot{x}_n) d\dot{x}_1 \qquad (9.73)$$

This value is the searched risk function based on the classical definition of stability. For the one-dimensional case formula, (9.73) gives the same result as (9.13) or (9.6).

9.2.3 Method of Energy Balance

We already considered the energy balance approach for capsizing in regular seas in subchapter 7.1.5. This method is very important, since it is the background of the weather criterion.

According to Boroday [1968] G.A. Firsov was the first, who attempted to determine the probability of a ship capsizing in beam seas using the energy balance method.

The series of research using the energy balance method for irregular seas was carried out by Boroday [1967, 1968]. Boroday [1968], Boroday and Nikolaev [1975], Boroday and Netsvetaev [1982] generalized the previous works for any course relative to waves.

This research considered the general hydrodynamic problem of forces and moments, which are acting on the ship from when she has a given course in irregular seas. The results were the statistic characteristics of the work of those forces and moments. Consideration of the balance of the work allowed formulating a method for calculation of capsizing probability.

Dudziak and Buczkowski [1978] proposed another approach: they considered all input values of the weather criterion as stochastic quantities.

Weather criterion suggests that the beginning of the wind gust coincides with the maximal roll angle on the windward side. The assumption was adopted, because such a situation is the most dangerous one: angular velocities caused by roll and by wind gust have the same directions in this situation. In reality, wind gust can start at any moment and is not related with the phase of the ship roll. But the danger of wind gust action will be decreased when the beginning of its action is shifted to the moment of roll to the leeward side.

Dudziak and Buczkowski [1978] proposed to calculate two probabilities: one when gust starts at the moment of windward roll phase (maximal danger) and another, when it coincides with leeward roll phase (minimal danger). Probability criterion is calculated as a mean value:

$$P(X) \approx \frac{1}{2}(P_W + P_L) \tag{9.74}$$

Here, $P(X)$ is probability of capsizing, P_W is conditional probability of capsizing when wind gust coincides in time with windward amplitude of roll, P_L is conditional probability of capsizing when wind gust coincides in time with leeward amplitude of roll.

Consider the conditional probability of capsizing if gust starts when roll angle reaches the amplitude value on the windward side. The initial conditions are:

$$\dot{\phi}_0 = 0 \quad ; \quad \phi_0 = \phi_{st} - \phi_{am} \tag{9.75}$$

Here, ϕ_{st} is constant angle of heel caused by constant component of the winds pressure, ϕ_{am} is a current stochastic amplitude of irregular roll. Since the ship has zero angular velocity at the initial moment, the kinetic energy is also zero:

$$K(\dot{\phi}_0) = K(0) = 0 \tag{9.76}$$

The energy balance equation can be written as :

$$P(t_0, t) = A_A(t_0, t) \tag{9.77}$$

Changes of potential energy can be expressed through the dynamic righting arm (or the area under GZ curve):

$$P(t_0, t) = \int_{\phi_0}^{\phi} mgGZ(\phi)d\phi = P(\phi_0, \phi) \tag{9.78}$$

The value of potential energy is a random variable, because the initial conditions are stochastic:

$$A_A(t_0, t) = (M_{Am} + M_{Af})(\phi - \phi_0) = A_n(\phi_0, \phi) \tag{9.79}$$

Here: M_{Am} is the value of wind heeling moment, caused by constant component of wind pressure: it is a deterministic value; M_{Af} is the additional wind heeling moment caused by wind gust. This variable has to be considered as random.

After substitution of (9.78) and (9.79) in (9.77) and taking into account (9.76), we obtain:

$$\int_{\phi_{st}-\phi_{am}}^{\phi_d} mgGZ(\phi)d\phi = (M_{Am} + M_{Af})(\phi_d - \phi_{st} + \phi_{am}) \tag{9.80}$$

Here: ϕ_d is dynamic angle of heel, which is associated with capsizing see fig. 9.11.

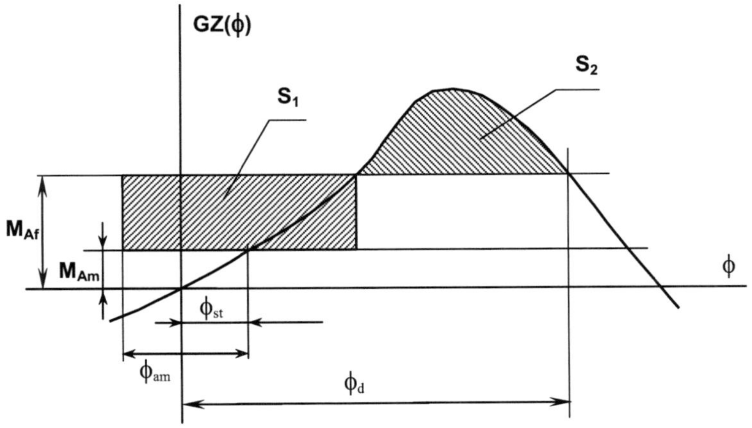

Fig. 9.11 Scheme of determination of dynamic angle of heel

The probability of capsizing can be calculated by the integration of the distribution $f(\phi_{am}, M_{Af})$ over the area S_C where the values of ϕ_{am} and M_{Af} cause capsizing:

$$P_W = \iint_{S_{C1}} f(\phi_{am}, M_{Af}) d\phi_{am} dM_{Af} \tag{9.81}$$

How do we define this area? Its boundary is M_{Af} vs. ϕ_{am}. The points of this boundary correspond to critical values of amplitude ϕ_{am} and additional heeling moment M_{Af}. These values provide satisfaction of weather criterion without surplus and lack. It means that for the given amplitude ϕ_{am} and additional heeling moment M_{Af}, the resulting dynamic heel angle ϕ_d is always equal to angle of vanishing stability to the other stability limit. The boundary $M_{Af}(\phi_{am})$ can be calculated by sequential solution of equation (9.80) relative to M_{Af}. For a number of given amplitudes ϕ_{am}, see fig 9.12.

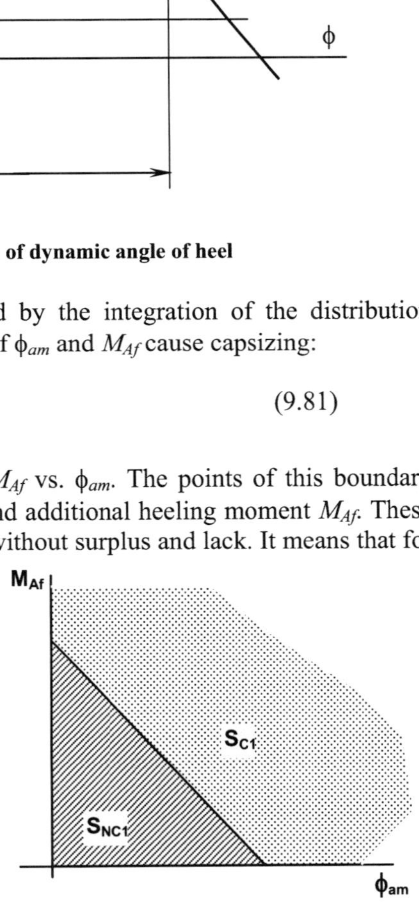

Fig. 9.12 Critical values of M_H versus ϕ_{am} capsizing S_{C1} and non-capsizing S_{NC1} area. [Dudziak and Buczkowski 1978]

It is more convenient to calculate the integral in (9.81) over the area S_{NC}, which corresponds to the non-capsizing domain in plane (ϕ_{am}, M_{Af}). So, the conditional probability of capsizing, if the beginning of a wind gust coincides in time with roll to the windward amplitude, can be calculated as:

$$P_W = 1 - \iint_{S_{NC1}} f(\phi_{am}, M_{Af}) d\phi_{am} dM_{Af} \tag{9.82}$$

Now, we consider the conditional probability of capsizing if the gust coincides with the roll leeward amplitude. The initial conditions in this case are as follows:

$$\dot{\phi}_0 = 0 \quad ; \quad \phi_0 = \phi_{st} + \phi_{am} \tag{9.83}$$

The energy balance equation is expressed as:

$$\int_{\phi_{st}+\phi_{am}}^{\phi_d} mgGZ(\phi)\,d\phi = (M_{Am} + M_{Af})(\phi_d - \phi_{St} - \phi_{am}) \tag{9.84}$$

Proceeding analogously to the previous case, the conditional probability of capsizing can be expressed as:

$$P_L = \iint_{Sc2} f(\phi_{am}, M_{Af})\,d\phi_{am}\,dM_{Af} \tag{9.85}$$

Area S_{C2} and its border can be obtained from the energy balance equation (9.84) in the same manner as that in the previous case, see fig. 9.13.

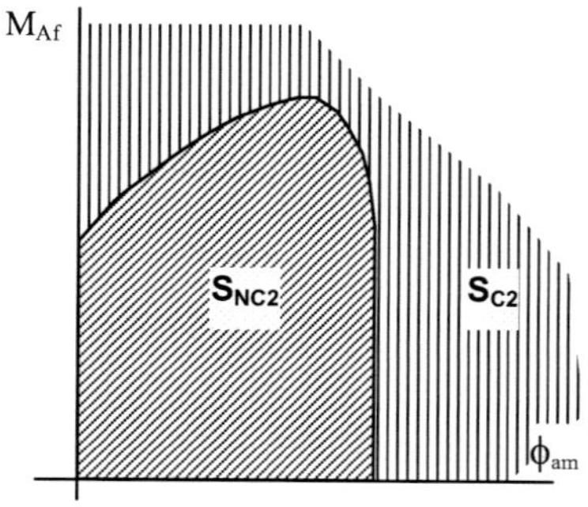

Fig. 9.13 Critical values of M_H versus ϕ_{am} capsizing S_{C2} and non-capsizing S_{NC2} area [Dudziak and Buczkowski 1978]

It is easier to calculate this probability by integrating over the non-capsizing area:

$$P_L = 1 - \iint_{NC2} f(\phi_{am}, M_{Af})\,d\phi_{am}\,dM_{Af} \tag{9.86}$$

The next step is evaluation of the joint distribution of roll amplitude and additional heeling moment. Dudziak and Buczkowski [1978] assumed them independent:

$$f(\phi_{am}, M_{Af}) = f(\phi_{am}) f(M_{Af}) \tag{9.87}$$

We have seen from subchapter 8.6.2 that the distribution of roll angles may be a non-Gaussian distribution. If this is the case, the problem of distribution of roll amplitudes requires additional consideration. If this is not the case, and roll distribution is Gaussian, then a Raleigh distribution can be used for amplitudes.

Consider the distribution of an additional heeling moment caused by a wind gust. Wind velocity is expressed as a sum of mean value u_{Am} and fluctuating part $u_{Af}(t)$:

$$u_A(t) = u_{Am} + u_{Af}(t) \tag{9.88}$$

Fluctuating part u_A(t) is assumed to be a Gaussian stochastic process:

$$f(u_A) = \frac{1}{2\pi\sigma_A} \exp\left[-\frac{(u_A - u_{Am})^2}{2\sigma_A^2} \right] \tag{9.89}$$

Here, σ_A is standard deviation of wind velocity. Dudziak and Buczkowski [1978] considered the wind gust as a maximum of the stochastic process of wind velocities.

Since secondary maxima are not of concern here, the Raleigh distribution can be used for the wind velocity value in the gust:

$$f(u_{Af}) = \frac{u_{Af}}{\sigma_A^2} \exp\left(-\frac{u_{Af}^2}{2\sigma_A^2}\right) \tag{9.90}$$

The additional heeling moment caused by a wind gust can be expressed via gust wind pressure:

$$M_{Af} = p_A A_W z_A \tag{9.91}$$

Here, p_{Af} is wind gust pressure; A_W is windage area z_A is vertical distance between the centre of windage area and the centre of hydrodynamic lateral resistance.

More information on probabilistic qualities of the wind pressures can be found in subchapter 8.5.1. Further research in this direction was carried out by Bielanski [1994].

There is one more question to address: we have found the probability of capsizing, but during what time?

The above method was developed for calculation of capsizing probability in the first semi-period of oscillation after the action of a wind gust. There is an assumption here that if the ship did not capsize after the first semi-period of oscillation, she will not capsize under action of this particular gust. Therefore, the above method yields probability of capsizing during one random gust of wind. Then the risk function can be estimated by dividing the probability by the averaged time of the wind gust, or otherwise related to average roll period.

9.2.4 Piecewise Linear Method

We introduced the piecewise linear system as the simplest dynamical system that is capable to reproduce capsizing as a transition to another equilibrium. Despite its non-smooth restoring term, the piecewise linear system has all the same qualities as that of a conventional model of roll and capsizing. We looked at its free motions in subchapter 4.1.2 and found a steady state solution under periodic excitation in subchapter 4.2.6. We have seen how the stability of this solution can be checked in subchapter 4.4.4 and that bifurcations are the same as a conventional nonlinear system (subchapter 4.5.4). Most important, we found out how simple it is to describe capsizing in a piecewise linear system under periodic excitation (subchapter 5.2). Also, the piecewise linear system helped us to understand how the classical definition of stability works (subchapter 5.2.3). Now, we are looking into capsizing in the piecewise linear system under the action of irregular seas [Belenky, 1989, 1993].

We have seen that capsizing in a piecewise linear system could be associated with the event of upcrossing of the boundary between ranges 0 and 1 (in fig. 9.14) with the simultaneous satisfaction of condition of the arbitrary constant A being positive (subchapter 5.2.2):

$$P_T(X) = P_T(\phi > \phi_{m0}) P(A > 0) \tag{9.92}$$

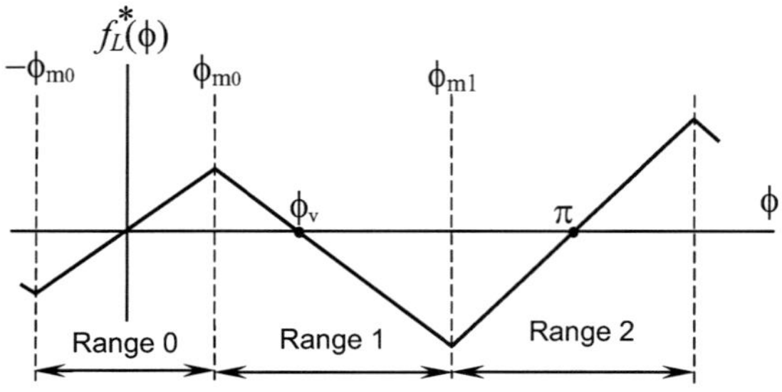

Fig. 9.14 Piece-wise linear restoring term

We mean under $P_T(\phi > \phi_{m0})$ the probability of at least one upcrossing of level ϕ_{m0} during time T. We assume that the level ϕ_{m0} is such that the upcrossing of this level is a rare event. It means that enough time passes between two neighboring upcrossings for the autocorrelation function of roll to be small. This assumption allows us to suppose that the flow of upcrossings is a Poisson flow and consequently the probability we are searching for can be calculated in accordance with exponential law (see subchapter 9.1 for details):

$$P_T(\phi > \phi_{m0}) = 1 - \exp(-\xi T) \tag{9.93}$$

Here, ξ is the averaged number of upcrossings - intensity of the flow of upcrossings. Its value depends on the distribution of roll angles and velocities (subchapter 9.1).

Consider the piecewise linear equation of roll under action of irregular waves:

$$\ddot{\phi} + 2\delta\dot{\phi} + \omega_\phi^2 f_L^*(\phi) = \sum_{i=1}^{N} \alpha_i \sin(\omega_i t + \varphi_{0i}) \tag{9.94}$$

The response can be expressed as:

$$\phi(t) = \phi_a \exp(-\delta t)\sin(\omega_0 t + \varepsilon_0) + \sum_{i=1}^{N} q_i \sin(\omega_i t + \beta_{0i} + \varphi_{0i}) \quad \text{range 0} \tag{9.95}$$

$$\phi(t) = A\exp(\lambda_1 t) + B\exp(\lambda_2 t) + \sum_{i=1}^{N} p_i \sin(\omega_i t + \beta_{1i} + \varphi_{0i}) + \phi_v \quad \text{range 1} \tag{9.96}$$

$$\phi(t) = \phi_a \exp(-\delta t)\sin(\omega_2 t + \varepsilon_2) + \sum_{i=1}^{N} q_i \sin(\omega_i t + \beta_{2i} + \varphi_{0i}) + \pi \quad \text{range 2} \tag{9.97}$$

Considering distribution of motions of piecewise linear system, it was assumed that the process is stationary. It means that its probabilistic characteristics are independent of time. This assumption, however, contradicts our ultimate task: a roll process that includes capsizing cannot be considered as stationary. To overcome this contradiction, we consider the roll process before capsizing separately and assume that enough time passes from the start point until capsizing to judge if the process is stationary.

The piecewise roll process before capsizing (9.95)-(9.96) consists of many segments corresponding to range 0 and range 1. Each of these segments has its initial conditions and its own free oscillations that are being generated after every crossing of the boundary between ranges, see fig. 9.15, where these points are marked.

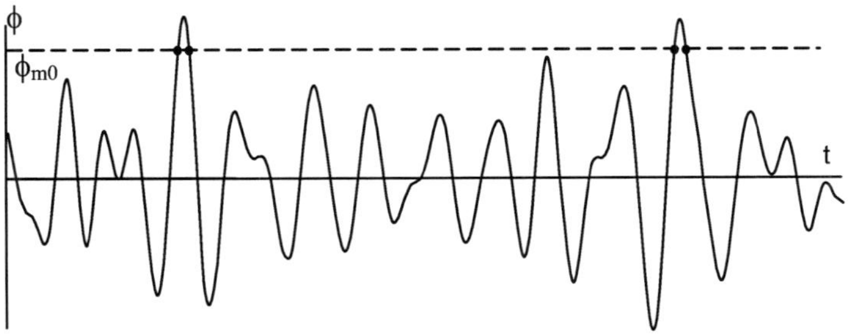

Fig. 9.15 Stochastic piecewise linear response

We assumed that the event of upcrossing of level ϕ_{m0} is rare. So, there is enough time between crossings for the correlation function to be small and for free oscillation in (9.95) to die that the system (9.94) "forgets" about any previous crossings.

So, we can assume that probabilistic characteristics of roll oscillation before capsizing can be determined by the partial solution only. Consequently the distribution of piecewise linear roll is Gaussian:

$$f(\phi) = \frac{1}{\sqrt{2\pi V_\phi}} \exp\left[-\frac{\phi^2}{2V_\phi}\right] = f(q) = \frac{1}{\sqrt{2\pi V_q}} \exp\left[-\frac{\phi^2}{2V_q}\right] \tag{9.98}$$

The variance also can be calculated by forced roll solution $q(t)$:

$$V_\phi = V_q = \frac{1}{2}\sum_{i=1}^{N} q_i^2 \tag{9.99}$$

Considering the piecewise linear roll response as a normal process, it is easy to express rate of upcrossings of level ϕ_{m0} as follows (see subchapter 9.1.2):

$$\xi = \frac{1}{2\pi}\sqrt{\frac{V_{\dot\phi}}{V_\phi}} \exp\left(-\frac{\phi_{m0}^2}{2V_\phi}\right) \tag{9.100}$$

The variance of roll velocity $V_{\dot\phi}$ also can be calculated by forced roll solution $q(t)$:

$$V_{\dot\phi} = V_{\dot q} = \frac{1}{2}\sum_{i=1}^{N} q_i^2 \omega_i^2 \tag{9.101}$$

The next term that defines probability of capsizing is $P(A > 0)$. It means the probability that arbitrary constant A is positive if upcrossing of level ϕ_{m0} has occurred.

Analyzing formula (4.113) we can see that this arbitrary constant depends on eigenvalues $\lambda_{1,2}$, initial conditions $\phi_1, \dot\phi_1$ and values of partial solution and its derivative in the moment of upcrossing $p_1, \dot p_1$. Three of these figures are stochastic values: $p_1, \dot p_1$ and $\dot\phi_1$. Other values are deterministic: eigenvalues are determined by piecewise linear term and damping and $\phi_1 = \phi_{m0}$. So, arbitrary constant A is a deterministic function of three random arguments:

$$A(\dot{\phi}_1, \dot{p}_1, p_1) = \frac{(\dot{\phi}_1 - \dot{p}_1) - \lambda_2(\phi_1 - p_1)}{\lambda_1 - \lambda_2} \tag{9.102}$$

Let us consider each of these random arguments in details:

$\dot{\phi}_1$ is a value of roll rate that is determined at the moment of upcrossing. In accordance with our hypothesis that upcrossings are rare events, we can substitute it by the value of the first derivative of the forced roll solution $\dot{q}_1$ determined at the moment of upcrossing. The processes of forced roll solution and its first derivative have a Gaussian distribution and are not correlated. So, upcrossing can take place with any positive roll rate and the distribution of $\dot{q}_1$ can be assumed to be a Rayleigh distribution (absolute value of normal variable has Rayleigh distribution):

$$f(\dot{\phi}_1) = f(\dot{q}_1) = \frac{\dot{q}_1}{V_{\dot{q}}} \exp\left[-\frac{\dot{q}_1^2}{2V_{\dot{q}}}\right] \tag{9.103}$$

p_1 is a value of partial solution $p(t)$ that is fixed at the moment of upcrossing. The process $p(t)$ is correlated with the process $q(t)$ because these processes are generated by the same excitation. The correlation coefficient can be expressed as follows (subchapter 8.3.2):

$$r_{pq} = \frac{1}{2\sqrt{V_q V_p}} \sum_{i=1}^{N} p_i q_i \cos(\beta_1 - \beta_0) \tag{9.104}$$

The joint distribution of both partial solutions is:

$$f(p,q) = \frac{1}{2\pi\sqrt{V_q V_p (1 - r_{pq}^2)}} \times$$
$$\times \exp\left\{-\frac{1}{2\sqrt{1 - r_{pq}^2}}\left[\frac{q^2}{V_q} - \frac{2q(p - m_p) r_{pq}}{\sqrt{V_q V_p}} + \frac{(p - m_p)^2}{V_p}\right]\right\} \tag{9.105}$$

The conditional distribution of $p(t)$, if upcrossing has taken place or $q = \phi_{m0}$, is:

$$f(p \mid q) = \frac{1}{2\pi\sqrt{V_q V_p (1 - r_{pq}^2)}} \exp\left\{-\frac{1}{2\sqrt{1 - r_{pq}^2}}\left[\frac{p - m_p}{\sqrt{V_p}} - r_{pq}\frac{q}{\sqrt{V_q}}\right]^2\right\} \tag{9.106}$$

Where $m_p = \phi_V$. Equation (9.106) can be presented in the more compact form:

$$f(p_1) = \frac{1}{\sqrt{2\pi V_{p1}}} \exp\left[-\frac{(p_1 - m_{p1})^2}{2V_{p1}}\right] \tag{9.107}$$

Here:

$$m_{p1} = \phi_v + \frac{\sqrt{V_p}}{V_q}\phi_{m0}; \quad V_{p1} = V_p (1 - r_{pq}^2) \tag{9.108}$$

As can be seen from (9.108), there is a difference between probabilistic characteristics of p_1 and the value of stochastic process $p(t)$ at any moment of time.

$\dot{p}_1$ is a value of the first derivative of partial solution $\dot{p}(t)$ that is determined at the moment of upcrossing. The process $\dot{p}(t)$ is correlated with the process $q(t)$, nevertheless the processes $q(t)$ and $\dot{q}(t)$ are not correlated:

$$r_{\dot{p}q} = \frac{1}{2\sqrt{V_q V_{\dot{p}}}} \sum_{i=1}^{N} p_i q_i \omega_i \sin(\beta_0 - \beta_1) \qquad (9.109)$$

The distribution of $\dot{p}_1$ can be derived analogously to the above distribution of p_1:

$$f(\dot{p}_1) = \frac{1}{\sqrt{2\pi V_{\dot{p}1}}} \exp\left[-\frac{(\dot{p}_1 - m_{\dot{p}1})^2}{2V_{\dot{p}1}} \right] \qquad (9.110)$$

Here:

$$m_{\dot{p}1} = \frac{\sqrt{V_{\dot{p}}}}{V_q} \phi_{m0}\,; \quad V_{\dot{p}1} = V_{\dot{p}}(1 - r_{\dot{p}q}^2) \qquad (9.111)$$

Considering the contribution of each term in formula (9.102), we have seen in subchapter 5.2.3 that partial solution $p(t)$ and its first derivative $\dot{p}(t)$ cannot have significant influence because their variances are much smaller in comparison with $q(t)$ and $\dot{q}(t)$ correspondingly, see fig 5.8.

So, variances of $\dot{p}(t)$ and $p(t)$ are significantly smaller than $V_{\dot{\phi}}$. This allows avoiding consideration of arbitrary constant A as a function of three random arguments that is a complex problem (a detailed consideration of it can be found in [Belenky, 1993]) and substitute p_1 and $\dot{p}_1$ by their mean values. We obtain a linear deterministic function for one random argument:

$$A(\dot{\phi}_1) = \frac{(\dot{\phi}_1 - m_{\dot{p}1}) - \lambda_2(\phi_{m0} - m_{p1})}{\lambda_1 - \lambda_2} = a_1\dot{\phi}_1 + a_2 \qquad (9.112)$$

Here:

$$a_1 = \frac{1}{\lambda_1 - \lambda_2} > 0 \quad \text{and} \qquad (9.113)$$

$$a_2 = \frac{m_{\dot{p}1} - \lambda_2(\phi_{m0} - m_{p1})}{\lambda_1 - \lambda_2} < 0$$

It should be mentioned that:

$$m_{\dot{p}1} = \frac{\sqrt{V_{\dot{p}}}}{V_q} \phi_{m0} \approx 0 \qquad (9.114)$$

$$m_{p1} = \phi_v + \frac{\sqrt{V_p}}{V_q}\phi_{m0} \approx \phi_v \tag{9.115}$$

To find probability that the arbitrary constant is positive it is enough to find such a value of roll rate at upcrossing (called further critical roll rate) that turns (9.112) into zero.

$$\dot{\phi}_{cr} = \dot{\phi}_1(A=0) = -a_2 / a_1 \tag{9.116}$$

All the upcrossings with the roll rate above the critical will lead to capsizing. Using distribution of the roll rates at upcrossing (9.103):

$$P(A>0) = P(\dot{\phi}_1 > \dot{\phi}_{cr}) = \int_{\dot{\phi}_{cr}}^{\infty} f(\dot{\phi}_1) d\dot{\phi}_1 \tag{9.117}$$

As it was shown in Chapter 1, it is not easy to work with probability of capsizing. Risk function λ, which is an average number (rate) of events per unit of time, is more practical.

Piecewise linear method allows considering capsizing as a subset of upcrossings. Then rate of upcrossing can be expressed as:

$$\lambda = \xi \cdot P(A>0) \tag{9.118}$$

Here ξ is an upcrossing rate (averaged number of upcrossings per unit of time) defined by formula (9.100). Expression (9.118) allows rewriting formula for probability of capsizing (9.92) expressing capsizing as an event of Poisson flow:

$$P_T(X) = 1 - \exp(-\lambda T) \tag{9.119}$$

The risk function λ derived here should be considered as a short term value; it is related to a particular sea state and loading condition. To emphasize this relation it is shown as $\lambda(X|S_i,L_j)$ – a function of particular realizations of vectors of loading conditions and assuming situations in the formula (1.66).

We have completed probabilistic consideration of capsizing in the piecewise linear system. As we previously mentioned, a "pure" piecewise linear model was introduced for rather theoretical purposes. The problem of practical calculation based on a piecewise linear model is addressed in the next subchapter.

9.2.5 Combined Piecewise-Linear-Numerical Method

The "triangle" *GZ* curve used above is sufficient for the theoretical study and understanding of the physics of capsizing phenomenon, however a real *GZ* curve is required for comparison of capsize probabilities of real ships. The distribution density for the roll process may be different from the Gaussian. As we have seen in subchapter 8.6.2, this difference strongly depends on the shape of the initial part of the *GZ* curve, because the most frequently encountered roll angles are in this range and their statistical weight has a large influence on the result.

Therefore, it is desirable to have a method for practical calculation that should not be dependent on the particular type of roll distribution. A combined numerical/piecewise linear method was proposed in [Belenky, 1994], it is based upon the following assumptions:

- Range No. 0 remains nonlinear. The boundary of range No. 0 (or nonlinear range) is located at the maximum of the *GZ* curve.
- A broken line is used to present the rest of the *GZ* curve (see fig. 9.16).
- We look for initial angular velocity, which satisfies the condition $A \approx 0$ in the beginning of the last range of the broken line.

Further, we call this initial velocity "critical" since the initial angular velocities exceeding it would lead to capsizing. All the velocities less than the critical one would not lead to capsizing because of the small influence of the excitation on the motion when roll angles correspond to the decreasing part of the *GZ* curve.

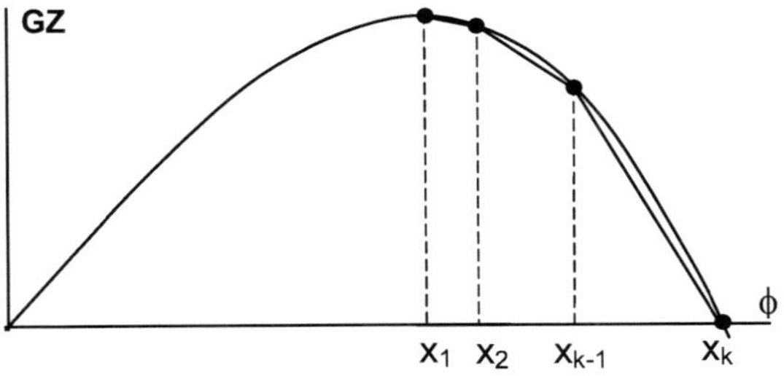

Fig. 9.16 Piecewise presentation of the decreasing part of *GZ* curve

The probability of exceeding the critical angular velocity can be easily found with the given distributions for roll angles and velocity. The rate of flow of upcrossings (e.g. average number of upcrossings per unit of time) ξ can be calculated with formula (9.13).

$$\xi = \int_0^\infty \dot{\phi} \cdot f(\phi = \phi_{\max}, \dot{\phi})\, d\dot{\phi} \tag{9.120}$$

Let us consider the algorithm for calculation of the critical velocity. We approximate the decreasing part of the *GZ* curve with a broken line with angular coefficients a_i and free terms b_i. Calculation of the critical velocity can be done with any iteration method, which is used for solution of nonlinear algebraic equations. An auxiliary function $A^{**}(\dot{\phi})$ is introduced for this purpose:

$$A^{**}(\dot{\phi}) = \begin{cases} A_{k-1} & \text{if process reaches } x_{k-1} \\ -1 & \text{if process does not reach } x_{k-1} \end{cases} \tag{9.121}$$

Here, x_{k-1} is the point at the beginning of the last range. The procedure for calculation of this function is the following.

We start from range 0 and proceed to the range $k-1$. $\dot{\phi}_i$ is the initial velocity at the entrance of the current range *i*.

Does the roll process reach the end of range *i* or not? First, we check the sign of arbitrary constant A_i. If A_i is positive, the angular velocity will increase without any limits and the roll process will reach the end of the range *i*. If A_i is negative, we calculate a time, for which it is necessary for roll velocity $\dot{\phi}(t)$ to reach zero. The following equation yields the required figure:

$$\dot{\phi}(t) = 0 \qquad (9.122)$$

Where:

$$\dot{\phi}(t) = \lambda_{1i} A_i \exp(\lambda_{1i} t) + \lambda_{2i} B_i \exp(\lambda_{2i} t) \qquad (9.123)$$

With:

$$A_i = \frac{\dot{\phi}_i - \lambda_{2i} x_i + \lambda_{2i}(b_i / a_i)}{\lambda_{1i} - \lambda_{2i}} \quad \text{and} \quad B_i = -\frac{\dot{\phi}_i - \lambda_{1i} x_i + \lambda_{1i}(b_i / a_i)}{\lambda_{1i} - \lambda_{2i}} \qquad (9.124)$$

$$\lambda_{1i,2i} = -\delta \pm \sqrt{a_i + \delta^2} \qquad (9.125)$$

The equation (9.122) can be solved with any iteration method. Once the time $t(\dot{\phi}_i = 0)$ was found, we check the following condition:

$$\phi(t_{\dot{\phi}_i = 0}) > x_{i+1} .$$

If this is true, the roll process will reach the end of the range *i*.

If this condition is not true, the roll process will never reach the end of the range and capsizing is impossible. Therefore:

$$A^{**}(\dot{\phi}_i) = -1$$

We calculate a time, which is necessary for the roll process to reach the end of the range *i*:

$$\phi(t) = x_{i+1} \qquad (9.126)$$

Where:

$$\phi(t) = A_i \exp(\lambda_{1i} t) + B_i \exp(\lambda_{2i} t) + b_i / a_i \qquad (9.127)$$

We calculate the roll velocity when the boundary between ranges *i* and $i+1$ is reached:

$$\dot{\phi}_{i+1} = \dot{\phi}(t_{\phi = x_{i+1}}) \qquad (9.128)$$

Calculations should be repeated until *i* reach value *k*-1.

When *i* will reach value *k-1*, only the value of arbitrary constant $A_{k\text{-}1}$ is calculated, which becomes the value for auxiliary function $A^{**}(\dot{\phi})$.

An initial angular velocity of roll, which makes a zero value for the auxiliary function, is searched. The critical initial value of angular velocity is expressed as:

$$A^{**}\left(\dot{\phi}_{cr}\right) = 0 \qquad (9.129)$$

The probability of capsizing finally is calculated as:

$$P_T(X) = \left(1 - \exp\left(-\left(\xi \cdot \int_{\dot{\phi}_{cr}}^{\infty} f(\dot{\phi}_1) d\dot{\phi}_1\right) \cdot T\right)\right) \qquad (9.130)$$

This method requires the distribution of roll angles and velocities as an input. These distributions may be the results of or results of direct Monte-Carlo simulation.

The last method seems to be preferable. The combination of numerical simulation without limitation on complexity of the model and piecewise linear model, which can treat really rare events, promises to be the most efficient solution for calculation of capsizing probability during a given time.

For further application of the piecewise linear method for calculation of capsizing probability see [Islander, *et al*, 2000, 2001], [Islander and Umeda, 2001, 2001a], [Paroka, *et al*, 2006], [Paroka and Umeda, 2006].

9.2.6 Methods Based on Motion Stability

We have considered stability of roll motion in subchapter 4.4. As we have seen, loss of motion stability means that a small deviation will increase, taking the system to another stable regime of motion. However, we did not consider motion stability when we were looking into capsizing in beam seas (Chapter 5).

The major drawback of the motion stability approach to capsizing is that we, in fact, linearized the system and, as a result, we do not know where the system goes if it is unstable; we just learn that this particular roll motion mode cannot exist for a long time. Loss of motion stability in regular seas does not necessarily constitute capsizing; the system may find another regime near original equilibrium, as with one with a doubled period of oscillation. Loss of stability of steady state mode oscillation is necessary, but not a sufficient condition of capsizing; however it definitely can be used as an indicator of approaching danger. The major advantage of the motion stability approach is we do not need actually to solve the roll equation in order to determine the stability of its solution.

This advantage becomes important when we consider capsizing in irregular seas. At the same time, the possibility to get the capsizing indicator based on a linearized solution is rather an advantage for the problem in irregular seas. Price [1975], Haddara [1975] and Nekrasov [1978, 1994] developed methods that combine the motion stability approach and stochastic description of seaway.

The main point of these methods is to estimate stability of the statistical characteristics of roll. Loss of stability of these values is associated with capsizing. External conditions that lead to such loss are recognized as dangerous.

The Markov process analysis is used by Haddara [1975] and Nekrasov [1978]. Haddara [1975] used the Fokker-Plank-Kholmogorov equation to obtain ordinary differential equations relative to mean value and variance. Analyzing the stability of these equations gives the required evaluation of safety against capsizing in an irregular seaway. Nekrasov [1978] also used Markov processes with the statistical moment technique. Price [1975] used impulse functions to determine stability of mean and mean square.

Zelenin [1989, 1989a] considered ship rolling and capsizing in irregular seas by taking into account heaving, nonlinear roll damping and gusty wind:

$$\ddot{\phi}+\delta\dot{\phi}+\delta_3\dot{\phi}^3+[\omega_\phi^2+a_1 z(t)]\phi-a_3\phi^3=f_E(t)+f_A(t) \qquad (9.131)$$

Here: $z(t)$ is heaving, f_E is wave excitation and f_A is wind excitation. The last value is presented as:

$$f_A(t)=\frac{C_P}{(I_{xx}+a_{44})}[2u_{Am}u_{Af}(t)+u_{Am}^2] \qquad (9.132)$$

Here: C_P is dimensional aerodynamic coefficient, u_{Am} is mean value of wind velocity, u_{Af} is fluctuating part of wind velocity. Canonical presentation is used for all stochastic processes only:

$$f_E(t)=\sum_{i=1}^{N}[\alpha_{Eci}\cos\omega_i t+\alpha_{Esi}\sin\omega_i t] \qquad (9.133)$$

$$f_A(t)=\sum_{i=1}^{N}[f_{ci}\cos\omega_i t+f_{si}\sin\omega_i t] \qquad (9.134)$$

$$z(t)=\sum_{i=1}^{N}[z_{ci}\cos\omega_i t+z_{si}\sin\omega_i t] \qquad (9.135)$$

Here, amplitudes of harmonics of Fourier series are non-correlated random values. They have a Gaussian distribution and their variance can be calculated as follows:

$$V_{\alpha i}=\int_{\omega_i-\frac{\Delta\omega}{2}}^{\omega_i+\frac{\Delta\omega}{2}} s_a(\omega)d\omega\approx s_\alpha(\omega_i)\cdot\Delta\omega \qquad (9.136)$$

Here: $s_\alpha(\omega)$ is spectral density of wave slope angles and $\Delta\omega$ is frequency step. Similar formulae can be written for all other stochastic amplitudes.

Then, the nonlinear equation of roll was linearized and solved. The solution was also searched in a form of canonical presentation:

$$\phi(t)=\phi_m+\sum_{i=1}^{N}[\phi_{ci}\cos\omega_i t+\phi_{si}\sin\omega_i t] \qquad (9.137)$$

Then, this solution is substituted into the original equation (9.131) with some small deviation ξ, exactly as we did for regular excitation, see subchapter 4.4.2. After simplification, we obtain the variation equation. We also introduce a new variable to get eliminate the damping term:

$$\varsigma=\xi\cdot\exp(-\delta_L t/2) \qquad (9.138)$$

As result, the variation equation is now presented in the standard form of the Hill equation:

$$\ddot{\varsigma}+\left[\theta_0+\sum_{i=1}^{2N}\left(\theta_{ic}\cos\omega_i t+\theta_{is}\sin\omega_i t\right)\right]\varsigma=0 \qquad (9.139)$$

The Hill equation is a generalization of the Mathieu equation (see subchapter 6.2.2): it has a Fourier series where the Mathieu equation has just one sine or cosine function.

The Hayashi [1964] criterion expresses the condition of boundedness for this form of the Hill equation. This criterion consists of two inequalities:

$$\begin{cases} \theta_0 > 0 \\ H(\omega_i) > \dfrac{\theta_i^2}{4} \end{cases} \tag{9.140}$$

Here:

$$\theta_i = \sqrt{\theta_{ci}^2 + \theta_{si}^2}$$

$$H(\omega_i) = \left[\theta_0 - \left(\frac{\omega_i}{2}\right)^2\right] + \frac{\delta_L^2}{2}\left[\theta_0 + \left(\frac{\omega_i}{2}\right)^2\right] + \frac{\delta_L^2}{16}$$

The probability of loss of motion stability at any moment of time can be expressed as only:

$$P_L = 1 - P_1 \cdot P_2 \tag{9.141}$$

Where P_1 and P_2 are probabilities of satisfaction of the first and the second inequalities of (9.140) correspondingly:

$$P_1 = 1 - \exp\left(\frac{\omega_\phi^2 - \delta_L^2}{3a_3(V_\phi + \phi_m^2)}\right) \tag{9.142}$$

$$P_2 = \prod_{i=1}^{2N}\left[1 - \exp\left(\frac{2H(\omega_i)}{V_{\theta i}}\right)\right] \tag{9.143}$$

Here, $V_{\theta i}$ is the variance of stochastic harmonic amplitudes θ_i.

To find the probability of capsizing per unit of time, Zelenin [1989, 1989a] proposed multiplying the probability of loss of motion stability at any time by the mean frequency of roll:

$$\lambda = \frac{1}{2\pi}\sqrt{\frac{V_{\dot\phi}}{V_\phi}}P_L \tag{9.144}$$

Here, $V_{\dot\phi}$ and V_ϕ are variances of roll velocities and angles.

9.2.7 Methods Based on Nonlinear Dynamics[1]

We have seen in subchapter 5.3, that increasing danger of capsizing in regular waves comes with a number of nonlinear phenomena: factorization of the safe basin is one of them. We have seen that the Melnikov function may be used as an indicator of these phenomena. The Melnikov function expresses distance between stable and unstable invariant manifolds. Once they intersect, the system slips into chaos and the likelihood of capsizing increases dramatically.

---

[1] The author is grateful to Prof. A.W. Troesch for fruitful discussion of the materials of this subchapter.

Hsieh, Troesch and Shaw [1994] considered behavior of the Melnikov function for the case of stochastic excitation and proposed a probabilistic measure of capsizing danger. The following equation of roll was used for the study:

$$\ddot{\phi} + \varepsilon\delta_1\dot{\phi} + \varepsilon\delta_2\dot{\phi}\,|\,\dot{\phi}\,| + \phi - a_3\phi^3 = \varepsilon\alpha_m(\tau) \tag{9.145}$$

Here, time is non-dimensional; it is expressed in terms of natural frequency:

$$\tau = \omega_\phi t \tag{9.146}$$

Here, ε is a small value. It is important to notice that this small parameter is not present in the nonlinear restoring term. This is the principal difference with Melnikov analysis and classical asymptotic methods. If the latter ones have a linear equation as a limit when $\varepsilon \to 0$, while equation (9.145) has an undamped and unforced Duffing equation, which keeps all major nonlinear properties including capsizing:

$$\ddot{\phi} + \phi - a_3\phi^3 = 0 \tag{9.147}$$

This also means that Melnikov analysis can be considered as an asymptotic extension of the classical definition of stability: in this sense, it is also similar to weather criterion (relationship between weather criterion and classical definition of stability is examined in subchapter 7.1.5).

Consider the Melnikov function for the equation (9.145) (we considered Melnikov in general vector form in subchapter 5.3.4):

$$M(\tau_0) = \int_{-\infty}^{\infty} \dot{\phi}_M(\tau)\left(-\delta_1\dot{\phi}_M - \delta_2\dot{\phi}_M\,|\,\dot{\phi}_M\,| + \alpha_E(\tau+\tau_0)\right)d\tau \tag{9.148}$$

Index "M" indicates that integration is performed along the invariant manifold.

Excitation α_E is the only stochastic figure in the formula (9.148). It can be seen from this formula that the Melnikov function can be presented in a form of a sum of constant and fluctuating parts:

$$M(\tau_0) = M_m + M_f(\tau_0) \tag{9.149}$$

The constant part is associated with damping and can be expressed in closed form (compare with formula (5.71)):

$$M_m(\tau_0) = \int_{-\infty}^{\infty} \dot{\phi}_M(\tau)\left(-\delta_1\dot{\phi}_M - \delta_2\dot{\phi}_M\,|\,\dot{\phi}_M\,|\right)d\tau = \frac{2\sqrt{2}}{3a_3} + \frac{8}{15}\frac{\delta_3}{\sqrt{a_3^3}} \tag{9.150}$$

The fluctuating part is related with the excitation:

$$M_f(\tau_0) = \int_{-\infty}^{\infty} \dot{\phi}_M(\tau)\cdot\alpha_m(\tau+\tau_0)\,d\tau \tag{9.151}$$

The fluctuating part of the Melnikov function is a stochastic process. Its probabilistic characteristics were studied in [Hsieh, *et al*, 1993, 1994]. It was found that:

- It is a stationary and ergodic process;
- It has a Gaussian distribution.

The last statement also means that mean value and variance determine distribution completely. It can be easily seen that mean value of the fluctuating part of the Melnikov function is zero:

$$m[M_f(\tau_0)] = \int_{-\infty}^{\infty} \dot{\phi}_M(\tau) \cdot m[\alpha_E(\tau+\tau_0)]d\tau = 0 \tag{9.152}$$

The variance of the fluctuating part of the Melnikov function can be expressed through its spectrum:

$$V[M_f(\tau_0)] = \int_0^{\infty} S_M(\omega)d\omega \tag{9.153}$$

This spectrum can be expressed with a Fourier transform in integral form:

$$\begin{aligned} S_M(\omega) &= 2\pi\left(\frac{1}{2\pi}\int_{-\infty}^{\infty} M_f(\tau_0)\exp(-i\omega\tau_0)d\tau_0\right)\times CC = \\ &= 2\pi\left(\frac{1}{2\pi}\int_{-\infty}^{\infty}\int_{-\infty}^{\infty}\dot{\phi}_M(\tau)\cdot\alpha_E(\tau+\tau_0)\ \exp(-i\omega\tau_0)d\tau_0 d\tau\right)\times CC \end{aligned} \tag{9.154}$$

CC is a complex conjugate. We need to multiply by the complex conjugate to eliminate the imaginary values. This is analogous to getting amplitude and eliminating phase when we were working with the discrete Fourier presentation, see subchapter 8.1.5.

Here, we introduce a new variable [Hsieh, *et al*, 1994]:

$$\tau_1 = \tau + \tau_0 \tag{9.155}$$

The new variable allows expressing spectrum (9.154) as two Fourier transformations:

$$\begin{aligned} S_M(\omega) &= 2\pi\left(\frac{1}{2\pi}\int_{-\infty}^{\infty}\int_{-\infty}^{\infty}\dot{\phi}_M(\tau)\exp(i\omega\tau)\alpha_E(\tau_1)\ \exp(-i\omega\tau_1)d\tau_0 d\tau\right)\cdot CC = \\ &= \int_{-\infty}^{\infty}\dot{\phi}_M(\tau)\exp(i\omega\tau)d\tau\cdot\int_{-\infty}^{\infty}\alpha_E(\tau_1)\ \exp(-i\omega\tau_1)d\tau_1 \times \\ &\times \int_{-\infty}^{\infty}\dot{\phi}_M(\tau)\exp(-i\omega\tau)d\tau\cdot\int_{-\infty}^{\infty}\alpha_E(\tau_1)\ \exp(i\omega\tau_1)d\tau_1 = \\ &= 2\pi S_{\dot{\phi}M}(\omega)S_\alpha(\omega) \end{aligned} \tag{9.156}$$

Here, $S_\alpha(\omega)$ is a known spectrum of roll excitation $S_{\dot{\phi}M}(\omega)$ which can be derived through Fourier transform of the invariant manifold:

$$S_{\dot{\phi}M}(\omega) = \frac{\pi}{a_3}\left(\frac{\omega}{\sinh(\pi\omega/\sqrt{2}\,)}\right) \tag{9.157}$$

We have completely defined the probabilistic qualities of the Melnikov function. Now we need to build a probabilistic measure of capsizing danger.

We have seen in subchapter 5.3 that once the invariant manifold is crossed, the safe basin becomes fractal, see fig. 5.10. This means that initial conditions leading to capsizing are mixed with safe initial conditions. This is a "still" picture for regular waves. In case of stochastic excitation, this picture is changing all the time.

Once the invariant manifold is crossed, some areas of the safe basin become unsafe, so we can consider it as "transport" of initial conditions out of the safe basin. The value related to the amount of transported area was proposed as the stability criterion in regular waves [Rainey, *et al*, 1990] [Rainey and Thompson, 1991].

This value becomes a random variable for irregular seas. Therefore, Hsieh, Troesch and Shaw [1994] proposed to use its averaged value as a criterion. It can be calculated using the Monte-Carlo method; however, this is a very large amount of calculations, because all initial conditions have to be checked at each time step. Since the safe basin becomes fractal, we may need a fine mesh to reach necessary accuracy, so the calculation procedure may be expensive.

Instead, the Melnikov function can be used. It gives the distance between invariant manifolds; when the Melnikov function is positive, the area, enveloped by it got transported out of the safe basin. If we integrate the positive parts of the Melnikov function over the time, we obtain a measure of capsizing danger for this particular ship in this specific environment. : This criterion also has a geometrical interpretation: it is a rate of flux in the phase plane. Following derivations in [Hsieh, *et al*, 1994] we write:

$$\Phi = \lim_{T\to\infty} \frac{\varepsilon}{2T} \int_{-T}^{T} f_{pos}\left(M(\tau_0)\right) d\tau_0 \tag{9.158}$$

Auxiliary function $f_{pos}(x)$ is introduced to pick up only the positive part of the Melnikov function:

$$f_{pos}(x) = \begin{cases} x & \text{if} \quad x \ge 0 \\ 0 & \text{if} \quad x < 0 \end{cases} \tag{9.159}$$

As we already mentioned, Hsieh, *et al* [1993] has shown that the Melnikov function is an ergodic process, so expression (9.158) can be considered as an average of the positive part of the Melnikov function and can be expressed not only over time, but also using the probability distribution:

$$\begin{aligned} m\left[f_{pos}\left(M(\tau_0)\right)\right] &= \varepsilon \int_{-M_m}^{\infty} (M - M_m) f(M) dM = \\ &= \varepsilon \int_{-M_m}^{\infty} (M - M_m) \frac{1}{\sqrt{2\pi V_M}} \exp\left(-\frac{(M - M_m)}{2V_M} \right) dM = \\ &= \varepsilon \left[\sqrt{V_m} f_G\left(\frac{M_m}{\sqrt{V_m}} \right) + M_m F_G\left(\frac{M_m}{\sqrt{V_m}} \right) + M_M \right] \end{aligned} \tag{9.160}$$

Here, f_G and F_G are Gaussian probability densities and their integrals are as follows:

$$f_G(x) = \frac{1}{\sqrt{2\pi}} \exp\left(-\frac{x^2}{2}\right); \quad F_G = \int_{-\infty}^{x} f_G(x)dx \qquad (9.161)$$

Jiang, Troesch and Shaw [1996] carried out further development of this technique, generalizing it for the case of a biased ship.

The Melnikov function was used to study chaotic roll response in irregular seas [Lin and Yim, 1996]. The probability distribution was obtained with the Fokker-Plank-Kholmogorov equation.

Vishnubhota, Falzarano and Vakakis [2000, 2001] developed another method to predict capsizing in irregular seas. This method is based on direct calculation of stable and unstable invariant manifolds, using the perturbation technique with the Duffing equation (9.147) as a zero-order expansion. A probabilistic study based on the Melnikov function is described in Chapter 7 and 8 of [Shlesinger and Swean, 1998].

Further developments of application of Nonlinear Dynamic for study of capsizing in irregular seas included consideration of the Lyapunov exponent, see [McCue, *et al*, 2006].

9.2.8 Markov Processes Application

We briefly introduced Markov processes in subchapter 8.6.4, while considering the probabilistic qualities of nonlinear roll. The main motivation of application of Markov processes was an opportunity to get distribution density as a solution of the Fokker-Planck-Kholmogorov equation. There is another aspect of Markov processes that may be applicable for the probabilistic study of capsizing - distribution of time to reach the boundary.

As we discussed in subchapter 9.1.1 the crossing problem is directly related with probability of capsizing. Crossing theory (subchapter 9.1.2) gives the mean value for the number of crossings without special assumptions. To get the time before crossing (subchapter 9.1.4), we had to assume that the crossings are a Poisson flow (subchapter 9.1.3), which imposes additional assumptions. The theory of Markov processes allows getting the time before reaching the boundary without additional assumptions; it is sufficient to consider the "carrier" process as a Markov process [Sveshnikov, 1968].

Let us look for the probability that within time interval $[t; t+\tau]$, the Markov process U would not reach the boundary that is located at u_1 and u_2:

$$u_1 < x(t) < u_2 \qquad (9.162)$$

Consider the distribution density of probability that the process $x(t)$ belongs to the range $[x, x+dx]$ and during given interval $[t; t+\tau]$ never has approached the boundaries u_1 or u_2: $p(T, x)$. The probability that the process would never reach the boundaries during time period T is expressed then with the following integral:

$$P(T) = \int_{u_1}^{u_2} p(T, x)\, dx \qquad (9.163)$$

The fact that we are interested in the time to reach the boundary cannot change probabilistic qualities of the process. So, before the process crosses the boundary, both distribution $p(\tau, x)$ and conditional distribution $f(t, x; \tau, y)$ (see subchapter 8.6.4) are governed by the same Fokker-Planck-Kholmogorov equation (8.264):

$$\frac{\partial f}{\partial \tau} + \frac{\partial}{\partial y}[a(\tau, y) \cdot f] - \frac{1}{2}\frac{\partial^2}{\partial y^2}[b(\tau, y) \cdot f] = 0$$

Once the boundary is crossed, $p(\tau, x)$ becomes zero, since the process is no longer within the given boundaries. This is actually a new boundary condition for (8.264):

$$p(\tau, u_1) = p(\tau, u_2) = 0 \quad \text{for} \quad \tau > t \tag{9.164}$$

Initial conditions for $p(\tau, x)$ may be defined as a delta-function if we need to start from a specific value of x, or just any given distribution, if vice versa.

This approach, also known as the "first passage" method, was used by several researchers. Cai, Yu and Lin [1994] considered total energy of the system as a Markov process and found the time before capsizing. Shen and Huang [2000] studied the same problem, but they were considering rolling as a Markov process. A comprehensive study of the subject can be found in Chapter 5 of [Shlesinger and Swean, 1998].

9.3 Probability of Capsizing in Following Seas and Risk Caused by Breaking Waves

9.3.1 Classical Definition of Stability and Pure Loss of Stability

The classical definition of stability in a probabilistic sense is associated with upcrossing of the separatrix by the multidimensional stochastic process (subchapter 9.2.2). Association with an upcrossing event is important, because it allows relating capsizing probability with time through the Poisson flow of random events.

This problem becomes more complex in following and quartering seas: a changing *GZ* curve in time actually adds one more stochastic process into the roll model. As a result, the separatrix also becomes stochastic. :

We already introduced the model of roll in following and quartering irregular seas (subchapter 8.4.3), equation (8.209). Here, we just added wind action F_A:

$$(I_{xx} + a_{44})\ddot{\phi} + 2N_\phi\dot{\phi} + mg \cdot GZ(\phi, t, \xi_G) = mg \cdot GM \cdot \alpha_W(t) + F_A(t) \tag{9.165}$$

Here N_ϕ is the linearized roll damping coefficient.

While surging, that influences the *GZ* curve, is described in subchapter 8.4.2 by the equation (8.200):

$$(M + m_{11})\ddot{\xi}_G + R(v_s + \dot{\xi}_G) - T(v_s + \dot{\xi}_G) = F_w(t, \xi_G)$$

Following [Umeda, *et al*, 1990] [Umeda and Yamakoshi, 1993], we use the effective wave concept, which was described in subchapter 8.4.3. As a result, the *GZ* curve now depends only on the amplitude of the effective wave, which is also a stochastic figure. This significantly simplifies the problem: otherwise we would have to consider at least three stochastic processes: height of the wave, length of the wave and ship position relative to the wave:

$$(I_{xx} + a_{44})\ddot{\phi} + 2N_{\phi}\dot{\phi} + mg \cdot GZ(\phi, \zeta_{eff}(t)) = mg \cdot GM \cdot \alpha_E(t) \tag{9.166}$$

As it is known, surging may increase the danger of capsizing in following and quartering seas. Usually, when a ship reaches the wave crest, surging velocity experiences a minimum. As a result, the ship spends more time on the wave crest, where stability is decreased. Umeda, *et al* [1990] proposed to approximate effective wave modulation with rectangle pulses, see fig. 9.17.

Duration of the pulse is approximated as:

$$T_C(\dot{\xi}_G) = \frac{\pi}{\omega_L - (\omega_L^2 / g)(v_s + \dot{\xi}_G)\cos\chi} \tag{9.167}$$

Here:

$$\omega_L = \sqrt{\frac{2\pi g}{L\cos\chi}} \tag{9.168}$$

The approximation of modulation is:

$$\zeta_{eff}(t) = \begin{cases} \zeta^*_{eff} & 0 \le t \le T_C \\ 0 & T_C < t \end{cases} \tag{9.169}$$

Here, ζ^*_{eff} is the effective wave height as we have described it in subchapter 8.4.3.

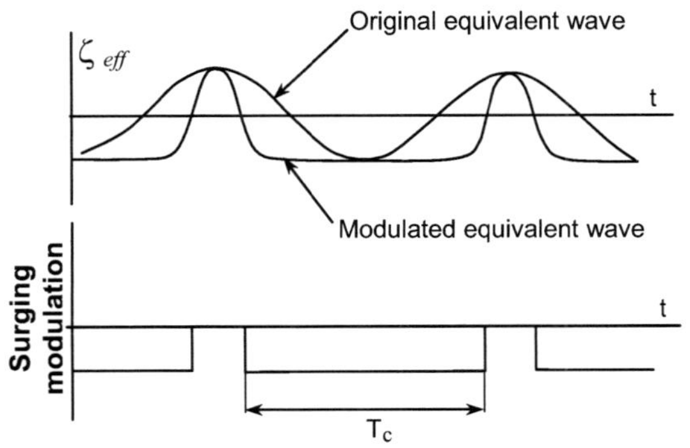

Fig. 9.17 Approximation for surging influence on effective wave [Umeda, 1990]

The classical definition of stability associates capsizing with crossing of the separatrix by the phase trajectory of roll motion. The influence of the following/quartering irregular seas is stochastic changes of the *GZ* curve and therefore, the separatrix becomes a stochastic process. The *GZ* curve depends on two random figures: effective wave height and surging velocity, so the separatrix also depends on them. As a result, capsizing is defined by a combination of four stochastic processes: roll angle and velocity, surging velocity and effective wave:

$$P(X) = 1 - \int_{-\infty}^{\infty}\int_{-\infty}^{\infty}\iint_{Sb(\phi,\dot{\phi},\zeta_{eff},\dot{\xi}_G)} f(\zeta_{eff}, \dot{\xi}_G, \phi, \dot{\phi},)\, d\phi, d\dot{\phi}, d\zeta_{eff} d\dot{\xi}_G \tag{9.170}$$

To calculate the probability of capsizing, we need to know the joint distribution of all the above stochastic processes. As we have seen from subchapter 8.6.2, the joint distribution of roll angle and velocity can be assumed Gaussian if the *GZ* curve does not have an S-shaped form. The distribution of effective wave and surging velocity also can be assumed normal [Umeda, *et al*, 1990], [Umeda and Yamakoshi, 1993]. Therefore the joint distribution $f(\zeta_{eff}, \dot{\xi}_G, \phi, \dot{\phi})$ also can be assumed Gaussian and then the only thing we need to know is the covariation matrix. Elements of the covariation matrix can be easily estimated if we use any linearization technique for roll angles, surging and roll velocities. The linearization technique allows presenting these processes as a Fourier series and effective wave is already in this form (subchapter 8.4.3). Once stochastic processes are presented in Fourier form, calculation of their correlation is easy, see subchapter 8.3.2.

Then, integration of the formula (9.170) should not meet any theoretical difficulties. However, the result may be too close to zero to be used for practical purposes. The reason is that a ship meets serious danger of capsizing only when she encounters a wave crest and her stability is significantly decreased. Therefore, the conditional probability of capsizing on the wave crest is more practical. Umeda, *et al* [1990], proposed a method to estimate this probability.

9.3.2 Piecewise Linear Method

Here, we try to extend the piecewise linear method (see subchapter 9.2.4) for calculating the probability of capsizing in following and quartering seas [Belenky, 1999a, 2000b]. Consider roll in following and quartering irregular seas (subchapter 8.4.3), equation (8.209). Here, surging influence is not included, trying to simplify the model for qualitative study.

$$(I_{xx} + a_{44})\ddot{\phi} + 2N_{\phi}\dot{\phi} + mg \cdot GZ(\phi, t) = mg \cdot GM \cdot \alpha_W(t) \qquad (9.171)$$

Here N_{ϕ} is the linearized roll damping coefficient.

The *GZ* curve here is a two-dimensional stochastic process. Its piece linear presentation therefore, is a surface that consists of many flat panels. The simplest way to perform such penalization is by using peaks of the *GZ* curve and angles of vanishing stability. These figures are, indeed, stochastic processes: a sample of their realization is presented in fig. 9.18.

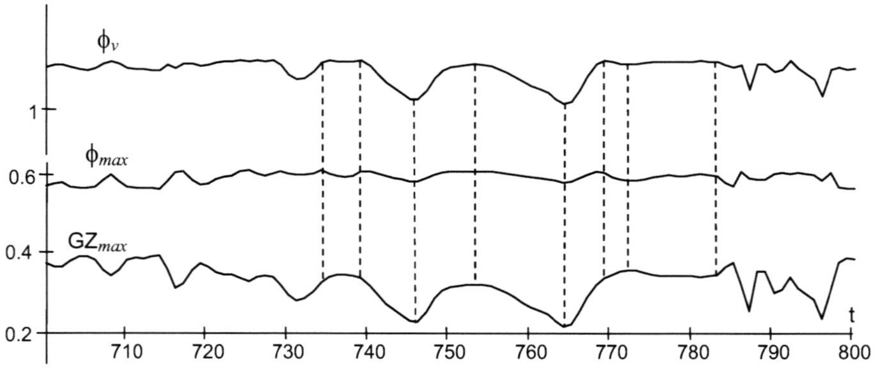

Fig. 9.18 Elements of *GZ* curve as stochastic processes [Belenky, 2000b]

Following the idea of combined piecewise linear – numerical method (subchapter 9.2.5), we make a piecewise linear presentation for the decreasing part of the *GZ* curve only, leaving the rest in its original nonlinear form:

$$GZ(\phi,t)=\begin{cases} GM(t)\phi & \phi<\phi_{\max}(t) \\ \begin{cases} a_1\phi+b_1t+c_1 & t\in[t_0;t_1[\\ a_2\phi+b_2t+c_2 & t\in[t_1;t_2[\\ \dots & \dots \\ a_n\phi+b_nt+c_n & t\in[t_{n-1};t_n[\end{cases} & \phi\geq\phi_{\max}(t) \end{cases} \tag{9.172}$$

The peak of the *GZ* curve is presented as a broken line:

$$\phi_{\max}(t)=\begin{cases} l_1t+m_1 & t\in[t_0;t_1[\\ l_2t+m_2 & t\in[t_1;t_2[\\ \dots & \dots \\ l_nt+m_n & t\in[t_{n-1};t_n[\end{cases} \tag{9.173}$$

The resulting 2-dimensional piecewise linear presentation is shown in fig. 9.19.

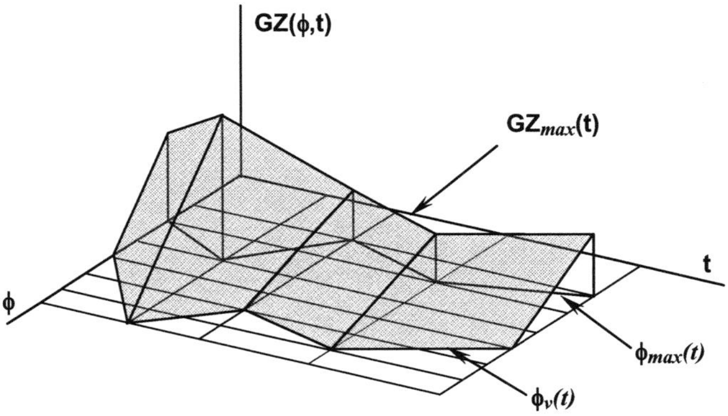

Fig. 9.19 Piecewise linear presentation of decreasing part of *GZ* curve in following or quartering seas [Belenky, 2000b]

All coefficients in formulae (9.172) and (9.173) are discrete stochastic processes. Their probabilistic characteristics were considered in [Belenky, 2000b]. It was found that they may be ergodic processes; distributions, however, are not Gaussian.

Let us focus on the roll motions at decreasing part of the GZ curve:

$$(I_{xx}+a_{44})\ddot{\phi}+2N_\phi\dot{\phi}+mg\cdot(a_i\phi+b_it+c_i)=mg\cdot GM\cdot\alpha_W(t) \tag{9.174}$$

The solution of (9.174) does not meet any difficulties:

$$\phi(t)=A_i\exp(\lambda_{1i}t)+B_i\exp(\lambda_{2i}t)+p(t)+q_{1i}t+q_{2i} \tag{9.175}$$

Here *p(t)* is a partial solution corresponding to wave excitation. Let us try to formulate capsizing conditions for (9.175). To do this, let us introduce the "line of no return" (LNR) that is the border between the decreasing part and another increasing part adjacent to upside – down equilibrium.

The solution (9.175) has three possibilities: immediate capsizing that is associated with crossing of LNR (trajectory 1 at Fig. 9.20), returning and downcrossing (trajectory 2) and escape from this range (trajectory 3). The same possibilities exist at the next range (trajectory 4 for capsizing, trajectory 5 for downcrossing and trajectory 6 for the next escape).

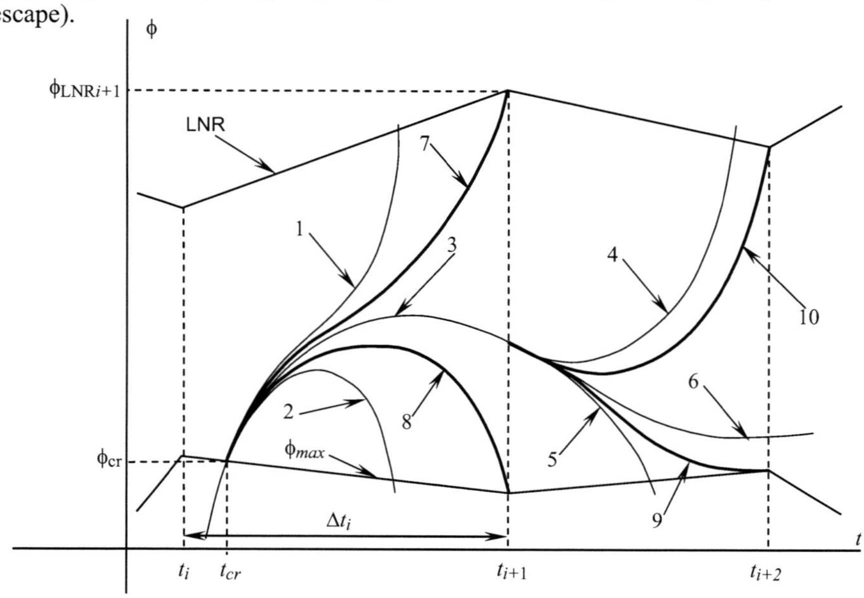

Figure 9.20 Possible variants of behavior at decreasing part of the *GZ* curve [Belenky, 2000b]

Following Belenky [1993], let us consider the sign of arbitrary coefficient A of solution (9.175). However, a positive sign for A is not enough for capsizing, because the vertical border of the range also exists. So, the system can cross it and escape capsizing even having a positive coefficient A. Therefore, an additional capsizing condition is necessary. Let us introduce three instances of time: t_{LNR} is a time necessary to reach the "line of no return" regardless of any other borders crossed. We consider t_{ESC} as a time necessary to escape from the range through the vertical border calculated with the above condition. Finally, we define t_{DCR} in the same way as a time necessary to reach level $\phi_{\max}$ and downcross it. We note that the first figure makes sense only if the coefficient A is positive, the third one if it is negative, and the second one – in both cases. Using this time, the capsizing condition can be formulated in the following manner:

$$
\begin{aligned}
A \geq 0 \quad & \begin{cases} t_{LNR} \leq t_{ESC} & \text{Immediate capsizing} \\ t_{LNR} > t_{ESC} & \text{Capsizing is possible} \end{cases} \\
A < 0 \quad & \begin{cases} t_{DCR} \geq t_{ESC} & \text{Capsizing is possible} \\ t_{DCR} < t_{ESC} & \text{Capsizing is impossible} \end{cases}
\end{aligned}
\tag{9.176}
$$

Then, probability of capsizing after time t can be expressed as:

$$P(X,t) = P(\phi > \phi_{\max}, t) \times \times \left[P(Y_i) + P(E_i)P(Y_{i+1} \mid E_i) + P(E_{i+1})P(Y_{i+2} \mid E_{i+1}) + \ldots\right] \quad (9.177)$$

Here Y_i is an event of immediate capsizing at the range $[t_i; t_{i+1}[$; E_i is event of escape from this range through a vertical border. The probability of capsizing after upcrossing is expressed by a theoretically infinite series. However, the series seems to converge quickly because of exponential functions in the solution (9.175) that try to take the system away from the decreasing part of the *GZ* curve. The probability of immediate capsizing after upcrossing formally can be presented as:

$$P(Y_i) = P(A > 0)P(t_{LNR} \le t_{ESC}) \quad (9.178)$$

Figures A, t_{LNR} and t_{ESC} are random values, they can be considered as deterministic functions of random initial conditions at upcrossing $(\phi_{cr}, \dot{\phi}_{cr})$ if we assume partial solution $p(t)$ is small in comparison with the other terms in (9.175) (subchapter 8.2.4).

Let us introduce random vector $\vec{B} = (l, m, \Delta t)$. To proceed with probability (9.178) we need the joint conditional distribution of initial conditions at upcrossing, assuming the last one as a random event that $\{\phi(t_{cr}) < l \cdot t_{cr} + m)\}$ and $\{\phi(t_{cr} + dt) < l \cdot (t_{cr} + dt) + m\}$, where $t_{cr} < \Delta t$ is a time of crossing. Having in mind that dt is small, the probability of this event can be expressed as:

$$P\{l \cdot t_{cr} + m - (\dot{\phi} - l)dt < \phi(t_{cr}) < l \cdot t_{cr} + m\} = = \int_0^{\infty} \int_{l \cdot t_{cr}+m+(\dot{\phi}-l)dt}^{l \cdot t_{cr}+m} f(\phi,\dot{\phi})d\phi d\dot{\phi} = dt \int_0^{\infty} f(lt_{cr} + m, \dot{\phi})(\dot{\phi} - l)d\dot{\phi} \quad (9.179)$$

Taking into account that $\phi_{cr} = \phi(t_{cr}) = l \cdot t_{cr} + m$:

$$P_U(t = t_{cr}) = P_U(\phi = \phi_{cr}) = \frac{d\phi_{cr}}{|l|} \int_0^{\infty} f(\phi_{cr}, \dot{\phi})(\dot{\phi} - l)d\dot{\phi} = f(\phi_{cr})d\phi_{cr} \quad (9.180)$$

Having in mind independence of roll angle and velocity and that upcrossing can happen with any positive roll velocity:

$$f(\phi_{cr}, \dot{\phi}_{cr} \mid \vec{B}) = f(\phi_{cr}) f(\dot{\phi})\Big|_{\dot{\phi}>0} = 2f(\phi_{cr}) f(\dot{\phi}_{cr}) = = \frac{2f(\dot{\phi}_{cr})}{|l|} \int_0^{\infty} f(\phi_{cr}) f(\dot{\phi})(\dot{\phi} - l)d\dot{\phi} \quad (9.181)$$

We can see from fig. 9.20 that some special trajectories can be found, - they are shown in bold. Trajectory 7 originates with the lowest upcrossing velocity that leads to capsizing defined here as $\dot{\phi}^u_{cr}(\phi_{cr})$ and can be easily found from (9.175) using the condition $\phi(t_{LNR}, \phi_{cr}, \dot{\phi}^u_{cr}) = \phi_{\text{LNR}\,i+1}$:

$$\dot{\phi}^u_{cr} = \{(\phi_{\text{LNR}i+1} - q_1 t_{LNR} + q_2)(\lambda_1 - \lambda_2) - [-q_1 - \lambda_2(\phi_{cr} - q_2)] \cdot e^{\lambda_1 t_{LNR}} + + [-q_1 - \lambda_1(\phi_{cr} - q_2)] \cdot e^{\lambda_2 t_{LNR}}\} / e^{\lambda_1 t_{LNR}} - e^{\lambda_2 t_{LNR}} \quad (9.182)$$

With: $t_{LNR}(\phi_{cr}) = \Delta t - t_{cr}(\phi_{cr}) = \Delta t - (\phi_{cr} - m)/l$. Then probability (9.178) can be easily found as:

$$P(Y_i) = \int_{\phi_{\max i}}^{\phi_{\max i+1}} \int_{\dot{\phi}^u_{cr}(\phi_{cr})}^{\infty} f(\phi_{cr}, \dot{\phi}_{cr}) d\dot{\phi}_{cr} d\phi_{cr} \tag{9.183}$$

The probability of escape through vertical border $P(E_i)$ can be found analogously, using trajectory 8 at fig. 9.20 to find the lowest possible initial upcrossing velocity to escape. To complete our study of conditional probability of capsizing after upcrossing at i^{th} range, we need to consider behavior of the system at range $i+1$ after it escapes through the vertical border.

This problem is analogous to the previous one; the only difference is the distribution of initial conditions at moment Δt $f(\phi^{II}_{cr}, \dot{\phi}^{II}_{cr} \mid \vec{B})$.

Consider random vectors $\vec{C} = (\phi_{cr}, \dot{\phi}_{cr})$ and $\vec{G} = (\phi^{II}_{cr}, \dot{\phi}^{II}_{cr})$. Then consider a vector valued function $F(\vec{C}) = \vec{G}$, and we can formulate a problem of multivariate probability transformation since the distribution of $\vec{C}$ is known from formula (9.181):

$$f(\vec{G}) = \det\left(J(F^{-1}(\vec{G}))\right) \cdot f(F^{-1}(\vec{G})) \tag{9.184}$$

Where $J(..)$ is the Jacobean determinant and F^{-1} denotes the inverse of the vector valued function F. The inverse problem here does not have an analytical solution, but a simple numerical one exists. We consider solution (9.175) back in time: starting at the moment Δt with initial conditions $(\phi^{II}_{cr}, \dot{\phi}^{II}_{cr})$ and going back in time to t_{cr} and $(\phi_{cr}, \dot{\phi}_{cr})$:

$$\begin{aligned} \phi_{cr} &= \phi^{II}_{cr}(-t_{LNR}(\phi_{cr})) = \\ &= A(\phi^{II}_{cr}, \dot{\phi}^{II}_{cr}) e^{-\lambda_1 t_{LNR}} + B(\phi^{II}_{cr}, \dot{\phi}^{II}_{cr}) e^{-\lambda_2 t_{LNR}} - q_1 t_{LNR} + q_2 \end{aligned} \tag{9.185}$$

Again:

$$t_{LNR}(\phi_{cr}) = \Delta t - (\phi_{cr} - m)/l \tag{9.186}$$

We can find ϕ_{cr} from (9.185) using any numerical method for nonlinear algebraic equations. The initial point could be found from (9.185) by expanding exponential functions into a power series.

As soon as ϕ_{cr} is found, calculation of t_{LNR} and $\dot{\phi}_{cr}$ (just differentiation of (9.185)) is no problem. The Jacobean matrix can be deduced analytically from (9.185) and its derivative:

$$\begin{aligned} J(F^{-1}(\vec{G})) &= \frac{1}{\lambda_1 - \lambda_2} \times \\ &\times \begin{pmatrix} -(\lambda_2 e^{-\lambda_1 t_{LNR}} - \lambda_1 e^{-\lambda_2 t_{LNR}}) & e^{-\lambda_1 t_{LNR}} - e^{-\lambda_2 t_{LNR}} \\ -\lambda_1 \lambda_2 (e^{-\lambda_1 t_{LNR}} - e^{-\lambda_2 t_{LNR}}) & \lambda_1 e^{-\lambda_1 t_{LNR}} - \lambda_2 e^{-\lambda_2 t_{LNR}} \end{pmatrix} \end{aligned} \tag{9.187}$$

The Jacobean matrix does not depend on initial conditions: this is a corollary of linearity of solution (9.175) or (9.185) vs. initial conditions. Further derivations in order to find

$P(Y_{y+1} \mid E_i)$ are analogous to $P(Y_i)$ – just using trajectories 9 and 10 of fig.9.20 to find the limits of integration.

The last problem to consider is the probability of upcrossing $P(\phi > \phi_{\max}, t)$. So far, we used a Poisson distribution for beam seas (subchapters 9.2.4 and 8.2.5). However, this could be questionable for quartering seas, because decreasing of stability upcrossing could happen more often and after-action influence could be significant. Some options are considered in [Belenky, 2000b].

9.3.3 Probability of Surf-Riding

As we have seen from Chapter 6, broaching is another possible cause of capsizing in following and quartering seas. Some (not all) broaching scenarios presume that a ship first experiences surf-riding. Here, we consider a method of estimation for the probability of surf-riding developed by Umeda [1990].

There are two modes of surf-riding in regular waves (subchapter 6.3): surf-riding with any initial conditions and co-existence of periodic surging and surf-riding. The latter case also can be called as surf-riding with specific initial conditions.

It means that a random event of surf-riding can happen in two ways: an encounter with such a wave that leads to surf-riding with any initial conditions. Another way is to encounter a wave that leads to surf-riding only with specific initial conditions while having them:

$$P(Y) = P_1(SR) + P_2(SR \mid \xi_G, \dot{\xi}_G) \cdot P(\xi_G, \dot{\xi}_G) \qquad (9.188)$$

Umeda [1990] used an envelope presentation of irregular waves assuming the spectrum to be narrow banded. We considered an envelope presentation in subchapter 8.1.6 and its application for waves in subchapter 8.2.4:

$$\zeta_W(t) = A(t)\cos(\omega_a t + \varphi(t)) \qquad (9.189)$$

Presentation (9.189) enables us to assume that surf-riding in irregular seas would happen when a ship heading with a speed corresponding to Fn_{cr2} (see subchapter 6.3.5) would encounter the wave with the length L_{WS} and height H_{WS} that satisfy surf-riding condition in regular seas.

Umeda [1990] used distributions of wave periods and heights [Longuet-Higgins, 1983] and approximated local wave period as:

$$T(t) = \frac{2\pi}{(\omega_a + \dot{\varphi}(t))} \qquad (9.190)$$

With corresponding length through the well-known formulae of small wave theory:

$$L_W(t) = \frac{2\pi g}{(\omega_a + \dot{\varphi}(t))} \qquad (9.191)$$

Local wave height, then can be presented as a doubled amplitude:

$$H_W(t) = 2A(t) \qquad (9.192)$$

Now, let us express the elements of the envelope presentation through characteristics of the wave:

$$\dot{\varphi} = \frac{2\pi g}{L_W} - \omega_a \tag{9.193}$$

$$A = 0.5H_W\,; \quad \dot{A} = 0.5\dot{H}_W \tag{9.194}$$

These formulae allow expressing distribution of elements of the envelope presentation in terms of characteristics of the wave (formula (8.38) from subchapter 8.1.6):

$$f_W(H_W, \dot{H}_W, L_W) = f\left(0.5H_W, 0.5\dot{H}_W, \frac{2\pi g}{L_W} - \omega_a\right) \tag{9.195}$$

The probability of surf-riding with any initial condition, then can be expressed as:

$$P_1(SR) = \int_{L_{\min}}^{L_{\max}} \int_{-\infty}^{\infty} \int_{H_{Wc}(L_W)}^{\infty} f_W(H_W, \dot{H}_W, L_W)\, dH_W\, d\dot{H}_W\, dL_W \tag{9.196}$$

Here, $L_{\max}$ and $L_{\min}$ are the largest and smallest wave length that can cause surf-riding. $H_{Wc}(L_W)$ is the critical height for the wave of given length that causes surf-riding with any initial conditions.

A description of the method of calculation of probability of surf-riding under specific initial conditions can be found in [Umeda, 1990].

9.3.4 Risk of Capsizing Caused by Breaking Waves

We have considered the action of a breaking wave on a ship in subchapter 7.3. Due to a lack of theoretical knowledge of the hydromechanics of breaking waves, a model test is still the main tool to judge upon stability in breaking waves.

Based on extensive model testing followed the capsizing of M/V "Helland Hansen" [Dahle and Kjaerland, 1980], Dahle, Myrhaug and Dahl [1988] developed a method to estimate risk of capsizing caused by breaking waves. Risk is measured with the probability of capsizing per year:

$$P(X) = P(W)P(X \mid W) \tag{9.197}$$

Where $P(W)$ is the probability to encounter a dangerous wave and $P(X|W)$ is the conditional probability of capsizing if such a wave is encountered.

The first figure is expressed as:

$$P(W) = \sum_j \sum_k P_{1jk} P_{2jk} W_{jk} P_{3kj} P_{4jk} \tag{9.198}$$

Here:

Index j is for significant wave heights. Dahle and Myrhaug [1993] used 12 groups.

Index k is for zero-crossing period, 11 used in the above reference.

P_{1jk} is a probability of dangerous wave direction. Value $P_{1jk} = 0.5$ was used for the sample calculation in [Dahle and Myrhaug, 1993], [Dahle, *et al*, 1995].

P_{2jk} is yearly fraction of time to exposure sea state defined by the indexes j and k.

W_{jk} is a value defined with the following formula:

$$W_{jk} = \begin{cases} t_{jk} / \tau_{jk} & \text{if} \quad t_{jk} \le \tau_{jk} \\ 1 & \text{if} \quad t_{jk} > \tau_{jk} \end{cases} \tag{9.199}$$

With t_{jk} is time of exposure in sea state defined by the indexes j and k; τ_{jk} is duration of the sea state. It is defined with the following formula:

$$\tau_{jk} = H_{Sj}^{-1.87}, \quad \text{hrs} \tag{9.200}$$

P_{3jk} is the conditional probability of steep (crest front steepness more than 0.25) and high waves for the sea state defined by the indexes j and k. This value can be found from wave statistics and it is region-specific. Sample data for the Norwegian sea [Dahle and Myrhaug, 1993], Baltic sea and North East Pacific [Dahle, *et al*, 1995] are given in tables 9.4 and 9.5 correspondingly.

P_{4jk} is the joint probability of the sea state defined by the indexes j and k. Such data is available from wave statistics [Hogben, *et al*, 1986].

Table 9.4 Probability of occurrence per season of steep ($\varepsilon > \varepsilon_c = 0.25$) and high ($H > H_c$) waves for Norwegian seas [Dahle and Myrhaug, 1993]

Season H_c, m	2	3	4
March to May	0.00419	0.00256	0.00156
June to August	0.00202	0.000819	0.000336
September to November	0.00286	0.00166	0.000943
December to February	0.00284	0.00171	0.00102
Annual	0.00298	0.00169	0.000965

Table 9.5 [Dahle, *et al*, 1995]

Area	Season				
5 Baltic sea	Dec. to March	April to May	June to August	Sept. to Nov.	Annual
	0.00048	0.00046	0.00027	0.00049	0.00043
19 North Pacific Ocean (Between Sakhalin and Hokkaido)	Dec. to Feb.	March to May	June to August	Sept. to Nov.	Annual
	0.0011	0.0012	0.00044	0.00090	0.00091
20 North Pacific Ocean (South of Kamchatka peninsula)	Dec. to Feb.	March to May	June to August	Sept. to Nov.	Annual
	0.0013	0.0010	0.00042	0.0011	0.00096

The second term in the equation (9.197) is defined as:

$$P(X|W) = P_5 P_6 P_7 \quad (9.201)$$

Here:

P_5 is the probability of encounter by a dangerous wave during the part of the roll period when a ship is the most susceptible. A value $P_5 = 0.5$ was used for the sample calculation in [Dahle and Myrhaug, 1993].

P_6 is the conditional probability of capsizing, when the vessel encountered the breaking wave. Model test data has to be used to determine this value. Dahle and Myrhaug [1993, 1997], Dahle, *et al* [1995] used results of the model test [Dahle and Kjaerland, 1980], see fig. 9.21. If the area under the *GZ* curve makes a point in a safe area on fig. 9.21, then it is assumed that ship is "safe" and $P_6 = 0$. Otherwise, the ship is unsafe and $P_6 = 1$.

P_6 is the conditional probability of good seamanship. Compliance to the requirements should decrease the risk. This is the place to put a numerical estimate of influence of human factors.

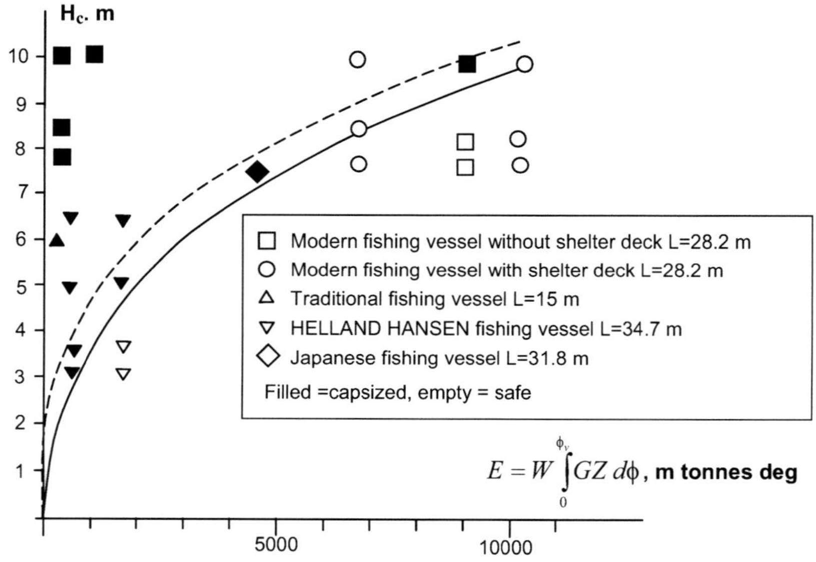

Fig. 9.21 Safe and unsafe regions from model tests [Dahle and Kjaerland, 1980]

There are a number of sample calculations available from [Dahle and Myrhaug, 1993] – for the Norwegian Sea, [Dahle, *et al*, 1995] for the Baltic Sea and Northeast Pacific and [Dahle and Myrhaug, 1997] for the Black Sea. The cited references also contain some suggestions for risk management. In general, these works show good examples of complete and simple methods for capsizing risk analysis for ships in real seas.

Appendix I

Nechaev Method[1]

The method [Nechaev, 1978, 1989] is based on a series of model tests conducted at the National Laboratory of Seakeeping of Fishing Vessels at Kaliningrad Institute of Technology (Russia). Results of these experiments were presented in a form of regression polynomials. Using these polynomials, it is possible to estimate GZ curves on the wave crest and at the wave trough, if the length of the wave is not very far from the length of the ship. Usage of the method is limited by the following ship parameter values:

$$3.2 \le L/B \le 8.00 \qquad 0.5 \le CB/CW \le 0.92$$
$$2.0 \le B/d \le 4.2 \qquad 0.5 \le CB/CM \le 0.85$$
$$1.05 \le D/d \le 2.2 \qquad 0.15 \le Fn \le 0.45$$

Here: L is length (between perpendiculars), B is breadth molded, d is draft, D is depth, CB is block coefficient, CW is water plane coefficient, CM is midship coefficient and Fn is Froude number. The GZ curve when the ship is located on the wave crest is expressed as:

$$GZ_C(\phi) = GZ(\phi) + B\left(F_C(\phi) + \sum_{i=1}^{17} A_i f_{Ci}(\phi) \right) \qquad \text{(A.1)}$$

Here:

$F_C(\phi)$ are the basic values for wave crest, see tables A.1-A.4;

$f_{Ci}(\phi)$ are the influence functions, see tables A.9-A.12;

A_i is a set of numbers expressing influence of ship parameters that differ from the basic model:

$A_1 = L/B - 4.82$	$A_7 = A_1^2$	$A_{13} = A_2\ A_4$
$A_2 = B/T - 2.67$	$A_8 = A_2^2$	$A_{14} = A_1\ A_6$
$A_3 = H/T - 1.3$	$A_9 = A_3^2$	$A_{15} = A_1^3$
$A_4 = C_B/C_W - 0.7$	$A_{10} = A_5^2$	$A_{16} = A_3^3$
$A_5 = C_B/C_M - 0.692$	$A_{11} = A_6^2$	$A_{17} = A_5^3$
$A_6 = Fn - 0.28$	$A_{12} = A_1\ A_2$	

[1] The author is grateful to Prof. Nechaev for fruitful discussions of the materials in this Appendix and for supplying a corrected version of the regression coefficients. Additional corrections were made by Prof. Nechaev during preparation of the second edition, his help is greatly appreciated.

The GZ curve, when the ship is on the wave trough, can be expressed as:

$$GZ_T(\phi) = GZ(\phi) + B\left(F_T(\phi) + \sum_{i=1}^{17} A_i f_{Ti}(\phi) \right) \quad \text{(A.2)}$$

Here:

$F_T(\phi)$ are the basic values for wave trough, see tables A.5-A.8;
$f_{Ti}(\phi)$ are the influence functions, see tables A.13-A.16;
Basic values for wave crest and trough are given in Tables A.1-A.8 for different values of wave steepness. The latter is defined as:

$$st = h_w / \lambda_w \quad \text{(A.3)}$$

Here h_w is wave height and λ_w is wave length. If the wave length is different from $L\cos\beta$, were β is wave direction, the following correction has to be used:

$$Corr = 1 + k_1 \cdot \bar{\lambda} - k_2 \cdot \bar{\lambda}^2 - k_3 \cdot \bar{\lambda}^3 \quad \text{(A.4)}$$

Here, λ_w is the wave length, $\bar{\lambda} = \lambda_w / L - 1$ is the non-dimensional relative wave length, k_i are coefficients from table A.17.

Example calculations are shown in tables A.18 and A.19.

Table A.1 Basic values for wave crest, $F_C \cdot 10^2$, wave direction $\beta = 0^0$

Wave Steepness	Angle of heel					
	10°	20°	30°	40°	50°	60°
0.03	-0.38	-0.58	-0.80	-0.92	-0.86	-0.75
0.04	-0.42	-0.74	-1.11	-1.30	-1.20	-1.05
0.05	-0.48	-0.92	-1.50	-1.70	-1.60	-1.42
0.06	-0.60	-1.12	-1.78	-1.98	-1.87	-1.70
0.07	-0.70	-1.31	-1.97	-2.15	-2.06	-1.90
0.08	-0.77	-1.45	-2.13	-2.30	-2.33	-2.06
0.09	-0.85	-1.56	-2.28	-2.42	-2.35	-2.20
0.10	-0.90	-1.65	-2.41	-2.55	-2.48	-2.35
0.11	-0.95	-1.75	-2.50	-2.65	-2.59	-2.43

Table A.2 Basic values for wave crest, $F_C \cdot 10^2$, wave direction $\beta = 15^0$

Wave Steepness	Angle of heel					
	10°	20°	30°	40°	50°	60°
0.03	-0.40	-0.53	-0.67	-0.74	-0.67	-0.57
0.04	-0.53	-0.68	-1.00	-1.08	-1.00	-0.80
0.05	-0.66	-0.86	-1.27	-1.37	-1.27	-1.04
0.06	-0.75	-1.08	-1.46	-1.60	-1.46	-1.22
0.07	-0.87	-1.24	-1.61	-1.68	-1.61	-1.40
0.08	-0.97	-1.36	-1.75	-1.90	-1.75	-1.53
0.09	-1.05	-1.47	-1.88	-2.02	-1.88	-1.65
0.10	-1.10	-1.56	-1.97	-2.12	-1.97	-1.78
0.11	-1.03	-1.67	-2.06	-2.21	-2.06	-1.85

Table A.3 Basic values for wave crest, $F_C \cdot 10^2$, wave direction $\beta = 30^0$

Wave Steepness	Angle of heel					
	10^0	20^0	30^0	40^0	50^0	60^0
0.03	-0.48	-0.42	-0.50	-0.60	-0.46	-0.48
0.04	-0.60	-0.54	-0.79	-0.80	-0.68	-0.60
0.05	-0.70	-0.67	-1.00	-0.93	-0.80	-0.70
0.06	-0.80	-0.80	-1.17	-1.09	-0.93	-0.80
0.07	-0.88	-0.90	-1.30	-1.22	-1.03	-0.88
0.08	-0.93	-1.03	-1.42	-1.37	-1.12	-0.93
0.09	-1.02	-1.14	-1.50	-1.43	-1.20	-1.02
0.10	-1.10	-1.28	-1.60	-1.48	-1.24	-1.10
0.11	-1.18	-1.41	-1.67	-1.50	-1.29	-1.18

Table A.4 Basic values for wave crest, $F_C \cdot 10^2$, wave direction $\beta = 45^0$

Wave Steepness	Angle of heel					
	10°	20°	30°	40°	50°	60°
0.03	-0.34	-0.41	-0.50	-0.45	-0.37	-0.34
0.04	-0.42	-0.49	-0.71	-0.63	-0.51	-0.42
0.05	-0.53	-0.70	-0.82	-0.78	-0.64	-0.53
0.06	-0.62	-0.83	-0.98	-0.91	-0.75	-0.62
0.07	-0.70	-0.91	-1.10	-1.00	-0.82	-0.70
0.08	-0.80	-1.00	-1.20	-1.08	-0.90	-0.80
0.09	-0.84	-1.08	-1.30	-1.17	-0.96	-0.84
0.10	-0.90	-1.17	-1.37	-1.22	-1.00	-0.90
0.11	-0.98	-1.20	-1.42	-1.26	-1.03	-0.98

Table A.5 Basic values for wave trough, $F_T \cdot 10^2$, wave direction $\beta = 0^0$

Wave Steepness	Angle of heel					
	10°	20°	30°	40°	50°	60°
0.03	0.12	0.32	0.43	0.30	0.04	-0.18
0.04	0.23	0.45	0.62	0.42	0.08	-0.24
0.05	0.31	0.54	0.74	0.51	0.11	-0.31
0.06	0.38	0.61	0.83	0.58	0.12	-0.40
0.07	0.43	0.64	0.91	0.61	0.11	-0.48
0.08	0.46	0.71	0.96	0.63	0.08	-0.57
0.09	0.48	0.76	1.02	0.64	0.04	-0.66
0.10	0.49	0.78	1.07	0.65	0.00	-0.79
0.11	0.50	0.79	1.10	0.68	-0.05	-0.88

Table A.6 Basic values for wave trough, $F_T \cdot 10^2$, wave direction $\beta = 15^0$

Wave Steepness	Angle of heel					
	10°	20°	30°	40°	50°	60°
0.03	0.18	0.30	0.38	0.25	0.03	-0.15
0.04	0.27	0.40	0.52	0.28	0.06	-0.17
0.05	0.36	0.47	0.63	0.32	0.09	-0.19
0.06	0.41	0.52	0.72	0.38	0.09	-0.22
0.07	0.46	0.60	0.81	0.41	0.09	-0.28
0.08	0.50	0.63	0.84	0.43	0.06	-0.32
0.09	0.54	0.68	0.89	0.45	0.03	-0.41
0.10	0.57	0.72	0.93	0.48	0.00	-0.50
0.11	0.59	0.76	0.97	0.50	-0.04	-0.60

Table A.7 Basic values for wave trough, $F_T \cdot 10^2$, wave direction $\beta = 30^0$

Wave Steepness	Angle of heel					
	10°	20°	30°	40°	50°	60°
0.03	0.22	0.22	0.29	0.17	0.02	-0.01
0.04	0.31	0.31	0.38	0.21	0.04	-0.12
0.05	0.37	0.37	0.46	0.23	0.06	-0.14
0.06	0.42	0.42	0.51	0.27	0.06	-0.16
0.07	0.46	0.46	0.58	0.30	0.06	-0.18
0.08	0.48	0.48	0.61	0.32	0.04	-0.20
0.09	0.52	0.52	0.63	0.33	0.02	-0.22
0.10	0.53	0.53	0.66	0.34	0.00	-0.28
0.11	0.55	0.55	0.67	0.35	-0.03	-0.34

Table A.8 Basic values for wave trough, $F_T \cdot 10^2$, wave direction $\beta = 45^0$

Wave Steepness	Angle of heel					
	10°	20°	30°	40°	50°	60°
0.03	0.17	0.20	0.20	0.14	0.14	-0.08
0.04	0.20	0.28	0.28	0.18	0.18	-0.09
0.05	0.23	0.33	0.33	0.20	0.20	-0.10
0.06	0.30	0.38	0.38	0.22	0.22	-0.11
0.07	0.34	0.42	0.42	0.23	0.23	-0.12
0.08	0.38	0.46	0.46	0.24	0.24	-0.14
0.09	0.40	0.50	0.50	0.25	0.25	-0.18
0.10	0.42	0.52	0.52	0.26	0.26	-0.22
0.11	0.43	0.54	0.54	0.27	0.27	-0.27

Table A.9 Influence functions for wave crest $f_{Ci} \cdot 10^2$, wave direction β= 0⁰

Index of function		Angle of heel					
		10°	20°	30°	40°	50°	60°
1		-0.2	-0.36	-0.45	-0.38	-0.3	-0.25
2		-0.35	-0.85	-1.03	-0.73	-0.36	-0.3
3		1.25	2.0	2.2	1.56	1.1	0.9
4		0.4	0.7	1.0	1.48	1.8	2.0
5		1.92	2.68	3.58	4.5	4.96	4.4
6	*Fn*<0.28	2.6	4.7	5.0	4.2	3.6	3.4
	Fn>0.28	-2.74	-3.6	-2.86	-2.2	-1.86	-1.74
7		0.05	0.09	0.106	0.070	0.0016	0.0
8		-0.08	-0.13	-0.096	-0.032	-0.006	0.0
9		-0.8	-1.5	-1.98	-1.710	-1.26	-1.120
10		-0.4	-3.8	-12.0	-26.0	-40.0	-38.8
11		1.83	2.37	2.02	1.67	1.48	1.4
12		-0.5	-0.9	-1.08	-0.8	-0.52	-0.4
13		-2.44	-2.7	-1.8	-0.8	-0.26	0.0
14		-0.44	-0.8	-0.88	-0.64	-0.5	-0.48
15		-0.0136	-0.0221	-0.0188	-0.0119	-0.0055	-0.0035
16		0.092	0.107	0.58	0.67	0.46	0.32
17		30	100	260	350	380	345

Table A.10 Influence functions for wave crest $f_{Ci} \cdot 10^2$, wave direction β = 15⁰

Index of function		Angle of heel					
		10°	20°	30°	40°	50°	60°
1		-0.232	-0.331	-0.392	-0.304	-0.210	-0.165
2		-0.406	-0.782	-0.896	-0.584	-0.252	-0.198
3		1.450	1.840	1.914	1.248	0.770	0.594
4		0.464	0.644	0.870	1.184	1.216	1.320
5		2.227	2.466	3.115	3.600	3.472	2.904
6	*Fn*<0.28	3.016	4.324	4.350	3.360	2.520	2.244
	Fn>0.28	-3.178	-3.312	-2.488	-1.760	-1.302	-1.148
7		0.058	0.083	0.092	0.056	0.011	0
8		-0.093	-0.120	-0.083	-0.026	0.004	0
9		-0.812	-1.380	-1.723	-1.368	-0.882	-0.739
10		-0.464	-3.496	-10.440	-20.80	-26.04	-26.928
11		2.123	2.180	1.757	1.336	1.036	0.924
12		-0.580	-0.830	-0.939	-0.640	-0.364	-0.264
13		-2.830	-2.484	-1.566	-0.640	-0.182	0
14		-0.510	-0.736	-0.766	-0.512	-0.350	-0.317
15		-0.0158	-0.0203	-0.0164	-0.0095	-0.0039	-0.0023
16		0.107	0.098	0.5	0.536	0.322	0.211
17		34.8	92	226.2	280	266	227.7

Table A.11 Influence functions for wave crest $f_{Ci} \cdot 10^2$, wave direction $\beta = 30^0$

Index of function		Angle of heel					
		10°	20°	30°	40°	50°	60°
1		-0.212	-0.263	-0.270	-0.194	-0.126	-0.095
2		-0.371	-0.621	-0.618	-0.372	-0.184	-0.114
3		1.325	1.460	1.320	0.796	0.462	0.342
4		0.424	0.511	0.620	0.755	0.790	0.760
5		2.035	1.956	2.148	2.300	2.083	1.678
6	*Fn*<0.28	2.756	3.431	3.000	2.142	1.840	1.292
	Fn>0.28	-2.904	-2.628	-1.716	-1.122	-0.781	-0.661
7		0.053	0.066	0.064	0.036	0.007	0
8		-0.085	-0.095	-0.058	-0.016	-0.003	0
9		-0.742	-1.095	-1.188	-0.872	-0.660	-0.426
10		-0.424	-2.774	-7.200	-13.26	-15.620	-15.500
11		1.940	1.730	1.212	0.852	0.622	0.532
12		-0.530	-0.657	-0.648	-0.408	-0.218	-0.152
13		-2.586	-1.971	-1.080	-0.408	-0.109	0
14		-0.466	-0.584	-0.528	-0.326	-0.210	-0.184
15		-0.0144	-0.0161	-0.0113	-0.0061	-0.0023	-0.0013
16		0.0975	0.0781	0.4	0.341	0.193	0.122
17		31.8	73	156	178.5	159.6	131.1

Table A.12 Influence functions for wave crest $f_{Ci} \cdot 10^2$, wave direction $\beta = 45^0$

Index of function		Angle of heel					
		10°	20°	30°	40°	50°	60°
1		-0.180	-0.227	-0.216	-0.152	-0.102	-0.075
2		-0.315	-0.536	-0.494	-0.292	-0.122	-0.090
3		1.125	1.260	1.056	0.624	0.374	0.270
4		0.360	0.441	0.480	0.592	0.639	0.600
5		1.728	1.688	1.718	1.800	1.686	1.320
6	*Fn*<0.28	2.340	2.961	2.400	1.680	1.224	1.020
	Fn>0.28	-2.466	-2.270	-1.373	-0.880	-0.632	-0.522
7		0.045	0.057	0.051	0.028	0.005	0
8		-0.072	-0.082	-0.046	-0.013	-0.002	0
9		-0.630	-0.945	-0.950	-0.684	-0.428	-0.336
10		-0.360	-2.394	-5.760	-10.40	-12.650	-12.240
11		1.647	1.493	0.970	0.668	0.503	0.420
12		-0.450	-0.567	-0.518	-0.320	-0.177	-0.120
13		-2.196	-1.701	-0.864	-0.320	-0.088	0
14		-0.396	-0.504	-0.422	-0.256	-0.170	-0.144
15		-0.0122	-0.0139	-0.0090	-0.0048	-0.0019	-0.0011
16		0.0828	0.0674	0.383	0.268	0.156	0.096
17		27	63	124.8	140	129.2	103.5

Table A.13 Influence functions for wave trough $f_{Ti} \cdot 10^2$, wave direction $\beta = 0^0$

Index of function		Angle of heel					
		10°	20°	30°	40°	50°	60°
1		0.016	0.04	0.078	0.138	0.164	0.15
2		0.11	0.137	0.09	-0.03	-0.07	-0.12
3		0.7	1.3	1.82	2.18	2.56	2.7
4		-0.7	-1.1	-1.26	-1.03	-0.8	-0.7
5		-0.3	-0.6	-0.78	-0.7	-0.56	-0.5
6		0.57	1.12	1.58	1.57	1.18	1.0
7		0	-0.005	-0.01	-0.025	-0.037	-0.035
8		0.05	0.032	0.014	0.006	0.003	0
9		0.18	0.78	1.3	1.64	1.8	1.78
10		0	0	0	0	0	0
11		-1.16	-1.63	-1.84	-1.83	-1.6	-1.46
12		-0.1	-0.7	-1.1	-1.35	-2.2	-3.65
13		0	0	0	0	0	0
14	*Fn*<0.28	-1.47	-2.29	-2.54	-2.53	-2.3	-2.15
	Fn>0.28	0.78	1.55	1.93	1.75	1.28	1.02
15		0	0	0	0	0	0
16		-0.65	-1.75	-2.37	-2.26	-1.5	-0.87
17		0	0	0	0	0	0

Table A.14 Influence functions for wave trough $f_{Ti} \cdot 10^2$, wave direction $\beta = 15^0$

Index of function		Angle of heel					
		10°	20°	30°	40°	50°	60°
1		0.020	0.038	0.069	0.115	0.130	0.113
2		0.140	0.130	0.080	-0.025	-0.055	-0.090
3		0.889	1.235	1.620	1.809	2.022	2.025
4		-0.889	-1.045	-1.121	-0.855	-0.632	-0.525
5		-0.381	-0.570	-0.694	-0.581	-0.442	-0.375
6		0.724	1.064	1.406	1.303	0.932	0.750
7		0	-0.004	-0.008	-0.015	-0.020	-0.018
8		0.064	0.030	0.012	0.005	0.002	0
9		0.229	0.741	1.157	1.361	1.422	1.335
10		0	0	0	0	0	0
11		-1.473	-1.549	-1.638	-1.519	-1.264	-1.095
12		-0.127	-0.665	-0.979	-1.121	-1.738	-2.738
13		0	0	0	0	0	0
14	*Fn*<0.28	-1.868	-2.176	-2.261	-2.100	-1.817	-1.613
	Fn>0.28	0.991	1.473	1.718	1.453	1.011	0.765
15		0	0	0	0	0	0
16		-1.20	-1.66	-2.11	-1.88	-1.19	-0.65
17		0	0	0	0	0	0

Table A.15 Influence functions for wave trough $f_{Ti} \cdot 10^2$, wave direction $\beta = 30^0$

Index of function		Angle of heel					
		10°	20°	30°	40°	50°	60°
1		0.020	0.032	0.052	0.080	0.084	0.072
2		0.135	0.110	0.060	-0.017	-0.036	-0.058
3		0.861	1.040	1.219	1.264	1.306	1.296
4		-0.861	-0.880	-0.844	-0.597	-0.408	-0.336
5		-0.369	-0.480	-0.523	-0.406	-0.286	-0.240
6		0.701	0.896	1.059	0.911	0.602	0.480
7		0	-0.003	-0.006	-0.011	-0.016	-0.014
8		0.062	0.024	0.009	0.003	0.002	0
9		0.221	0.624	0.871	0.951	0.918	0.854
10		0	0	0	0	0	0
11		-1.427	-1.304	-1.233	-1.061	-0.816	-0.701
12		-0.123	-0.560	-0.800	-1.083	-1.122	-1.752
13		0	0	0	0	0	0
14	*Fn*<0.28	-1.808	-1.832	-1.702	-1.467	-1.173	-1.032
	Fn>0.28	0.959	1.240	1.293	1.015	0.653	0.490
15		0	0	0	0	0	0
16		-0.80	-1.40	-1.58	-1.31	-0.77	-0.42
17		0	0	0	0	0	0

Table A.16 Influence functions for wave trough $f_{Ti} \cdot 10^2$, wave direction $\beta = 45^0$

Index of function		Angle of heel					
		10°	20°	30°	40°	50°	60°
1		0.018	0.029	0.044	0.065	0.067	0.060
2		0.131	0.099	0.051	-0.014	-0.029	-0.048
3		0.770	0.936	1.037	1.025	1.203	1.080
4		-0.770	-0.792	-0.718	-0.484	-0.328	-0.280
5		-0.330	-0.432	-0.445	-0.329	-0.263	-0.200
6		0.627	0.806	0.901	0.738	0.555	0.400
7		0	-0.002	-0.004	-0.007	-0.012	-0.009
8		0.055	0.023	0.008	0.003	0.001	0
9		0.198	0.562	0.741	0.771	0.846	0.712
10		0	0	0	0	0	0
11		-1.276	-1.174	-1.049	-0.860	-0.752	-0.584
12		-0.110	-0.504	-0.628	-0.635	-1.034	-1.460
13		0	0	0	0	0	0
14	*Fn*<0.28	-1.617	-1.649	-1.448	-1.189	-1.081	-0.860
	Fn>0.28	0.858	1.116	1.100	0.823	0.602	0.408
15		0	0	0	0	0	0
16		-0.72	-1.26	-1.35	-1.06	-0.71	-0.35
17		0	0	0	0	0	0

Table A.17 Correction coefficients for wave length not equal to $L\cos\beta$

	k_1	k_2	k_3
wave trough	0.87	1.20	0.21
wave crest	0.90	1.50	0.39

Table A.18 Data for example calculations

Length, B.P., m	40
Breadth molded, m	9
Draft, m	3
Depth,	4.2
Block coefficient *CB*	0.6
Water plane coefficient, *CW*	0.8
Midship coefficient, *CM*	0.85
Froude Number	0.2
Wave length λ_w, m	70
Wave height h_w, m	5
Wave direction β, deg	5

Table A.19 Example calculations

Heel angle, deg.	10	20	30	40	50	60
GZ curve in calm water, m	0.169	0.307	0.38	0.358	0.208	-0.101
$F_C 10^2$ $\beta=0^0$, steepness 0.07	-0.7	-1.31	-1.97	-2.15	-2.06	-1.9
$F_C 10^2$ $\beta=15^0$, steepness 0.07	-0.87	-1.24	-1.61	-1.68	-1.61	-1.4
$F_C 10^2$ $\beta=5^0$, steepness 0.07	-0.757	-1.29	-1.85	-1.99	-1.91	-1.73
$F_C 10^2$ $\beta=0^0$, steepness 0.08	-0.77	-1.45	-2.13	-2.3	-2.33	-2.06
$F_C 10^2$ $\beta=15^0$, steepness 0.08	-0.97	-1.36	-1.75	-1.9	-1.75	-1.53
$F_C 10^2$ $\beta=5^0$, steepness 0.08	-0.837	-1.42	-2	-2.17	-2.14	-1.88
$F_C 10^2$ $\beta=5^0$, steepness 0.0714	-0.768	-1.31	-1.87	-2.02	-1.94	-1.75
$\Sigma A_i f_{Ci} 10^2$, $\beta=5^0$	0.381	0.445	0.433	0.405	0.403	0.365
$(F_C + \Sigma A_i f_{Ci})10^2$, $\beta=5^0$	-0.387	-0.86	-1.44	-1.61	-1.54	-1.39
$B\,(F_C + \Sigma A_i f_{Ci})$	-0.0348	-0.0774	-0.129	-0.145	-0.139	-0.125
$\Delta GZ_C = B\,(F_C + \Sigma A_i f_{Ci})\,Corr$	-0.0309	-0.0688	-0.115	-0.129	-0.123	-0.111
$GZ_C = GZ + \Delta GZ_C$	0.138	0.238	0.265	0.229	0.0849	-0.212
$F_T 10^2$ $\beta=0^0$, steepness 0.07	0.43	0.64	0.91	0.61	0.11	-0.48
$F_T 10^2$ $\beta=15^0$, steepness 0.07	0.46	0.6	0.81	0.41	0.09	-0.28
$F_T 10^2$ $\beta=5^0$, steepness 0.07	0.44	0.627	0.877	0.543	0.103	-0.413
$F_T 10^2$ $\beta=0^0$, steepness 0.08	0.46	0.71	0.96	0.63	0.08	-0.57
$F_T 10^2$ $\beta=15^0$, steepness 0.08	0.5	0.63	0.84	0.43	0.06	-0.32
$F_T 10^2$ $\beta=5^0$, steepness 0.08	0.473	0.683	0.92	0.563	0.0733	-0.487
$F_T 10^2$ $\beta=5^0$, steepness 0.0714	0.445	0.635	0.883	0.546	0.099	-0.424
$\Sigma A_i f_{Ti} 10^2$, $\beta=5^0$	0.055	0.137	0.169	0.178	0.318	0.493
$(F_T + \Sigma A_i f_{Ti})10^2$, $\beta=5^0$	0.5	0.771	1.05	0.725	0.417	0.0695
$B\,(F_T + \Sigma A_i f_{Ti})$	0.045	0.0694	0.0947	0.0652	0.0375	0.00625
$\Delta GZ_T = B\,(F_T + \Sigma A_i f_{Ti})\,Corr$	0.03	0.0463	0.0631	0.0435	0.025	0.00417
$GZ_T = GZ + \Delta GZ_T$	0.199	0.353	0.443	0.401	0.233	-0.097

Appendix II

Basic Statistics and Ergodicity of Stochastic Process

A2.1 Statistical Estimates of Stochastic Process as Random Numbers

Estimate of variance of roll motions and their confidence intervals were used in subchapter 8.6.1 to evaluate the difference between realizations. This appendix gives a brief review of confidence interval techniques for stochastic processes and using this technique to evaluate ergodicity of the process.

Any averaged value based on a finite amount of statistical material, including results of Monte-Carlo simulation, is a random number. The randomness of these values results in uncertainty of any judgment based on statistics. Confidence interval is a tool to evaluate the level of statistical uncertainty.

If a realization of stochastic process is presented in a form of array of time history data points x_i, the estimates are calculated as follows:

$$m_X^* = \frac{1}{N}\sum_{i=1}^{N}(x_i) \tag{A2.1}$$

$$V_X^* = \frac{1}{N}\sum_{i=1}^{N}\left(x_i - m_X^*\right)^2 \tag{A2.2}$$

Here m_X^* and V_X^* are estimates of mean value and variance respectively; asterisk superscript shows that these figures are only estimates of true or theoretical values. In fact, formulae (A2.1) and (A2.2) are discretized version of the expressions (8.13). If the process is ergodic these estimates can be used for the entire process.

The autocorrelation function can be estimated as:

$$C_j^* = \frac{1}{N-j}\sum_{i=1}^{N-j}(x_i - m_X^*)(x_{i+j} - m_X^*) \tag{A2.3}$$

This formula is a discretized version of (8.15).

Further consideration is focused on the estimates (A2.1) and (A2.2). They are random variables and can be characterized by probability distribution as any other random variables; normal distribution is used when the number of data points is large, which is usually true for stochastic processes. To define a normal distribution, the mean value and variance are needed.

The mean value of the mean value estimate equals itself as the mean value is an unbiased estimate [Bendat and Piersol 1986].

$$M\{m_X^*\} = m_X^* \qquad (A2.4)$$

The mean value of the variance estimate requires correction of bias:

$$M\{V_X^*\} = V_X^* \cdot \frac{N}{N-1} \qquad (A2.5)$$

The variance of the mean value estimate depends on the autocorrelation function.

$$V\{m_X^*\} = \frac{1}{N} C_0^* + \frac{2}{N} \sum_{i=1}^{N-1} C_i^* - \frac{2}{N^2} \sum_{i=1}^{N-1} i C_i^* \qquad (A2.6)$$

A formula for the variance of the variance estimate is only available for a process that has the same relationship between the second and the fourth statistical moment as a process with a Gaussian distribution:

$$V\{V_X^*\} = \frac{4}{N} \sum_i (C_i^*)^2 \left(1 - \frac{i}{N}\right) \qquad (A2.7)$$

A2.2 Confidence Interval of Statistical Estimates

Once the mean value and variance of the estimates are found, their distributions are defined. To evaluate the confidence interval the confidence probability P_β needs to be set up. The confidence probability is the probability that the true or theoretical value is within the confidence interval, see fig. A2.1.

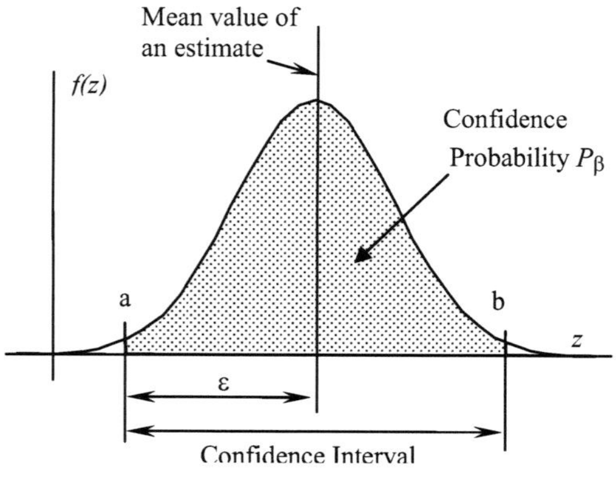

Fig. A2.1 On the definition of confidence interval and confidence interval

Boundaries b1 and b2 of confidence interval can be found from the following equation:

$$P_\beta = \int_a^b f(z)dz \qquad (A2.8)$$

Here *z* is the estimate for which the confidence interval is searched. Since the distribution of this estimate was assumed normal and normal distribution is symmetric relative to the mean value of the estimate, the equation (A2.8) can be re-written:

$$P_\beta = \int_{m_z-\varepsilon}^{m_z+\varepsilon} f(z)dz = 2\int_{m_z}^{m_z+\varepsilon} f(z)dz\,; \quad a = m_z - \varepsilon\,; \quad b = m_z + \varepsilon \tag{A2.9}$$

Using cumulative probability distribution instead of the distribution density:

$$P_\beta = 2\int_{m_z}^{m_z+\varepsilon} f(z)dz = 2\big(F(m_z+\varepsilon) - F(m_z)\big) \tag{A2.10}$$

Here $F(z)$ is cumulative probability distribution of the estimate z. Substitution of normal distribution yields:

$$F(z) = \int_{-\infty}^{z} f(z)dz = \frac{1}{\sqrt{2\pi V_z}}\int_{-\infty}^{z} \exp\left(-\frac{(z-m_z)^2}{2V_z}\right)dz = \Phi\left(\frac{z-m_z}{\sqrt{V_z}}\right) \tag{A2.11}$$

Here Φ is integral error function, defined as:

$$\Phi(z) = \frac{1}{\sqrt{2\pi}}\int_{-\infty}^{z} \exp\left(-0.5\cdot z^2\right)dz \tag{A2.12}$$

Taking into account that for the case of normal distribution $F(m_z) = 0.5$ and using definitions (A2.11) and (A2.12) in the equation (A2.10):

$$\Phi\left(\varepsilon/\sqrt{V_z}\right) = 0.5(P_\beta + 1) \quad \Leftrightarrow \quad \varepsilon = \sqrt{V_z}\cdot inv\Phi(0.5(P_\beta + 1)) \tag{A2.13}$$

Here *inv*Φ is inverted integral error function. Its value for typical confidence probabilities are given below:

$$\begin{aligned} &P_\beta = 0.9973\,; \qquad inv\Phi(0.5(P_\beta+1)) = 3.0 \\ &P_\beta = 0.95\,; \qquad inv\Phi(0.5(P_\beta+1)) = 1.959964 \end{aligned} \tag{A2.14}$$

The upper line in the formulae (A2.14) corresponds to the commonly used "six-sigma rule", while 95% of confidence probability is most frequently used for engineering calculations.

A2.3 Measure of Ergodicity

Relative measure of ergodicity was proposed in Belenky [2004]. It is based on the comparison of confidence intervals for variance estimated for several realizations.

Consider two variance estimates with confidence intervals calculated from two independent realizations of a stochastic process, see fig. A2.2.

Consider variance V_1 estimated over the realization 1 and variance V_2 estimated over the realization 2 with their respective confidence intervals defined with the confidence probability P_β:

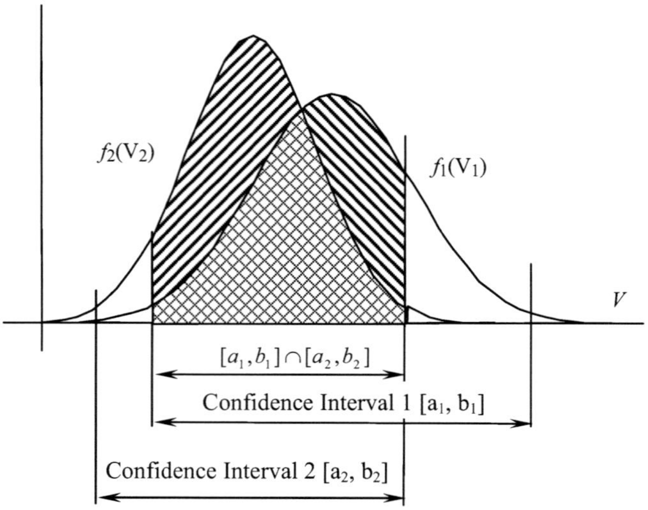

Fig. A2.2 Two confidence intervals

$$P\left(V_1 \in [m_1\{V_X^*\} - \varepsilon_1; m_1\{V_X^*\} + \varepsilon_1]\right) = P_\beta \quad \text{(A2.15)}$$

$$P\left(V_2 \in [m_2\{V_X^*\} - \varepsilon_2; m_2\{V_X^*\} + \varepsilon_2]\right) = P_\beta \quad \text{(A2.16)}$$

The probability that the difference between these two estimates is caused by statistical factors only (with the given confidence probability) is to be calculated over the subset of the confidence intervals of these estimates, which is defined as:

$$[a_1; b_2] = [m_1\{V_X^*\} - \varepsilon_1; m_1\{V_X^*\} + \varepsilon_1] \cap [m_2\{V_X^*\} - \varepsilon_2; m_2\{V_X^*\} + \varepsilon_2] \quad \text{(A2.17)}$$

The probability that true variances belong to the above interval is calculated with a Gaussian distribution assumed above for the estimates:

$$P_1 = P(V_1 \in [a_1, b_2]) = \frac{1}{\sqrt{2\pi \cdot V_1\{V_X^*\}}} \int_{a_1}^{b_2} \exp\left(-\frac{(x - m_1\{V_X^*\})}{V_1\{V_X^*\}}\right) dx \quad \text{(A2.18)}$$

$$P_2 = P(V_2 \in [a_1, b_2]) = \frac{1}{\sqrt{2\pi \cdot V_2\{V_X^*\}}} \int_{a_1}^{b_2} \exp\left(-\frac{(x - m_2\{V_X^*\})}{V_2\{V_X^*\}}\right) dx \quad \text{(A2.19)}$$

The probability that the true value of both belong to the interval $[a_1, b_2]$ is calculated with simple multiplication of the probabilities P_1 and P_2 as the realizations 1 and 2 are assumed to be independent:

$$P_{Erg}(2) = P\left((V_1 \in [b_1; b_2]) \cap (V_2 \in [b_1; b_2])\right) = P_1 P_2 \quad \text{(A2.20)}$$

If the process is ergodic, both estimates have the same limit equal to the theoretical value:

$$\begin{aligned}&\lim_{T\to\infty} m_1\{V_X^*\} = \lim_{T\to\infty} m_2\{V_X^*\} = V_X\\&\lim_{T\to\infty} \varepsilon_1 = \lim_{T\to\infty} \varepsilon_2 = 0\end{aligned} \qquad \text{(A2.21)}$$

Here, V_X is a theoretical value. At this limit, variance of the variance estimate equals to zero as both estimates produce a true value.

In the ideal case, variance estimated for both realizations are very close to each other and have very similar boundaries of the confidence interval:

$$m_1\{V_X^*\} \approx m_2\{V_X^*\}\,; \qquad \varepsilon_1 \approx \varepsilon_2 \qquad \text{(A2.22)}$$

Therefore for the ideal case:

$$P_1 \approx P_2 \approx P_\beta \qquad \text{(A2.23)}$$

In a real case, even two realization of the ergodic process would not produce exactly equal estimates, so the probabilities calculated by formulae (A2.18) and (A2.19) are always less than the accepted confidence probability:

$$P_1 < P_\beta\,; \quad P_2 < P_\beta \qquad \text{(A2.24)}$$

Therefore

$$P_1 P_2 < P_\beta^2 \;\Leftrightarrow P_\beta^2 - P_1 P_2 > 0 \qquad \text{(A2.25)}$$

A positive value K_{U2} is defined as

$$K_{U2} = P_\beta^2 - P_1 P_2 \qquad \text{(A2.26)}$$

This value could be used as a measure of difference between estimates obtained from two different realizations of the same stochastic process.

The value K_{U2}, however, is a random number because it depends on statistical estimates. To decrease influence of random factors, it is better to average this value over n realizations:

$$P_{Erg}(n) = M\left\{\begin{pmatrix} P_1P_1 & P_1P_2 & \dots & P_1P_n \\ P_2P_1 & P_2P_2 & \dots & P_2P_n \\ \dots & \dots & \dots & \dots \\ P_nP_1 & P_nP_2 & \dots & P_nP_n \end{pmatrix}\right\} = \frac{1}{(n-1)\cdot n}\left(\sum_{i=1}^{n}\sum_{j=1}^{n} P_iP_j - \sum_{i=1}^{n} P_i^2\right) \qquad \text{(A2.27)}$$

Finally the measure of ergodicity

$$K_{Un} = P_\beta^2 - P_{Erg}(n) \qquad \text{(A2.28)}$$

It is important to note that the measure defined by the formula (A2.28) has a relative character. Even if the process is ergodic, the measure may not be equal to zero, as the different realizations are unlikely to yield exactly equal estimates.

References

Adee, B. and Caglayan, I. (1982). "The effect of free water on deck on the motions and stability of vessels", *Proc. of STAB'82: 2nd International Conference on Stability of Ships and Ocean Vehicles*, Tokyo, pp. 413-426.

Alman, P.R., Minnick, P. V., Sheinberg, R. and Thomas III, W., L. (1999). "Dynamic capsize vulnerability: reducing the hidden operational risk", *SNAME Transactions*, Vol. 107.

Amagai, K., Kimura, N. and Ueno, K. (1994). On the practical evaluation of shallow water effect in large inclination for small fishing boats, *Proc. of STAB'94: 5th International Conference on Stability of Ships and Ocean Vehicles*, vol. 3, Melbourne, Florida.

Amersdorffer, R. (1998). "Parametrisch erregte Rollbewegungen in längslaufendem Seegang" *Schiff & Hafen*, No. 10/98 and No. 11/98, Hamburg.

Ananiev, D. M. (1964). "On directional stability of a ship in waves", *Transactions of Krylov Society*, Vol. 54, Leningrad (in Russian).

Ananiev, D. M. (1966). "On surf-riding in following seas", *Transactions of Krylov Society*, Vol. 73, Leningrad, (in Russian).

Ananiev, D. M. (1981) "On stability of forced rolling of a vessel with given GZ curve", *Transactions of Kaliningrad Institute of Technology*, Vol. 93, "Seakeeping of Ships", Kaliningrad, (in Russian).

Ananiev, D. M. and Savchuck, S. V. (1982). "Experimental approval of applicability of exponent distribution to capsizing probability", *Transactions of Russian Register Shipping*, Vol.12, Leningrad, pp. 115-119 (in Russian).

Ananiev, D. M. (1989). "Ship rolling with negative initial stability", in *Transactions of Kaliningrad Institute of Technology* "Seakeeping and Design of Ships" Kaliningrad, pp. 4-12 (in Russian).

Ananiev, D. M. (1994). "Determination the boundaries of surf-riding domain analysing surging stability", *Proc. of STAB'94: 5th International Conference on Stability of Ships and Ocean Vehicles*, Vol. 5, Melbourne, Florida.

Ananiev, D. M. (1995) "On the excitation forces acting on ship in horizontal plane during her motion with drift and rotation" *Proc. of International Symposium - Ship Safety in a Seaway: Stability, Manoeuvrability, Nonlinear Approach (Sevastianov Symposium)*, Vol. 1, paper No 12, Kaliningrad.

Andronov, A. A., Vitt A. A. and Khaikin, S. E, (1966) "Theory of oscillators", Pergamon Press Oxford, New York.

Angelides, D. C., Veneziano, D., and Shyam Sunder (1981). "Random sea and reliability of offshore foundations". -*J. Eng. Mech. Div.,* Vol.107, No 1, pp. 131-148.

Apollinariev, V. I. and Sevastianov N. B. (1992). "Stability Estimation based on Simulative Modelling", *Proc. of PRADS'92: International Conference on Practical Design and Mobile Units*, Elsevier Science, New Castle upon Tyne, Vol. 2, pp. 1161-1172.

Apostol, T. M. (1997). "Linear algebra. The first course with application to differential equations", Wiley Interscience, New York.

Armenio, V., Francescutto, A. and La Rocca, M. (1996). "On the roll motion of a ship with partially filled unbaffled and baffled tanks. Part 1 and 2". *International Journal of Offshore and Polar Engineering,* Vol. 6, No 4.

Arndt B., and Roden S. (1958). Stabilität bei vor- und achterlichem Seegang", *Schiffstechnik*, Vol. 5, No 29, Hamburg.

Arndt, B., Kastner, S. and Roden, S. (1960). "Stabilitätsmessung auf Schiffen", *Hansa*, No. 47/48, Hamburg.

Arndt, B. (1965). "Ausarbeitung einer Stabilitaetsvorschrift fuer die Bundesmarine", *Transactions STG*, Vol.59, p.594.

Arndt, B., Brandl H. and Vogt K. (1982). "20 Years of experience - stability regulations of the West-German Navy". *Proc. of STAB'82: 2nd International Conference on Stability of Ships and Ocean Vehicles*, Tokyo, p.765.

Arndt, B. (1984). "Marine computers for ship handling and ship operation", IMAEM, Athens.

Balitskaya, E. O. (1965). "Some results on the ship behaviour in shallow breaking waves", *Transactions of Russian Register Shipping* "Theoretical and Practical Problems of Stability and Survivability", Transport, Moscow-Leningrad, pp.147-165 (in Russian).

Baker, G. L. and Gollub, J. P. (1996). "Chaotic dynamics: an introduction", (2nd ed.), Cambridge University Press.

Beck, R.F. and Reed, A. M. (2001). "Modern computational methods for ships in seaway", *SNAME Transactions*, Vol. 109, pp. 1-48.

Belenky, V. L. (1989). "A new method of statistical linearisation in severe rolling and capsizing problem", *Proc. of 18th SMSSH: Scientific and Methodological Seminar on Ship Hydrodynamics*, Vol. 1, Paper No 39, Bulgarian Ship Hydrodynamic Centre, Varna.

Belenky, V. L. (1990). "Applicability of the theoretical model large-amplitude rolling motion", *Proc. of 19th SMSSH: Scientific and Methodological Seminar on Ship Hydrodynamics*, Vol.1, Paper No 45, Bulgarian Ship Hydrodynamic Centre, Varna.

Belenky, V. L. (1991). "Practical method for capsizing probability determination by "double" linearisation", *Proc. of HADMAR'91: International Symposium on Hydro- and Aerodynamics in Marine Engineering,* Vol. 2, Paper No 54, Bulgarian Ship Hydrodynamic Centre, Varna.

Belenky, V. L. (1993). "A capsizing probability computation method", *Journal of Ship Research*, Vol. 37, No. 3, pp. 200-207.

Belenky, V. L. (1993a). "On a probabilistic balance of IMO stability regulations", Technical University of Gdansk, Report of Dept. of Ship Hydromechanics No 6/93.

Belenky, V. L. (1994). "Piecewise linear methods for the probabilistic stability assessment for ship in a seaway", *Proc. of STAB'94: 5th International Conference on Stability of Ships and Ocean Vehicles,* Vol. 5, Melbourne, Florida, 1994.

Belenky V. L. (1995). "Analysis of probabilistic balance of IMO stability regulation by piecewise linear method", *Marine Technology Transactions*, Polish Academy of Sciences, Branch in Gdansk, Vol. 6, p 5-55.

Belenky, V. L. (1995a). "On the dynamics of piecewise linear system" *Proc. of International Symposium - Ship Safety in a Seaway: Stability, Manoeuvrability, Nonlinear Approach (Sevastianov Symposium)*, Vol. 1, Paper No 6, Kaliningrad Institute of Technology, Kaliningrad.

Belenky, V. L., Degtyarev, A. B. and Boukhanovsky, A. V. (1995). "Probabilistic qualities of severe rolling", *Proc. of International Symposium - Ship Safety in a Seaway: Stability, Manoeuvrability, Nonlinear Approach (Sevastianov Symposium)*, Vol. 1, Paper No 7, Kaliningrad Institute of Technology, Kaliningrad.

Belenky, V. L., Degtyarev, A. B. and Boukhanovsky, A. V. (1997). "Probabilistic qualities of severe ship motions", *Proc. of STAB'97: 6th International Conference on Stability of Ships and Ocean Vehicles*, Varna, Vol. 1, pp.163-172.

Belenky V. L. and Umeda N. (1997). "Contemporary remarks on classic weather criteria", *Proc. of STAB'97: 6th International Conference on Stability of Ships and Ocean Vehicles*, Varna, Vol. 2, pp. 325-332.

Belenky, V. L., Degtyarev, A. B., and Boukhanovsky, A. V. (1998). "Probabilistic qualities of nonlinear stochastic rolling" *Ocean Engineering*, Vol. 25, No 1, pp. 1-25.

Belenky, V. L. (1999). "Piecewise linear approach to nonlinear dynamics of ships", *Bulletin of National Research Institute of Fisheries Engineering*, No 20, Hasaki, Ibaraki, pp. 21-40.

Belenky, V. L. (1999a). "Probabilistic assessment of ship stability in quartering seas", *Proc. of JAKOM'99: 4th Japan-Korea Joint Workshop on Ship and Marine Hydrodynamics*, Fukuoka, pp. 291-297.

Belenky, V. L. (2000). "Probabilistic Approach for Intact Stability Standards", *SNAME Transactions*, Vol. 108, pp. 123-146.

Belenky, V. L. (2000a). "Piecewise linear approach to nonlinear ship dynamics", in "*Contemporary Ideas of Ship Stability*", D. Vassalos, M. Hamamoto, A. Papanikolaou and D. Moulyneux (Editors), Elsevier Science, pp. 149-160.

Belenky, V. L. (2000b). "Piecewise linear approach to probabilistic stability in quartering seas", *Proc. of STAB'2000: 7th International Conference on Stability of Ships and Ocean Vehicles,* Vol. 1, Launceston, Tasmania, pp. 503-510f.

Belenky, V. L. Suzuki, S. and Yamakoshi, Y. (2001). "Preliminary results of experimental validation of practical non-ergodicity of large amplitude rolling motions",

Proc. of 5th International Workshop on Ship Stability, Paper No 4.1 University of Trieste, Trieste.

Belenky, V.L., Liut, D., Weems, K. and Shin Y.S., (2002) "Nonlinear ship roll simulation with water-on-deck" *Proceeding of 6th International Ship Stability Workshop*, Webb Institute, New York.

Bendat, J.S. and Piersol, A.G. (1986) "Random data (Analysis and measurement procedures)", 2nd edition, John Wiley & Sons, New York.

Bielanski, J. (1994) Ship rolling under action of beam wind and sea as a background of comparison IMO stability criteria, *Proc. of OTRADNOE'93: International Workshop on the Problems of Physical and Mathematical Stability Modelling,* Vol.2, Paper No 7, Kaliningrad Institute of Technology, Kaliningrad.

Blagoveshchensky, S. N. (1932). "On the method of stability standardisation", *Transactions of Shipbuilding Research Institute, (NISS, later Krylov Research Centre)*, United Science and Technology Publishing (ONTI), Vol. 12, (in Russian).

Blagoveshchensky, S. N. (1951). "On stability regulations of sea vessels", *Transactions of Central Maritime Institute (TSNIIMF)*, Vol.8, (in Russian).

Blagoveshchensky, S. N. (1962). "Theory of Ship Motions", in two volumes, Dover Publications, New York.

Blagoveshchensky, S. N. (1965). "On the wind load on vessel", *Transactions of Russian Register of Shipping* "Theoretical and practical problems of stability and Survivability", Transport publishing, Moscow-Leningrad, pp, 100-146 (in Russian).

Blocki, W. (1980). "Ship safety in connection with parametric resonance of the roll", *International Shipbuilding Progress*, Vol.27, No. 306, pp.36-53.

Bogolubov, N. N., and Mitropolsky Y. A. (1961). "Asymptotic methods in the theory of nonlinear oscillation", Gordon and Breach, New York.

Boroday, I. K. (1967). "Statistical characteristics of stability and the probability of capsizal of a ship running on any course in irregular seas", Documents of the USSR Expert of the IMCO, Working Group on Stability of Fishing Vessels, IMO, London.

Boroday, I.K. (1968). "Statistical characteristics and probability of capsizing of ship heading with arbitrary course in irregular seas", *Transactions of Russian Register of Shipping* "Theoretical and Practical Problems of Stability and Survivability", Transport publishing, Moscow-Leningrad, pp.21-48 (in Russian).

Boroday, I. K. and Netsvetaev, Yu. A. (1969). "Ship motion in a seaway", Sudostroenie publishing, Leningrad (in Russian).

Boroday, I. K. and Nikolaev, E. P. (1975). "Methods of estimating the ship's stability in irregular seas", *Proc. of STAB'75: 1st International Conference on Stability of Ships and Ocean Vehicles*, Glasgow.

Boroday, I. K. and Netsvetaev, Yu. A. (1982). "Seakeeping of ships", Sudostroenie publishing, Leningrad. (In Russian).

Boroday, I. K. and Morenshildt, V. A., (1986). "Stability and parametric roll of ships in waves", *Proc. of STAB'86: 3rd International Conference on Stability of Ships and Ocean Vehicles*, Gdansk, pp. 19-26.

Boroday, I. K., Morenshildt, V. A., Vilensky, G. V., Dubitsky, V. M. and Smirnov, B. N. (1989) "Applied dynamic of ship in a seaway", Sudostroenie publishing, Leningrad, (in Russian).

Boukhanovsky, A. V. and Degtyarev, A. B. (1996). "Nonlinear stochastic ship motion stability in different wave regime" *Proc. of CRF'96: 3rd International Conference in Commemoration of the 300-th Anniversary of Creating Russian Fleet by Peter the Great*, Vol. 2, St. Petersburg, pp. 296-306.

Boukhanovsky, A. V., Lopatoukhin, L. J. and Ryabinin, V .E. (1998). "Evaluation of the highest wave in a storm" WMO/TD - No. 858, 18 p.

Boukhanovsky, A. V., Degtyarev, A. B., Lopatoukhin, L. J. and Rozhkov, Y. V. (1998a). "Probabilistic modelling of wave climate", *Physics of Atmosphere and Ocean*, Vol. 34, No 2, pp. 261-266 (in Russian).

Boukhanovsky, A. V., Degtyarev, A. B., Lopatoukhin, L. J. and Rozhkov, Y. V. (2000) "Stable states of wave climate: Applications for risk estimation*Proc. of STAB'2000: 7th International Conference on Stability of Ships and Ocean Vehicles*, Vol. 2, Launceston, Tasmania, pp. 831-846.

Boukhanovsky A., Rozhkov V., Degtyarev A. (2001) "Peculiarities of computer simulation and statistical representation of time–spatial metocean fields", in Computational Science - ICCS 2001, LNCS 2073, part I , Springer, pp.461-472.

Bronshtein, I. N. and Semendyayev, K. A. (1997). "Handbook of mathematics", (3rd ed.), Springer, Berlin.

Buckley, W. H. (1988). "Extreme and climatic wave spectra for use in structural design of ships". *Naval Engineers Journal*, Vol. 100, No 5, ASNE, pp. 36-57.

Buckley, W. H. (1992). "Matching Vehicles Characteristics to Seaway Environments", ASNE, June.

Caglayan, I.H. and Storch, R.L. (1982) "Stability of fishing vessel with water on deck: a review", *Journal of Ship Research*, Vol. 26, No. 2.

Caldeira-Saraiva, F. (1986). "A stability criterion for ships using Lyapunov's method", *Proc. of Safeship Conference.*

Caldeira-Saraiva, F. (1986a). "The boundness of roll motion of a ship by Lyapunov's method", *Proc. of STAB'86: 3rd International Conference on Stability of Ships and Ocean Vehicles*, Vol. 1, Gdansk, pp 71-75.

Calisal S.M., Rudman M. J., Akinturk A. and Tasevski (2000) "Water discharge from an opening in ships", in "*Contemporary Ideas of Ship Stability*", D. Vassalos, M. Hamamoto, A. Papanikolaou and D. Moulyneux (Editors), Elsevier Science, pp. 473-483.

Cai, G. Q., Yu, J. S. and Lin, Y. K. (1994). "Ship rolling in random sea", *Stochastic Dynamics and Reliability of Nonlinear Systems* DE 77 ASME.

Cardo, A., Francescutto, A. and Nabergoj, R. (1981) "Ultraharmonics and subharmonics in the rolling motion of ship" steady state solution", *International Shipbuilding Progress*, Vol. 28, No 326, pp. 234-251.

Cardo, A., Francescutto, A. and Nabergoj, R. (1982) "On the maximum amplitudes in nonlinear rolling", *Proc. of STAB'82: 2nd International Conference on Stability of Ships and Ocean Vehicles*, Tokyo.

Cardo, A., Francescutto, A. and Nabergoj, R. (1984). "Subharmonic oscillations in nonlinear rolling", *Ocean Engineering*, Vol. 11, No. 6, pp. 553-669.

Cardo, A., Francescutto, A. and Nabergoj, R. (1985). "Deterministic nonlinear rolling: a critical review", *ATMA - Association Technique Maritime et Aeronautique*, Paris.

Caughey, T. K. (1963). "Derivation and application of the Fokker-Plank equation to discrete nonlinear systems subjected to white random excitation", *J. Acoust. Soc. Am.*, Vol. 35, No. 11, pp. 1683-1692.

Caughey, T. K. (1964). "On the response of a class on nonlinear oscillators to stochastic excitation", *Proc. Colloq. Int. Cent. Nat. Rech. Sci.*, Vol. 148.

Čerka, J. and Batuev, A. D. (1985). "Influence of superstructures and deck houses on capsizing probability", *Transaction of Nikolaev Shipbuilding Institute*, Nikolaev, pp.48-56 (in Russian).

Chang, M.S. (1977) "Computation of three-dimensional ship-motions with forward speed", *Proc. of 2nd International Conf. Num. Ship Hydrodynamics*, Berkeley pp.124-135.

Choi, Y. S. and Noah, S. T. (1988) "Forced periodic vibration on unsymmetric piecewise linear systems" *Journal of Sound and Vibration*, Vol. 121, No 1, pp.117-126.

Chernetsky, V. I. (1968). "Analysis of accuracy of nonlinear control systems", Mashinostroenie publishing, Moscow, (in Russian).

Dahle, E. A. and Kjaerland, O. (1980). "The capsizing of M/S HELLAND HANSEN", *Naval Architect*, RINA, Vol. 122, pp 51-70.

Dahle, E. A., Myrhaug, D. and Dahl, S. J. (1988). "Probability of capsizing in steep waves and high waves from the side in open sea and coastal waters", *Ocean Engineering* Vol. 15, pp. 139-151.

Dahle, E. A. and Myrhaug, D. (1993). "Risk analysis applied to capsize of smaller vessels in breaking waves", *Presented at spring meeting of RINA.*

Dahle, E. A., Enerhaug, B. and Myrhaug D. (1995). "Risk analysis applied to capsize of smaller vessels in breaking waves with examples from Russian waters" *Proc. of International Symposium - Ship Safety in a Seaway: Stability, Manoeuvrability, Nonlinear Approach (Sevastianov Symposium)*, Vol.1, Paper No 2, Kaliningrad Institute of Technology, Kaliningrad.

Dahle, E.A. and Myrhaug, D. (1997). "Capsize risk of a typical fishing vessel in bleak sea" *Proc. of STAB'97: 6th International Conference on Stability of Ships and Ocean Vehicles,* Vol. 2 Varna, pp. 201-209.

Davidan, I. N., Lopatoukhin, L. J. and Rozhkov, V. A. (editors) (1974). "Wind and waves in the oceans and seas. A reference book". Russian Register of shipping, Transport publishing, Leningrad, (in Russian).

Davidan I. N (editor) (1995). "Problems of Investigation and Mathematical Modelling of Wind Waves", Hydrometeoizdat publishing, St. Petersburg, (in Russian).

Davidson, K. S. M. (1948). "A note on the steering of ships in following seas", *Proc. of 7th Int. Congress Applied Mech.*, London.

Davies, H. G. and Rajan, S. (1988). "Random superharmonic and subharmonic response: multiple time scaling of a Duffing equation", *Journal of Sound and Vibration*, Vol. 126, No 2, pp. 195-208.

Degtyarev, A. B. (1994) "Distribution of nonlinear roll motion" *Proc. of OTRADNOE'93: International Workshop on the Problems of Physical and Mathematical Stability Modelling* Vol. 2, Paper 9, Kaliningrad Institute of Technology, Kaliningrad.

Degtyarev A.B. and Boukhanovsky A.V. (1995) "On the estimation of the ship motion stability in real sea" *Proc. Int. Symp. on Ship Safety in a Seaway: Stability, Manoeuvrability, Nonlinear Approach (SEVASTIANOV SYMPOSIUM)*, Kaliningrad, vol.2, paper 8.

Degtyarev A. B. and Boukhanovsky A.V. (2000) "Peculiarities of motions of ship with low buoyancy on asymmetrical random waves" *Proc. of STAB'2000: 7th International Conference on Stability of Ships and Ocean Vehicles,* Launceston, Australia, vol. 2, pp 665-679.

Dillingham, J. T. and Falzarano, J. M. (1986) "Three dimensional numerical simulation of green water on deck". *Proc. of STAB'86: 3rd International Conference on Stability of Ships and Ocean Vehicles*, Vol. 1, Gdansk, pp 57-64.

Dimentberg, M.F. (1980). "Nonlinear stochastic problems of mechanical oscillations", Nauka publishing, Moscow (in Russian).

Dragan, Ya. G., Rozhkov, V. A. and Yavorsky I. N. (1987). "Methods of probabilistic analysis of the oceanographic process rhythmic" "Gidrometeoizdat" publishing Leningrad (in Russian).

Du Cane, P. and Goodrich, G. J., (1962). "The following sea, broaching and surging", *RINA Transactions.*

Dudziak, J., Buczkowski, A. (1978). "Probability of ship capsizing under the action of the beam wind and sea as a background of stability criteria", Polish Register of Ships, Prace Studialno-Roswojowe, Zeszyt Nr. 13, Gdansk.

Elis, Ya. M. (1980). "Added masses and damping of transverse section of heeled vessel", *Transactions of Kaliningrad Institute of Technology*, Vol. 90, "Seakeeping of Fishing Vessels", Kaliningrad, pp. 30-41 (in Russian).

Elis, Ya. M. (1980a). "Hydrodynamic coefficients of heeled ship section", *Transactions of Kaliningrad Institute of Technology*, Vol. 90, "Seakeeping of Fishing Vessels", Kaliningrad, pp. 42-93 (in Russian).

Esparza, I. and Falzarano J., M. (1993). "Nonlinear rolling motion of a statically biased ship under the effect of external and parametric excitations", *ASME*, DE-Vol. 56 "*Dynamics and vibration of time–varying system and structures*".

Ericson, A., Jianbo, H. and Rutgersson, O. (2000) "Mixed seas and cargo shifting onboard" *Proc. of STAB'2000: 7th International Conference on Stability of Ships and Ocean Vehicles*, Vol. 2, Launceston, Tasmania, pp. 802-818.

Ericson, A., Person, J. and Rutgerson O. (1997). "On the use of formal safety assessment when analyzing the risk for cargo shift in rough seas", *Proc. of RINA International Conference on Design and Operation for Abnormal Conditions*, Glasgow.

Falzarano, J. M. and Troesch, A. W. (1990). "Application of the modern geometric methods for dynamical systems to the problem of vessel capsizing with water-on-deck", *Proc. of STAB'90: 4th International Conference on Stability of Ships and Ocean Vehicles*, Naples.

Falzarano, J. M., Shaw, S. W. and Troesch, A.W. (1992) "Application of global methods for analysing dynamical systems to ship rolling motion and capsizing", *International Journal of Bifurcation and Chaos,* Vol. 2, No 1, pp. 101-115.

Falzarano, J. M., Esaprza, I. and Taz Ul Mulk, M. (1995) "A combined steady-state and transient approach to study large amplitude ship rolling motion and capsizing", *Journal of Ship Research,* Vol. 39, No. 3, pp. 213-224.

Fediaevsky, K.K. and Firsov, G.A. (1957) Ship heeling under wind action, *Sudostroenie*, No. 12, Leningrad (in Russian).

Ferrant, P. (1998) "The pressure of water waves upon fixed a obstacle", *Proc. Royal Soc. of London,* Series A; 175:(963), pp. 409-421.

France, W.G., Levandou, M., Treakle, T. W., Paulling, J.R., Michel, R. K. and Moore, C. (2003). "An investigation of head seas parametric rolling and its influence on Container Lashing Systems", *Marine Technology*, Vol. 40, No. 1, pp 1-19.

Francescutto, A. and Nabergoj, R. (1990). "A stochastic analysis of nonlinear rolling in narrow band sea", *Proc. of 18th Symposium on Naval Hydrodynamics*, National Academy Press, Washington D.C.

Francescutto, A. (1991). "On the nonlinear motion of ships and structures in narrow band sea", in "Dynamics of Marine vehicles and Structure in Waves" W.G. Price, P. Temarel and A.J. Keane (editors), Elsevier Science, pp. 291-303.

Francescutto, A. (1991a). "On the probability of large amplitude rolling and capsizing as a consequence of bifurcations", *Proc. of OMAE'91: 10th International Conference on Offshore Mechanics and Arctic Engineering*, ASME, Vol. 2 pp. 91-96.

Francescutto, A. (1992). "Stochastic modelling of nonlinear motions in the presence of narrow band excitation" *Proc. of the 2nd International Offshore and Polar Engineering Conference*, , Vol.3, San Francisco, pp.554-558.

Francescutto, A. and Contento, G. (1994). "An experimental study of the coupling between roll motion and sloshing in a compartment", *Proc. of the 4th International Offshore and Polar Engineering Conference*, Osaka, Vol. 3, pp. 283-291.

Francescutto, A. (1998). "On the statistical distribution of stochastic nonlinear rolling", in "*Risk and reliability in marine technology*", C. Guedes-Soares (Editor), A.A. Balkema publishers, Rotterdam, pp.107-116.

Fujino, M., Yamasaki, K. and Ishi, Y. (1983), "On the stability derivatives of a ship travelling in the following waves", *Journal of Society of Naval Architects of Japan,* Vol. 130, pp. 167-179 (in Japanese).

Garkavy, V. V. and Ponomarenko, A. D. (1977). "Experimental investigation of motion of low-built vessel model and condition of her capsizing in regular seas", *Transactions of Kaliningrad Institute of Technology* "Seakeeping of fishing vessels", Vol.61, Kaliningrad, pp.3-8 (in Russian).

Garkavy, V. V. (1979). "Froude-Krylov part of the excitation forces of asymmetrical rolling and heaving being accompanied with the deck immersion", *Transactions of Kaliningrad Institute of Technology*, Vol.81 "Seakeeping of fishing vessels", Kaliningrad, pp. 23-37 (in Russian).

Garkavy, V. V., Batuev, A. D. and Kovalenko, N. I. (1982). "Some results of experimental estimation of value and point of application of drift resistance force", *Transactions of Kaliningrad Institute of Technology*, Vol. 99 "Seakeeping of fishing vessels", Kaliningrad, pp. 31-37 (in Russian).

Garkavy, V. V. (1985). "On stability of steady state rolling being accompanied with the deck immersion", *Transactions of Kaliningrad Institute of Technology* "Seakeeping and Design of Fishing Vessels", Kaliningrad, pp. 56-62, (in Russian).

Garkavy, V. V. and Kovalenko, N. (1989). "Serial model test on drift heeling moment and drag", *Transactions of Kaliningrad Institute of Technology* "Seakeeping and Design of Ships", Kaliningrad, p. 121-127 (in Russian).

Garkavy, V. V. (1991). "Deterministic chaos in the task of the roll motion ship with a small freeboard", *Proc. of HADMAR'91: International Symposium On Hydro- And Aerodynamics In Marine Engineering*, Paper 49, Vol. 2, Bulgarian Ship Hydrodynamics Centre, Varna.

Gerasimov, A. V. (1973). "Energy principle of linearisation roll damping for irregular seas", *Transactions of Krylov Research Institute*, Vol. 269 (in Russian).

Gerasimov, A. V. (1974). "Energy characteristics of stationary random oscillation and their use in statistical linearisation" *Proc. of International Symposium of the Dynamics of Marine Vehicles and Structures in Waves*, Paper 41, London.

Gerasimov, A. V. (1979). "Energy-statistical theory of nonlinear irregular ship motion", Sudostroenie publishing, Leningrad (in Russian).

Gorski, J. (2002). "A perspective on the role of RANS codes for predicting large amplitude ship motions", *Proceeding of 6th International Ship Stability Workshop*, Webb Institute, New York.

Grim, O. (1952). "Rollschwingungen, Stabilität und Sicherheit im Seegang", *Schiffstechnik*, No. 1, Schiffahrts-Verlag Hansa, Hamburg.

Grim, O. (1961). "Beitrag zu dem Problem der Sicherheit des Schiffes im Seegang", *Schiff und Hafen*, Helft 6, pp.490-491.

Grochowalski, S. (1979). "Calculation of the amount of water shipping on deck in shallow water beam seas" *Tech. University of Gdansk, Institute of Naval Architecture*, Report No 1210/MR-324/79, Gdansk (in Polish and German).

Grochowalski, S. (1981). "Theoretical prediction of mass of water shipping on deck in irregular waves" *Tech. University of Gdansk, Institute of Naval Architecture*, Report No 1969/MR-514/81, Gdansk (in Polish).

Grochowalski, S., Rask, I. and Soderberg, P. (1986). "An experimental technique for investigation into physics of ship capsizing". *Proc. of STAB'86: 3rd International Conference on Stability of Ships and Ocean Vehicles*, Vol. 2, Addendum 1, Gdansk.

Grochowalski, S. (1989). "Investigation into the physics of ship capsizing by combined captive and free-running model tests", *Transactions SNAME*, Vol. 97, pp. 169-212.

Grochowalski, S. (1990). "Hydrodynamic phenomenon generated by bulwark submergence and its influence on ship susceptibility to capsizing", *Proc. of STAB'90: 4th International Conference on Stability of Ships and Ocean Vehicles*, Vol. 2, Naples.

Grochowalski, S. and Lee, C. K. (1990a). "Time analysis of bulwark submergence and instantaneous wave crest position in the capsizing model test" *NRC/IMD* Report No TR-1995-40.

Grochowalski, S. (1993). "Effect of bulwark and deck edge submergence in dynamics of ship capsizing", *Proc. of US Cost Guard Vessel Stability Symposium*, US Coast Guard Academy, New London, Connecticut, pp.94-109.

Grochowalski, S. (1993a). "Ship in quartering seas - a case which determines stability safety", *Proc. of OTRADNOE'93: International Workshop on the Problems of Physical and Mathematical Stability Modelling*, Vol. 2, Paper No 12, Kaliningrad Institute of Technology, Kaliningrad.

Grochowalski, S., Archibald, J. B., Connolly, F. J. and Lee, C. K. (1994). "Operational factors in stability safety of ships in heavy seas" *Proc. of STAB'94: 5th International Conference on Stability of Ships and Ocean Vehicles*, Vol. 4, Melbourne, Florida.

Grochowalski, S. (1997). "Experimental investigation of the hydrodynamic forces and pressure distribution on the submerged part of deck in wave", *NRC/IMD Report* No TR-1997-09.

Grochowalski, S., Hsiung, C.C., Huang, Z.J. and Cong L.Z. (1998). "Theoretical modelling of ship motions and capsizing in large and steep waves" *Transactions SNAME,* Vol. 106, pp. 241-267.

Grochowalski, S. (2000). "Experimental investigation of ship dynamics in extreme waves" "*Contemporary Ideas of Ship Stability*", D. Vassalos, M. Hamamoto, A. Papanikolaou and D. Moulyneux (Editors), Elsevier Science, pp. 3-13.

Guckenheimer, J. and Holmes, P. (1983). "Nonlinear oscillations, dynamical system and bifurcation of vector fields", Springer-Verlag New York, Berlin and Heidelberg.

Gumbel, E. J. (1958). "Statistics of extremes". Columbia University Press.

Guralnik, B. and Kulagin, V. (1995). "Load changes of the ship in light due to her ageing and their correction in planning stability trials" *Proc. of International Symposium - Ship Safety in a Seaway: Stability, Manoeuvrability, Nonlinear Approach (Sevastianov Symposium)*, Vol. 1, Paper 16, Kaliningrad Institute of Technology, Kaliningrad.

Gurgenidze, A.T. and Trapeznikov, Yu.A. (1988). "Probabilistic models of wind induced waves", in "*Theoretical Foundation and Calculation Methods of Wind Induced Waves*", Gidrometeoizdat publsihing, pp.8-23 (in Russian).

Haddara M. R., Kastner S., Magel L. F., Paulling J. R., Perez y Perez, L. and Wood, P. D. (1972). "Capsizing experiments with a model of a fast cargo liner in San Francisco Bay", Dept. of Transportation, U.S. Coast Guard, Technical Report.

Haddara, M.R. (1974). "A modified approach for the application of Fokker-Plank equation to nonlinear ship motion in random waves", *International Shipbuilding Progress*, Vol. 21, No 242, pp. 283-288.

Haddara, M. R. (1975). "A study of the mean and variance of rolling motion in random waves", *Proc. of STAB'75: 1st International Conference on Stability of Ships and Ocean Vehicles*, Glasgow.

Haddara, M.R. and Nassar, M.A. (1986). A stochastic model for the analysis of rolling motion in a realistic seaway, *International Shipbuilding Progress* Vol. 33, pp.144-150.

Haddara, M.R. and Zhang, Y. (1994). "On the joint probability density function of nonlinear rolling motion", *Journal of Sound and Vibration*, Vol. 169, No 4, pp.562-569.

Hamamoto, M. (1973). "On the hydrodynamic derivatives for the directional ship stability of ships in following seas, part 2" *Journal of Society of Naval Architects of Japan* Vol. 133, pp.133-142 (in Japanese).

Hamamoto M., Fujino M. and Kim Y. S. (1994). "Dynamic stability of a ship in quartering seas", *Proc. of STAB'94: 5th International Conference on Stability of Ships and Ocean Vehicles*, Vol. 4, Melbourne, Florida.

Hayashi, Ch. (1964), "Nonlinear oscillation in physical systems", McGrow-Hill.

Hogben, N. and Lumb, F. E. (1967). "Ocean wave statistics". Her Majesty's Stationary Office, London.

Hogben, N., Dacunha, N. M. C. and Oliver, G.F. (1986). "Global wave statistics". Compiled and edited by British Maritime Technology Ltd, Unwin Brothers.

Hooft, J. P. (1987). "Mathematical description of manoeuvrability of high speed surface ships", MARIN Report No 47583-1-MOA.

Hsieh, S.-R., Shaw, S.W and Troesch A.W (1993). "A predictive method for vessels capsize in random seas" in *Nonlinear Dynamics of Marine Vehicles* ASME vol. DSC-51 (OMAE-1).

Hsieh, S.-R. Troesch, A.W. and Shaw, S. W (1994). "A nonlinear probabilistic method for predicting vessel capsizing in random beam seas" *Proc. R. Soc. London*, No 446, pp. 195-211.

Huang, Z. J. (1995). "Nonlinear shallow water flow on deck and its effect on ship motions", *Ph.D. Thesis*, Tech. University of Nova Scotia, Halifax.

Huang, Z. J. and Hsiung, C. C. (1996). "Nonlinear shallow-water flow on deck", *Journal of Ship Research*, Vol. 40, No 4, pp. 303-315.

Huang, Z. J. and Hsiung, C. C. (1997). "Dynamic simulation of capsizing for fishing vessels with water on deck", *Proc. of STAB'97: 6th International Conference on Stability of Ships and Ocean Vehicles*, Vol. 1, Varna, p. 287-299.

Huang, Z. J., Cong, L., Grochowalski, S. and Hsiung, C. C. (1999). "Capsize analysis for ships with water shipping on and off the deck" *Proc. of 22nd Symposium of Naval Hydrodynamics,* National academy press, Washington, D.C.

Ikeda, Y., Tanaka, N. and Himeno, Y. (1982). "Effect of hull form and appendage on roll motion of small fishing vessel", *Proc. of STAB'82: 2nd International Conference on Stability of Ships and Ocean Vehicles*, Tokyo.

Ikeda, Y. and Kawahara, Y. (1993). "Studies of roll characteristics of high-speed hardchine craft", *Proc. of 2nd Japan-Korea Joint Workshop on Ship and Marine Hydrodynamics*, pp. 37-44.

Ikeda, Y. (2002). "Prediction methods of roll damping of ships and their application to determine optimum stabilisation device" *Proceeding of 6th International Ship Stability Workshop*, Webb Institute, New York.

IMO, (1967). "Recommendations on intact stability of fishing vessels", Document STAB VII/11, Annex V.

IMO, (1967a). "Approximate evaluation of safety at sea for a small vessel among waves", Submitted by Soviet Union, Document PFV VI/22.

IMO, (1968). "Stability recommendation for passenger, cargo and fishing vessels", Resolutions A-167 and A168.

IMO, (1969). "Results of experimental investigation of ship motions and the possibility of capsizing of an undamaged low built vessel in regular waves", Submitted by Soviet Union, Document PFV IX/3/2.

IMO, (1984). "Intact stability of fishing vessels in breaking waves. Results obtained on small trawlers", Submitted by Soviet Union, Document SLF/28.

Inglis, R.B. and Price, W.G. (1982) "A three-dimensional ship motion theory: the hydrodynamic coefficient with forward speed", *Trans. RINA*, Vol. 124, pp.141-157.

Ishida, S. and Takaishi, Y. (1990) "A capsizing experiment of a small fishing boat in breaking waves". *Proc. of STAB'90: 4th International Conference on Stability of Ships and Ocean Vehicles*, Vol. 1, Naples.

Iskandar, B. H., Umeda, N. and Hamamoto M. (2000). "Capsizing probability of an Indonesian Ro-Ro passenger ship in irregular beam seas", *Journal of the Society of Naval Architects of Japan*, Vol. 188 (December), pp.183-189.

Iskandar, B. H., Umeda, N. and Hamamoto M. (2001). "Capsizing probability of an Indonesian Ro-Ro passenger ship in irregular beam seas (Second Report)", *Journal of the Society of Naval Architects of Japan*, Vol. 189 (June), pp.31-37.

Iskandar, B. H. and Umeda, N. (2001). "Some examinations of capsizing probability calculation for an Indonesian Ro-Ro passenger ship in waves", *Journal of Kansai Society of Naval Architects*, No. 236 (September), pp. 81-86.

Iskandar, B. H. and Umeda, N. (2001a). "Capsizing probability of an Indonesian Ro-Ro passenger ship in irregular beam seas (Third Report)", *Journal of the Society of Naval Architects of Japan*, Vol. 190 (December), pp.211-216.

Jiang, Ch., Troesch, A. W. and Shaw, S. W. (1996). "Highly nonlinear rolling motion of biased ships in random beam seas", *Journal of Ship Research,* Vol. 40, No 2, pp. 125-135.

Jonson, R. A. and Wichern, D. W. (1992). "Applied multivariate statistical analysis" (3rd ed.). Prentice Hall Inc.

Kan, M. (1990). "Surging of large amplitude and surf-riding of ships in following seas" *Naval Architecture and Ocean Engineering* The Society of Naval Architects of Japan, Vol. 28.

Kan, M., Taguchi, H., (1991) Chaos and fractal in capsizing of a ship, *Proc. of HADMAR'91: International Symposium on Hydro- and Aerodynamics in Marine Engineering*, Vol.1, Varna, pp.81-88.

Kan, M., Saruta, T. and Taguchi, H. (1992). "Capsizing of a ship in quartering waves", *Naval Architecture and Ocean Engineering,* The Society of Naval Architects of Japan Vol. 29.

Kan, M. (1992). "Chaotic capsizing", *Proc. of ITTC.* SCR-KFR Osaka Meeting on Seakeeping Performance.

Kan, M., and Taguchi, H. (1992). "Chaos and fractals in loll type capsizing equation" *Transactions of West-Japan Society of Naval Architects* No 83, pp. 131-149. (in Japanese).

Kan, M. and Taguchi, H. (1993). "Chaos and fractals in nonlinear rolling and capsize of a damaged ship", *Proc. of OTRADNOYE'93: International Workshop on the Problems of Physical and Mathematical Stability Modelling* Vol. 2, Paper No 2, Kaliningrad Institute of Technology, Kaliningrad.

de Kat J.O. and Thomas, W. L. (1999). "Extreme rolling, broaching, and capsizing – model tests and simulations of a steered ship in waves", *Proc. of 22nd Symposium of Naval Hydrodynamics,* National academy press, Washington, D.C.

de Kat J.O. (2000) "Dynamics of a ship with partially flooded compartment" in Contemporary Ideas of Ship Stability, D. Vassalos, M. Hamamoto, A. Papanikolaou and D. Moulyneux (Editors), Elsevier Science pp. 249-264.

Karr, D.G., Troesch, A. W. and Levi, R. (1995). "Some effect of threshold singularities with intermittent contact and breakage", *Journal of Sound and Vibration* Vol. 185, No 4, pp. 609-625.

Kim, M.H. and Yue, D.K.P. (1989) "The second-order diffraction solution for an axisymmetric body – Part 1. Monochromatic incident waves" *J. Fluid Mech.* Vol. 200, pp. 235-264.

Kinoshita, M. and Okada, S. (1957). "Heeling moment due to wind pressure at small vessels", *Proc. of Symposium on the Behaviour of Ship in Seaway*, Washington D.C.

Kjeldsen, S. P. and Myrhaug, D. (1978). "Kinematics and dynamics of breaking waves", Part 4 of Report "Ship in Rough Seas", River and Harbour Laboratory at Norwegian Institute of Technology, Report No STF60 A78100, Trondheim.

Knuth D.E. (1997). "The art of computer programming", v.2 "Semi-numerical algorithms", (3rd edition) Addison-Wesley, Reading, Massachusetts.

Khaskind, M. D. (1973). "Hydrodynamic theory of ship motion", Nauka publishing, Moscow, (in Russian).

Kim, Y. B. and Noah, S.T. (1991). "Stability and bifurcation analysis of oscillations with piece-wise linear characteristics: a general approach", *Journal of Applied Mechanics -- ASME TRANS.*, Paper No 91-APM-22.

Kholodilin, A. N. (1963). "On small ship stability regulations" *Transactions of Russian Register Shipping* "Theoretical and Practical Problems of Stability and Survivability of Sea Vessels", Morskoy Transport publishing, pp.45-51 (in Russian).

Kholodilin, A. N. and Tovstikh, E.V. (1969). "The model experiments for the stability of small ships on erupting waves". *Proc. 12th ITTC*, Rome, pp. 795-797.

Kholodilin, A. N. and Mirokhin, B. V. (1972). "The action of the erupting shallow water waves upon the models of ships", *Proc. of 13th ITTC* .

Kerwin, J.E. (1955), "Notes on rolling in longitudinal waves", *International Shipbuilding progress,* Vol. 2, No 16, pp. 597-614.

Kochin, N. E., Kibel, I. A. and Roze, N. B. (1948). Theoretical hydromechanics, in 2 volumes, Vol.1, OGIZ publishing, Leningrad-Moscow (in Russian).

Komuro, M. (1988). "Normal forms of continuous piecewise linear vector fields and chaotic attractors. Part I: Linear vector field with a section", *Japan Journal of Applied Mathematics*, Vol. 5, No. 2, pp. 257-304.

Komuro, M. (1988a). "Normal forms of continuous piecewise linear vector fields and chaotic attractors. Part II: Chaotic attractors", *Japan Journal of Applied Mathematics*, Vol. 5, No. 3, pp. 503-549.

Komuro, M. (1992). "Bifurcation equations of continuous vector fields" *Japan Journal of Industrial and Applied Mathematics*, Vol. 9, No. 2, pp. 269-312.

Komuro, M. (1994). "Bifurcation equations and standard forms of continuous piecewise vector fields" in "Towards the Harnessing of Chaos", Yamaguti, M. (editor) Elsevier, pp. 257-271.

Korn G.A. and Korn T.M. (1968). "Mathematical handbook", Book *McGraw Hill*, New York.

Korpus, R, and Falzarano, J. (1996) "Prediction of ship roll damping by unsteady Navier-Stokes technique," *Proc. of The 15th International Conference on Offshore Mechanics and Artic Engineering*, Florence.

Kramer, H. and Leadbetter, M.R. (1967). "Stationary and related stochastic processes", John Wiley, New York.

Krylov, A. N. (1958). "Selected papers", Published by Academy of Science of the USSR, Moscow, (in Russian).

Kuo, C. and Odabasi, A. Y. (1975). "Application of dynamic system approach to ship and ocean vehicles stability", *Proc. of STAB'75: 1st International Conference on Stability of Ships and Ocean Vehicles*, Glasgow.

Leadbetter M., Lindgren G., Rootzen, H. (1986) "Extremes and related properties of random sequences and processes" book Springer-Verlag, N.Y.

Lee, Ai-Kuo and Adee, B. (1994). "Numerical analysis of a vessel's dynamic responses with water trapped on deck", *Proc. of STAB'94: 5th International Conference on Stability of Ships and Ocean Vehicles*, vol. 2, Melbourne, Florida.

Lewis, E. V. (Editor) (1989). "Principles of Naval Architecture. Second Revision", SNAME.

Lin, H. and Yim, S. (1995). "Chaotic roll motion and capsize of ships under periodic excitation with random noise" *Applied Ocean Research*, Vol. 17, pp.185-204.

Lin, W.M. and Yue, D.K.P. (1990). "Numerical Solutions for Large-Amplitude Ship Motions in the Time-Domain," *Proceedings of the Eighteenth Symposium of Naval Hydrodynamics*, The University of Michigan, 1990.

Lin, W. M. and Salvesen, N. (1998) "Nine years of progress with LAMP – the Large Amplitude Motion Program", SAIC Report No. 97/1079.

Longuet-Higgins, M. S. (1983). "On the joint distribution of wave period and amplitudes in random wave field" *Proc. Royal Soc. London* A389, p. 241.

Lopatoukhin, L. J., Rozhkov, V. A. and Trapeznikov Yu. A. (1990). "Spectral structure of waves". In "Results of oceanographic studies in the eastern part of the Pacific Ocean tropical zone", "Gidrometeoizdat" publishing, Leningrad, (in Russian).

Lopatoukhin, L. J., Boukhanovsky, A. V., Rozhkov, V. A. and Divinsky, B. V. (1999) "Climatic wave spectra of the Black Sea," *Proc. of MEDCOAST: International Conference "Wind and wave climate of the Mediterranean and the Black sea"*, Antalya,, pp. 97-109.

Lopatoukhin, L. J., Rozhkov, V. A., Ryabinin, V. E., Swail, V. R., Boukhanovsky, A. V. and Degtyarev, A. B. (2000). "Estimation of extreme wind wave heights". WMO/TD - No. 1041.

Lugovsky, V. V. (1966). "Nonlinear problems of seakeeping of ships", Sudostroenie publishing, Leningrad (in Russian).

Lugovsky, V. V. (1971). "Theoretical background of stability standards of sea vessels", Sudostroenie publishing, Leningrad, 248 p. (in Russian).

Lugovsky, V. V. (1976). “Dynamics of sea”, Sudostroenie publishing, Leningrad, (in Russian).

Lugovsky, V. V. (1980). “Hydrodynamics of Ship Non-Linear Motions” Sudostroenie publishing, Leningrad, (in Russian).

Lyapunov, A.M. (1954) “On steady helical motions of a rigid body in a fluid”, in Collected Works, Moscow, (in Russian).

Magnus, K. (1976). “Schwingungen”, B. G. Teubner publishing, Stuttgart.

Makov, Yu. L. (1969). “Some results of theoretical analysis of surf-riding in following seas”, *Transactions of Krylov Society*, "Manoeuvrability and Seakeeping of Ships" Vol. 126, Sudostroenie publishing, Leningrad, pp.124-128 (in Russian).

Makov, Yu. L. (1985). “Forecasts of capsizing of damaged vessels”, *Transactions of Kaliningrad Institute of Technology* "Seakeeping and design of fishing vessels", Kaliningrad, pp. 84-97, (in Russian).

Makov, Yu. L., Prosuryakov, Yu. L., Ortiz, A. G. and Sevastianov, N. B. (1987). “Ship heeling at steady wind drifting”, *Sudostroenie*, No. 3, pp. 7-10 (in Russian).

Makov, Yu. L. and Sevastianov, N. B. (1993). “Physical modelling of damaged vessels capsizing under influence of wind and wave action”, *Proc. of OTRADNOE'93: International Workshop on the Problems of Physical and Mathematical Stability Modelling*, Vol. 1, Paper No 13, Kaliningrad Institute of Technology, Kaliningrad.

Malenica, S. and Molin, B. (1995) “Third harmonic wave diffraction by a vertical cylinder”, *J. Fluid Mech*, Vol. 302, pp. 349-373.

Martin, J., Kuo, C. and Welaya, Y. (1982) “Ship stability criteria based on time varying roll restoring moments”, *Proc. of STAB’82: 2nd International Conference on Stability of Ships and Ocean Vehicles*, Tokyo.

Mauro, H. (1967) “Application of the slender body theory to the longitudinal motion of ships along waves”, *Bulletin of the Faculty of Engineering*, Yokohama National University, Vol. 16, pp. 29-61.

Mc Taggart K. and De Kat, J. O. (2000) “Capsize risk of intact frigates in irregular seas” *Transcations SNAME*, Vol. 108, pp. 147-177.

Mei, C., C. (1983) “The applied dynamics of ocean surface waves”, Wiley-Interscience, New York.

Meylunas, V. F. (1971). “Some statistic data on the distribution of weight cargo coefficient and specific volume of cargo”, *Transactions of Russian Register of Shipping* Vol. 1, Leningrad, pp. 241-250 (in Russian).

Meylunas V. F. and Braslavskaya A. I. (1976). “Probability field for displacement and height of centre of gravity of seagoing vessels in operation”, *Transactions of Russian Register of Shipping*, Vol.4, Leningrad, pp. 46-58 (in Russian).

Michell, J. H. (1893). “On the highest waves in water”, Phil. Mag., 5th Series 36, p. 430-437.

Moisseyeva, M. E. (1971). "Analysis of loading conditions and stability of big trawler Mayakovsky", *Rybnoe khozaistvo*, No 7, pp.27-30 (in Russian).

Moisseyeva, M. E. (1976). "Compatibility of subdivision, stability and seakeeping requirements", Leningrad, Sudostroenie publishing (in Russian).

Moon, F. C. (1987) "Chaotic vibrations: an introduction for applied scientists and engineers", John Wiley and sons, New York.

Mordachev, S. V., Sevastianov, N.B. and Belenky, V.L. (1994). "Assumed situation in probabilistic norms of stability", *Proc. of STAB'94: 5th International Conference on Stability of Ships and Ocean Vehicles*, Vol. 6, Melbourne, Florida.

Mordachev, S. V. (1995). "Assumed situations and human factor in probabilistic approach to stability assessment", *Proc. of OTRADNOE'93: International Workshop on the Problems of Physical and Mathematical Stability Modelling*, Vol.1, Paper 8, Kaliningrad Institute of Technology, Kaliningrad.

Möckel, W. (1960). "Behaviour of trawlers at sea", *Fishing boats of the world*, No 2, pp. 404-417.

Motora, S., Fujino, M. and Fuwa, T. (1982). "On the mechanism of broaching-to phenomena" *Proc. of STAB'82: 2nd International Conference on Stability of Ships and Ocean Vehicles*, Tokyo.

Murashige, S., Komuro, M. and Aihara, K. (1998) "Nonlinear roll motion and bifurcation of a Ro-Ro ship with flooded water in regular beam seas" *Proc. of R. Soc.*

Naito, S. and Sueyoshi, M. (2001). "A numerical analysis of violent free surface flow on flooded car-deck using particle method" *Proc of 5th International Workshop "Stability and Operational Safety of Ships"*, University of Trieste, pp 5.1.1-4.

Nayfeh, A. H. (1973). "Perturbation methods", Wiley-Interscience, New York.

Nayfeh, A. H. (1981). "Introduction to perturbation technique", Wiley-Interscience, New York.

Nayfeh, A. H. and Khdeir, A. A. (1986). "Nonlinear rolling of ships in regular beam seas", *International shipbuilding progress* Vol. 33, No. 379, pp. 40-49.

Nayfeh A. H. and Khdeir A. A. (1986a). "Nonlinear rolling of biased ships in regular beam waves", *International shipbuilding progress* Vol. 33, No. 381, p.84-93.

Nayfeh, A. H., Sanchez, N. E. (1990). "Stability and complicated rolling response of ships in regular beam seas", *International Shipbuilding Progress*, Vol. 37, No 412, pp. 331-352.

Nechaev, Yu. I. (1972). "Measuring of ship rolling amplitude in longitudinal irregular waves in parametric resonance regime", *Transactions of Nikolaev Shipbuilding Institute*, Vol. 58, pp. 52-59 (in Russian).

Nechaev, Yu. I. (1978). "Stability of ships in following seas", Sudostroenie publishing, Leningrad (in Russian).

Nechaev, Yu. I. (1989). "Modelling of ship stability in waves", Sudostroenie, publishing Leningrad, (in Russian).

Nechaev, Yu. I. and Degtyarev, A. B. (2000). "Account of peculiarities of ship's nonlinear dynamics in seaworthiness estimation in real time intelligence systems". *Proc. of STAB'2000: 7th International Conference on Stability of Ships and Ocean Vehicles*, Vol.2, Launceston, Tasmania, pp.688-701.

Nechaev, Yu. I., Degtyarev, A. B. and Boukhanovsky, A. V. (2001). "Complex Situation Simulation When Testing Intelligence System Knowledge Base" in "Computational Science - ICCS 2001", Part 1, LNCS 2074, Springer, , pp.453-462.

Nekrasov, V. A. (1978). "Probabilistic problems of seakeeping", Sudostroenie publishing, Leningrad, (in Russian).

Nekrasov, V. A. (1994). "Stochastic stability theory of ship motion" *Proc. of STAB'94: 5th International Conference on Stability of Ships and Ocean Vehicles*, Vol. 5, Melbourne, Florida.

Newman, J. N. (1977). "Marine hydrodynamics", Massachusetts Institute of Technology Press. Cambridge.

Newman, J. N. (1978). "Theory of ship motions", *Advanes in Applied Mechanics*, Vol. 18.

Odabasi, A. Y. (1982). "A morphology of mathematical stability theory and its application to intact stability assessment". *Proc. of STAB'82: 2nd International Conference on Stability of Ships and Ocean Vehicles*, Tokyo.

Okhusu, M. (1986). Prediction of wave forces on a ship running in following waves with very low encounter frequency. *Journal of Society of Naval Architects of Japan*, Vol. 159, pp. 129-138.

Ortiz, A. F. G. (1985). On aerodynamic forces on hull of trawler in beam position" *Transactions of Kaliningrad Institute of Technology* "Seakeeping and design of fishing vessels", Kaliningrad, pp. 3-17, (in Russian).

Paulling J. R. and Rosenberg R.M. (1959). "On unstable ship motions resulting from nonlinear coupling", *Journal of Ship Research*, Vol. 3, No 1, pp. 36-46.

Paulling, J. R. (1961). "The transverse stability of a ship in a longitudinal seaway". *Journal of Ship Research*, vol. 4, no. 4, pp. 37-49.

Pavlenko, G. E. (1949). "Problems of ship static", Morskoy Transport publishing, Leningrad (in Russian).

Pham Ngock Hoeh, (1981). "Evaluation of capsizing risk function of vessels affected by irregular seas and gusty wind", *Transactions of Kaliningrad Institute of Technology*, Vol. 93, "Seakeeping of ships", Kaliningrad, pp. 45-53 (in Russian).

Phillips, O. M. (1977). "Dynamics of the upper ocean", 2nd edition, Oxford University Press, London

Phillips, S. R. (1986). "Determining the roll stability of a vessel using Lyapunov methods", *Safeship Conference*.

Phillips, S. R., (1986a). Applying Lyapunov methods to investigate roll stability, *Proc. of STAB'86: 3rd International Conference on Stability of Ships and Ocean Vehicles,* Vol. 1, Gdansk, pp. 65-69.

Phomenko, Yu. I., (1967). "Seakeeping trials of m/s Krasnogradsk" *Transactions of Russian Register of Shipping* "Theoretical and Practical Problems of Stability and Survivability of Sea Vessels", Leningrad, Transport, pp. 313-349 (in Russian).

Price, W.G., (1975). "A stability analysis of the roll motion of a ship in an irregular seaway", *International Shipbuilding Progress* Vol.22, No 247, pp. 103-112.

Rahola, J. (1935). "The judging of the stability of ships", Transactions INA.

Rahola, J. (1939). "The judging of the stability of ships and the determination of the minimum amount of stability", Ph.D. Thesis, Helsinki.

Rainey, R. C. T, Thompson, J. M. T., Tam, G. W. and Noble, P. G. (1990). "The transient capsize diagram - a route to soundly-based new stability regulations", *Proc. of STAB'90: 4th International Conference on Stability of Ships and Ocean Vehicles*, Naples.

Rainey, R. C. T. and Thompson, J. M. T. (1991). "The transient capsize diagram - a new method of quantifying stability in waves", *Journal of Ship Research*, Vol. 35, No.1, pp.58-62.

Rakhmanin, N. N. (1966). "On dynamic stability of ship with water on deck", *Transactions of Krylov Society* Vol. 73, Sudostroenie publishing, Leningrad (in Russian).

Rakhmanin, N. N. (1966a). "n dynamic stability of a ship with water on deck", IMCO doc. STAB/INF.27 10.10.66 Submitted by USSR.

Rakhmanin, N. N. (1971). "Approximate assessment of safety of small vessel in seas", *Transactions of Russian Register Shipping*, Leningrad, pp.12-24 (in Russian).

Rakhmanin, N. N. (1995). "Roll damping when deck is submerging into seawater" *Proc. of International Symposium - Ship Safety in a Seaway: Stability, Manoeuvrability, Nonlinear Approach (Sevastianov Symposium)* Vol. 2, Paper 1, Kaliningrad Institute of Technology, Kaliningrad.

Rakov A. I. and Sevastianov N. B. (1981). "Design of fishing vessels", Sudostroenie publishing, Leningrad (in Russian).

Register of USSR (1974). "Handbook data: Wind and waves in oceans and seas", Transport publishing, Leningrad (in Russian).

Renilson, M. (1982). "An investigation into the factors affecting the likelihood of broaching-to in following seas" *Proc. of STAB'82: 2nd International Conference on Stability of Ships and Ocean Vehicles*, Tokyo, pp. 551-564.

Roberts, J. B. (1980). "A stochastic theory for nonlinear ship rolling in irregular seas" Report of National Maritime Institute, NMI R 99.

Roberts, J. B. (1980a). "The effect of parametric excitation on ship rolling motion in random waves", Report of National Maritime Institute, NMI R 100.

Roberts, J. B. (1982). "A stochastic theory for nonlinear ship rolling in irregular seas" *Journal of Ship Research* Vol. 26, No 4, pp. 229-245.

Roberts, J. B. (1982a). "Effect of parametric excitation on ship rolling motion in random waves" *Journal of Ship Research* Vol. 26, No 4, pp. 245-253.

Rosenberg, R. M. (1954). "On the stability of nonlinear nonautonomous systems", *Proc. of 2nd U.S. Nat. Congr. of Applied Mechanics*, Ann Arbor, Michigan.

Rayleigh, J. W. S. (1896). "The theory of sound", Reprinted by Dover, New York.

Sanchez, N. E. and Nayfeh, A. H. (1990). "Nonlinear rolling motions of ships in longitudinal waves", *International Shipbuilding Progress*, Vol. 37, No 411, pp.247-272.

Scavounos, P.D and Nakos, D.E. (1988) "Stability analysis of panel methods for free surface flows with forward speed" *Proc. of 17th Symp. on Naval Hydrodynamics*s

Semenov-Tian-Schansky, V. V., Blyagoveschensky, S. N. and Kholodilin, A. N. (1969). "Ship motions", Sudostroenie publishing, Leningrad.

Sevastianov, N. B. (1963). "On probabilistic approach to stability standards" *Transactions of Kaliningrad Institute of Technology*, Vol. 18, Kaliningrad, pp. 3-12 (in Russian).

Sevastianov, N. B. (1968a). "Comparison and evaluation of different systems of stability standards", *Sudostroenie*, No 6, Leningrad, pp. 3-5 (in Russian).

Sevastianov, N. B. (1970). "Stability of fishing vessels", Sudostroenie publishing, Leningrad (in Russian).

Sevastianov, N. B. (1977). "Mathematical experiment on ship capsizing under combined action wind and waves", Part 2 of "Investigation of Possibilities of Practical Realisation of Probabilistic Stability Standard" Report of Kaliningrad Institute of Technology, No 77-2.1.6 (in Russian).

Sevastianov, N. B. (1978). "On possibility of practical implementation of probabilistic stability regulations", *Sudostroenie* No 1, Leningrad, pp.13-17, (in Russian).

Sevastianov, N. B. (1979). "The advantages of probabilistic approach to stability regulations as compared deterministic one", *Transactions of Kaliningrad Institute of Technology* "Seakeeping of Fishing Vessels", Vol. 81, Kaliningrad, pp. 3-9, (in Russian).

Sevastianov, N. B. and Pham Ngock Hoeh (1979). "Boundary between the domains of stable and unstable free ship motion in drift-rolling regime", *Transactions of Kaliningrad Institute of Technology* "Seakeeping of Fishing Vessels", Vol. 81, Kaliningrad, pp. 17-25 (in Russian).

Sevastianov, N. B. (1982). "Probabilistic stability regulation as a problem of reliability theory", *Transactions of Russian Register Shipping*, Vol. 12, Leningrad, pp. 94-100, (in Russian).

Sevastianov, N. B. (1982a). "An algorithm for averaging of risk function over the sets of stochastic situations and conditions of loading". *Transactions of Russian Register Shipping*, Vol.12, Leningrad, pp. 100-105, (in Russian).

Sevastianov, N. B. (1982b). "Assumed situations in stability estimation and physical pictures of ships' capsizing" *Trans. of Kaliningrad Institute of Technology* "Seakeeping of Ships", Vol. 99, Kaliningrad, pp. 110-122, (in Russian).

Sevastianov, N.B. (1984). "Practical and scientific aspects of stability problem from small fishing vessels", *Proc. of International Conference on Design Considerations for Small Crafts*, RINA, London, pp.13-15.

Sevastianov, N. B. (1992). "Reliability integral and its interpretation in the task of stability estimation". *International Workshop on Safety Problems Related to Stability of Ships*, Ilawa.

Sevastianov, N. B. (1993). "Theoretical and practical models for probabilistic estimation of vessels stability" *Proc. of OTRADNOYE'93: International Workshop on the Problems of Physical and Mathematical Stability Modelling*, Vol. 1, Paper 5, Kaliningrad Institute of Technology, Kaliningrad.

Sevastianov, N. B. (1994). "Distribution of aerodynamic pressures of gusty wind" *Transactions of Kaliningrad Institute of Technology* "Design and Seakeeping Standards of Ships", Kaliningrad, pp. 94-107 (in Russian).

Sevastianov, N. B. (1994a). "An algorithm of probabilistic stability assessment and standards" *Proc. of STAB'94: 5th International Conference on Stability of Ships and Ocean Vehicles*, Vol. 5, Melbourne, Florida.

Shen, D. and Huang, X. (2000). "The study of lasting time before capsize of a ship under irregular wave excitation" *Proc. of STAB'2000: 7th International Conference on Stability of Ships and Ocean Vehicles* Vol. 2, Launceston, Tasmania, pp.710-723.

Shlesinger, M. F. and Swean, T. (editors) (1998). "Stochastically excited nonlinear ocean structures", World Scientific, Singapore.

Shin, C. I. (2000). "Prediction of shipping water on deck of ships", *Proc. of STAB'2000: 7th International Conference on Stability of Ships and Ocean Vehicles*, Vol. 1, Launceston, Tasmania, pp. 190-205.

Shin, Y.S., Chung, J.S., Lin, W.M., Zhang, S. and Engle, A. (1997). "Dynamic loadings for structural analysis of fine form ships based on non-linear large amplitude motions and loads method" *SNAME Transactions* pp. 127-154.

Skomedal, N. G. (1982). "Parametric excitation of roll motion and its influence on stability", *Proc. of STAB'82: 2nd International Conference on Stability of Ships and Ocean Vehicles*, Tokyo.

Smolnikov, L. P. and Byohkov, Yu. A. (1972). "Calculation of piecewise linear systems", Energia publishing, Leningrad, (in Russian).

Soliman, M. S. (1990). "An analysis of ship stability based on transient motion", *Proc. of Proc. of STAB'90: 4th International Conference on Stability of Ships and Ocean Vehicles*, Naples.

Spyrou, K. (1995). 'Surf-riding, yaw instability and large heeling of ships in following/quartering waves", *Ship Technology Research/Schiffstechnik*, Band 42, Helft 2, pp.103-111.

Spyrou, K. (1995a). "Surf-riding and oscillation of a ship in quartering waves", *Journal of Marine Science and Technology*, Vol. 1, pp.24-36.

Spyrou, K. (1996). "Dynamic instability in quartering seas: the behaviour of a ship during broaching" *Journal of Ship Research*, Vol. 40, No 1, pp. 46-59.

Spyrou, K. (1996a). "Dynamic instabilities in quartering seas – part II: analysis of ship roll and capsize for broaching", *Journal of Ship Research*, Vol. 40, No 4, pp. 326-336.

Spyrou, K. (1997). "Dynamic instabilities in quartering seas – part III: nonlinear effects on periodic motions", *Journal of Ship Research*, Vol. 41, No 3, pp. 210-223.

Spyrou, K. (2000). "The nonlinear dynamics of ships in broaching", *Annals of Marie Curie Fellowship Association,* Vol. 1.

St. Denis, M. and Pierson W. J. (1953). "On the motions of ships in confused seas", *Transactions SNAME* Vol. 61.

Stokes, G. G. (1880). "Mathematical and Physical papers". Vol. 1, 314, Cambridge University Press.

Stoker, J. J. (1950). "Nonlinear vibrations in mechanical and electrical systems", New York.

Stratonovich, R. L. (1963). "Topics in the theory of random noise", Vol. 1, Gordon and Breach, New York.

Sveshnikov, A. A, (1968). "Applied methods of stochastic function theory", Nauka publishing, Moscow, (in Russian).

Terao, Y (1980). "The hydrodynamic force acting on the ship in following sea [1st report]", *Journal of Society of Naval Architects of Japan,* Vol. 148, pp. 77-85 (in Japanese).

Thompson, J. M. T. and Stewart, H. B. (1986). "Nonlinear dynamics and chaos", John Wiley, Chichester.

Tikhonov, V. I. (1982). "Statistical radio engineering", Radio i sviaz publsihing, Moscow (in Russian).

Trincas, G. (1986) "Influenza del profilo ondoso sulla stabilita' transversale delle navi", *Tecnica Italiana*, N. 1/2, pp.11-20.

Troesch, A. W., Karr, D. G. and Beier, K.-P. (1992). "Global contact dynamics of an ice-structure interaction model", *International Journal of Bifurcation and Chaos* Vol.2 No.3 pp. 607-620.

Umeda, N., Yamakoshi, Ya. and Tsuchiya, Ts. (1990). "Probabilistic study on ship capsizing due to pure loss of stability in irregular quartering seas", *Proc. of STAB'90: 4th International Conference on Stability of Ships and Ocean Vehicles*, Naples.

Umeda, N. (1990). "Probabilistic study on surf-riding of a ship in irregular following seas", *Proc. of STAB'90: 4th International Conference on Stability of Ships and Ocean Vehicles*, Naples.

Umeda, N. Ikeda, Y. and Suzuki, I. (1992). "Risk analysis applied to the capsizing of high-speed craft in beam seas," *Proc. of PRADS'92: International Conference on Practical Design and Mobile Units*, (New Castle upon Tyne), Elsevier Science.

Umeda, N. and Renilson, M. R. (1992). "Broaching – a dynamic behaviour of a vessel in following and quartering seas", in Manoeuvring and Control of Marine Craft, Wilson P.A. (editor), Computational Mechanics Publication, Southampton, pp. 533-543.

Umeda, N. and Renilson, M. R. (1992a). "Wave forces of a ship running in quartering seas - a simplified calculation method", *Proc. of 11th Australian Fluid Mechanics Conference.*

Umeda, N. and Yamakoshi, Ya. (1993). "Probability of ship capsizing due to pure loss of stability in irregular quartering seas", *Naval Architecture and Ocean Engineering* Vol.30, Soc. Naval Arch. Japan.

Umeda, N. and Renilson, M. R. (1994). "Broaching of a fishing vessel in following and quartering seas nonlinear dynamical system approach", *Proc. of STAB'94: 5th International Conference on Stability of Ships and Ocean Vehicles*, Vol. 3, Melbourne, Florida.

Umeda N., Yamakoshi Y. and Suzuki S. (1995). "Experimental study for wave forces on a ship running in quartering seas with very low encounter frequency" *Proc. of International Symposium – Ship Safety in a Seaway: Stability, Manoeuvrability, Nonlinear Approach (Sevastianov Symposium),* Vol. 1, Paper 14, Kaliningrad Institute of Technology, Kaliningrad.

Umeda, N. and Vassalos, D. (1996). "Nonlinear periodic motions of a ship running in quartering and following seas". *Journal of Society of Naval Architects of Japan*, Vol. 179, pp. 89-101.

Umeda, N., Vassalos, D. and Hamamoto, M. (1997). "Prediction of ship capsize due to broaching in following / quartering seas", *Proc. of STAB'97: 6th International Conference on Stability of Ships and Ocean Vehicles*, Vol. 1, Varna, pp. 45-54.

Umeda, N. (1999). "Nonlinear Dynamics on Ship Capsize due to Broaching in Following and Quartering Seas" *Journal of Marine Science and Technology*, Vol. 4, pp. 16-26.

Umeda, N., Matsuda, A., Hamamoto, M. and Suzuki, S. (1999). "Stability assessment for intact ships in the light of model experiments" *Journal of Marine Science and Technology*, Vol.4, pp. 45-57.

Umeda, N. and Hamamoto, M. (1999). "Capsizes of ship model in following /quartering waves – physical experiments and nonlinear dynamics", in The Nonlinear Dynamics of Ships, Thompson J. M. T. (Editor), Theme issue, Royal Society.

Umeda, N. and Matsuda, A. (2000). "Broaching in following and quartering seas-theoretical attempts and new prevention device" *Proc. of STAB'2000: 7th International Conference on Stability of Ships and Ocean Vehicles*, Vol.1, Launceston, Tasmania, pp.460-470.

Vassalos, D. (1994). "A unified treatment of ship stability: a heuristic approach using recent finding in nonlinear system dynamics", National Research Institute of Fisheries Engineering (Japan).

Vassalos, D. (2000). " The water on deck problem of damaged RO-RO ferries", in Contemporary Ideas of Ship Stability, D. Vassalos, M. Hamamoto, A. Papanikolaou and D. Moulyneux (Editors), Elsevier Science, pp. 163- 186.

Vassilopoulos, L. (1971). " Ship rolling at zero speed in random beam seas with nonlinear damping and restoration". *Journal of Ship Research*, Vol. 15, No 4, pp.289-294.

Virgin, L. N. (1987). "The nonlinear rolling response of a vessel including chaotic motions leading to capsize in regular seas", *Applied Ocean Research*, Vol. 9, No 2, pp.89-95.

Vishnubhota, S., Falzarano, J. and Vakakis, A. (2000). "A new method to predict vessel/platform critical dynamics in a realistic seaway", *Phil. Trans. R. Soc.*, Vol. 358 London, pp. 1967-1981.

Vishnubhota, S., Falzarano, J. and Vakakis, A. (2001). "A new method to predict vessel/platform critical dynamics in a realistic seaway", *Proc. of 5th International Workshop: Stability and Operational Safety of Ships*, University of Trieste, paper No 4.3.

Voitkunsky, Y. J. (editor) (1985). "Handbook on Ship Hydrodynamics", (Vols. 1-3) Sudostroenie publishing, Leningrad, (in Russian).

Wahab, R., and Swaan, W. A. (1964). "Coursekeeping and broaching of ship in following seas", *Journal of Ship research*, Vol.4, No 7, pp. 1-15.

Wellicome, J. F. (1975). "An analytical study of the mechanism of capsizing", *Proc. of STAB'75: 1st International Conference on Stability of Ships and Ocean Vehicles*, Glasgow.

Whitham, G.W. (1974). "Linear and nonlinear waves" Wiley-Interscience, New York.

Wright, J. H. G. and Marshfield, W. B. (1980). "Ship roll response and capsize behaviour in beam seas", *Naval Architect*, No 23, pp. 129-144.

Yamagata, M. (1959). "Standard of stability adopted in Japan", *Transactions INA*, pp. 417-443.

Zelenin, L. K. (1989). "Probability characteristic of ship capsizing in irregular waves", *Proc. of 18th SMSSH: Scientific and Methodological Seminar on Ship Hydrodynamics*, Vol. 2, Paper 38, Bulgarian Ship Hydrodynamic Centre, Varna.

Zelenin, L.K. (1989a). "Estimation of reliability of ship stability characteristics in storm conditions", *Transactions of Kaliningrad Institute of Technology*, "Seakeeping and Design of Fishing Vessels", Kaliningrad, pp. 87-93 (in Russian).

Zinkovsky-Gorbatenko, V.G. (1965). "Some results of full scale of wind and its utilisation for ship stability assessment", *Transactions of Russian Register of Shipping* 'Theoretical and Practical Problems of Stability and Survivability of Fishing and Cargo Vessels", Transport publishing, Moscow-Leningrad, pp. 87-106, (in Russian).

Zuo, Q. H. and Hjelmstad, K. D. (1998). "Piecewise linear wrapping theory for multilayered elastic beams", *Journal of Engineering Mechanics*, Vol. 124, No 4 pp. 377-384.

References for the Second Edition

American Bureau of Shipping (2004). "Guide for the assessment of parametric roll resonance in the design of container carriers", Houston.

American Bureau of Shipping (2006). "Guide for vessel maneuverability", Houston.

Belenky, V. L., Weems, K.W., Luit, D. and Shin, Y. S. (2003) "Nonlinear Roll With Water-On-Deck: Numerical Approach" *Proc. of STAB'2003: 8th International Conference on Stability of Ships and Ocean Vehicles*, Madrid, pp.59-80.

Belenky, V. (2004) "On risk evaluation at extreme seas", *Proc. of 7th International Workshop on Ship Stability, Shanghai Jiao Tong University* (Also available form ABS Technical Papers, 2004, Houston).

Belenky, V. (2005) "On long numerical simulations at extreme seas" *Proc. of 8th International Workshop on Ship Stability,* Istanbul Technical University, Istanbul. (Also available form ABS Technical Papers, 2005, Houston).

Belenky, V., Yu, H. and Weems, K. (2006) "Numerical Procedures and Practical Experience of Assessment of Parametric Roll of Container Carriers" *Proc. of STAB'2006: 9th International Conference on Stability of Ships and Ocean Vehicles,* Vol. 1, Rio-de-Janeiro, pp. 119-130.

Brizzolara, S. and Rizzuto, E. (2006). "Wind heeling moments on very large ships. some insights through CFD", *Proc. of STAB'2006: 9th International Conference on Stability of Ships and Ocean Vehicles*, Rio-de-Janeiro, Vol. 2 pp. 781-793.

Brunswig, J., Pereira, R. and Kim, D. (2006). "Validation of parametric roll motion predictions for a modern containership design", *Proc. of STAB2006: 9th International Conference on Stability of Ships and Ocean Vehicles,* Vol. 1, pp. 157-169.

Davenport, A. G. (1964). "Note on the distribution of the largest value of a random function with application to gust loading". *Proc. of Royal Soc.*

Ikeda, Y., Munif, A., Katayama, T. and Fujuwara, T. (2005). "Large parametric rolling of large passenger ship in beam seas and role of bilge keel in its restraint" *Proc. of 8th International Workshop on Ship Stability,* Istanbul Technical University, Istanbul.

de Kat, J.O. and Paulling, J. R. (1989). "The Simulation of Ship Motions and Capsizing in Severe Seas", *Transaction SNAME*, Vol. 97 pp. 139-168.

de Kat, J.O., Brouwer, R., McTaggarrt, K. and Thomas, W. L. (1994). "Intact ship survivability in extreme waves: new criteria from research and navy perspective", *Proc. of STAB'94: 5th International Conference on Stability of Ships and Ocean Vehicles,* Vol. 1, Melbourne, Florida.

Kobylinski, L. K. and Kastner S. (2003) "Stability and safety of ships. Volume 1: regulation and operation", Elsevier Ocean Engineering Book series Vol. 9, Elsevier, Amsterdam.

Lin, W.M. and Yue, D.K.P. (1993). "Time-domain analysis for floating bodies in mild-slope waves of large amplitude", *Proceedings of the Eighth International Workshop on Water Waves and Floating Bodies*, Newfoundland, Canada.

McCue, L., Bassler, C. and Belknap, W. (2006). "Real-time identification of behavior leading to capsize", *Proc. of STAB'2006: 9th International Conference on Stability of Ships and Ocean Vehicles*, Rio-de-Janeiro, Vol. 1, pp. 373-383.

Munif, A., Ikeda, A., Fujiwara, T. and Katayama, T. (2006). "Parametric roll resonance of a large passenger ship in dead ship condition in all heading angles", *Proc. of STAB'2006: 9th International Conference on Stability of Ships and Ocean Vehicles*, Rio-de-Janeiro, Vol. 1, pp. 81-87.

Nechaev, Yu.I., Degrtayrev, A.B. and Anischenko, O. (2006). "Ships dynamic on wave-breaking condition", *Proc. of STAB'2006: 9th International Conference on Stability of Ships and Ocean Vehicles*, Rio-de-Janeiro, Vol. 1, pp. 409-418.

Neves, M. A. S and C. A. Rodríguez (2006) "An investigation on roll parametric resonance in regular waves", *Proc. of STAB'2006: 9th International Conference on Stability of Ships and Ocean Vehicles*, Rio-de-Janeiro, Vol. 1, pp. 99-108.

Paroka, D., Okura, Y. and Umeda, N. (2006). "Analytical prediction of capsizing probability of a ship in beam wind and waves", *Journal of Ship Research*, Vol. 50, No. 2, pp. 187-195.

Paroka, D. and Umeda, N. (2006). "Capsizing probability prediction of the large passenger ship in irregular beam wind and waves: comparison of analytical and numerical methods" *Journal of Ship Research*, Vol. 50, No. 4, pp. 371-377.

Paulling, J. R., Kastner, S. and Schaffran, S. (1972). "Experimental studies of capsizing of intact ships in heavy seas". *US Coast Guard, Technical Report*. (Also IMO Doc. STAB/7, 1973).

Paulling, J. R., Oakley O.H., and Wood, P. D. (1974). Ship motions and capsizing in astern seas. *Proc. 10th Symposium on Naval Hydrodynamics*, ACR-204, Office of Naval Research, MIT.

Paulling, J. R., Oakley O.H., and Wood, P. D. (1975). "Ship capsizing in heavy seas: the correlation of theory and experiments", *Proc. of STAB'75: 1st International Conference on Stability of Ships and Ocean Vehicles*, Glasgow.

Roberts, J.B. and Spanos, P. (2003) "Random vibration and statistical linearization", Dover Publications.

Shin, Y.S, Belenky, V.L., Lin, W.M., Weems, K.M. and Engle, A.H. (2003). "Nonlinear time domain simulation technology for seakeeping and wave-load analysis for modern ship design" *SNAME Transactions* Vol. 111, pp. 557-578.

Shin, Y.S, Belenky, V.L., Paulling, J.R., Weems, K.M. and Lin, W.M. (2004). "Criteria for parametric roll of large containerships in longitudinal seas", *SNAME Transactions* Vol. 112, pp. 14-47.

Stern, F., Carrica, P., Kandasamy, M., Gorski, J., O'Dea, J., Hughes, M., Miller, R., Hendrix, D., Kring, D., Milewski, W., Hoffman, R. and Cary, Ch. (2006) "Computational hydrodynamic tools for high-speed transports", *SNAME Transactions* Vol. 114, pp. 55-81.

Tikka, K. K. and Paulling, J. R. (1990). "Prediction of critical wave conditions for extreme vessel response in random seas", *Proc of STAB'90: 4th International Conference on Stability of Ships and Ocean Vehicles*, Naples.

Subject Index